Partnership in Space
The Mid to Late Nineties

A History of Human Space Exploration

Other Springer-Praxis books by Ben Evans in this Series

Escaping the Bonds of Earth – The Fifties and the Sixties
2009
ISBN: 978-0-387-79093-0

Foothold in the Heavens – The Seventies
2010
ISBN: 978-1-4419-6341-3

At Home in Space – The Late Seventies into the Eighties
2011
ISBN: 978-1-4419-8809-6

Tragedy and Triumph in Orbit – The Eighties and Early Nineties
2012
ISBN: 978-1-4614-3429-0

Ben Evans

Partnership in Space

The Mid to Late Nineties

Published in association with
Praxis Publishing
Chichester, UK

Ben Evans
Space Writer
Atherstone
Warwickshire
UK

SPRINGER–PRAXIS BOOKS IN SPACE EXPLORATION

ISBN 978-1-4614-3277-7   ISBN 978-1-4614-3278-4 (eBook)
DOI 10.1007/978-1-4614-3278-4
Springer New York Heidelberg Dordrecht London

Library of Congress Control Number: 2012947009

© Springer Science+Business Media New York 2014
This work is subject to copyright. All rights are reserved by the Publisher, whether the whole or part of the material is concerned, specifically the rights of translation, reprinting, reuse of illustrations, recitation, broadcasting, reproduction on microfilms or in any other physical way, and transmission or information storage and retrieval, electronic adaptation, computer software, or by similar or dissimilar methodology now known or hereafter developed. Exempted from this legal reservation are brief excerpts in connection with reviews or scholarly analysis or material supplied specifically for the purpose of being entered and executed on a computer system, for exclusive use by the purchaser of the work. Duplication of this publication or parts thereof is permitted only under the provisions of the Copyright Law of the Publisher's location, in its current version, and permission for use must always be obtained from Springer. Permissions for use may be obtained through RightsLink at the Copyright Clearance Center. Violations are liable to prosecution under the respective Copyright Law.
The use of general descriptive names, registered names, trademarks, service marks, etc. in this publication does not imply, even in the absence of a specific statement, that such names are exempt from the relevant protective laws and regulations and therefore free for general use.
While the advice and information in this book are believed to be true and accurate at the date of publication, neither the authors nor the editors nor the publisher can accept any legal responsibility for any errors or omissions that may be made. The publisher makes no warranty, express or implied, with respect to the material contained herein.

Cover design: Jim Wilkie
Project copy editor: David M. Harland
Typesetting: BookEns, Royston, Herts., UK

Printed on acid-free paper

Springer is part of Springer Science+Business Media (www.springer.com)

Contents

Author's preface vii
Acknowledgements xi

1. **Rise from the ashes** 1
    The odyssey of Ulysses 1
    The odyssey of the Shuttle 5
    A difficult summer 16
    Secret sentinels 30
    Lost mission 43
    Four powerful eyes 54
    "We're back!" 56
    How to save a 'Great Observatory' 67
    Milkshakes and doggie biscuits 86
    Three women or three doctors? 99
    First medical research flight 110
    "Even the girls…" 120
    Ozone watcher 128
    Infrared eyes 135

2. **The last Soviet citizen** 151
    A new start 151
    Perestroika and *glasnost* 157
    New arrivals, new challenges 158
    "No experience necessary" 183
    An extended flight 190
    End of an era 197

3. **The International Space Years** 201
    Journeys to the edge 201
    Meeting of minds 216
    Success and failure 231

"Train like you fly" 245
"Weird science" 261
International mission 277
A glass half-empty? 286
Telescience on trial 295
Under the 'veil' 297
A changed mission 303
"Red lights in the cockpit" 309
"We didn't have a clue" 326
Abort! 336
Long road to SLS-2 347
"You and the rest!" 356

4. **From foes to friends** 375
"Pinch me!" 375
An awkward start 376
Cool crew 395
Radar love 401
The majesty of 'Columbia' 412
"Safing in progress" 422
Eyes on the Earth, on the stars, on the future 430

5. **Dawn of a new era** 453
A gloomy road to freedom 453
Seeds of friendship 461
The partnership evolves 468
Footsteps to the future 480

Bibliography 483
Index 493
About the author 497

Author's preface

When I set out to write a five-volume series to commemorate the first 50 years of our adventure in space, it seemed a big project, though relatively straightforward and something that I have always wanted to do. An obsessive space enthusiast for as long as I can remember, I received my first space book at the age of five, was given a toy Space Shuttle for a birthday present soon afterwards and by the time I reached my seventh birthday I had watched in childish astonishment as Enterprise – mounted atop a Boeing 747 carrier aircraft – hurtled over my primary school in Birmingham, during sports day. It caused me to drop the egg from my spoon, unfortunately, but the sight was so spectacular and awe-inspiring that it hooked me for life. The Moon landings excited me – and still do – beyond compare and I began writing articles at the age of 15, for the British Interplanetary Society's *Spaceflight* magazine and, later, for *Countdown* and *Astronomy Now*. As I grew older, it became a goal of mine to someday write a 'meaty' history of the human exploration of space, which I continue to believe firmly is one of the greatest adventures ever undertaken by our species, but everything I read seemed to 'lack' something. Some were overloaded with facts and figures, others were devoted solely to the 'popular' audience, while still more simply lacked the detail and human interest factor. I cannot promise the reader that my series fulfils each of these gaps, but what I *can* say with certainty is that I have spent an enjoyable three years researching the history of our adventure, told through news sources, books, the memoirs of those involved, magazines, press kits and oral histories, and have learned an immense amount. The reader may love or hate my book – they may find it hard to put down or may simply find use for it as an expensive additional castor for their sofa – but I have derived great joy from researching and writing it.

It has been impossible to track entire decades within the pages of each volume. The first, *Escaping the Bonds of Earth*, had to take into account some of the achievements of the 1950s, as a prerequisite to focusing on 'its' decade, the 1960s. In a similar vein, the second volume, *Foothold in the Heavens*, needed the focus to fall in considerable depth on some of the most remarkable achievements of the Space Age – Apollo 11 being the obvious example – at the expense of covering an entire decade. The third instalment, *At Home in Space*, tackled the 1980s, the devastating tragedy of Challenger and the triumphs of Salyut 7 and Mir. Only with the fourth volume did it become clear that I was punching above my weight. My determination to cover each mission with the level of detail that it deserved, including biographies of each spacefarer, turned this book into something much longer; so long, in fact, that I had barely covered the 1980s and the extent was already rapidly approaching 600 pages. As a result, with the approval of Clive Horwood, the series has expanded from five into *six* volumes, to cover the first half-century in its entirety, whilst maintaining the kind of depth that the reader would expect.

As the work progressed into this volume, *Partnership in Space*, the depth has been easier to fulfil in some areas than others. The Russians, even in the late 1980s and early 1990s, with the advent of *glasnost* (openness) and *perestroika* (restructuring), and the eventual collapse of the Soviet Union, proved notoriously secretive about their activities and in several cases it has been impossible to offer biographies of a handful of their cosmonauts in more than just a few words or paragraphs. That secrecy extended to the West, too, when the Americans staged a handful of military Shuttle missions, many of whose precise objectives remain classified to this day. I have attempted, using the sources and contacts at my disposal, to shine a meagre light on these shadowy flights and I fervently hope that a few years from now they will emerge from the shadows to take their rightful place in history.

I have learned much about the human space programmes of the United States and Russia and, equally importantly, have learned a great deal about the political events which shaped their progress. Starting with Yuri Gagarin's pioneering voyage in April 1961, the journey has carried me through a handful of dramatic decades, punctuated by conflict and reconciliation, meddling and political manoeuvring, and has seen the first men land on the Moon, the first men occupy an Earth-circling space station, the first men pilot a reusable vehicle beyond the atmosphere, the first men from other nations – Czechoslovakia, Poland, Vietnam and Mongolia, to name but a few – and, of course, the first members of *womankind* to carry their dreams and aspirations into the heavens. My intention has always been for something a little more than a basic log of crewed expeditions into space, but as time has rolled on, the project evolved into something much larger and more complex than I had envisaged. The *human* side of the story has always been profoundly important to me. Take Sergei Korolev: a man who endured so much physical and psychological trauma in his youth – dreaming, even as he was transported to a living death in the Kolyma gulag, of one day sending a rocket into space. Deke Slayton is another example: a man chosen in his prime as one of America's finest, the 'Mercury Seven', whose hopes were cruelly dashed by a devastating heart condition, yet who sprang back as the man who guided the astronaut corps to the Moon, chose Neil Armstrong to take the first historic steps on its dusty surface, and eventually overcame every obstacle in his path to fly into orbit. Their remarkable stories and their individual trials and tribulations, from childhood to the grave, carry just as much weight and drama and excitement and adventure as the missions they flew. Yuri Gagarin pissing on the wheel of the bus as he prepared to board his space capsule, Wally Schirra conceiving of another legendary 'gotcha', Alexei Leonov and his love of cowboy hats and boots, John Glenn and his competitive yearning to be first, Alexander Serebrov and his penchant for fast cars, Christa McAuliffe and her passion to carry education to the final frontier and Shannon Lucid and Svetlana Savitskaya, who both refused to accept that their gender should tie them to Earthbound pursuits and who both strove for the stars.

The story of our adventure in space is not simply about who spent the most time there, who performed the most spacewalks or who flew the most missions. It is a collection of *stories*: the stories of how a few hundred remarkable people, all of whom achieved an uncommon goal, forever changed our perspective on the world in which we live and fixed our eyes and our minds and our imagination on the Universe around us.

Ben Evans
Atherstone, May 2013

Acknowledgements

This book would not have been possible without the support of a number of individuals, to whom I am enormously indebted. I must firstly thank my wife, Michelle, for her constant love, support and encouragement throughout the time it has taken to plan, research and write this manuscript. As always, she has been uncomplaining during the weekends and holidays when I sat up late, typing on the laptop, or poring through piles of books, old newspaper cuttings, magazines, interview transcripts, press kits or websites. It is to her, with all my love, that I would like to dedicate this book. My thanks also go to Clive Horwood of Praxis for his enthusiastic support and to David Harland for reviewing the manuscript and offering a wealth of advice and guidance; I deeply appreciate not only their support, but also their patience in what has been an overdue project and one which has proven more difficult to write than I had imagined. Additional thanks go to Ed Hengeveld, who has been enormously gracious with his time in identifying suitable illustrations for this book, including many 'unfamiliar' ones which surely bolster the text. Others to whom I owe a debt of gratitude include Sandie Dearn and Malcolm and Helen Chawner. To those friends who have encouraged my fascination with all things 'space' over the years, many thanks: to Andy Salmon and Andy Rowlands and to Dave Evetts and Mike Bryce and to Rob and Jill Wood. Our two golden retrievers – the ever-hungry Rosie and the attention-seeking Milly – have provided a ready source of light relief and a regular opportunity for me to leave the laptop and either play with them or give them a biscuit.

1

Rise from the ashes

THE ODYSSEY OF ULYSSES

Human hands have stretched far into the cosmos during our half-century of exploring the final frontier. Men and women have circled hundreds of kilometres above the protective gaseous veil of Earth's atmosphere and a handful of men have ventured further and left their footprints on the flat plains and undulating hills of our closest celestial neighbour, the Moon. Someday, our direct presence will expand to Mars and other more exotic places. For now, though, efforts to develop the capacity to transport us beyond low-Earth orbit remain in their infancy. Still, many space machines crafted by human hands have been sent deep into the Solar System ... and several of those were delivered, personally, by human crews. In June 2009, one such machine fell silent after two decades exploring the polar regions of the Sun. The joint US-European Ulysses mission, now defunct, continues to orbit our parent star, completing a full circuit every six years or so, and its legacy stands testament to the ingenuity of the scientists, engineers, visionaries and thinkers who laboured to put it there. From its launch in October 1990 to the end of its life, Ulysses pushed the boundaries of knowledge about our Sun and fundamentally altered our understanding of how it functions.

The Sun is, quite literally, the reason that life exists on Earth. Most scientists accept that the Solar System formed from a vast cloud of gas and dust, around 4.5 billion years ago, with immense temperatures and pressures serving to form a protostar and an enormous disk which eventually coagulated into the primordial versions of the planetary attendants that exist today. On the third of those attendants, the largest of the innermost, 'rocky' planets, life eventually arose; life which would someday build and despatch Ulysses to learn more about the star which had given it life. Ulysses was the first spacecraft to significantly depart the 'ecliptic plane' – the plane of Earth's orbit – to directly explore the Sun's northern and southern poles. In doing so, it enabled physicists to study the star in three dimensions and provide an accurate assessment of the total solar environment, across a full range of heliographic latitudes. Since the ecliptic plane differs from the solar equatorial

2 Rise from the ashes

plane by only 7.25 degrees, it was previously only possible to directly sample the solar wind from low solar latitudes. To investigate the solar wind at higher inclinations demanded a prohibitively large launch vehicle, but by utilising the enormous gravity of the planet Jupiter a significant 'plane change' could be effected, enabling travel outside of the ecliptic.

More than four decades ago, consideration was given to launching a Pioneer spacecraft in 1974 for precisely this purpose, but it failed to gain approval. A seed of interest had been sown, however, and ultimately bore fruit as the International Solar Polar Mission (ISPM). In its original incarnation, this was a truly 'international' endeavour, employing two separate spacecraft – one built by America's National Aeronautics and Space Administration (NASA), the other by the European Space Agency (ESA) – to travel towards Jupiter. One would hurtle 'beneath' the giant planet's south pole, using its gravity to deflect it northwards, out of the ecliptic, towards northern solar latitudes. Meanwhile, the other craft would do the reverse, travelling 'above' Jupiter's north pole to bend its trajectory southwards to explore southern solar latitudes. The result would be a pair of *in-situ* instruments to provide simultaneous measurements of both solar hemispheres for mapping, measurements of magnetic fields and observations of the 'solar wind', a stream of charged particles known to emanate from the Sun at 400 km/sec in the ecliptic plane.

The ISPM was formally approved in 1976, its scientific instruments were agreed the following year and work on the spacecraft began in October 1978. Both would be launched on a single Space Shuttle mission in February 1983 and were to be boosted towards Jupiter by an attached, solid-fuelled rocket, known as the Inertial Upper Stage (IUS). However, the limited capability of this rocket, built by Boeing for the US Air Force, was already raising eyebrows in scientific and political circles and there was doubt that it was powerful enough to deliver the twins as far as Jupiter. As a result, in April 1980 the ISPM was split into two halves and rescheduled for separate launches in 1985. The IUS woes continued, however, because the infant Shuttle drew voraciously on NASA's funds. In February 1981, the space agency was forced to slow the development of its ISPM craft and the IUS was dropped in favour of a more powerful, liquid-fuelled booster, built by General Dynamics. It was called the Centaur-G Prime – part of a family of upper stages originally developed for the Atlas rocket in the 1960s – and its selection for the ISPM pushed the launch back still further to May 1986. It also opened an entirely new can of worms.

The Centaur carried an enormous load of cryogenic hydrogen and oxygen – totalling more than 16,500 kg – and came to be nicknamed a 'balloon tank', since it required total pressurisation in order to become fully rigid. In fact, if it was not fully pressurised, a single push from a finger could literally *flex* its metal walls. Right from the start, the Centaur was viewed warily by NASA's safety officials, whose rule of thumb dictated that no single failure should ever be capable of endangering the Shuttle or her crew. Disturbingly, the Centaur's pressure regulation hardware lacked a backup facility and, worse, a failure of its internal bulkhead had the potential to rupture the walls of both of its propellant tanks. Moreover, the dangers of these propellants 'sloshing' risked a whole range of controllability problems for the Shuttle itself … but, balanced against these enormous risks was the promise that the

Centaur was powerful enough to boost the ISPM and other deep space probes, including the Galileo mission to Jupiter. In the end, that was not enough and the Centaur was removed from consideration in favour of the less powerful, but safer IUS.

Potential disaster hit the ISPM in September 1981, when NASA was forced by the House Appropriations Committee to terminate the production of its spacecraft. However, ESA pressed on with its own craft, which, at 370 kg, was small and light enough to be reassigned back onto a two-stage IUS in January 1982. (In fact, it was so small, said astronaut Dick Richards, that it could quite easily be fitted onto the back of a pickup truck.) By this time, the absence of the Centaur was creating massive financial consequences: Galileo was a hugely important voyage and to be launched by an IUS meant that its journey time to Jupiter would double, the duration of its mission at its target would effectively be halved and its overall scientific harvest would be seriously compromised. Within a matter of months, plans changed yet again. A groundswell of support for the Centaur, spearheaded by New Mexico Senator and former Moonwalker Harrison 'Jack' Schmitt, led to its reinstatement. In spite of the cost of changing boosters *again* and the lingering safety fears, the reduced journey times and increased scientific bounty which the Centaur could offer Galileo and the ISPM were deemed worthy of the risk. In July 1982, President Ronald Reagan himself approved the change.

The ISPM, therefore, reverted back to the Centaur and, since both it and Galileo needed to travel to Jupiter, they were scheduled for separate Shuttle missions during the same 'launch window' in May 1986. In the meantime, by 1983, the Europeans had completed the fabrication of their spacecraft – a small, boxy machine, some $3.2 \times 3.3 \times 2.1$ m in size, with a 1.4 m dish antenna on a fixed mounting and a NASA-provided Radioisotope Thermoelectric Generator (RTG) power unit. It would be spin-stabilised at five revolutions per minute and its attitude would be managed by four pairs of hydrazine thrusters. Ten scientific instruments were manifested, half of them provided by ESA and half by NASA, to explore radio wave emissions from solar plasmas, together with measurements of magnetic fluxes and observations of electrons, ions, neutral gas, dust and cosmic rays.

As the ISPM changed, so too did its name. One leading contender was 'Odysseus', to honour the mythical Greek hero of the Trojan War, whose ten-year journey back home to reclaim his kingdom of Ithaca and his suitor-pestered wife, Penelope, has made his name a synonym for a voyage with many changes of fortune. The name was entirely fitting. In the same way that the ISPM would follow an indirect path to explore an uncharted destination, so the mythical Odysseus had taken many unexpected twists and turns before concluding his journey. At length, the Latinised version of Odysseus' name – Ulysses – was picked instead. It had been proposed by an Italian physicist, Bruno Bertotti of the University of Pavia, whose gravitational wave experiment was conducted during the mission. In Bertotti's mind, the name drew upon not only the rich cultural heritage of Troy, but also offered a nod toward other more recent writings: Alfred, Lord Tennyson's epic poem, James Joyce's novel and Dante Alighieri's *Inferno*. In the latter, Ulysses famously guided his crew westwards into the unexplored waters beyond the Strait of Gibraltar. His terrified

4 Rise from the ashes

men mutinied, but Ulysses calmed them and encouraged them "to follow after knowledge and excellence".

Unlike its Homeric namesake, the spacegoing Ulysses would face no witches or cyclops, no cannibals or sea monsters, no suitors or sirens, and nor would it risk being transformed into swine, but its long and arduous odyssey would face equally intimidating obstacles as engineers and trajectory specialists sought to chart a route into regions of the Sun's realm never previously encountered. Its nine scientific instruments – or *ten*, if one also counts the spacecraft's radio – were designed to investigate the strength and direction of solar magnetic fields, the basic properties of ions and electrons in the solar wind, X-ray and gamma-ray emissions from the Sun, gravitational waves, cosmic dust, neutral helium atoms and high-energy cosmic rays entering the Solar System from the interstellar medium.

As with all voyages of exploration, Ulysses and its Centaur booster required massive preparation on the ground. Challenger, the vehicle assigned to deliver the spacecraft into low-Earth orbit, underwent extensive modifications for the purpose. Extra plumbing and emergency dumping vents were installed into the Shuttle to load and drain the Centaur's propellants, control panels were fitted in the flight deck, an S-band telemetry antenna was added and a huge Centaur Integrated Support Structure in the payload bay served to position the 'stack' for deployment.

According to the crew activity plan for Challenger's mission, released by NASA in mid-January 1986, the Shuttle and its crew of four – astronauts Rick Hauck, Roy Bridges, Mike Lounge and Dave Hilmers – would launch at 4:10 pm Eastern Standard Time (EST) on 15 May. Assuming an on-time liftoff, the Shuttle carried provisions for a four-day flight and the sheer weight of the Centaur was such that a number of crew provisions, including the galley, had to be removed. Hauck's crew would have entered a relatively low orbit of just 170 km and had just nine hours to get Ulysses out of the payload bay and on its way. The Centaur was required to periodically dump its boiled-off gaseous hydrogen to keep tank pressures within their mandated limits and, beyond nine hours, it would have 'bled' so much propellant that the remainder would have been insufficient to perform the engine burn for Jupiter. After deployment, the Centaur's twin engines would have ignited to carry Ulysses towards a rendezvous with the giant planet in July 1987, thus setting the spacecraft up for an encounter with the Sun's polar regions in 1989-1991.

All of these plans ground brutally to a halt on 28 January 1986, when Challenger exploded during liftoff, carrying one of NASA's own communications satellites, killing her entire crew. The resultant investigation uncovered many safety flaws in the reusable Shuttle, several of which related directly to the Centaur, and in June the tempestuous booster was formally cancelled by NASA Administrator Jim Fletcher. While both Ulysses and Galileo waited out the lengthy delay for the Shuttle to resume flying, other options were pursued to deliver them to Jupiter. At this point, the IUS returned to the fore and in April 1987 a firm launch target of October 1990 was established for Ulysses. It would be a narrow 'window', just two weeks long, and achieving it would be critical if the spacecraft was to properly rendezvous with Jupiter in February 1992 and then go on to explore the solar poles later in the decade. The happiness of the successful mission which Ulysses would become seemed a world away in the dismal days after Challenger. In fact, the Shuttle itself had suffered enormously, from delay and disaster to trial and tribulation, virtually from conception.

THE ODYSSEY OF THE SHUTTLE

To say that the summer of 1990 had been a difficult one for NASA is something of an understatement. As Space Shuttle Discovery sat on Pad 39B at the Kennedy Space Center (KSC) on Merritt Island, Florida, in the pre-dawn hours of 6 October, even the media commentators were remarking that no previous mission had gotten this close to launch for more than five months. The reusable fleet of orbiters now numbered just three – Discovery, Columbia and Atlantis – and had been operating for ten years, but it had been a decade of triumph and tragedy. The Ulysses deployment mission, designated STS-41, was the 36th flight of the Space Transportation System, yet only the eleventh since the September 1988 return to normal operations after Challenger. Many improvements had been implemented – redesigned Solid Rocket Boosters (SRBs), better brakes and tyres, improved computers, an escape pole for high-altitude bailout and hyperbaric protection for the astronauts, in the form of bright-orange pressure suits – but the fact remained that the three-part Shuttle was the most complex machine ever launched.

That complexity had been amply demonstrated in the summer of 1990, when both Columbia and Atlantis were grounded for several months by a persistent spate of hydrogen leaks. This focused more negative public attention on NASA itself, which was already smarting from an embarrassing problem which had hit its scientific showpiece, the $1.5 billion Hubble Space Telescope. Toted as being capable of looking further into the Universe than ever before, and unlocking many secrets of the early cosmos, a test of Hubble's optics within weeks of launch had shown that its optics could not focus correctly. A Shuttle repair and servicing mission was scheduled, but the telescope quickly turned NASA into a butt of jokes on late-night television shows.

This unhappy time must have led some NASA old-timers to reflect on the similar unhappiness which had plagued much of the Shuttle's own development, some two decades earlier. The reusable orbiter would be entirely different from any previous manned spacecraft. In fact, from the moment it was conceived, the sheer notion of changing the United States' spacegoing philosophy from small capsules, lofted initially atop ballistic missiles and later by the giant Saturn V, to something the size and shape of a conventional airliner – ferried into orbit attached to a large fuel tank and a pair of solid-propelled rocket boosters – was a vast effort and posed monumental, and frequently maddening, obstacles. Naturally, the idea for the Shuttle did not appear from nowhere; it could trace its development across several decades and many of its systems had been pioneered by a series of aerospace projects organised by NASA and its forerunner, the National Advisory Committee for Aeronautics (NACA). As early as the 1930s, the German aerospace engineer Eugen Sänger developed early concepts for rocket-propelled aircraft, some of which

6 Rise from the ashes

envisaged velocities as high as ten times the speed of sound and altitudes up to 70 km. In his later work, Sänger showed how the addition of wings could enhance a spacegoing rocket: although its initial range would be quite modest, following an arc-like trajectory, much akin to an artillery shell, the *lift* generated by those wings during re-entry would carry it *upward*, allowing it to 'skip' off the atmosphere repeatedly, thus opening up new possibilities for a craft which could circle the globe and return to its launch site. Within a few years, such dreams rose from the drawing boards and into hardware: in 1947, Chuck Yeager flew the Bell X-1 rocket-propelled aircraft through the sound barrier for the first time and in 1951 the Douglas Skyrocket and its pilot, Scott Crossfield, reached an impressive Mach 2.

At these velocities, the issue of aerodynamic overheating was not yet a significant obstacle, but as the US military focused more attention on the prospect of delivering heavy nuclear warheads to Moscow, speed and stability became more important. In January 1952, NACA was urged to pursue a manned research aircraft, capable of exceeding Mach 5, with early reports even suggesting a *commercial* hypersonic vehicle. If such a plan were to be realised, it would need to overcome issues of aerodynamic overheating and stability. The latter problem was resolved by Charles McLellan, an aerodynamicist at NACA's Langley Aeronautical Laboratory in Hampton, Virginia, who proposed that small, wedge-shaped vertical fins and horizontal stabilisers would be much more effective than conventional thin surfaces. Overheating proved a harder nut to crack. If the aircraft re-entered the atmosphere with its nose pointing in the direction of flight, its streamlined shape would subject it to disastrous overheating and destructive aerodynamic stress. A re-entry with the nose positioned at a higher angle, with the flat undersurface presented to the hypersonic airflow, would be more manageable, permitting the craft to lose speed in the upper atmosphere and easing overheating and lowering aerodynamic loads.

This realisation that overheating could be reduced did not detract from the reality that such vehicles would be subjected to re-entry temperatures of significantly higher severity than had previously been encountered. Bell Aircraft had already begun to explore temperature-resistant materials, including a chrome-nickel alloy called 'Inconel-X' and stainless steel 'shingles', capable of radiating heat away from the airframe, coupled with techniques for water-cooling areas along the leading edges of the wings. In October 1954, the Aircraft Panel of the Scientific Advisory Board expressed its belief that the next decade's research and development goal should be to focus on the field of hypersonic flows. "This is one of the fields in which an ingenious and clever application of the existing laws of mechanics is probably not adequate," the panel reported to Air Force Chief of Staff Nathan Twining. "It is one in which much of the necessary physical knowledge still remains unknown at present and must be developed before we arrive at a true understanding and competence." Summing up, the panel considered the time ripe for the construction of a new aircraft to surpass Mach 5 and reach altitudes as high as 150 km. The project received overwhelming support from within the Air Force and NACA and in the spring of 1955 it also received a name: the X-15. In time, it would become the fastest and highest-flying aircraft prior to the arrival of the Shuttle. In August 1963, with NASA test pilot Joe Walker at the controls, the X-15 reached a record altitude of 107 km, and a few years later, the Air Force's William 'Pete' Knight achieved a speed of Mach 6.72 (some 7,270 km/h). The Air Force decided to award its own 'astronaut wings' to pilots who flew higher than 80 km, but the Fédération Aéronautique Internationale (FAI) decreed that only flights which exceeded 100 km officially reached the threshold of space. The FAI's ruling was based on the so-called 'Kármán Line', named after Theodore von Kármán, the Hungarian-American engineer who first described it as the altitude at which the atmosphere is too thin for aeronautical purposes, since a vehicle would need to travel faster than orbital velocity to achieve sufficient lift. Following these two rulings, between July 1962 and August 1968, *thirteen* X-15 missions by eight men exceeded the Air Force's limit, but only *two* of those missions (both flown by Joe Walker) passed the Kármán Line and were considered official 'space flights' by the FAI. Years later, however, NASA honoured the 80 km Air Force limit and in August 2005 awarded 'astronaut wings' to its three civilian X-15 pilots: posthumously in the cases of Walker and Jack McKay and in person to Bill Dana. (The five Air Force X-15 pilots had long since been awarded such wings by their service.) Journalist Tim Furniss has referred to this baker's dozen of X-15 sorties as 'astro-flights', thereby differentiating them from actual orbital or suborbital 'space flights'.

America's space ambitions in the 1960s were driven by the desire to land a man on the Moon, but in the opening months of President Richard Nixon's administration in 1969 a Space Task Group was established to chart possible courses for the immediate and long-term future. It was chaired by Vice President Ted Agnew and, sadly, his lack of real clout in the White House meant that only one of his group's four recommendations – the Shuttle – was actually endorsed by Nixon. Plans for a large space station, advanced lunar bases and a manned voyage to Mars fell on deaf ears. The president liked the 'heroic' image of astronauts, but his enthusiasm for space travel itself was low. Unlike John Kennedy and, to a lesser extent, Lyndon Johnson, Nixon saw little political, scientific or technological value in exploring the heavens. The American Institute of Aeronautics and Astronautics (AIAA) also endorsed the Shuttle above the other three options, noting that a low-cost manned spacecraft for delivering medium to large payloads into orbit would enable it to "effectively compete with present expendable boosters" and its members collectively felt that "commitment to an entirely new space station is less urgent than commitment to a new logistics system". The President's Science Advisory Committee (PSAC) went further, suggesting that the Shuttle – "a reusable space transportation system with an early goal of replacing *all* existing launch vehicles" – would not only allow for the deployment and recovery of satellites *and* the orbital assembly of a variety of different structures, but would also lead to a "radical reduction in unit cost of space transportation". In general, the concept of a reusable vehicle was gaining popularity, supported by the AIAA, the PSAC, NASA itself and the Air Force. Since it was expected to be one of the Shuttle's main users, the Air Force became a key player in its final design.

"During the year that followed the landing of Apollo 11 in the Sea of Tranquillity," wrote Thomas Heppenheimer in *The Space Shuttle Decision*, "NASA received a cold bath in the Sea of Reality." A pessimistic remark, perhaps, but appropriate, since 1970 was a time of retreat from lofty goals of permanent lunar bases and expeditions to Mars and saw a refocusing of national priorities: ending the war in Vietnam, beginning the war on poverty, improving the health care system ... and sharply cutting back the space budget. Although Congress *would* ultimately agree that the Shuttle *was* the most important 'next step', and would appropriate $110 million to start the work in Fiscal Year 1971, the chances of this reusable winged spaceplane ever getting off the ground seemed slim. Its initial funding had but barely survived in the House and had only won by four votes in the Senate; clearly even this low-cost approach to future space exploration was vulnerable to the slightest change in anti-space sentiment.

The physical form of what would become the Shuttle was also steadily evolving into a delta-winged spacecraft, because this would produce considerably more lift at hypersonic speeds and permit flying substantial distances to the left or right of the nominal track in order to attract the interest of the Air Force. In 1967, Air Force attempts to acquire detailed satellite imagery of troop movements during the Six-Day War in the Middle East and, a year later, during the Soviet invasion of Czechoslovakia had been thwarted by the limitations of the system then employed by reconnaissance satellites and by the time required for photographic films to be returned and developed. 'Real-time' reconnaissance required not only new technologies, such as electronic imaging devices, but also the ability of the Shuttle to launch into space, perform a single orbit and return to base with film exposed less than an *hour* earlier. Other plans included using the craft to quickly snare Soviet spy satellites from orbit. For the Shuttle to receive Congressional approval, NASA *needed* the Air Force on its side, because such important missions of national security would benefit from the capabilities of a delta-winged, high-cross-range vehicle. "These Air Force leaders knew that they held the upper hand," wrote Heppenheimer. "They were well aware that NASA needed a Shuttle programme and therefore needed both the Air Force's payloads *and* its political support. The payloads represented a tempting prize, for that service was launching over *two hundred* reconnaissance missions between 1959 and 1970."

Unfortunately, the Air Force did *not* need NASA to launch their satellites; they could rely perfectly well on expendable boosters, such as the Titan III. Their support for the civilian space agency would come at a cost. The Air Force wanted to launch its Shuttles from Vandenberg Air Force Base in California and land there after a single-orbit, 90-minute mission, which would require a 1,700 km cross-range capability – more than twice as large as NASA had envisaged. NASA's Max Faget, who wanted a straight-winged Shuttle, optimised for subsonic performance, was disappointed when the delta wing shape was accepted, but it turned out to be a small price to pay for the Air Force's support. In order to achieve the larger cross-range capability, the Shuttle would glide hypersonically, thereby increasing both the *rate* of aerodynamic heating and the *duration* of that heating ... which, in turn, would require the additional weight (and cost) of a beefed-up thermal protection system.

Linked to the requirements of *both* NASA and the Air Force was the issue of the size of the payload bay. NASA wanted it to measure about 7.5 m in diameter and 10 m long, but, as Heppenheimer explained, "the Air Force needed length, for its

1-01: Pictured during the STS-41 launch, the Space Shuttle stack represented an unusual marriage of technical practicality and political compromise. A pair of Solid Rocket Boosters (SRBs) provided 80 percent of the impulse necessary to deliver the DC-3-sized orbiter into space. Meanwhile, the bulbous External Tank fed liquid oxygen and hydrogen propellants into the Shuttle's cluster of three main engines.

reconnaissance satellites amounted to orbiting telescopes and these had to be long to yield the sharpest images. Moreover, such satellites were growing markedly in length. The Corona spacecraft of the 1960s, each with an attached Agena upper stage, had *started* at [5.5 m] and quickly grown to [8 m]." Already, the CIA was preparing for new generations of satellites, known as 'Big Bird' and 'Kennan'. The latter was only in the preliminary design stage, but was expected to have a focal length of around 20 m. As a result, the Air Force told NASA that they would require the Shuttle's payload bay to accommodate a 20 m satellite. Any reduction in size would be simply unacceptable: even a 14 m bay would mean that less than *half* of the Air Force's satellites could fly on the Shuttle. "This requirement," they told NASA, "is *still* considered valid. Should you elect to develop the Shuttle with a [smaller] payload compartment, it will preclude our full use of the potential capability and operational flexibility ... Also, if a portion of the present expendable launch vehicle stable must be retained to satisfy some mission requirements, then the potential economic attractiveness and the utility of the Shuttle ... is severely diminished." For the civilian space agency, the bay would enable the Shuttle to carry components for a modular space station, to be launched, piecemeal, in the early 1980s ... but that was a mere side note, compared to the needs of the Air Force and the Department of Defense. In terms of weight, the Shuttle *had* to be capable of delivering an 18,000 kg payload into low polar orbit; Big Bird was known to be around 13,000 kg and future satellites would be even heavier. With few quibbles from NASA, the military got its way. One anecdote from this period concerns an attempt made by Max Faget, one of the world's foremost aerospace engineers, designer of the Mercury spacecraft and NASA's then-head of engineering and development in Houston, to slightly reduce the *diameter* of the payload bay (a dimension in which he thought the military had little interest) from 5 m to 4 m. "It took the Air Force only three days to put him in his place," wrote Heppenheimer, "with a reply that read: *The USAF fully supports and stands firm on the present Level 1 requirement for a payload diameter of 5 m and a length of 20 m*. The reply came from one Patrick Crotty, whose rank was no higher than Major." NASA can have been under no illusions that the Air Force was in control of virtually every detail of the Shuttle. This control was underlined, once and for all, in January 1971, when NASA formally presented the requirements of the Shuttle to all contractors involved in the design studies at a meeting in Williamsburg, Virginia ... and gave the Air Force all it demanded: the cross-range capability, the payload bay length *and* diameter, the delivery capacity to polar orbit; *everything*. "One sometimes hears that when two parties are in a relationship, the one that wants it *more* is the weaker," remarked Heppenheimer. "NASA certainly had been pursuing support for the Shuttle with unmaidenly eagerness and the Williamsburg rules were the result. The agency was now promising to build a bigger and heavier Shuttle than it had *wanted* for its own uses, with considerably more thermal protection. It was also prepared to treat the Shuttle as a *national asset*, which meant the Air Force would not pay for its development or production, yet would receive the equivalent of exclusive use of one or more of these vehicles, entirely *gratis*." The Air Force would be responsible for the development of its own Shuttle launch facility at Vandenberg Air Force Base, but setting that to one side Heppenheimer is not alone in comparing the relationship to a treaty between a superpower and a banana republic ...

As the Shuttle danced to the Air Force's tune, other changes were ongoing in the design of the rocket booster. Potential contractors had considered expendable or only partially-reusable elements, but from August 1969, NASA turned its gaze to a fully-reusable booster. "Partially-reusable designs had represented an effort to meet economic goals by seeking a Shuttle that would cost less to develop than a fully-reusable system," wrote Heppenheimer, "even while imposing higher costs per flight. This approach had held promise prior to the spring of 1969, when the Shuttle had been considered largely as a means of providing space station logistics. Now, its intended uses were broadening to include launches of automated spacecraft, which meant it might fly far more often. The low cost per flight of a fully-reusable [system] now made it attractive and encouraged NASA to accept its higher development cost." Flying more often, it was hoped, would gradually help to reduce overall launch costs and NASA anticipated it might also cut the cost of the *payloads* themselves. As more payloads entered orbit, frequently and reliably, they might stimulate new uses for space, thereby encouraging contractors to build further satellites, with the Shuttle serving much the same role as commercial aviation had done, achieving enormous growth by cutting the prices of its passengers' tickets. Its payload bay offered a vast volume, which eased restrictions on weight and size, and the satellites to be placed into orbit could be fault-checked by the Shuttle crew, *in orbit*, thus lessening the mass of paperwork on the ground. Subsystems aboard these satellites could be standardised and when they started to fail in orbit after a few years of service, the Shuttle would have the capability to retrieve, refurbish and deploy them for a fraction of the cost of a full replacement.

Processing of the Shuttle and its booster was also required to contribute to substantial cost savings. In the early part of 1971, the booster was expected to be longer than a Boeing 747 and considerably fatter in the fuselage, and it would be swung into an upright position and mated with the orbiter. An enormous, diesel-powered 'crawler' would then transport it from the assembly building to the pad. Launch would occur under the combined thrust of maybe a dozen main engines – the contract for which went to Rocketdyne, a division of North American Rockwell – and the booster and Shuttle would part company in the high atmosphere, with the former returning to a conventional runway, under the power of its own jet engines. The Shuttle would return to a similar style of landing after completing its mission in space. Processing of the booster and orbiter between flights was expected to require two weeks. Mathematical and economic studies from this period suggest that NASA was confident that each mission could be launched for around $4.6 million – a million dollars less than for an expendable Delta rocket, one of the smallest rockets then in use – and a full order of magnitude lower than the $55 million cost of the Saturn IB booster, used to launch Skylab crews. Early development cost outlays, to be fair, *would* be significant, but within a few years the Shuttle would begin to show massive savings over expendable vehicles; perhaps as high as a billion dollars per annum. This dramatic reduction in costs was expected to spark the new revolution in spacecraft design, but it would not come to pass, for in Thomas Heppenheimer's words, "the *first* statement was a speculation [and] the *second* then amounted to a speculation that *rested* on a speculation".

The final approval of the Shuttle came in April 1972. NASA had drummed up a list of possible names for the new craft – Pegasus, Hermes, Astroplane, Skylark, to name a few – but it was Richard Nixon himself who ultimately made the final choice, by referring to it simply as "the Space Shuttle". It would thus break with previous practice of actually *naming* the project, although, as time would tell, individual orbiters *would* receive their own names. Of course, Nixon's acceptance of the Shuttle was rooted securely in down-to-Earth politics and his motivation was to strengthen employment within the aerospace industry as he entered election year for his second term in office. NASA Administrator Jim Fletcher had told the White House in November 1971 that an early start on the Shuttle "would lead to a direct employment of 8,800 by the end of 1972 and 24,000 by the end of 1973". In terms of employment and the need to pacify America's 'battleground states', the space programme was recognised to have an importance which was out of proportion to its budget.

On 26 July 1972, out of a pool of proposals which also included Lockheed, McDonnell Douglas and Grumman, the Nixon administration selected North American Rockwell, based in California, with its 55-vote monopoly in the Electoral College, to receive the $2.6 billion contract to build the Shuttle. This prompted Jean Westwood, chair of the Democratic National Committee, to harshly criticise Nixon's "calculated use of the American taxpayers' dollars for his own pre-election purposes". Westwood queried the 'help' that five of the corporation's directors provided to Nixon's 1968 election campaign and requested "a full airing" of the background to the contract award. Nevertheless, there *were* solid engineering reasons for not selecting the others. Lockheed's craft, for example, was heavy, "unnecessarily complex", according to Jim Fletcher, and it left a minute-long gap during ascent when there was no emergency abort provision. (In retrospect, this one minute proved a strange criticism, given that the design ultimately adopted imposed a *two-minute* phase, on the Solid Rocket Boosters, during which time there was no viable means of escape.) McDonnell Douglas' proposal was technically deficient and weak, whilst Grumman – which had earlier built the Apollo lunar module for NASA – came second to North American Rockwell. Grumman's presentation was impressive, identifying fundamental problems and offering good solutions, but fell down in terms of costs and overall project management. Even North American Rockwell itself, whose cost estimate was the *lowest* of the four, had a number of weaknesses in its design, not least a crew cabin which would be difficult to build.

High scores and low costs, therefore, were the reason behind the decision to pick North American Rockwell to build the Shuttle. In fact, when NASA Deputy Administrator George Low asked the three losing bidders to comment on the fairness of the competition, all three felt that it was the best and fairest competition that they had ever participated in. The situation was quite the opposite for the leading contenders to build the Shuttle's main engines, with Rocketdyne having been chosen for the $450 million contract over Pratt & Whitney. The latter had confidently *expected* to win, even taking out adverts in major aerospace magazines and declaring themselves ready to get the project started. Pratt & Whitney even lodged a complaint against NASA, alleging that the space agency had performed unfair acts in selecting Rocketdyne, but in March 1972 the Comptroller-General, Elmer Staats, upheld the decision. He slammed Pratt & Whitney, telling their lawyers that it was *also* unfair on NASA, which had helped them "to bring [an] original, *inadequate* proposal up to the level of other, *adequate* proposals, by pointing out those weaknesses which were the result of [Pratt & Whitney's] *own* lack of diligence, competence or inventiveness".

As the contracts rolled out, North American Rockwell (which became 'Rockwell International' in March 1973) proved itself more than honourable, by offering important subcontracts to its rivals. Grumman would take charge of building the orbiter's delta-shaped wings *and* the Gulfstream II Shuttle Training Aircraft, for example, and an unsuccessful attempt had earlier been made to interest Pratt & Whitney in sharing the main engine development programme. McDonnell Douglas would take responsibility for the Shuttle's tail-mounted Orbital Manoeuvring System (OMS) engines. In August 1973 NASA chose Martin Marietta to build a large External Tank which would carry cryogenic propellants for the main engines and in June 1974 picked Morton Thiokol for a pair of reusable, solid-fuelled rocket boosters to provide 80 percent of the thrust needed for the orbiter to reach space. By this time, of course, the shape, design and definition of the Shuttle had evolved into the form that we can recognise today. The orbiter itself was similar in dimensions to the DC-9 airliner, roughly 36 m long with its wings spanning 24 m from tip to tip. The first set of wings, built by Grumman, were among the earliest sections to be completed and were delivered to Rockwell's Palmdale plant in California in April 1975. The habitable area of the Shuttle, meanwhile, consisted of a two-tiered cockpit – a 'flight deck' for operations and a 'middeck' for experimental work, eating and sleeping – which backed onto the 20 m long payload bay and an aft compartment to house the three main engines and two OMS pods and support a vertical stabiliser tail fin. Forty-four tiny Reaction Control System (RCS) thrusters in the Shuttle's nose and tail would provide attitude control and additional manoeuvrability whilst in space. The graphite-epoxy payload bay doors were the largest aerospace structures yet built from composite material and *had* to be opened within a couple of hours of reaching orbit, to enable the radiators lining their interior faces to shed excess heat from electrical systems into space. The five-piece doors were hinged at either side of the mid-fuselage, mechanically latched at the forward and aft bulkheads and thermally sealed at the centreline. Ordinarily, they were driven 'open' and 'closed' by electromechanical power, but if they were unable to be opened, it was declared that the vehicle must return to Earth at the earliest opportunity. Conversely, if the doors did not *close* properly at mission's end, two crew members were trained to operate the mechanism manually on a spacewalk. Each orbiter was designed to make a hundred missions or a full decade of use (whichever came sooner) before major refurbishment would become necessary, although of the surviving vehicles even the fleet leader, Discovery, had made only 39 voyages when it was finally retired in March 2011.

Preparing for each Shuttle flight required several years, but the actual bringing together of the components began with setting up the Solid Rocket Boosters (SRBs) on a Mobile Launch Platform in the gigantic Vehicle Assembly Building (VAB) at the Kennedy Space Center. A booster comprised six blocks, called 'segments', each of which was positioned by overhead cranes with pinpoint accuracy, one atop the next and joined by a ring of bolts. To prevent a leakage of searing gases whilst operating, a series of rubberised O-rings sealed the joints between the segments. After propelling the Shuttle and External Tank to an altitude of about 45.7 km, about two minutes after launch, pyrotechnics separated the boosters, auxiliary rockets at their nose and tail pushed them away and parachutes were deployed to lower them to a gentle splashdown in the Atlantic Ocean. They were then recovered, stripped down and refurbished for reuse. When the assembly of the SRBs was complete, the External Tank was moved into position between them and connected by a series of spindly, but strong, attachment struts. After checks to verify its mechanical and electrical compatibility, the Shuttle itself was moved from the nearby Orbiter Processing Facility (OPF), tilted by crane onto its tail and mated to the tank. The transfer of the 1.8 million kg stack from the VAB to either Pad 39A or 39B – a distance of 5.6 km – took six hours, with the aptly named 'crawler' inching the precious, $2.2 billion national asset along a track made from specially imported Mississippi river gravel. Once the stack was 'hard down' on the pad surface, the crawler withdrew and a servicing structure rotated into place. Further checks were conducted, payloads installed and the crew participated in a Terminal Countdown Demonstration Test. This was essentially a full dress rehearsal of the final part of the countdown, after which there was a simulated main engine shutdown and emergency evacuation exercise.

The unusual appearance of the Shuttle 'stack' on the launch pad has been described by astronaut Story Musgrave as "like bolting a butterfly onto a bullet". The *bullet* was the External Tank, which resembled an enormous aluminium zeppelin, standing on end, some 46.6 m tall. It comprised two tanks for liquid oxygen and liquid hydrogen, separated by an 'inter-tank' for instrumentation and umbilicals. The oxygen tank at the top housed up to 542,640 litres of oxidiser and the hydrogen tank held some 1.4 million litres of fuel, both of which were fed through a pair of 43 cm lines into the Shuttle's aft compartment and thence into the combustion chambers of the three high-performance main engines.

At the start of each Shuttle mission, the main engines burned for about eight minutes and were shut down a few seconds before the External Tank was jettisoned, right on the edge of space. Each engine measured 4.2 m in length, weighed 3,400 kg and was 'throttleable' at one-percent incremental steps from 65 percent to 104 percent of its rated peak performance. The throttle was controlled by the Shuttle's General Purpose Computers and throttling back reduced stress on the vehicle during periods of maximum aerodynamic turbulence and also served to limit the G-loads in the final phase of ascent. Development of the main engines was undertaken at the National Space Technology Laboratory in Mississippi, a NASA facility previously used for testing the Saturn booster, with work getting underway in 1974. A full-thrust chamber test of an integrated subsystem demonstrator was conducted in the summer of the following year and in March 1976 the engine was fired successfully for

42 seconds at 65 percent rated performance. These tests uncovered a number of potentially serious problems, including failures of its high-pressure fuel and oxidiser turbopumps, which prompted modifications and the addition of more monitoring instrumentation. Twenty-five further tests got underway in April 1977, with no serious difficulties, although turbopump problems *would* continue to plague the effort throughout its development. Two years later, in July 1979, the main fuel valve experienced a major fracture during a test firing which allowed hydrogen to leak into a mockup aft compartment; the system was commanded to shut down, but not before it suffered serious structural damage. Four months later, the oxidiser turbopump failed again, less than ten seconds into a full-flight-duration, 510-second firing of a three-engine cluster. Triumph followed hard on the heels of failure in the following months: a perfect static test in December 1979, then a premature shutdown in April 1980, then another success ... and another shutdown. Two missions prior to the loss of Challenger would suffer on-the-pad main engine aborts and one of those, in July 1985, would also endure a hair-raising engine failure during ascent. In the early years of operating the Shuttle, a main engine failure was the anomaly most feared by the astronauts.

Despite the immense power generated by each engine and the colossal amount of

1-02: The flight deck of the Shuttle was the location from which the bulk of payload activities were conducted. In this view from STS-41, astronaut Bruce Melnick (left) floats against the aft control panel, from which RMS operations were conducted, whilst crewmate Bill Shepherd floats between the commander and pilot stations, with a bank of switches overhead.

propellant needed to run them, they provided only about 20 percent of the muscle to reach space. The remainder came from the two 45.4 m tall SRBs, the first solid-fuelled rockets ever to be used in conjunction with a manned spacecraft. Loaded with a powdery aluminium fuel, mixed with an oxidiser of ammonium perchlorate, the boosters, built by Morton Thiokol (now ATK Thiokol) of Utah, were mounted like a pair of big Roman candles on either side of the External Tank. When the decision to use solid-fuelled rockets was made in March 1972, Jim Fletcher explained that they would help to reduce the Shuttle's development costs from $5.5 billion to around $5.15 billion. Typically, during pre-flight preparations, the SRBs were paired in matching sets and filled with propellant ingredients from identical 'batches' to minimise the risk of thrust imbalances during ascent. The SRBs were to be capable of flying 25 times apiece (although they would need to be stripped down and refurbished between missions), but the External Tank was to be discarded and burn up in the atmosphere. It would have been more costly to recover and reuse the tank than to simply build a new one for each mission. The boosters represented relatively new technology for NASA. The Air Force had been using solid-fuelled rockets for some time, but those for the Shuttle were rather larger. The first tests of an empty casing took place in September 1977 at Thiokol's facility in Utah to verify structural integrity and fracture mechanics and to permit analysis of the growth and development of cracks. Four static firings of the motors were conducted between July 1977 and February 1979, followed by three qualification tests from June 1979 to February 1980, which validated the design. Parachute drops were conducted in El Centro, California, and historian Dennis Jenkins noted that the development of the SRBs over a total of just seven test firings was dwarfed by more than *seven hundred* to qualify the main engines.

Given the enormous difficulties – engineering and political – faced by the Shuttle during its genesis, it is unsurprising that the maiden launch of the vehicle was postponed by three years and finally occurred in April 1981. Even in the wake of the Challenger tragedy, and the benefits that the enforced down time provided in terms of attending to systemic flaws and implementing improvements, the first two years after the resumption of flight operations turned into a rocky road. There would be more trouble ahead.

A DIFFICULT SUMMER

"... T-minus 31 seconds ... we've had a Go for Autosequence Start ... Discovery's four redundant computers have primary control of vehicle critical functions ..."

Dawn was breaking over the marshy landscape of Florida's Space Coast on 6 October 1990 as NASA prepared to launch its first Shuttle mission in more than five months. It should not have been that way; in fact, at the start of the year, 1990 was expected to be a banner 12 months, with as many as nine flights scheduled to take place. Three of those occurred without incident – NASA's Long Duration Exposure Facility had been triumphantly recovered from orbit by Columbia in January, a classified Department of Defense payload had been delivered by Atlantis in

February and the gigantic Hubble Space Telescope had been launched by Discovery in late April – and more success was to follow. In mid-May, on STS-35, Columbia would undertake the first dedicated science mission of the post-Challenger era, carrying a battery of four ultraviolet and X-ray telescopes into orbit for astronomical research. The mission, called ASTRO-1, had been long-delayed; in fact, it was to have been the next Shuttle flight after the ill-fated Challenger and had been planned for March 1986 to observe, primarily, Halley's Comet, as the latter made its 75-yearly passage through the inner Solar System. Four years later, with Halley long gone, ASTRO-1 was finally ready to observe a catalogue of astronomical targets in unprecedented detail.

"... T-minus 15 seconds ..."

Following ASTRO-1, it was planned for Atlantis to fly a classified Department of Defense assignment (STS-38) in mid-July, after which Columbia would undertake a dedicated life sciences Spacelab mission in early September. Discovery would carry Ulysses aloft in its critically brief October launch window and Atlantis and Columbia would round out the year in November and December by flying NASA's Gamma Ray Observatory and another Spacelab science mission, known as the International Microgravity Laboratory. It was to be a challenging year and by the beginning of May NASA had made an exceptionally good start. Then came the issue with the disconnects.

"... T-minus ten, nine ..."

Deep inside the belly of the Shuttle were a pair of 43 cm propellant lines, through which thousands of litres of liquid oxygen and hydrogen flowed from the External Tank into the cluster of three main engines. Both 'sides' possessed mechanical disconnect fixtures and shortly before the separation of the External Tank, on the edge of space, a pair of flapper valves would be commanded to close by pneumatic helium pressure in order to prevent further propellant discharge and contamination. The criticality of the disconnect valves and their flapper valves cannot be underestimated, for any inadvertent closure whilst the main engines were firing would have prevented propellant flow from the External Tank and precipitated a catastrophic failure. In the days preceding the launch of STS-35, these disconnects took centre stage in one of the most difficult engineering challenges to face the Shuttle.

"... We have a Go for main engine start ..."

Before each mission, a Flight Readiness Review scrutinised all pertinent documentation, before formally reaching an agreement on a target launch date. At the beginning of May 1990, Columbia was provisionally scheduled to fly on the 16th, but a problem was noted with a proportioning valve on her Freon coolant loop, necessitating its replacement. A second review was convened and settled on the 30th for launch. By this time, the delays were expected to have a significant effect on the remaining flights of the year, which included two more Spacelab missions by Columbia, in late August and mid-December. Then, during the lengthy process to load liquid hydrogen into the External Tank, a tiny leak was detected, close to the tail service mast on the Mobile Launch Platform. Further investigation revealed a much more extensive – and far more worrisome – leak, apparently coming from the

18 Rise from the ashes

disconnect hardware in Columbia's belly. The launch attempt was immediately called off and the External Tank emptied and inerted. Since the crew would be working in two 12-hour teams in orbit, they had already begun 'sleep-shifting' and the 'blue' team was awake when the news of the launch delay came through. Blue team member Jeff Hoffman recalled waking up his 'red' team counterparts with the unwelcome news.

"... *six, five, four* ..."

A few days later, on 6 June, a miniature 'tanking test' was conducted to identify the exact location of the leak. It soon became clear that Pad 39A was not an appropriate place for exploratory work to be undertaken and on the 12th the STS-35 stack was rolled back to the VAB. Columbia was removed from the stack and returned to the OPF for repairs on the 15th. Launch was provisionally rescheduled for mid-August, which inevitably pushed Columbia's next two flights, STS-40 and STS-42, into December 1990 and the spring of 1991, respectively. The Shuttle 'side' of the disconnect hardware was replaced with a set borrowed from the new orbiter, Endeavour, which was then undergoing final construction as Challenger's replacement, and apparatus for the External Tank 'side' arrived shortly thereafter from its assembly facility at Michoud in New Orleans. Certainly, Jeff Hoffman remembered that it was a tough time on his crewmates, but not so much for himself. He had already been assigned to *another* mission, STS-46 with an Italian-built tethered satellite, and there was plenty of work for him during the enforced down time. "I carried on working with the tethered satellite," he said, "in addition to the maintenance training that we had to do for ASTRO." For the others, it was harder, since they were all fully trained, at the peak of mental and physical readiness and yet forced to sit through endless simulator sessions with no new launch date.

As this work was being undertaken, Atlantis progressed smoothly – or so it seemed – towards her own launch on STS-38 in July. As a precautionary measure, shortly after the STS-38 stack arrived at Pad 39A on 18 June, NASA decided to perform a tanking test on Atlantis, to verify that she was not similarly affected. Two modes of propellant loading were followed; the first, termed the 'slow fill', served to chill down the pipework and the structure of the tank, so that when the cryogenic hydrogen was pumped in at a higher rate, known as the 'fast fill', it would not boil and generate excessive amounts of gas. On the 29th, liquid hydrogen was duly pumped into her External Tank ... and to engineers' dismay the *same* problem appeared: after the fuelling process moved from 'slow fill' to 'fast fill', concentrations of gas were found in the vicinity of her disconnect hardware. It was small and was described by NASA as "both temperature and flow-rate-dependent", but the agency was convinced that the two leaks in Columbia and Atlantis represented nothing more than coincidence.

"... *three, two, one* ..."

To identify the source, more instrumentation was fitted around the disconnect and a second tanking test was performed on 13 July. Sealants were added in a bid to stop the leak, but a third test on the 25th revealed that the problem persisted. Two weeks later, in the pre-dawn darkness of 9 August, the STS-38 stack rolled back to the VAB, providing a unique photo opportunity in the process, as the STS-35 stack passed it on its way back out to Pad 39A. Despite the similarities between the problems, NASA managers were convinced that the hydrogen leaks were unrelated. "Incorrect torqueing of bolts around the flange interface between the tank and the orbiter caused the Atlantis mishap," *Flight International* told its readers on 8 August. "The Columbia leak was caused by a faulty seal in the drive mechanism used to close the flapper valve in the disconnect." STS-35 was rescheduled for launch on 1 September. Misfortune, though, was far from finished with the unlucky flight; two days before launch, an avionics box on the astronomy payload failed and required replacement, prompting a delay of several days. Then, on the evening of the 5th, technicians began to load propellants into Columbia's External Tank ... only to discover hydrogen gas *again* leaking into the aft compartment of the orbiter. The maximum allowable rate was 660 parts per million; but the actual rate of leakage was upwards of 6,500 ppm ...

"... *Ignition* ..."

It was now apparent that *two* separate leaks had evolved; the data indicated that the leakage from the disconnect was gone, but a new one had emerged, somewhere in Columbia's aft engine compartment. Initial diagnostics placed it close to the orbiter's recirculation pump package inset or the manifold. Time to resolve the problem was short, for Discovery *had* to launch Ulysses between 6-23 October. In fact, so urgent was the need to get the solar probe launched that an *alternate* crew had been chosen. Throughout the summer of 1990, astronauts Loren Shriver, Jim Wetherbee and Dave Hilmers trained in tandem with the primary crew, in case illness should preclude any of them from flying. The decision was taken following the illness of STS-36 commander John Creighton, whose severe cold in February 1990 had caused his mission to be delayed by several days. Any such delay to the Ulysses launch had to be avoided at all costs. If STS-35 did not fly soon, ASTRO-1 would need to wait until after the Ulysses mission. A package of three hydrogen recirculation pumps in Columbia's aft compartment were removed and replaced, together with a damaged Teflon seal on one of the main engines, but to no avail: in the hours preceding another launch attempt on 18 September leaking gas was *again* detected, emerging at a rate of 6,700 ppm. The looming Ulysses window meant that NASA was now out of options. STS-35 would be indefinitely postponed until the hydrogen problem was resolved. Space Shuttle Program Manager Bob Crippen – himself a former astronaut – assembled a 'tiger team' to investigate the disconnects and totally 'retorque' Columbia's entire liquid hydrogen system.

Crippen assigned a veteran engineer, Bob Schwinghamer, from the Marshall Space Flight Center (MSFC) in Huntsville, Alabama, to lead the investigation. In his oral history, Schwinghamer remembered NASA Deputy Administrator J.R. Thompson telling him, without a hint of humour, that he had a *one-way* ticket to Florida; he was *not* to return to Alabama until the hydrogen leak was solved. Schwinghamer's team spent three months at KSC, from September until December, setting up an intricate fault tree and co-ordinating a huge number of personnel, spread across several NASA centres. By the time they completed the final tanking test on 30 October 1990, Schwinghamer could confidently declare that Columbia was now "the soundest leak-free orbiter at that time in the fleet". The explosive nature of hydrogen meant that a leak of any sort could not be tolerated – even though the main propulsion system was designed to overpower leaks with a nitrogen purge – and it certainly surprised Vance Brand, the commander of STS-35, that an orbiter was being grounded for such a long period. It marked a change in attitude from the way in which NASA management had worked before the Challenger accident.

The September 1990 delay was clearly remembered by Jeff Hoffman, many years later. Rather than risk having their children out of school for a lengthy spell, several of the crew decided to pay, personally, for air tickets from Houston to Florida and asked neighbours to take them to the airport. In the hours preceding the 18 September attempt, the Hoffman children – aged 11 and 15 at the time – were seated aboard the aircraft, with one engine in the process of starting up, when the Continental Airlines departures desk received a call from Hoffman himself. Another leak had emerged; there was no point in the children flying down to Florida. At the last possible moment, the engines fell quiet, the doors opened and the disappointed children got off. Not until December would they see their father launched into space. By this time, the effort to fix the hydrogen leaks had cost NASA $3.8 million and the agency suspected that Columbia's problems originated from a complete disassembly of the main engines following the STS-32 mission in January 1990 to remove polishing grit from her fuel lines. When the engines were reassembled, the seals were imperfectly fitted and minute glass beads contaminated the disconnect hardware. From the beginning of June, the attention of engineers was drawn solely to the disconnect leak, which posed a more serious problem, and the seals were overlooked. "As a result," explained *Flight International* in mid-November, "NASA has introduced a new processing programme in which key engine components will be checked for leaks before the engines are finally assembled."

In late November 1990, NASA announced that a new hydrogen dispersion apparatus would be added to the Mobile Launch Platform for future missions, beginning with STS-39. The system would provide a nitrogen-rich air flow around the disconnect hardware, thus helping to disperse hydrogen concentrations. As for the disconnects themselves, NASA was already working on plans with Rockwell International to develop an upgraded system, somewhat narrower at 35.5 cm in diameter, and contracts worth upwards of $27.6 million were awarded in February 1991. It was expected that the new disconnect hardware would help to prevent inadvertent closure of the flapper doors during ascent, which threatened catastrophe. Unfortunately, the new disconnect project was cancelled by NASA in 1993, although much of the technology behind it was employed to improve the safety and reliability of the existing 43 cm hardware. Furthermore, the space agency instituted more rigorous rules around the issue of hydrogen leakage, embedding them more firmly within its Launch Commit Criteria.

"*... and liftoff of Discovery and the Ulysses spacecraft, bound for the polar regions of the Sun!*"

Having been dogged by such cruel luck for so many months, it was a profound relief when Discovery thundered perfectly into orbit at 7:47 am Eastern Daylight Time (EDT) on 6 October, following a brief 12-minute delay. She was now the unrivalled record-holder in terms of number of flights, for this was her eleventh voyage into orbit. Discovery's construction had begun more than a decade earlier, in February 1979, under a $1.9 billion contract between NASA and Rockwell International. She was named in honour, primarily, of Captain James Cook's HMS *Discovery*, but also offered a respectful nod to vessels commanded by Henry Hudson to search out the Northwest Passage, by George Nares to reach the North Pole and by Robert Falcon Scott to conquer Antarctica. The spacefaring Discovery's assembly began in August 1979 and the orbiter was structurally complete by February 1983. Several months of testing followed and she was finally rolled out of Rockwell's Palmdale facility in California in October for an overland trek to Edwards Air Force Base and delivery to the Kennedy Space Center atop a Boeing 747 Shuttle Carrier Aircraft on 9 November. Thanks to manufacturing changes in the internal structure of her airframe and the inclusion of newer, lighter thermal protection materials – including quilt-like Advanced Flexible Reusable Surface Insulation (AFRSI), in place of tiles at various points on the upper wings, fuselage, payload bay doors and vertical stabiliser fin – Discovery's dry weight of 67,100 kg

1-03: During the summer of 1990, two of NASA's orbiters – Columbia and Atlantis – were grounded during preparations for their respective missions, due to hydrogen leakage issues. In August, as the STS-35 stack (left) returned to Pad 39A after its first period of repairs, it passed the STS-38 stack (right), newly returned from the pad and temporarily 'parked' outside the Vehicle Assembly Building (VAB) in readiness for its own repairs.

was some 300 kg less than her sister Challenger and more than 2,000 kg less than the queen of the fleet, Columbia.

In the summer and early autumn of 1990, Discovery continued to impress. Her disconnect hardware had shown none of the leakage problems which afflicted her two sisters. Yet the incident highlighted that the whole machine – orbiter, External Tank and boosters – remained very much 'experimental' in the minds of those who processed them for flight and more than one veteran astronaut has remarked that simply getting the stack into a position whereby it could launch into space, with everything ticking in harmony, was as much a miracle as any of the technology involved. It was also tremendously dangerous. From the commander's seat aboard Discovery, astronaut Richard Noel Richards knew instinctively that flying the Shuttle was riskier and more complex than anything his years as a naval aviator, flying off aircraft carriers and testing new jets, had offered.

The sandy-haired Richards came from Key West, Florida, where he was born on 24 August 1946, but received much of his schooling in Missouri, before entering the University of Missouri to study chemical engineering on a scholarship from the Reserve Officers Training Corps. With a father who had served as a submariner, it seemed inevitable that Richards would follow in his footsteps and join the Navy. "I'd already sort of committed that I was going to do four years in the Navy," he explained, "and at the time, that was the Vietnam War going on, so the draft was in vogue, so any young male at that point knew he *had* to deal with it." Richards opted to tackle the issue head-on and get part of his college fees paid by the military. After receiving his degree in 1969, followed by a master's in aeronautical systems from the University of West Florida in 1970, he opted for naval aviation and began flight training in Pensacola, Florida, finally earning his aviator's wings from Corpus Christi, Texas. He never saw action in Vietnam, but served with a shore-based squadron until 1973, with the bonus of flying the F-4 Phantom fighter.

Richards next deployed to the North Atlantic aboard the USS *America* and then to the Mediterranean aboard the USS *Saratoga*, again flying the F-4. Despite the ongoing conflict in Vietnam, he reflected, "we had to maintain, because of our NATO commitments, two aircraft carriers in the Mediterranean all of that time, because of the perceived Russian threat over there." He returned to the United States to report to test pilot school at Patuxent River, Maryland, graduated in 1976 and worked for over three years as a project pilot for automatic carrier landing systems for the F-4. Shortly before his selection as an astronaut, he served as a carrier suitability project officer for the F/A-18 Hornet, performing its first shipboard catapults and arrested landing tests aboard the *America*. Richards had been drawn to the astronaut programme in 1977, when John Young visited Pax River to encourage the test pilots to apply. He was unsuccessful in his first attempt, but "got enough encouragement to try to reapply" and was duly selected two years later. Although he remained an active duty naval officer – eventually reaching the rank of captain – Richards admitted that he probably donned his military uniform on only a handful of occasions after 1980. "That was effectively the end of my Navy career," he said, "and the start of my NASA career."

At the time of the Challenger disaster in January 1986, Richards was assigned as pilot of the original ASTRO-1 flight, scheduled for launch just five weeks later in March. Not surprisingly, he was stood down from training for several years and did not make it into orbit for his first flight until August 1989, when he served as pilot on STS-28, a classified Department of Defense mission. Six weeks after his return, at the end of September, he was named to command STS-41. It was an incredibly rapid turnaround and a 'plum', of sorts, for Richards had waited an unenviable *nine years* for his first flight ... longer than any of his astronaut classmates. In fact, when he did his debriefing after STS-28, he described himself as "the plank-holder", for having waited the longest time for a mission, and expressed his fervent hope that no one else would be subjected to the same. "I guess management felt like they owed it to me to make it up to me," Richards told the NASA oral historian, "and so they had turned me around and got me ready for my first command on STS-41, right away."

His crew for the planned four-day flight were relatively inexperienced; to such an extent that this mission was the first in which none of them had flown in the pre-Challenger period. Pilot Bob Cabana, a Marine Corps aviator who went on to become chief astronaut and command the first International Space Station construction flight, sat alongside Richards on Discovery's flight deck, with former Navy SEAL Bill Shepherd – on his second Shuttle mission – perched behind and between them as the flight engineer, offering an extra set of eyes to monitor the instruments during the eight-and-a-half-minute climb to orbit. Two other mission specialists, an Air Force test engineer named Tom Akers and the first active-duty astronaut from the US Coast Guard, Bruce Melnick, were making their first flights. As the commander, Richards saw it as his duty to ensure that his crew was fully trained; "I had the luxury of nine years getting ready to go fly," he said, but "they didn't have that much time". To make them as confident with the Shuttle as possible, he decided to put together a crash course in systems knowledge – they ended up giving each other weekly lectures from their perspective – and although Richards admitted that the move was both popular and unpopular, the end justified the means. "I spent a lot of time worrying about their systems knowledge," he said, "and ship basics, because of the lack of their shelf life. By the time we got done on that crew, we knew the vehicle backwards and forwards."

It helped, of course, that all five men were incredibly smart, focused and self-motivated individuals *and* that all were active military personnel. They knew the chain of command, they understood the importance of duty and single-minded devotion to accomplishing The Mission, and performed admirably. Robert Donald Cabana was born on 23 January 1949 in Minneapolis, Minnesota, the oldest of two sons of Ted and Annabell Cabana. After high school in his home town, Cabana entered the US Naval Academy and received a degree in mathematics in 1971, then joined the Marine Corps. He completed Basic School at Marine Corps Base Quantico in Virginia, then entered naval flight officer training at Naval Air Station Pensacola in Florida. Cabana flew as a bombardier and navigator in the A-6 Intruder attack aircraft, based at Marine Corps Air Station Cherry Point in North Carolina, and at that service's base at Iwakuni in Japan. Returning to Pensacola in 1975, he started pilot training and was designated a naval aviator in September of the following year. As the top Marine to complete pilot training, Cabana was awarded the Daughters of the American Revolution Award. Graduation from the Naval Test Pilot School at Patuxent River, Maryland, followed in 1981 and Cabana served as programme manager for the A-6 and project officer for the X-29 advanced technology demonstrator. He was serving as assistant operations officer of Marine Aircraft Group 12 in Iwakuni when he was selected as a Shuttle pilot candidate by NASA in June 1985.

At the time of writing, in 2013, Cabana serves as director of the Kennedy Space Center, from which his four Shuttle missions originated. His reputation as a top-notch – and often terrifying – aviator quickly attained legendary status, as recalled by fellow astronaut Tom Jones in his memoir, *Skywalking*. On one occasion, a few weeks after STS-41, Cabana took Jones up in a T-38 Talon – one of a fleet of training jets maintained by NASA to keep its astronauts' flying skills sharp – and mimicked for the young rookie what a Shuttle ascent was like: he put the aircraft into a near-supersonic dive, bottomed out at an altitude of 3,000 m and hauled back the stick to commence a *vertical* climb at more than 950 km/h. When they hit a ceiling of 7,500 m, and by now rapidly running out of airspeed, Cabana pushed the stick forward and the two men *floated*, momentarily weightless, in their seats! On another occasion, Cabana and Jones flew out over the Gulf of Mexico and performed more than a dozen aileron rolls. Even Jones, a veteran bomber pilot, found it difficult to keep the contents of his breakfast in his violently churning stomach. Cabana's desire to "set his hair on fire" in this manner was far from unique; in fact, he was cut from the exact same cloth that has produced every military pilot astronaut for more than half a century.

On the morning of 6 October 1990, as Cabana blazed into orbit for the first time, the experiences were already well-known to Dick Richards and to the flight engineer, William McMichael Shepherd. Today, Shepherd is best known for having commanded the first crew to live aboard the International Space Station, but his pre-NASA work as an underwater demolition expert and Navy SEAL brought an interesting anecdote into his astronaut career: that he had *killed* a man with his bare hands, whilst on a covert military operation. Totally inaccurate, Shepherd later told an interviewer. "The story was *with knives*," he deadpanned. The tall tale started making its rounds in Houston soon after his selection in May 1984. "I heard about it," he said, "and I thought, this is just really *too good* to deny, so I just wouldn't comment on it for a long time, which made people *really* wonder!" Shepherd came from Oak Ridge, Tennessee, where he was born on 26 July 1949, the son of a naval aviator who had flown in combat during the Second World War. It is perhaps hardly surprising that Shepherd also followed a naval career, but by his own admission he had grown up with a natural love of boats and the water. After high school in Phoenix, Arizona, he entered the Naval Academy and received a degree in aerospace engineering in 1971. "I wanted to be a pilot," he told an interviewer, "but unfortunately could not pass my eye exam, so I became a diver." Shepherd subsequently commenced training as a member of Basic Underwater Demolition/SEAL (BUD/S) and later qualified as a Navy SEAL. In 1978, he obtained a master's degree in mechanical engineering and the degree of Ocean Engineer from MIT. Shepherd unsuccessfully tried for NASA in 1980 and was finally selected to join the next class in May 1984. Four years later, in December 1988, he became the first of his class to fly into space on STS-27, a classified mission for the Department of Defense.

The remaining members of the crew – Bruce Edward Melnick of the Coast Guard and Thomas Dale Akers of the Air Force – achieved similar recognition to Shepherd, in that they were the first members of *their* astronaut class to fly into space. Selected three years earlier, in June 1987, they were part of the first astronaut group chosen in the wake of the Challenger disaster and had earned themselves the nickname 'The GAFFers', for 'George Abbey's Final Fifteen'. The moniker came from the fact that they were the final astronaut class chosen under the direction of George Abbey, head of Flight Operations and Flight Crew Operations at the Johnson Space Center (JSC) in Houston, Texas, from 1976-87. Bruce Melnick, born on 5 December 1949 in New York City, received much of his later schooling in Florida and attended the Coast Guard Academy. A nautical career had long beckoned him; as a teenager, he worked as a deckhand, guide and lifeguard on Clearwater Beach. He earned a degree in engineering from the academy and his first duty within the Coast Guard was as a deck-watch officer aboard a cutter stationed in St Petersburg, Florida. Melnick won his pilot's wings at Naval Air Station Pensacola, gained a master's degree in aeronautical systems from the University of West Florida and later worked with the Coast Guard in areas as far afield as Texas, Alaska, Michigan and Cape Cod.

During his 20-year military career, Melnick served as operations officer and chief test pilot at the Coast Guard's Aircraft Program Office in Grand Prairie, Texas, participating extensively in the development and acceptance trials of the HH-65 Dolphin helicopter; in fact, so extensive was his involvement that Melnick actually *wrote* the flight manual for his fellow pilots. The four-crew HH-65 had been selected in 1979 as a short-range recovery air-sea-rescue helicopter and almost a hundred were procured by the Coast Guard for a wide range of service, including homeland security patrols, drug enforcement, ice-breaking, pollution control and general military readiness. Its Fenestron tail rotor and autopilot meant that it could complete unaided approaches to hover just 15 m above water. Melnick's work during this period and later in his career earned him a pair of Distinguished Service Medals from the Department of Defense, a pair of Distinguished Flying Crosses and the Secretary of Transportation Heroism Award. A long-time fisherman and angler, Melnick today maintains two boats – one inshore, the other offshore – and has won numerous competitions on the east and west coasts of Florida.

If Melnick was drawn to the sea, then Tom Akers was drawn to the air. Born in St Louis, Missouri, on 20 May 1951, he earned undergraduate and master's degrees in mathematics, both from the University of Missouri-Rolla. In 1973, the year after receiving his first degree, he found work as a National Park ranger at Alley Springs, and remained there each summer until 1976. By the time his ranger service ended, he had gained his master's degree and spent several years as a high school principal in his hometown of Eminence, before entering the Air Force in 1979. After initial office training, Akers was assigned to Eglin Air Force Base in Florida as an air-to-air missile data analyst and during this period he also taught evening classes in mathematics and physics for Troy State University. Selected in 1982 for the Air Force's Test Pilot School at Edwards Air Force Base in California, he graduated as a flight test engineer the following year and subsequently worked on a variety of advanced weapons programmes. Interestingly, his association with the University of Missouri-Rolla endured after both his Air Force and NASA careers and in 1999, after departing both services, he taught mathematics there.

On STS-41, Akers was primarily responsible for overseeing the deployment of Ulysses, which was mounted, uniquely, atop an IUS and a PAM-S booster. The Air Force's IUS had originally been developed as a temporary substitute for a reusable 'space tug' and was initially dubbed the 'Interim Upper Stage', before changing to 'Inertial' in recognition of its sophisticated internal guidance system. Losing this 'interim' status also reflected a growing awareness, when the space tug was cancelled in late 1977, that the IUS' services would be needed throughout the 1980s. In fact, not until the early years of the present century did it fly for the last time as a 'standalone' booster. Prime contractor for the IUS was Boeing, which began developing the two-stage vehicle in August 1976 and supported its first launch aboard a Titan 34D rocket six years later. Measuring 5 m long and a little under 3 m in diameter and weighing some 14,740 kg, the cylindrical booster – made from Kevlar-wound aluminium – was capable of delivering 2,270 kg payloads to geostationary altitudes. Its first stage carried 9,700 kg of solid propellant and a large motor, capable of firing for up to 145 seconds; this made it the longest burning solid-fuelled engine ever used in space applications. Meanwhile, the second stage carried 2,720 kg of propellant. Both the first and second stage nozzles, commanded by redundant electromechanical actuators, could steer the former by up to four degrees and the latter up to seven degrees. Although solid rockets were known to generate a harsh impulse, the separation mechanism between the first and second stages employed a low-shock ordnance device in order to avoid damaging its payload. Moreover, as has been seen in the IUS-Centaur debate, solid propellant was chosen over a liquid-fuelled booster because of its simplicity, safety, high reliability and low cost. Hydrazine-fed reaction control thrusters provided the IUS with additional stability during the 'coasting' phase between the first and second stage firings, as well as ensuring accurate roll control during the engine firings and assisting with the satellite's insertion into geostationary orbit on satellite-deployment missions. Situated between the two stages was an equipment section with avionics systems to provide guidance, navigation, control, telemetry and data management services to Ulysses. Importantly, most critical components, except the bellows for the gimbal actuator, were fully redundant to provide a reliability of more than 98 percent. In the early days of the IUS' development, Boeing even proposed adding a smaller third stage to propel planetary missions out of Earth orbit, although, with a payload on top, it would have been a tight squeeze in the payload bay. Since Ulysses was so small, it was possible to insert such a small motor betwixt the IUS and the spacecraft. This PAM-S was a special variant of McDonnell Douglas' Payload Assist Module, whose primary objective was to deliver the spacecraft out of Earth orbit and onto a trajectory towards Jupiter. Equipped with a Star-48B solid rocket motor, the PAM-S was designed to be spin-stabilised after separation from the final stage of the IUS.

After Discovery reached orbit, the payload bay doors were opened, allowing

A difficult summer 27

unfiltered sunlight to flood across Ulysses for the first time, and Akers took the lead in preparing the solar explorer for its voyage. Deployment was scheduled for six hours and one minute into the mission. Watching his movements, Dick Richards could not hide his admiration for the young rookie. "There was a time-critical bunch of steps," he recalled, none more so that the purging of coolant from Ulysses' plutonium-powered Radioisotope Thermoelectric Generator. "Tom had to get down on this switch panel, which was, for some reason, located in this obscure corner of the flight deck." Step by critical step, the instant of deployment drew closer: forward payload restraints were released, the aft frame of the IUS' support structure tilted the stack to an angle of 29 degrees, Richards and Cabana manoeuvred to the correct attitude and electrical power was switched from the orbiter to the IUS. Finally, the three-minute purge of RTG coolant occurred, minutes before deployment. As Akers worked, his crewmates anxiously eyed the clock, keenly aware that a few minutes

1-04: Against the intense blue of Earth and the ethereal blackness of space, Ulysses and its attached IUS and PAM-S boosters drifts away, *en-route* to two decades of solar exploration from beyond the ecliptic plane.

hence Ulysses would have to be *gone*. It brought back memories from a couple of their pre-flight simulations, in which Akers had been momentarily late with switch throws, but Richards trusted him implicitly to complete the job and for a few minutes left him alone. At length, however, the anxiety was pressing.

"Tom?" he asked. "How you doing?"

Akers looked up from his work and gave a broad grin. "Never had so much time!"

The tension in Discovery's cabin was thus broken and, precisely on time, the ordnance to separate the IUS umbilical cables was activated and the stack was tilted to its deployment position of 58 degrees above the payload bay. Seemingly in slow motion, the IUS stack drifted smoothly and serenely away from Discovery. Nineteen minutes later, Richards and Cabana fired the orbiter's thrusters to manoeuvre to a safe distance in anticipation of the firing of the IUS' first stage engine. *That* occurred three-quarters of an hour after deployment, unseen by the crew because Discovery had been oriented with her belly facing the direction of Ulysses to protect the orbiter's windows from the exhaust plume. The first stage burned out, as planned, after a 150-second firing and was jettisoned; whereupon the second stage ignited for almost two minutes, before separating itself. Next came the turn of the PAM-S. Firstly, it spun up Ulysses to 70 revolutions per minute for stability, then executed an 88-second burn to provide the final velocity increment and set the spacecraft on its way to Jupiter. After the burnout of the PAM-S, the spacecraft was 'yo-yo-despun' – by the principle of angular momentum conservation, using weights deployed at the end of cables – to less than 8 revolutions per minute. By now, travelling at an Earth escape velocity of 55,540 km/h, Ulysses became the fastest man-made object yet to leave the vicinity of the Home Planet; a record which would remain unbroken until the New Horizons spacecraft was boosted towards Pluto in January 2006.

For the astronauts, their involvement with Ulysses was now effectively at an end and responsibility passed to an army of flight controllers and trajectory specialists who would guide it towards a rendezvous with the Solar System's largest planet in February 1992 and, later, supervise its exploration of the Sun. Although the role of the STS-41 crew had been exclusively to launch Ulysses, they had undertaken several trips to Europe, and particularly Holland and Germany, where much of the contracting and project management was undertaken. On one occasion, Richards recalled, it gave him a slightly unsettling introduction to European culture. At the end of each working day, at 4:30 pm, the German team would reach the end of their day and open up their cooler and pull out several kegs of beer. "We'd all sit around there, next to Ulysses," he recalled, "toasting Ulysses and having beer. We *didn't* do that here in the United States, so *that* was different. I kinda liked it."

With Ulysses gone, the astronauts' attention shifted to their secondary experiments. Of particular visibility was the Intelsat Solar Array Coupon (ISAC), affixed to Discovery's Canadian-built Remote Manipulator System (RMS) mechanical arm. Six months before STS-41, the International Telecommunications Satellite Organisation had launched its Intelsat 603, but the Commercial Titan III booster failed to deliver it to geosynchronous transfer orbit, instead leaving it in an inoperable low-Earth orbit. Shortly afterwards, NASA contracted with the satellite's manufacturer, Hughes, to investigate the feasibility of staging a Shuttle mission to rendezvous with Intelsat, attach a new booster and deploy it into its correct orbit. Part of that feasibility study was ISAC, which exposed two solar array material specimens to the harsh atomic oxygen environment of low-Earth orbit for upwards of 23 hours to assess the level of degradation which the satellite itself might have suffered. The satisfactory results from ISAC led directly to a firm commitment to stage the recovery mission and a seven-member astronaut crew – which, incidentally, included STS-41 veterans Akers and Melnick – was announced in mid-December 1990.

Also in the payload bay was the Shuttle Solar Backscatter Ultraviolet (SSBUV) instrument, flying its second mission as part of ongoing efforts to calibrate the ozone-monitoring sensors aboard the National Oceanic and Atmospheric Administration's NOAA-9 and NOAA-11 satellites, together with NASA's Nimbus-7. Calibration 'drift' of solar backscatter ultraviolet data from these three satellites had proven a cause for concern, in terms of the reliability of their data over time, and SSBUV was designed to be flown at least once per year. Before each of its missions, SSBUV was calibrated to a laboratory standard and recalibrated after its return to Earth. Housed inside a pair of dustbin-sized Getaway Special (GAS) canisters, it flew annually from October 1989 until its final mission in January 1996. Its most alarming results described the marked depletion of stratospheric ozone since the mid-1980s – the much-publicised 'ozone hole' – due partly to solar effects and atmospheric dynamics, but chiefly to the release of man-made carbon fluorides. This could prove catastrophic for life on Earth, because the effects observed over a four-year period, if extrapolated over half a century, could lead to a 60-percent ozone depletion.

On Discovery's middeck, research was conducted to explore how genetic material in the cells of plant roots – responsible for the flowering process – responded to the microgravity environment. Flame formation and propagation over solid fuels was examined in one experiment, whilst another involved the carriage of several rats in self-contained animal enclosure modules to investigate the effectiveness of new therapies for bone and muscle wastage, organ tissue regeneration and immune-system disorders. Polymer membrane processing was undertaken as part of efforts to understand the role of convection-driven currents in their formation and Bill Shepherd and Bruce Melnick trialled an experimental voice command system for operating television cameras. A single military payload, the Department of Defense's Radiation Monitoring Experiment, consisted of a hand-held device to measure levels of ionising radiation to which the astronauts were exposed in flight.

STS-41 ended, as planned, after just four days – a woefully short period of time, according to Richards, who felt the monumental effort to simply get there could have been exploited to spend more time aloft and undertake more experiments – and the orbiter returned to a desert landing on Runway 22 at Edwards Air Force Base in California at 6:57 am Pacific Daylight Time (PDT) on 10 October. Edwards had been selected by NASA very early in the Shuttle development process; the added safety margins of its broad runways and virtually year-round good weather made it an obvious choice to land the orbiters. These margins were particularly evident in the nature of the two primary Edwards landing strips: Runway 17 crossed a salt flat and measured 13 km long, whilst the all-concrete Runway 22 was 4.5 km long and was extended by a 500 m asphalt overrun at each end. For Dick Richards, the challenge of landing the Shuttle for the first time in his career was enormous. "I decided to use a different technique than the rest of the astronaut office was using at that point," he told the NASA oral historian, "and that turned out to work wonderfully. Basically, fly the vehicle down to about five feet and then *don't touch it* anymore, because the vehicle's got so much wingspan that once it gets down to [the] ground effect there's a natural cushioning. In fact, you'll end up landing at maybe about one foot per second. I think the astronaut office liked that."

By the time Discovery landed, Ulysses had already traversed more than 1.6 million km in its journey towards Jupiter. On 8 February 1992, it utilised the planet's gravitational influence to increase the inclination of its orbit relative to the ecliptic plane by 80.2 degrees and bend its trajectory southwards to encounter the solar south pole in June-October 1994. Its journey carried it for the first time over the solar north pole in June-September 1995 and a series of mission extensions provided further pairs of solar passes in 2000-2001 and 2007-2008. Its observations profoundly altered our knowledge of the Sun, demonstrating the dynamic nature of solar magnetism and highlighting the strength of the solar wind. From its unique vantage point, Ulysses was also employed to observe Jupiter and Comet Hale-Bopp from afar, as well as examining gamma ray bursts and interstellar dust from beyond the Solar System. For a mission which came so close to cancellation, Ulysses transformed itself into one of the greatest success stories and one of the grandest adventures of scientific exploration ever undertaken in the annals of human history.

SECRET SENTINELS

'High profile' are words which can easily be applied to the astronaut career of Dick Covey. During his 16-year tenure, he rendezvoused with a crippled satellite, piloted the first post-Challenger Shuttle mission and commanded the singular flight which restored NASA's image in the eyes of politicians and public alike: the repair of the Hubble Space Telescope. Yet there is another aspect of Covey's career which was far quieter. Early in 1989, he replaced fellow astronaut Brewster Shaw as the lead representative on classified Shuttle missions for the Department of Defense. On paper, the assignment looked relatively non-descript. In reality, it exposed him to the innermost details of the Shuttle's most secret flights ... flights which, even today, remain shrouded in myth, mystery and rumour.

Richard Oswalt Covey was the last of his 35-strong astronaut class to fly into space. "I got that distinction," he said, "and that was hard to take." At the time of his assignment, in November 1983, he knew that he would not fly until at least the end of the following year. "At the time, I didn't realise I was going to be the very *last* one, but I knew I was going to be somewhere down there." Born in Fayetteville, Arkansas, on 1 August 1946, Covey was the son of a fighter pilot who had flown in the Second World War and Korea. It was clear that the young boy would follow suit.

After high school in Florida, he entered the Air Force Academy and graduated in June 1968 with a degree in engineering sciences, majoring in astronautical engineering. In January of the following year he got a master's in aeronautics and astronautics from Purdue. Covey would pay tribute to Professor Jack Arnold, a family friend and lecturer at the Air Force Academy, for having steered him towards astronautical engineering and, years later, would wonder if it was *that* decision which ultimately guided him to NASA.

Flight training at Williams Air Force Base in Arizona came next, followed by operational work in the F-100 Super Sabre, the A-37 Dragonfly and the A-7 Corsair, then two deployments to Vietnam, during which time Covey flew 339 combat missions in the final years of the bitterly divisive conflict, right up to the end of planned US aerial activity in August 1973. Flying the Super Sabre was a real thrill. "It was an older fighter aircraft at that time," he recalled, "single-engine, single-seat … and single-engine just seemed to imply more danger." Returning home, he was discouraged from entering test pilot school by his superiors, who felt that he would be ruining his career, but Covey submitted his application and was accepted in 1974, at the relatively young age of 28. "I was just six years out of the Air Force Academy," he said, "and again it was one of those things where somebody was in the right place at the right time." Graduation from Edwards' famed school was followed by test work in the F-4 Phantom and the A-7D at Eglin Air Force Base in Florida and, later, as joint test force director the electronic warfare testing of the new F-15 Eagle fighter. It was whilst at Eglin that Covey applied for the astronaut programme in 1977 and received an invitation to an interview in Houston. This posed a problem, for he had already planned a week's vacation with his wife, Meredith. Whereas most applicants were *jumping* at the interview, Covey was concerned that asking for a delay would jeopardise his chances. Still, he asked, and was told to come to Houston at a later date … but spent his vacation worrying about whether NASA would call him back. "It was a terrible vacation," he recalled, "a *terrible* vacation!" Towards the end of it, he received the call to Houston and in January 1978 he was chosen as part of the first group of Shuttle astronauts. Eleven years later, in May 1989, after flying two Shuttle missions as pilot, Covey was assigned his first command on STS-38.

In his oral history, he acquiesced that it did not surprise him to receive a Department of Defense flight, since his previous duty in the astronaut office had amply prepared him for the closeted, secretive world of 'black' operations in space. The insights which Covey received were greater than those to which even the astronauts assigned to the classified flights were normally privy. "Each crew was read into the particular DoD programme that they were supporting," he explained to NASA's oral historian, "but there needed to be someone who was aware of what all of those missions were going to be doing and working that interface with the appropriate agencies within the DoD to make sure that the crew issues that may cross all of those were being taken of." Covey's mandate was to manage the classified materials and the staff whose involvement with the missions demanded a level of security clearance beyond Top Secret. Many of the meetings attended by Covey were held in "special environments" and the crucial nature of his role is underlined by the fact that even the astronauts for one DoD mission did not necessarily know what their fellow astronauts, assigned to *another* DoD mission, might be working on.

From STS-27 in December 1988 until his own flight, STS-38, Covey was thus involved on an intimate level with the planning and preparation of these secret flights and was exposed to details of classified payloads whose existence continues to be officially denied or concealed. Many observers considered the missions 'boring', since so little was known about them. In truth, they were quite the reverse. "There were exciting things going on in DoD space back then," Covey recalled, cryptically. "There still are. And being a part of that, then, with my military background and being able to support all that was really very, very cool." Indeed, more than one astronaut has said that his involvement in these missions was the crowning glory of his career, as an astronaut, as a military officer and as a patriot, since each flight placed a new element of the United States' national security arsenal into orbit.

As technicians battled to resolve Columbia's hydrogen leak problems in the early autumn of 1990, the hurdles which faced Atlantis appeared to be closing in on a resolution and as Discovery flew successfully for four days in October it became clear that STS-38 would precede STS-35 into orbit. On the 8th of that month, shortly after the Ulysses launch, the STS-35 stack rolled over from Pad 39A to the now-vacant Pad 39B, only to be returned to the cover of the VAB the following day as a precautionary measure against the onslaught of Tropical Storm Klaus. Circumstances began to improve somewhat when the STS-38 stack arrived at Pad 39A on the 12th and sailed through a tanking test with flying colours on the 24th. Bolstered by the success of the Ulysses flight and the encouraging results from the test, NASA managers felt able to schedule Atlantis' launch for 9 November. 'Problems' with her classified payload effected a slight delay until the 15th, but as darkness fell over the Kennedy Space Center that evening, all seemed ready for the fifth Shuttle mission of the year. In keeping with the secrecy of all Department of Defense flights, the media were told that the launch would occur in a four-hour period between 6:30 pm and 10:30 pm EST; Atlantis finally rose from Pad 39A at 6:48 pm, turning night into day across the Space Coast.

For Dick Covey, launching in darkness presented few surprises; his first flight in August 1985 had begun on the cusp of daybreak and he was familiar with many of the sights and sounds and sensations involved. Most of Covey's family was in attendance that night, with the exception of his eldest daughter, Sarah, who was on the Clear Lake High School volleyball team, back in Houston, which had just won the region. She was on her way to Austin, Texas, to play in the state championship series and was on the bus when Atlantis launched; throughout the flight, Covey would receive uplinked teleprinter messages to keep him aware of the scores.

On STS-38, Covey and the mission's flight engineer, Bob Springer, were the only crew members who had flown before; for pilot Frank Culbertson and the other two mission specialists, Carl Meade and Charles 'Sam' Gemar, it was their first space voyage. As a crew, the men of STS-38 represented all four military services – Covey and Meade were from the Air Force, Culbertson was from the Navy, Springer was a

Marine Corps aviator and Gemar an Army helicopter pilot – and all were graduates of a military test pilot school. Flying with three rookies, though, presented its own unique challenges. On his last flight, STS-26, both Covey and his crewmates had all flown before and the same would be true on his final mission. On STS-38, the presence of first-timers brought good and bad news. "The good news is that we weren't flying a real long mission," he recounted, "so they didn't have to worry about a whole lot of things. The bad news was that our most critical operations were *all* on the first day, right between launch and the time we went to bed; so these guys were going to be adapting to space for a first time, with all of the 'gee-whiz' factors, and we had to do our most serious and significant work that first day in deploying a payload." As the mission commander, Covey saw it as his responsibility – like Dick Richards – to ensure that his crew was as finely schooled in the systems knowledge as possible. His approach was to keep operations as simple as possible; in fact, the most difficult part was dealing with the orbital lighting conditions for photography.

Carl Joseph Meade's Air Force heritage began from the moment of his birth, at Chanute Air Force Base in Illinois on 16 November 1950. He attended Randolph High School on Randolph Air Force Base in Texas and entered the University of

1-05: The STS-38 crew is pictured during a pre-flight simulation, clad in training versions of their orange launch and entry suits. Frank Culbertson (left) sits in the pilot's seat, whilst Dick Covey (right) occupies the commander's station. Behind and between them is Bob Springer, who fulfilled the role of flight engineer, with Carl Meade at his right shoulder. Standing is Sam Gemar, who sat downstairs on the middeck for ascent. This view offers a perspective of the flight deck, looking aft from the forward cockpit.

Texas at Austin to study electronics engineering. Meade received his degree in 1973 and went on to gain a master's qualification in electronics engineering as a Hughes Fellow at the California Institute of Technology, two years later. He then entered the Air Force at Laughlin Air Force Base in Texas, where he shone as the Distinguished Graduate of Undergraduate Pilot Training and went on to fly the RF-4C – an all-weather tactical reconnaissance version of the F-4 Phantom fighter – before being selected for test pilot school at Edwards Air Force Base in 1980. Meade earned the Liethen-Tittle Award as Outstanding Test Pilot for his work and remained at Edwards, undertaking research development and evaluation of the F-5E Freedom Fighter, RF-5E Tigergazer and F-20 Tigershark aircraft, as well as ground- and air-launched cruise missiles. He later flew performance trials of the F-16 Fighting Falcon, evaluating structural loads and weapons systems, and was reassigned as a test pilot school instructor at Edwards in 1985, only months before his selection by NASA. Like Bob Springer before him, Meade applied to the space agency as a pilot candidate, but was ultimately chosen in June 1985 as a mission specialist.

That all five STS-38 astronauts were qualified test pilots was amply demonstrated by the fact that, whenever they flew anywhere, they took *five* T-38 jets. Having said this, Covey recalled some humorous situations involving Sam Gemar, whose pre-NASA flying career had been almost exclusively in low-altitude helicopters. "He didn't have as much experience in flying high-performance jet aircraft as the rest of us did," Covey admitted, with a laugh, "and it *showed!* We always had to accommodate that and make sure we were watching out for Sam to make sure we didn't put him in the wrong situation." Still, despite a rather ugly 'stalking' episode, later in his astronaut career, following an extramarital affair, Gemar's credentials were impeccable and when selected by NASA in June 1985 he became only the fourth Army officer ever chosen for astronaut training. Charles Donald Gemar – "Sam was a college nickname", he once told this author – had been born in Yankton, South Dakota, on 4 August 1955, of Russian-German extraction. He graduated from Scotland Public High School in his home state in 1973 and enlisted in the Army later that same year, reporting for duty in June. Five months later, Gemar was assigned to the 18th Airborne Corps at Fort Bragg in North Carolina and it was from there that he received an appointment to the Military Academy Preparatory School at Fort Belvoir, Virginia. Later, a Department of the Army appointment admitted him to the Military Academy to study engineering and he earned his degree in 1979. Infantry training followed graduation and Gemar entered the flying world, completing rotary and fixed wing courses and serving with the 24th Infantry Division at Fort Stewart in Georgia until January 1985. In addition to his experience in helicopters, he also completed advanced training for aviation officers and a parachutist course.

On his crew, Covey heaped the greatest of praise. "They were all very much process-oriented people," he said, "because of their test piloting backgrounds and so they would go off and work the specific issues that they were assigned. I had smart people and I wanted them to be the smartest person on specific areas and I would rely on them to be able to keep *me* smart." Responsibility was a fundamental tenet and Covey saw *his* personal responsibility as fostering an appropriate relationship with the Air Force and the Department of Defense – STS-38's payload customers – and with flight directors and support teams to ensure that his crew was properly trained. That training was rendered more complex in view of the military secrecy which enshrouded the mission. Flight data files were developed and managed in a classified manner, Mission Control had to function in a classified manner, with encrypted communications, and special provisions were made for 'secure' meetings. Often, the astronauts had to file flight plans to false locations, simply because they were not permitted to tell anyone where they were going, even their wives.

It was under a metaphorical and literal cloak of darkness, therefore, that Atlantis rose to orbit on the night of 15 November 1990, entering space only hours before Carl Meade's 40th birthday. The mission marked Atlantis' sixth flight and of those half-dozen voyages, two-thirds had been Department of Defense assignments; indeed, she would fly more dedicated military missions than any of her siblings. Atlantis was named in honour of a two-masted ketch, operated by the Woods Hole Oceanographic Institute from 1930 until 1966, which became the first vessel built specifically for interdisciplinary research in marine biology and geology and physical oceanography. During her time at sea, Atlantis and her scientists made a number of impressive discoveries, not least of which was the first identification and description of an abyssal plain on the ocean floor – the 'Sohm Abyssal Plain', located off the continental margin, to the south of Newfoundland – in 1947. Today, she is owned by Argentina as a naval research vessel. The construction of the spacegoing Atlantis got underway in January 1979, following a contract award to Rockwell International to configure a structural test article into the future Challenger and build two additional orbiter vehicles, OV-103 and OV-104. The names 'Discovery' and 'Atlantis' were assigned to the new orbiters a few days after the award. In March 1980, engineers started the structural assembly of Atlantis' crew compartment and over the next few years the vehicle grew – construction of her aft fuselage began in November 1981, her wings arrived from contractor Grumman in June 1983 and she was complete by April 1984. Rollout from Rockwell's Palmdale plant in March 1985 was followed by an overland transfer to Edwards Air Force Base and arrival in Florida, atop the Boeing 747 Shuttle Carrier Aircraft, on 9 April. Two days after Discovery touched down from her 51I mission, on 5 September, Atlantis' three main engines burned at full power in a Flight Readiness Firing, one of the last milestones to prepare her for her maiden voyage. All told, Atlantis required less than half as much time to assemble as did the queen of the fleet, Columbia, and historian Dennis Jenkins has pointed to the greater use of thermal protection blankets, rather than tiles, on the youngest vehicle's airframe as one of the main reasons.

In his NASA oral history, Covey recounted that training for STS-38 was relatively straightforward in that his crew had no spacewalks for which to prepare, nor did they have any activities with the Shuttle's Canadian-built RMS mechanical arm and nor was there a rendezvous commitment. Most of their focus, therefore, was upon ascent and re-entry preparations and a Day One deployment of their classified payload. To a great extent, for the ascent contingencies, Covey trained closely with his pilot, Frank Culbertson, and the mission's flight engineer, Bob Springer ... who, but for a slight quirk of fate, might have been a Shuttle pilot himself.

Robert Clyde Springer, by now a colonel in the Marine Corps, was making his second Shuttle flight. He came from St Louis, Missouri, where he was born on 21 May 1942 and had actually been interviewed by NASA for its 1978 *and* 1980 astronaut classes ... though as a *pilot*, not a mission specialist. In fact, his interview as a pilot for the 1980 group came in April of that year and he was selected as a mission specialist, just a few weeks later. Springer completed high school in Ohio and entered the Naval Academy, receiving his degree in naval science in 1964 and a commission into the Marine Corps. After finishing Basic School in Virginia, he reported for flight training in Pensacola, Florida, and Beeville, Texas, and was designated as a naval aviator in August 1966. Springer flew in Vietnam, completing 300 combat missions in the F-4 Phantom and 250 combat missions in helicopters and he also served as an advisor to the South Korean Marine Corps in June 1968. Upon his return home, he completed a master's degree in operations research and systems analysis at the Naval Postgraduate School and in March 1971 was assigned to the Third Marine Aircraft Wing in El Toro, California, as a wing operations analysis officer. Springer later attended the Navy Fighter Weapons School – today's 'Top Gun' – and completed test pilot school at Patuxent River, Maryland, in 1975. Among his test pilot work were the first flights of the AHIT attack helicopter instrumented test programme and, after graduation from the Armed Forces Staff College in Norfolk, Virginia, in 1978, he assumed responsibility for joint operational planning for Marine forces in NATO and the Middle East. Springer was serving as *aide-de-camp* for the commanding general of the Fleet Marine Force-Atlantic when he was selected by NASA in May 1980.

Unlike Springer, naval aviator Frank Lee Culbertson Jr was making his first flight on STS-38. He was born in Charlestown, South Carolina, on 15 May 1949, the son of a Navy pilot father and a teacher mother, both of whom were educated to degree level. (In fact, after his military duty, Mr Culbertson Sr received a medical degree.) Not surprisingly, the young Culbertson grew up with an interest in aviation and his father's exciting stories of flying from ships at sea made naval aviation a particularly attractive option. From the time that Yuri Gagarin and Al Shepard ventured into space, Culbertson wanted to follow in their footsteps; one Saturday, aged only 12 or 13, he was given a book about the legendary Mach 2 pilot Scott Crossfield and the X-15 rocket-propelled aircraft. He was hooked. "And *that* inspired me to want to explore the aerodynamics and the edge of space and the technical side of things and to become a test pilot," Culbertson told a NASA interviewer. Keenly aware that many of the early US astronauts hailed from military academy backgrounds, he eventually applied to, and was accepted by, the Naval Academy at Annapolis, Maryland.

He received his degree in 1971 and served for a time aboard the USS *Fox* in the Gulf of Tonkin, before returning to the United States for flight instruction in Pensacola. Designated a naval aviator in May 1973, Culbertson flew the F-4 Phantom fighter at Naval Air Station Miramar in California and aboard the USS *Midway* in Yokosuka, Japan, serving as a weapons and tactics instructor. Subsequent work as a catapult and arresting gear officer on the USS *John F. Kennedy* was followed by selection for the Naval Test Pilot School in May 1981.

Graduation led to assignment to the Carrier Systems Branch of the Strike Aircraft Test Directorate, where Culbertson oversaw all testing of the F-4 and served as test pilot for automatic carrier landing systems and suitability trials. He was working on the new F-14 Tomcat fighter when selected by NASA as part of its tenth group of astronauts – nicknamed 'The Maggots' – in May 1984. Years later, Culbertson would acknowledge that "perseverance", bordering on "stubbornness", was a fundamental tenet of his personality and enabled him to focus on his goals with almost single-minded determination. "I think it's important for anybody who wants to achieve a goal," he said, "or to arrive at a certain destination or start a certain journey, even, to have perseverance to get ready to do that." That perseverance was amply demonstrated when Culbertson was accepted for astronaut training; his wife refused to go to Houston and the couple divorced shortly afterwards. Culbertson was linked romantically with fellow astronaut Judy Resnik in his early days at NASA and, according to Bryan Burrough in his book *Dragonfly*, the pair were on the verge of marriage. One day in January 1986, he went to Florida to watch her launch aboard Challenger. Seventy-three seconds later, Resnik's life was gone and Culbertson and every other astronaut was brought face to face with the harsh truth that the Shuttle could never be routine.

A quarter of a century has now elapsed since Resnik's fateful flight, and more than 20 years since Culbertson's first mission, but steadily, like the trickling of a tap, some details of what the STS-38 crew did during their five days in orbit have emerged. In the late 1990s, images of one classified Shuttle payload (a pair of military communications satellites) were released and photographs of a second-generation Lacrosse radar-imaging sentinel, under construction, also entered the public domain. Speculation over the physical appearance and capabilities of these payloads has been rife and excitable: sketches of gigantic 'Magnum' satellites, with their vast 'farms' of antennas, are readily available on the Internet, and observations of the peculiar 'disappearance' and 'reappearance' of the optically stealthy 'Misty' satellite have been made by professional astronomers and amateur skywatchers alike. Then, in 1999, came the first mutterings that STS-38 – thought to have carried a military telecommunications relay satellite for the National Reconnaissance Office – might also have launched a second, more covert payload, known only as 'Prowler'.

At the time of launch, the primary satellite for Covey's mission was dubbed Air Force Program-658 (AFP-658) by the media and there was suspicion that it was a Magnum electronic intelligence platform, deployed atop a Boeing-built IUS booster. Saddam Hussein had invaded Kuwait only months earlier and the emplacement of this new intelligence monitor was expected to track the movement of his forces, in addition to the old enemy, the Soviet Union. However, when on-orbit images of Atlantis' vertical stabiliser were revealed, many years later, they revealed no trace of the airborne support equipment – a doughnut-like 'tilt table' – known to have supported IUS-boosted satellites in the payload bay. More recently, it has become generally accepted that STS-38 deployed a member of the second-generation Satellite Data Systems (SDS-B) telecommunications relay satellites. It was not the first such launch from the Shuttle; in August 1989, the crew of Columbia were believed to have deployed a similar payload.

Imagery and videotapes of an SDS-B under construction were released by the National Reconnaissance Office in the spring of 1998, as was the name of its prime contractor, Hughes, and its heritage seemed obvious: it was strikingly similar in appearance to the Syncom 4 military communications satellites, five of which were deployed from the Shuttle in a sideways, Frisbee-like motion. In a 2009 article for *Air & Space* magazine, Michael Cassutt quoted an Air Force source who was deeply familiar with the SDS-B project. "It's strange," he told Cassutt, "to work on a secret project for ten years, then see it on network television!" The Air Force began to develop the first-generation SDS in 1973 to offer America's intelligence community with a network of orbiting relays, capable of transmitting real-time data and images from reconnaissance satellites which were out of range of ground stations. Another of their responsibilities was to support voice and data communications for covert military activities. The second-generation SDS-B operated in high-apogee and low-perigee orbits, ranging from as close as 480 km and as far as 38,000 km, at steep inclinations which achieved their highest point over the northern hemisphere. This enabled them to cover two-thirds of the globe, relay spy satellite data of the entire Soviet land mass and cover the entire north polar region in support of Air Force communications. Such wide coverage was not possible to geostationary-orbiting satellites.

The SDS-B (possibly codenamed 'Quasar') featured a pair of 4.5 m dish antennas and a third, smaller dish for Ku-band downlink. It is also believed to have carried the Heritage (Radiant Agate) infrared early-warning system for ballistic missile detection capability. Overall, the satellite measured 4 m long and 3 m wide in its stowed configuration, with a launch mass estimated at somewhere in the range of 2,300-3,000 kg. In total, three of these cylindrical SDS-Bs were deployed by the Shuttle, on STS-28 in August 1989, STS-38 in November 1990 and STS-53 in December 1992. Although it is unclear as to *how* they were deployed, some observers have assumed that they were 'rolled' out of the payload bay, like a Frisbee, in a similar fashion to the Hughes-built Syncom 4 satellites. Others have noted that the solid rocket booster used for the SDS-B was an Orbus-21, physically identical to the motor later fitted to Intelsat 603 by spacewalking astronauts during STS-49 in May 1992. This has prompted alternative suggestions that the SDS-B was deployed 'vertically' from a special cradle in the payload bay, more like Intelsat 603 than the Syncom 4s.

Irrespective of how SDS-B departed Atlantis, it is certain that the deployment was completed on 16 November 1990, about seven hours after launch, after which the orbiter performed a separation burn to move to a safe distance in anticipation of the firing of the satellite's motor. However, according to observer Ted Molczan, writing in February 2011, the delta-V of Atlantis' burn was less than a tenth of what it should have been for a motor attached to a payload the size of an SDS-B. Moreover, he noted that the satellite itself lingered for some time in low-Earth orbit, rather than initiating its climb to operational altitude at the next available ascending node.

Molczan explained that the second satellite, Prowler, was then deployed 22 hours later ... after which Atlantis' crew performed an unusual manoeuvre, by lowering their orbit, rather than raising it. "It also happened to arrest the separation from the SDS," Molczan wrote, "and initiate a very gradual overtaking, perhaps to create the impression of a rough station-keeping manoeuvre [by Atlantis] to keep Soviet attention focused on the SDS." It would also appear that the SDS-B finally fired its perigee kick motor during a 16.5-hour period which overlapped the firing of Prowler's own motor. Detection by the Soviet-operated signals intelligence station at Lourdes, near Havana, might have been circumvented, Molczan continued, by timing the deployment of Prowler very carefully to when Atlantis passed beyond the Cuban radar horizon. As for Prowler itself, even today the fiction and the speculation greatly outweigh the facts and the evidence. Due to its brightness, a case has been advanced that it was a Hughes HS-376 spacecraft 'bus' – very much like the cylindrical communications satellites launched on several Shuttle missions in the early 1980s – with an attached Payload Assist Module (PAM-D) to boost it into geosynchronous transfer orbit. Molczan suggested a total payload weight of around 4,500 kg – of which 1,300 kg was the satellite itself, together with 2,100 kg for the PAM-D and the remainder for the support hardware in Atlantis' bay – and argued that the ability of the Shuttle to carry both it and the SDS-B would have been well within its performance envelope.

Nonetheless, STS-38 was heavy, as evidenced by its orbital altitude, which did not venture much higher than about 250 km. "You can read a lot into that," Dick Covey admitted to NASA's oral historian. "We didn't go very high because we couldn't go

1-06: Surrounded by post-flight servicing vehicles, Atlantis sits on the KSC runway on 20 November 1990, after the first Shuttle landing back in Florida for more than five years.

very high, which says we probably had a heavy payload. That was the thing that was really unique about the whole mission." Of course, for much of the first decade after the conjectured launch of Prowler, its existence was unacknowledged. The presence of two spent rocket motors – codenamed '1990-097C' and '1990-097D' in a catalogue of orbital objects – could be explained simply as representing the expended first and second stages of an IUS booster, without raising suspicion.

As to Prowler's nature, labels such as 'geolocation platform', 'optically stealthy' and 'inspector' have been banded around over the years and the consensus seems to be that it was some kind of low-observable satellite, employed to rendezvous and secretly inspect other nations' satellites in geostationary orbit, 35,000 km above Earth. At face value, the mission seemed impossible. Some observers doubted that it was even possible, in 1990, to conduct unmanned rendezvous in geostationary orbit, although such exercises had long since been routinely performed by the Soviets in low-Earth orbit between Progress cargo craft and the Salyut and Mir space stations. Others countered that geostationary altitudes provide a more benign environment for telerobotic operations and relative motion control and even wondered if Prowler might have performed radio frequency blocking; literally parking itself in front of a satellite's antenna path to block its signals. Still others have gone further: that Prowler did attempt such blocking, albeit experimentally, on US communications satellites, and several analysts have argued that it could have positioned itself within 30 cm of a target.

More than two decades later, the Department of Defense missions performed by the Shuttle are far more than an insignificant footnote of history, for additional detail and rumour continues to surface with every passing year. The flights once denigrated as 'boring' might actually have been some of the most exciting voyages ever undertaken by the reusable orbiters. The sense of pride harboured by so many military astronauts as a result of having participated in these flights is tangible; a pride centred not only on completing an ambitious mission, or fulfilling an urgent national security need, but on placing into operation a new masterpiece of technological prowess. Dick Covey was "read into" many of these levels of classification and restricted knowledge, sometimes more so than his fellow crew members. Families were kept out of the loop, as were other astronauts, and to this day, Covey knows many 'things' about his flight that remain Top Secret. "From that standpoint," he said, "we still have a secret 'ring' [of people] and can nod at each other when we hear things or know things that we know that nobody else still knows about what we did and how we trained."

Landing of STS-38 was scheduled for 19 November, but was postponed by 24 hours due to unacceptable crosswinds at Edwards Air Force Base, which scrubbed all three opportunities for that day. In the years before the loss of Challenger, landings at the Kennedy Space Center – on a specially-built runway, known as the Shuttle Landing Facility (SLF) – were expected to become routine. In fact, when Shuttle preparations began in Florida in 1974, this runway was one of the first structures to be built. Right from the start, landing back at the launch site was regarded by NASA as a key milestone in achieving truly 'routine' Shuttle missions, as well as saving an estimated $1 million needed to fly the orbiter across country from California and the loss of five days' worth of processing time.

There was a significant down side, however, for Edwards exhibited far more stable weather conditions than tropical Florida. The SLF was officially opened in 1976 and resided a few kilometres north-west of the Vehicle Assembly Building. It measured 4.6 km and 91 m wide, with a 300 m overrun at each end, and was entirely concrete, with slopes from the centreline to facilitate drainage. Two options were typically available to Shuttle crews: they could approach from the south-eastern end of the landing strip (designated Runway 33) or the north-western end (Runway 15), a decision that would be taken based upon wind speed and direction. In February 1984, Challenger performed the first Shuttle landing in Florida, but the unpredictable weather conditions meant that it was not always possible to avoid a diversion to Edwards. Then, in April 1985, as Discovery landed on the SLF in a crosswind, the efforts of her pilots to keep her on the runway centreline caused her to suffer seized brakes and a burst tyre. In the wake of the Challenger disaster, NASA's new-found conservatism led it to assign the SLF as a backup landing site, with Edwards as the preferred choice.

All that changed in the final hours of STS-38.

The weather on 20 November showed no signs of improving and was not forecasted to improve. Atlantis carried enough consumables to remain aloft only until the 21st. It was with some surprise that Covey received a call from Entry Flight Director Lee Briscoe – normally, only fellow astronauts speak directly with the crew, through the 'capcom' console in Mission Control – and asked him if he was happy to divert to Florida. Although it had been more than five years since a Shuttle had made landfall on the East Coast, the answer was a no-brainer for Covey. He had flown so many simulated approaches to Florida that he was more than happy. The landing would occur in the late afternoon and fellow astronaut Mike Coats, flying the Gulfstream Shuttle Training Aircraft on weather reconnaissance, was responsible for determining whether conditions were optimum to receive Atlantis.

"This is the fall of the year," explained Covey, "and one of the things that they do in Florida during the fall is burn the underbrush in their pine forests; a very controlled type of burn, just to get everything down. They were doing that over on the west side of the [Banana] River in Florida and the winds were predominantly from the north-east ... so they were blowing that smoke out over central Florida, toward Orlando." Based upon this factor, Coats recommended that Covey should land on the south-eastern end of the SLF, on Runway 33. By the time Atlantis neared the time to fire her OMS engines for the irreversible de-orbit burn, the winds shifted. "The smoke was coming pretty much right across the southern half of the runway," Covey recalled, "and the northern half was clear."

Coats held off from advising a landing on the Runway 15 end – which would have required Covey to perform a left-hand manoeuvre, during final approach – and the astronauts were advised that the smoky conditions might prove problematic. Added to this was the fact that it was near sunset on the Space Coast and the refractive effect caused the smoke to appear thicker. A little after 4:30 pm EST, as Atlantis came within minutes of touching down on Runway 33, the astronauts could see very little through their windows. Covey and Culbertson had great confidence in the Shuttle's guidance capabilities, however, and as the vehicle rolled out on final, they spotted the Precision Approach Position Indicator (PAPI) lights on the runway perimeter, barely visible through the smoke. These gave Covey the visual reference that he needed, but as Atlantis descended lower, passing *into* the smoke, he could see nothing but the lights; the runway itself was invisible to him. At length, the smoke cleared and *there*, right ahead, lay the runway.

"Frank lowered the landing gear and we landed," he recounted, "and I think, technically, I get to log an *instrument* approach on that!" Touchdown came at 4:42 pm EST. Atlantis had flown through conditions of visibility which would ordinarily never have been sanctioned and for several years afterwards, Covey enjoyed reminding Coats of his recommendation, for the Runway 15 end of the SLF was totally clear of smoke by landing time! Two weeks later, an increasingly confident NASA announced its intent to resume 'operational' Shuttle landings in Florida, beginning with the maiden voyage of the Challenger-replacement orbiter, Endeavour, in May 1992. "Accumulation of data on brakes and crosswind landings will be required," *Flight International* cautioned in early December, "before the decision is made to resume operational landings at KSC", adding that after Endeavour's flight, *both* sites – Florida and California – would be given similar weightings in terms of priority levels during a mission. That situation would change markedly in 1991, when a second Shuttle crew was diverted to Florida and a decision would be made to commence routine KSC landings far sooner than planned.

For Bob Springer, climbing out of Atlantis that November evening in 1990 was the end of his final opportunity to experience space travel. In the months before launch, he had accepted a position with Boeing in Huntsville, Alabama. The hydrogen leaks which pushed the flight from July to November complicated these plans and, although Boeing allowed Springer to postpone his report date to them, Dick Covey was unsure for a time whether or not his crewmate would even stick around to fly STS-38. Like a good astronaut and dedicated military man, Springer remained, but his departure after the mission was swift. NASA's press release, announcing his departure, came on 12 December and he actually completed part of his post-mission crew report whilst in Alabama.

One of the primary focal points of the United States military in the months surrounding the STS-38 flight was Operation Desert Shield and the First Gulf War, which began in January 1991 following Iraqi dictator Saddam Hussein's invasion of Kuwait. The invasion had been triggered as a consequence of Iraq's bankruptcy in the aftermath of its war with Iran and the perception that Kuwait was harming its economy by driving down oil prices. On 2 August 1990, days after Ambassador April Glaspie had told Saddam that the United States had "no opinion" on Arab-Arab conflicts, the first Iraqi bombs were dropped on Kuwait City. In the days which followed, the Kuwaiti armed forces were overwhelmed by Saddam's Republican Guard and Emir Jaber Al-Ahmad Al-Jaber Al-Sabah was forced to flee to neighbouring Saudi Arabia. Months of sanctions, initiated by the United Nations Security Council, led to a demand at the end of November 1990 for all Iraqi forces to depart Kuwait by 15 January of the following year. When this deadline passed unheeded, a 34-nation coalition commenced bombing operations to destroy

Iraqi military assets and a subsequent ground assault, which began on 23 February, restored Kuwait's independence.

The coalition invasion was still some weeks into the future when Dick Covey's crew launched their classified payload in November 1990, but the astronauts could clearly see the after-effects of Saddam's deliberate burning of Kuwaiti oil wells, which cast long black smudges across the region. From Atlantis' windows, Covey remembered seeing bombers waiting at Diego Garcia in the Indian Ocean. To him, it was obvious that the storm clouds of war were on the horizon. "Everybody had a sense that we were going to war," he told the NASA oral historian, "that Desert Shield was more than just sabre-rattling, that they were really positioning for an invasion and a rejection of the Iraqis out of Kuwait." The resultant conflict focused the eyes of the world on Iraq and upon a dictator whose actions still reverberate to this day.

LOST MISSION

On 28 January 1986, seven astronauts clambered out of the Shuttle Mission Simulator at JSC in Houston, Texas, to watch the launch of their comrades, the 51L crew, on television. Seventy-three seconds later, as Challenger was lost with all hands in a fireball which ripped across the blue Florida sky, the men of Mission 61E knew that their own flight, just five weeks hence, had similarly vanished. "We were preparing to fly in 40 days to observe Halley's Comet," remembered 61E mission specialist Bob Parker. "Obviously we *didn't* fly 40 days later!"

Yet the bullish attitude of NASA management in the weeks and months leading up the Challenger tragedy had different priorities ... and those were governed almost exclusively by the need to meet schedule targets. The return of Halley's Comet to the inner Solar System in late 1985 and early 1986 had already provoked a plethora of interplanetary probes – the International Cometary Explorer (ICE), the European Giotto, the Soviet Vegas and the Japanese Suisei and Sakigake – to explore the fabled celestial wanderer, and the Shuttle was expected to carry cameras and telescopes on three separate missions to observe it. Of these, Mission 61E was the most important, for Columbia would be carrying a battery of three powerful ultraviolet telescopes, known as 'ASTRO-1', to observe the comet's progress. In order to complete these observations at the most optimum time, Columbia *had* to launch early in March 1986. The criticality of this launch date had already been picked up by the press and *Flight International* reported in December 1985 that the mission "must be launched by 10 March to achieve maximum science return", warning that "a slip to 20 March would result in the flight's cancellation". By the time of the Challenger accident, 61E was scheduled to begin at 5:45 am EST on 6 March. In the years that followed, Bob Parker would prove vocal in his astonishment at such definitive, immovable targets. "It's amazing," he said, "when you look back at that, and the *rate* at which we thought we had to keep pumping this stuff out." Parker had an expression: *The Sun kept rising and setting*. Schedule pressure meant nothing in the face of crew safety.

Had Challenger not been lost that frigid January morning, Parker is almost certain that his own mission might have fallen victim to the technical and managerial cancers which riddled the Shuttle at this time ... for conditions in Florida in the early hours of 6 March were even *colder*, and the effect of cold weather on O-ring seals in the Solid Rocket Boosters was later identified as a key factor in the accident. It is almost certain that NASA would have pressed ahead with the 61E launch and, if Columbia made it to orbit safely, it promised to be one of the most exciting scientific missions to date. The crew would have been split into two 12-hour shifts to keep the ASTRO-1 instruments running around the clock. Pilot Dick Richards, mission specialist Parker and payload specialist Sam Durrance would have formed the 'red' team, whilst mission specialists Dave Leestma and Jeff Hoffman and payload specialist Ron Parise were on the 'blue' team and commander Jon McBride's schedule was anchored across both. They would have spent nine days in orbit – the second-longest Shuttle mission to date – and would have returned to land in Florida on the 15th at 3:47 am EST.

With such an important scientific payload for astronomy, it was unsurprising that the crew included four professional astronomers: Hoffman, Parker, Durrance and Parise. Hoffman had been assigned to follow ASTRO-1's development in 1982 and he quickly concluded that the mission was sufficiently complex to warrant the inclusion of a pair of 'payload specialists'. This position had long since crystallised into something which described a non-career astronaut; someone who was chosen to represent a research institution or a space agency, an aerospace corporation or even a foreign government. For major science missions, such as 'Spacelab' flights, payload specialists were selected by an Investigators Working Group (IWG), whose panel included the principal investigators responsible for the experiments. It had been received badly in some areas of NASA, with many managers opposing the idea that Shuttle crew members should be selected and trained outside of its immediate jurisdiction.

In his report to George Abbey, the head of Flight Crew Operations, Hoffman was vocal in his conviction that payload specialists were needed for ASTRO-1. Privately, though, he worried if this very conviction might prove career-limiting, for Abbey's powerful ability to approve or veto the selection of astronauts for Shuttle crews was known to be immense. Still, even though he was a professional astronomer, Hoffman knew that he could spend two years in dedicated training and probably would not know the payload in as much depth as the team that developed it. Consequently, payload specialist candidates were selected from representatives of each of the telescopes. Hoffman was fortunate and in June 1984 he and Bob Parker were assigned as mission specialists for the ASTRO-1 flight. By the end of January 1986, the payload had completed its pre-launch processing and was ready for installation aboard Columbia for launch. Its three instruments – the Hopkins Ultraviolet Telescope (HUT), the Ultraviolet Imaging Telescope (UIT) and the Wisconsin Ultraviolet Photopolarimeter Experiment (WUPPE) – were attached to a pair of Spacelab pallets and an intricate device known as an Instrument Pointing System (IPS). Of all the pieces of hardware aboard ASTRO-1, the IPS had generated the most severe headaches.

1-07: With the blue and white of Earth offering a stunning backdrop, the ASTRO-1 payload represented one of the most complex sets of instruments ever flown aboard the Shuttle. Mounted atop an Instrument Pointing System and Spacelab pallet, with the cylindrical igloo visible in the foreground, ASTRO-1 focused three sets of powerful ultraviolet eyes on the cosmos.

In fact, before Challenger, a senior ASTRO-1 manager had told *Flight International* that if the IPS failed, "the whole mission is down the drain". It was not an idle statement, for the pointing system's technical complexity, organisational and schedule difficulties and cost escalation had contributed to more than a decade of troubles. When ESA named an industrial consortium, headed by ERNO and Fokker, to build the Spacelab system in June 1974, the German Dornier Satellitensysteme GmbH (now EADS Astrium GmbH) was contracted to undertake IPS design studies. The specifications called for a three-axis system, with a pointing accuracy of one arc-second; it should measure up to 2 m in diameter and up to 4 m in length and weigh no more than 2,000 kg. The sheer size of the yoke mounting for such a large device prompted Dornier to propose an end-mounted approach in which the three gimbal systems would be attached to a Spacelab pallet and would support a circular mounting frame onto which the instruments would be installed. Authorisation to proceed with the design was granted in December 1974, but within months ESA was discussing options to reduce project costs by holding the contractor to less stringent specifications. NASA management was also concerned that no single IPS design would satisfy all of its users' pointing requirements. By September 1975, Dornier's design had been rejected on the basis of unacceptable cost and schedule risks and in December ESA issued a new Request for Proposals. Early the following year, two new proposals were delivered: a joint bid from Dornier and MBB and a bid from ERNO which covered the integration of the IPS into Spacelab. In mid-1976, the Dornier-MBB proposal was accepted.

However, issues of cost, schedule and technical difficulty continued to escalate and, with them, tensions grew between NASA and its European partners until, in early 1977, some ESA managers suggested removing the IPS from its Spacelab effort entirely, in order to find another means of development. By June, the agency had finally agreed to continue with the project, contracting with Dornier to deliver the first flight unit by the middle of 1980. Yet the problems persisted. Evidence of the susceptibility of certain components to stress corrosion and uncertainty about the IPS' software were flagged in December 1977. When the European members refused additional funding for the pointing system, Dornier was forced to make modifications and delay the first flight. For its part, NASA was keen to demonstrate its confidence that ESA and Dornier would solve the problems and in May 1980 the agency purchased a second IPS for $20 million. Meanwhile, Dornier submitted a proposal for a redesigned IPS in April 1981 and the first flight unit was delivered to the Kennedy Space Center in November 1984. "The last few months of checkout," wrote Douglas Lord in *Spacelab: An International Success Story*, "were fraught with debates about the state of readiness of both the hardware and software and the adequacy of documentation and operating instructions." Harbouring reservations, NASA was torn between pushing for the completion of qualification testing in Europe and pushing for the early delivery of the hardware to Florida in order to begin payload integration for the IPS-dependent Spacelab-2 mission. Dornier offered the Americans "iron-clad assurances" that all open actions, missing data and tests *would* be completed before launch and on this basis NASA and ESA accepted the IPS.

Its advertised capabilities included providing precise guidance for a suite of instruments and pointing them to within two arc seconds, holding them on target to just 1.2 arc seconds. One end of the pointing system was mounted on the pallet and the other end to an 'integration ring', to which the three-axis gimbal for the instrument package was affixed. When operational, the 1,180 kg IPS was capable of manoeuvring telescopes and instruments backwards and forwards, from side to side and could even 'roll' them in a 22-degree arc around its 'straight up' position. Its movements were commanded from the Spacelab subsystem computer and a pair of Data Display Units (DDUs) on the Shuttle's aft flight deck. It could be operated in manual or automatic modes and was capable of spending long periods focused on single objects or conducting slow-scan mapping operations. Moreover, its reaction times were much better than those of the Shuttle's attitude control system. When compared to the orbiter's pointing precision of perhaps a tenth of a degree at best, the IPS' ability to achieve accuracies of one-thirty-six-*hundredth* of a degree has been likened to keeping an instrument on the steps of Washington's Capitol Building aimed at a coin on the Lincoln Memorial, some 3.5 km away! Even the effects of crew motions, equipment operations or Shuttle thruster firings could be compensated by accelerometers mounted on the IPS to maintain the scientific instruments on target. Astronomical research from the Shuttle was highly desirable for one fundamental reason: simply *being* above the turbulence of the 'sensible' atmosphere meant it could acquire imagery at wavelengths invisible on Earth.

If the ASTRO-1 instruments were mounted atop an IPS, then the IPS itself was affixed to a U-shaped Spacelab 'pallet' in the Shuttle's payload bay. Built by British Aerospace, the pallets measured 3 m long by 4 m wide and were covered with aluminium panels. Up to five pallets – three of them bolted together in a rigid 'train' – could fit into the Shuttle's payload bay and the versatile platforms continue in service in today's International Space Station era. The pallet train was held in place by five attachment fittings – four along the walls of the payload bay and one in the floor – and included aluminium ducts and trays on its port and starboard sides to route cables to and from experiments and subsystems. Thermal control was provided by multi-layered insulation and Spacelab's Freon-21 coolant loop, which collected excess heat from the pallet-mounted hardware through a series of 'cold plates' and rejected it into space via the Shuttle's heat exchanger. Although the pallets had been tested prior to being used operationally to support a dedicated scientific payload, their carriage aboard Spacelab-2 in July 1985 marked the first time that the 'train' and another device, the 'igloo', had been utilised for a 'full' scientific mission. The igloo was a 2.1 m tall aluminium alloy cylinder and was mounted vertically on a crossbeam at the forward end of the train, providing a temperature controlled container to hold subsystems and equipment for the instruments. Pressurised to 14.7 psi, the 660 kg igloo offered electrical power, cooling and command and data acquisition services for the pallet-mounted experiments.

The ASTRO-1 crew itself experienced a great deal of change from the first assignments in June 1984 to its eventual launch in December 1990. Hoffman and Parker were both professional astronomers before their selection into NASA's astronaut corps. In fact, almost anyone who met Jeffrey Alan Hoffman when he joined NASA in January 1978 would have described him as the stereotypical professor: bearded, riding a collapsible bicycle and carrying a lunch pail in hand were three of the attributes noted by his contemporary Mike Mullane in *Riding Rockets*. "To the very end," Mullane wrote, "Jeff remained an unpolluted scientist." Yet his life had encompassed far more than academia; he was a skydiver, an accomplished mountaineer and a skilled engineer. Hoffman came from Brooklyn, New York, where he was born into a Jewish family of physicians on 2 November 1944. "My parents took me all over the place to museums and concerts," he recalled in his NASA oral history, "and among the other places was the Hayden Planetarium." This quickly hooked the young boy on astronomy. He received his schooling in Scarsdale and entered Amherst College in Massachusetts to study astronomy, graduating *summa cum laude* – with highest honours – in 1966. He went to Harvard for his doctorate, which he received in 1971. His research focused on high-energy astrophysics, specifically cosmic gamma rays and X-rays, and he participated in the design, construction, testing and flight operations of a balloon-borne, low-energy gamma ray telescope. Years later, he believed that this probably attracted him to NASA and vice versa. "I did a lot of work with my hands," he said, "building electronics, machining stuff. That probably stood me in good stead with NASA, because when they're selecting astronauts, they want people who know how to work in a lab, who can fix things and build things." He then moved to England for three years to undertake post-doctoral work at the University of Leicester, serving as project scientist for the medium-energy X-ray experiment on the European Exosat mission. Whilst in England, he met his wife, Barbara, and became a father. The Hoffmans returned to the United States in 1975 so that he could take up a position at MIT as the project scientist in charge of the hard X-ray and gamma ray experiment aboard the first High Energy Astronomy Observatory. "That was probably the most interesting scientific time that I've ever spent, because ... we discovered a new phenomenon called X-ray bursts," he recalled. "I guess we wrote about thirty-five to forty papers about it. These are thermonuclear explosions on the surface of a neutron star; pretty wild stuff."

By his own admission, Hoffman had been drawn to astronomy and astrophysics through his fascination for space exploration, although the opportunity to do such things himself seemed out of reach. "The early astronauts were all military test pilots," he pointed out. "I was never particularly interested in that career. In fact, I wasn't particularly interested in airplanes, because they didn't go high enough or fast enough. I always liked rocket ships." The chance finally came in October 1977, when he was invited to Houston for a week of interviews and testing; at first, his wife thought he was joking when he told her about the astronaut application. It came as a shock to Hoffman that it would spell the end of his research career in astrophysics. "NASA made it very clear that they were *not* looking for people to come and be research astronomers," he explained. "They were looking for astronauts who had to be generalists, because there were a lot of different things we were going to have to learn how to do." It was disappointing, in a sense, but it marked a change in Hoffman's career.

"The most unusual thing about my application," he continued, was that "I very well could have been the only person who was selected as an astronaut, who admitted in their application to having been convicted of a *crime*." It had happened during his tenure in Leicester, when he and some friends took a converted coastal steamer across the North Sea to explore the Norwegian fjords. Unfortunately, the original captain of the trip cancelled at the last minute and Hoffman – despite lacking the proper certification – stepped in. Upon their return, the coast guard arrested them and charged them a £10 fine, which Hoffman's friend disputed. The case was upheld because not only did Hoffman lack the required certification, to captain a British flagged vessel it was necessary to possess British citizenship. At the time, Hoffman did not have this and the party were convicted. "We actually had to go to Crown Court," he said, "with the wigs and the whole deal." They were fined £250 and when Hoffman came to fill in his NASA application form, he hesitated before deciding to be honest and admit to his offence. Surely, NASA would not delve *that* deeply into his past. They did. Nor could the selection board prevent themselves from making light of the situation. At his interview in Houston, the first greeting from astronaut Joe Kerwin, on the selection panel, as Hoffman walked into the room was: "Here comes the criminal!" This *criminal* look was surely made complete by a beard, which Hoffman quickly needed to remove. "As soon as I got to the altitude chamber in preparation for T-38 flying, it became pretty clear that you can't make a good face seal with a full beard, so off came the beard. My wife *shrieked* when I walked through the door!" (He kept a moustache, however, throughout his astronaut career.)

On ASTRO-1, both Hoffman and Robert Allan Ridley Parker would be making their second Shuttle flights, but for the latter it had been a long and often frustrating wait for a launch into space. Parker had been selected by NASA in August 1967 and did not make his first mission into space until the end of 1983. It is perhaps fortuitous, therefore, that he was a born-and-bred New Yorker, with all the ingrained tenacity that it entailed. Parker was born there on 14 December 1936, but received schooling in Shrewsbury, a suburb of Worcester, Massachusetts. Education and learning was in Parker's blood: his paternal grandfather had been a high school teacher and librarian, his father the chairman of the physics department at Worcester Polytechnic Institute, whilst his twin brother would later teach physics at Yale University and his younger brother became a systems analyst. Parker received a degree in astronomy and physics from Amherst College in 1958 and a doctorate from the California Institute of Technology in 1962, after which he accepted a National Science Foundation fellowship and joined the Badger faculty of the University of Wisconsin at Madison. Whilst there, Parker and one of his colleagues were caught up in the excitement of America's drive to land a man on the Moon – "it was way, way, *way* beyond saving the whales or the rainforest or the ozone layer," he told the NASA oral historian – and a call for applicants for scientist-astronauts in the spring of 1965 captured their attention. Unfortunately, NASA's requirements were strict: six feet (1.8 m) was the maximum height allowable and perfect eyesight was mandatory. "I had an office mate at Wisconsin," Parker recalled. "He was over six feet; I wasn't. He had perfect eyes; but I *didn't*." It did not seem worth the effort to complete the stack of application paperwork and Parker thought no more of becoming an astronaut for two more years, until he heard of another selection in 1967, in which the restrictions on eyesight had softened. He soon found himself among a group of 69 finalists, summoned for medical screening at Brooks Air Force Base in San Antonio, Texas, interviewed by NASA in Houston and given his first flight in a high-performance T-38 jet. "As a matter of fact, I got *sick*," he recalled. It did not affect his selection in August and, years later, Parker would strongly suspect that the selection committee were more interested in finding out whether he enjoyed flying, rather than simply dismissing him on the basis of airsickness. And Parker certainly *loved* flying.

Flight school, at Williams Air Force Base in Arizona, would not begin until March 1968, so intense was the need for the US military to prepare its own pilots for Vietnam. A year later, Parker returned from Williams with his jet credentials and worked initially on Skylab issues, before an assignment to the support crew of the Apollo 15 lunar mission. During this time, he undertook trips to Iceland, Hawaii, the Grand Canyon and other exotic destinations in support of the geology training, serving as the capcom for the backup crew – Dick Gordon, Vance Brand and Jack Schmitt – and liaising with the scientific community. Parker extended this work in his next role, as the mission scientist for Apollo 17, the final lunar landing flight. He returned to work on Skylab and in May 1974 was assigned as chief of the astronaut office for NASA's Science and Applications Directorate. One of Parker's primary focuses was upon Spacelab and just six months into his tenure, he was approached by Philip Culbertson, the head of mission and payload integration at NASA Headquarters, and also by Owen Garriott, one of the Skylab scientist-astronauts, to participate in the first mission of a project known as the Airborne Science/Spacelab Experiment System Simulation (ASSESS). This was a joint ESA/NASA effort to closely examine the operational requirements for future Spacelab missions and consisted of five flights of a CV-990 Galileo II aircraft over a six-day period in June 1975, in which a four-man crew simulated high-altitude experiments in atmospheric physics and infrared astronomy. Each flight lasted around seven hours and after touchdown they completed scientific debriefings, then bedded down in specially-designed crew quarters, adjacent to the aircraft parking area. The ASSESS-I mission provided valuable scientific data, as well as raising important issues of crew training and demonstrating the ability of two international space agencies to work together. Certainly, it must have factored into the eventual decision to assign Parker to one of the two mission specialist spots on Spacelab-1.

The ASTRO-1 payload specialists were announced at around the same time as Hoffman and Parker, in June 1984. Original plans called for three flights, each with a pair of payload specialists, chosen from a pool of three candidates: Sam Durrance and Ron Parise would be aboard ASTRO-1, then Durrance and Ken Nordsieck would be aboard ASTRO-2 and Parise and Nordsieck would be aboard ASTRO-3. At the time of the Challenger disaster, it was planned that all three missions would have been completed by July 1987 ... and as for Hoffman and Parker, they would have been aboard *all three*!

The two payload specialists ultimately assigned to ASTRO-1 were both highly qualified in their fields. Samuel Thornton Durrance came from Tallahassee, Florida, where he was born on 17 September 1943, and earned bachelor's and master's degrees in physics from California State University at Los Angeles and a doctorate in astrogeophysics from the University of Colorado at Boulder. As a principal research scientist in the Department of Physics and Astronomy at Johns Hopkins University in Baltimore, Maryland, he was a co-investigator for the HUT instrument aboard ASTRO-1. In this capacity, Durrance was intimately involved in the development, optical and mechanical design, construction and integration of HUT into the payload. As a professional astronomer, he led the team which designed and built the Adaptive Optics Coronagraph, which detected the first cool brown dwarf circling a nearby star, and co-discovered changes in the planet-forming disk around Beta Pictoris.

1-08: Seated in the flight engineer's position, behind and between the commander and pilot, astronaut Mike Lounge works his checklists ahead of the STS-35 re-entry. To his right is fellow mission specialist Bob Parker.

Meanwhile, Ronald Anthony Parise came from Italian-American parentage and grew up with a desire to pursue his twin interests of aviation and astronomy. He was born in Warren, Ohio, on 24 May 1951 and became a licenced ham radio operator, aged just 11. In his teens, he was active in the Mahoning Valley Astronomical Society and built two telescopes, grinding the mirrors himself. After high school, Parise entered Youngstown State University, from which he earned a degree in physics, and moved to the University of Florida to pursue his master's and PhD in astronomy. In 1979, following the completion of his doctorate, he worked for a time on the development of failure mode analyses on several NASA missions, before becoming involved in the design and definition of the Ultraviolet Imaging Telescope. As circumstances transpired, both he and Durrance flew aboard ASTRO-1 and ASTRO-2 and Parise would devote a sizeable portion of his later career to the analysis of data from his two Shuttle missions. Yet for Parise, who died in 2008, the opportunity to fly into space encompassed far more than just carrying his profession beyond the atmosphere. "It's such a remarkable mixture of feelings when you are in space," he once told an interviewer. "There is the experience of weightlessness, the incredible views ... I can remember looking back at Earth, with all of its brilliant colours, and that pitch-black sky as a backdrop. There's *nothing* like it!"

By the time Hoffman, Parker, Durrance and Parise were assigned to ASTRO-1, the process of selecting Shuttle crews – particularly for scientific Spacelab missions – had settled into an established routine. Typically, operations would be conducted around-the-clock, with a mission specialist and a payload specialist serving on each shift, together with a member of the 'orbiter' crew, who would be responsible for performing manoeuvres and maintaining the Shuttle's systems. In January 1985, commander Jon McBride, pilot Dick Richards and flight engineer Dave Leestma were named as Mission 61E's orbiter crew. In the aftermath of Challenger, the entire crew was stood down indefinitely. For Jeff Hoffman, the decision to stick around and wait for ASTRO-1 was an easy one ... but the wait would be a long one. Several astronauts took sabbaticals, or pursued further education, and Hoffman took a master's degree in materials science at Rice University in Houston – surprising many of the younger students in his classes as "this old guy" got straight-As, every time, in his exams – before ASTRO-1 was resurrected as STS-35 in November 1988. By then, Richards and Leestma had been named to another flight, but McBride remained as commander of the new mission. Richards' and Leestma's places were taken, respectively, by Guy Gardner and Mike Lounge. A few months later, in May 1989, however, McBride retired from NASA and was replaced in command by veteran astronaut Vance DeVoe Brand.

In his NASA oral history, McBride rationalised his thinking. Shortly after his assignment to STS-35, rumours arose that ASTRO-1 was destined for cancellation. "My wife and I had bought a home in [my] native West Virginia, in the beautiful Greenbrier Valley," he explained, "and I was commuting." McBride saw his family every two or three weeks, until one morning something changed. "I looked out the back window and there were deer and pheasants and squirrels and rabbits," he said, "and seeing the Greenbrief River and the snow-capped peaks ... I [had] two choices. I can hang it up now and come back here to West Virginia ... or I can go down to

Houston and take a chance of training for two more years and never going anywhere." By his own admission, the choice was tough, but at length McBride called Don Puddy, then-head of Flight Crew Operations, with his decision. "I might be the only person in history who was assigned to a mission that pulled out of it," he admitted. As circumstances transpired, ASTRO-1 *did* fly – with Vance Brand in command – but, years later, McBride felt that he made the right decision. In West Virginia, he established a venture capital company, bringing industry and jobs to the area, which he felt enabled him to make a positive impact and open up business opportunities.

As for STS-35's replacement commander, Vance Brand had long since earned himself a reputation as one of NASA's most gifted fliers – "a hard-nosed aeronautical engineer and an experienced test pilot", according to Deke Slayton, who flew in space with him – and served as a backup crew member for no fewer than three missions. Born in Longmont, Colorado, on 9 May 1931, Brand had been an active Boy Scout during his formative years and achieved the title of Life Scout. After high school, he entered the University of Colorado at Boulder, receiving his business degree in 1953 and accepting a commission into the Marine Corps. For four years, he served as a naval aviator, which saw him assigned as a fighter pilot in Japan, then resigned from active duty to return to his *alma mater* for a degree in aeronautical engineering. A master's degree in business administration from the University of California at Los Angeles followed in 1964, by which time Brand was working for Lockheed as a flight test engineer for the Navy's P-3 Orion aircraft. His ties with the military remained intact, however: he continued to serve in the Marine Forces Reserve and with Air National Guard jet squadrons until 1966. Along the way, he graduated from the Naval Test Pilot School in 1963 and worked at West Germany's F-104G Flight Test Centre in Istres, France, as an experimental test flier.

After selection by NASA in April 1966, Brand focused on thermal vacuum chamber tests of the Apollo spacecraft and later served as a member of the support crews for the Apollo 8 and 13 lunar flights. During the latter, in April 1970, he was a capcom in Mission Control, providing a direct voice link with the astronauts during one of the most harrowing missions ever attempted. At around the same time, Brand was named as the backup command module pilot for Apollo 15 and, mindful of a well-established three-flight 'rotation' system, confidently expected assignment to Apollo 18 and his own journey into lunar orbit. After the cancellation of that flight in September 1970, Brand continued his Apollo 15 duties and later transferred to Skylab, serving as backup commander for its second and third missions. (In August 1973, Brand and fellow astronaut Don Lind came within days of flying a rescue mission to the station.) In the midst of his Skylab work, Brand was named as command module pilot for the Apollo-Soyuz Test Project, a joint flight with the Soviet Union in July 1975. Brand returned from his first space flight, wrote up his crew report, debriefed with the NASA managers and physicians ... and promptly lost his *parking space*. Reserved spaces outside Building 4 at JSC have become legendary; an indicator that an astronaut-in-waiting has finally received a flight assignment, but *after* a mission their importance rapidly deteriorates. "Up to the time of the mission," Brand told the NASA oral historian, "when you're first up to bat, as we used to say, *everyone* will pay attention to you and if there's something that you think might help the mission, why, there are people eagerly waiting to get your ideas. You're sort of on top of the world. Well, after you come *back* from a mission, whereas *before* you had a parking place, when you come back, you *lose* your parking place. That's the *first* thing that happens. After about two weeks post-mission, their interest starts to wane, because they're thinking of the *next* mission." By the time of his assignment to STS-35, Brand had completed three space missions, including two Shuttle flights, and the opportunity to regain that parking place was simply irresistible.

It must also have been a pleasure for Brand to be commanding Columbia, the orbiter he had previously flown in November 1982 on the first operational mission of the Shuttle era. Named in honour of a privately-owned sailing frigate, built in 1773, which became the first US vessel to circumnavigate the globe, Columbia was the result of a NASA decision to give *names*, rather than impersonal *numbers*, to each of its orbiters. In May 1978, John Yardley, the agency's Associate Administrator for Space Transportation Systems, sent a list of options to the Public Affairs Office. Each name – 'America', 'Independence', 'Liberty', 'Freedom', to name a few – was required to bear some symbolic relevance to the heritage of the United States. Later that year, Arnold Frutkin chaired a committee which broadened the scope of the list to include sailing vessels and even Native American tribes. In December, three categories of names were selected: (1) Explorers' Vessels, which included 'Enterprise', 'Endeavour' and 'Discovery', (2) American Tradition and Spirit, which included 'Columbia', 'Constitution' and 'Republic', and (3) Stars and Constellations, which included 'Orion', 'Arcturus' and 'Pegasus'. At length, the committee settled on Explorers' Vessels and even the *Star Trek*-themed 'Enterprise' had a proud naval heritage, having lent its name to the world's first nuclear-powered aircraft carrier.

The spacegoing Columbia was also notoriously complex to prepare for flight; as the first orbiter, she had been built to different specifications than her sisters, was considerably heavier and some technicians described her as the "hangar queen" or, somewhat more disparagingly, as "the pig". When she returned from Mission 61C on 18 January 1986, just ten days before the launch of Challenger, the task of turning her around for ASTRO-1 in early March seemed difficult, at best. Four years later, her preparations for STS-35 had proven equally difficult, with the incredibly intricate work to repair her hydrogen leaks not fully completed until October 1990. However, she was forced to wait until Atlantis had flown STS-38. A plan to launch on 30 November was called off when astronomers argued that the timing would impair the mission's observations, but as night fell on the evening of 2 December ASTRO-1 was finally ready to go.

FOUR POWERFUL EYES

The ASTRO-1 observatory can trace its genesis back to 1978, when NASA issued an Announcement of Opportunity for advanced astronomical instruments for carriage aboard the Shuttle. Three were ultimately selected: an ultraviolet telescope from

Johns Hopkins University, an ultraviolet photopolarimeter from the Space Astronomy Laboratory at the University of Wisconsin at Madison and an ultraviolet imaging telescope from NASA's own Goddard Space Flight Center in Greenbelt, Maryland. The project was initially managed by the Office of Space Science, but by 1982 control had passed to MSFC in Huntsville, Alabama, and was officially renamed 'ASTRO'. When the first mission of the series was timed to coincide with the arrival of Halley's Comet, a Wide Field Camera was also added to the payload and in January 1986 ASTRO-1 was in the Operations and Checkout Building at the Kennedy Space Center, awaiting launch.

For almost three years after the loss of Challenger, the payload was kept in storage and the telescopes were removed from their Spacelab pallet. Periodic health checks were carried out, but NASA decided to recertify them all before clearing them for flight. Hundreds of bolts were replaced, for example, in 1987, and the Wide Field Camera was deleted, as Halley's Comet was by now long gone. In its place was the Broad Band X-Ray Telescope (BBXRT), built by the Goddard Space Flight Center. This had also been selected for further design work in 1978 and was originally scheduled to form part of another mission, known as the Shuttle High Energy Astrophysics Laboratory (SHEAL). The primary focus of the BBXRT was to shed new light on a major supernova – known as '1987A' – which had been spotted in the Large Magellanic Cloud, 170,000 light years away. One of the aims of BBXRT was to examine the spectrum of the expanding cloud of gas in order to identify the differing chemical elements present. It was mounted atop a separate support structure, known as a Two-Axis Pointing System (TAPS), which could manoeuvre the BBXRT backwards and forwards and from side to side. Unlike ASTRO-1, it was controlled not by the astronauts, but remotely from the ground. This X-ray instrument had not been expected to fly until at least 1992, but its completion ahead of schedule made it easier to add to ASTRO-1. (The second SHEAL instrument, a Diffuse X-ray Spectrometer, was later flown on Shuttle mission STS-54.)

As work progressed on readying the BBXRT for flight, the ASTRO-1 instruments were also gradually removed from storage and prepared for their mission. Of these, the Hopkins Ultraviolet Telescope had been kept at KSC throughout the post-Challenger down period, although its spectrograph had been returned to Johns Hopkins in October 1988 for attention. Tests had shown that – although gaseous nitrogen ordinarily protected it from air or moisture contaminants – the telescope's sensitivity had nevertheless degraded and its spectrograph needed replacement. HUT, which measured 3.6 m long and 1.2 m wide and weighed 770 kg, was designed to observe objects such as quasars, active galactic nuclei and 'normal' galaxies at far and extreme ultraviolet wavelengths. This region of the electromagnetic spectrum was virtually inaccessible using Earth-based instruments and HUT's mirrors were coated with iridium to enhance their sensitivity. The second spectrograph also failed its tests and was replaced, as was an aging camera.

The other two instruments underwent a lengthy period of recalibration and retesting. WUPPE was not shipped back to the University of Wisconsin, but instead a portable vertical calibration facility was erected and delivered to Florida. The instrument passed its tests with flying colours in April 1989. Similarly, UIT remained at KSC and received a new power unit for its on-board image intensifier in the summer of that year. By December, all three of the ASTRO-1 experiments had been declared ready to fly and were installed onto the IPS. When in orbit, WUPPE would examine the ultraviolet 'polarisation' of hot stars, galactic nuclei and quasars, whilst UIT was designed to acquire wide field-of-view images of star clusters, planetary nebulae, supernova remnants and galactic clusters. The ASTRO-1 observatory as a whole had its own focal targets; the process by which ancient red giant stars steadily shed their outer layers to leave a dense 'ember', no bigger than Earth, known as a 'white dwarf', was imperfectly understood and an important area of scientific inquiry. As these white dwarf stars were known to emit much of their radiation at ultraviolet wavelengths, they were ideal candidates for ASTRO-1.

The instruments were also directed to examine suspected 'black holes' – stars whose density was so high that they literally collapsed under their own gravitational influence, to such an extent that not even light could escape. The three ultraviolet telescopes and the BBXRT were expected to be able to resolve hot gas swirling into a black hole's clutches. Other studies would focus upon 'binary systems', in which a pair of stars reside in close proximity to one another, and colossal stellar clusters. In visible light, it was difficult to discern individual stars in these clusters, but under ASTRO-1's ultraviolet and X-ray gaze they were expected to blaze brightly. More broadly, the observatory would analyse the 'interstellar medium' – the vast gulf of gas and dust *between* stars – and explore its physical properties and characteristics.

ASTRO-1 became the first scientific Shuttle mission to be managed from MSFC's new Spacelab Mission Operations Control Facility, which superseded the Payload Operations Control Center at JSC, used on earlier flights. The new facility could transmit commands directly to the orbiter, monitoring the instruments, directing their observations, receiving and analysing their data and adjusting schedules to take proper advantage of unanticipated events. In total, the ASTRO-1 and BBXRT combined payload weighed 11,750 kg and represented one of the heaviest scientific cargoes ever carried aloft by the Shuttle.

"WE'RE BACK!"

In some ways, the delays to ASTRO-1 yielded both fortune and misfortune; for astronomy as a science often produces unexpected phenomena which can neither be predicted nor planned. Halley's Comet *was* expected, and losing the chance to observe it with the ultraviolet telescopes was a crushing disappointment, but the delays would enable a study of Supernova 1987A ... and also of another celestial event, known as a 'blazar', which occurred late in November 1990, just a few days before Columbia and her seven-man crew set off. Blazars are quasars – distant, highly luminous objects, now known to be the cores of active galaxies – which suddenly flare several dozen times brighter than normal. Johns Hopkins astronomer Art Davidsen could hardly contain his excitement in the days before launch, as the blazar observations gained priority on the ASTRO-1 schedule.

For once, Columbia herself was also behaving and Launch Director Mike

Leinbach told journalists on 29 November that his team was "right on the timeline", with no problems being tracked as the formal three-day countdown commenced. Weather conditions at the time of launch on 2 December were expected to be 70 percent favourable, although forecasters kept a close eye on a tropical storm to the south of Cuba. When Vance Brand and his crew arrived in Florida in their fleet of T-38 jets on the evening of the 29th, they were in a jubilant mood. "We're back! We're ready!" yelled Brand, who had already ensnared a new record for the mission, by becoming the oldest human ever to journey into space. He had turned 59 years old the previous May and would beat the existing record-holder, Karl Henize, by a little under nine months. Some prophets of doom noted that STS-35 would be the 13th post-Challenger launch, but Sam Durrance was personally convinced that, after so many delays, their day had finally come. Jeff Hoffman, equally, was optimistic. He knew that comets were supposed to be bringers of ill-fortune and many comets had been associated with ASTRO-1: Halley being the obvious one, but a string of other, lesser known, celestial wanderers had made the rounds as STS-35 struggled to get ready to fly. By December 1990, there were no cometary targets available to observe. "We all know comets are harbingers of bad news," Hoffman rationalised. "This time, we have *no* comet. So we're going to go."

And go they did. At 1:49 am EST on the 2nd, Columbia roared aloft on her tenth flight, lighting up the darkened Florida skies for hundreds of kilometres. The launch suffered a brief, 21-minute delay, due to concern about cloud beneath the 2,430 m 'ceiling' needed to monitor the first two minutes of ascent, but was exceptionally smooth. Coming on the heels of Discovery's 6 October launch with Ulysses and Atlantis' 15 November flight for the Department of Defense, the beginning of STS-35 set NASA's second-best record for despatching three Shuttles within 57 days. This slightly missed the 54-day record set by three other missions in October-November 1985, but marked an exceptional growth in maturity as the space agency recovered from Challenger.

Sitting on Columbia's flight deck, directly behind the pilot, Jeff Hoffman would never forget the awe-inspiring sensation of blasting off at night. As the vehicle ascended from the pad, cleared the launch tower and began a computer-controlled roll program manoeuvre to establish itself on the proper azimuth for a 28.45-degree-inclination orbit, he remembered looking back over his shoulder through the overhead windows to see the ground *light up* at the instant of SRB ignition. Later, when the boosters burned out and were jettisoned, Hoffman and the other three men on the flight deck were astonished as the entire cockpit was bathed in light.

Even though this was his fourth launch into space, it was one to savour for Vance Brand, since it was his first at night. "You had the feeling you were lighting up all that part of Florida," he told the NASA oral historian. "It was like night flying in an airplane. You had to really have your lights adjusted and pay attention to your gauges. You couldn't really tell much about what was going on outside. You couldn't see the horizon very well." The night launch oriented Columbia to ensure that her passage through the so-called 'South Atlantic Anomaly' – where the Van Allen radiation belt 'dips' toward the ionosphere – occurred mainly during orbital daytime. High-energy particles were already known to adversely affect instrument

1-09: Jeff Hoffman's recollection that the ground lit up at the instant of SRB ignition comes to life in this astonishing view of the STS-35 launch.

performance and increased 'background' levels in sensitive scientific detectors. Since this natural background, which consists of scattered light and ultraviolet atmospheric airglow emissions, was also higher on the daylit portion of each orbit, it preserved the nighttime passes for ASTRO-1 to focus on its faintest celestial targets.

Observation of those targets began crisply and activation of the payload commenced almost as soon as the pilots had established the Shuttle in her 300 km orbit. The 'red' team, consisting of Bob Parker and Ron Parise, led by pilot Guy Gardner, took charge of activating ASTRO-1 and its support equipment. Meanwhile, the 'blue' team of Hoffman and Durrance and shift leader Mike Lounge bedded down for an abbreviated sleep period. They would awaken for their first 12-hour work duty at 1:00 pm EST on 2 December. Although not specifically attached to either shift, Vance Brand tended to anchor his schedule across both teams, which enabled him to maintain a 'big picture' of operations and transfer knowledge from one team to the other.

As the first use of the new Spacelab Mission Operations Control Facility at NASA's Marshall Space Flight Center, ASTRO-1 got off to an exceptionally smooth start when, at 7:56 am EST on the 2nd – just six hours after launch – Bob Parker opened his first scientific communications session with "Huntsville, this is ASTRO". Michelle Snyder, the crew interface co-ordinator in Huntsville, Alabama, quickly responded, telling Parker that everyone was "really excited" and "looking forward to a great ten-day mission and a lot of terrific astronomy". Aboard Columbia, Ron Parise piped up. "Michelle, we know there's a lot of people down there that did a lot of work on this mission and we're hoping to make it a real success for everybody. Let's get this show on the road."

By this time, the red team had switched on the BBXRT and at 7:36 am Parker received a go-ahead to unlatch and raise the IPS and ASTRO-1 instruments from their horizontal position in the payload bay. The process took less than seven minutes and by the time STS-35 was 16 hours old and Lounge's blue team had assumed duty, the telescopes had sailed through their initial checks and were ready for calibration. One observation of the star Beta Doradus, using the Hopkins telescope, was particularly lauded by Sam Durrance. Located in the constellation of Dorado (the Swordfish), the star had been chosen for its suitability when aligning and focusing the telescopes. The sighting was part of the so-called Joint Focus and Alignment process, whereby all three ASTRO-1 instruments were trained on a common target as a prelude for upcoming observations. Unfortunately, a computer failure in the WUPPE prevented it from participating in the alignment. Ground-based engineers diagnosed the problem as having been caused by an unactivated heater and, when this was powered up, early on the 3rd, the WUPPE's checkout got underway.

Overall, the mission had gotten off to a fine start. Typically, the 'duty' mission and payload specialists, situated on the Shuttle's aft flight deck, overlooking the payload bay, used a pair of Spacelab keyboards and the two DDUs to command the IPS and the telescopes. Closed circuit monitors provided images of the starfields under observation and enabled the astronauts to check the data. Observations typically required between ten minutes and a full hour to complete. Meanwhile, shift leaders Gardner and Lounge were responsible for firing Columbia's RCS thrusters to keep the orbiter properly oriented.

From his selection in May 1980 to his first flight in December 1988, Guy Spence Gardner had waited an unenviable eight and a half years, although that was nothing compared to the lengthier waits endured by Vance Brand and, particularly, Bob Parker. Gardner came from Altavista, Virginia, where he was born on 6 January 1948, and attended George Washington High School in Alexandria. He later entered the Air Force Academy and earned a degree in astronautics, mathematics and engineering sciences in 1969, followed up, a year later, by a master's credential in astronautics from Purdue University. Initial flight instruction at Craig Air Force Base in Alabama and training in the F-4 Phantom fighter led to his assignment to Thailand in 1972, where he flew 177 combat missions over Vietnam. Gardner returned to the United States as an F-4 instructor and operational pilot at Seymour Johnson Air Force Base in North Carolina. He completed test pilot school at Edwards Air Force Base in 1975 and served there as a test pilot and an instructor pilot. After unsuccessfully applying as a pilot candidate in the 1978 group of astronauts, Gardner was successful two years later. Had the Challenger accident not occurred, he would have piloted the first Shuttle mission into polar orbit from Vandenberg Air Force Base in July 1986.

Seated behind and between Gardner and Brand during the STS-35 ascent and re-entry was civilian astrogeophysicist John Michael Lounge, who had come to NASA's Payload Operations Division in 1978, following a lengthy spell in the Navy. He was born in Denver, Colorado, on 28 June 1946, and graduated from high school in Burlington, then entered the Naval Academy. He received his degree in 1969 and proceeded onto an immediate master's programme in astrogeophysics at the University of Colorado at Boulder. "If you got selected for, and got a scholarship, somewhere," Lounge told the NASA oral historian, "the Navy let you go and spend a year or 15 months getting a master's degree before you reported to your first duty station." As a naval ensign, he wore his uniform whilst at the civilian university and the nature of his master's credential was an overt statement of his intent to someday apply to join the astronaut corps. "To put it in perspective," he said, "two days after I reported to the University of Colorado ... is when Neil Armstrong stepped onto the Moon. Every young ensign in the Navy wanted to follow in his footprints, I'm sure, so I did that."

Lounge completed flight instruction in Pensacola, Florida, moved on to advanced training as a radar intercept officer – "the systems guy" – in the back seat of the F-4J Phantom and was deployed to Vietnam in 1972 aboard the USS *Enterprise*. He participated in 99 combat missions, then undertook a Mediterranean cruise aboard the USS *America*. On returning to the United States in 1974, Lounge became an instructor in the Naval Academy's physics department and entered the Space Projects Office in Washington, DC, in 1976, for a two-year tour as a staff project officer for reconnaissance satellites. On hearing of NASA's plans to hire astronauts he submitted his application, but was not selected in 1978. However, he did garner sufficient interest to be offered a role at the Johnson Space Center, working in mission operations. When he asked his superiors in the Navy if they would consider reassigning him to Houston he was refused because the Navy wanted to offer him a position aboard an aircraft carrier. Consequently, in 1978 he resigned from the service, with the rank of commander, and spent two years in Houston, working on Shuttle payloads and serving as a member of the Skylab Re-entry Flight Control Team. Lounge knew that committing to NASA in this way surely helped his chances of selection in the *next* astronaut class. Whilst there, he worked with two other engineers, a civilian named Bonnie Dunbar and Air Force officer Jerry Ross, and they all applied when the agency issued its next request for astronauts. In May 1980, all three were selected. "Our offices," Lounge recalled, "were within shouting distance, so you could *hear* the shouts!"

Those shouts were the prelude to an astronaut career which would bring triumph and sorrow: on his first mission in August 1985, Lounge proved instrumental in a plan to capture an errant communications satellite – packed with hydrazine – and return it to its correct operational orbit. His second mission should have occurred a few months later, in May 1986, and deployed Ulysses, but the destruction of Challenger put that flight indefinitely on hold. In early 1987, Lounge was assigned to the crew of the first post-51L mission, STS-26, and described his preparations for *that* flight as being quite different from the first. "My first flight was a festive, picnic atmosphere," he told an interviewer for a BBC documentary, *The Diary of Discovery*, in the weeks preceding STS-26. "A good percentage of my home town came down to that launch and all my high school classmates that could make it, and it was a big social event. I think the next time won't be quite so festive, until we're safely into orbit." Lounge's words marked a distinct shift in attitudes towards the reusable Shuttle; a shift which had been thrown into sharp focus. Now, on his third mission, he had already decided that STS-35 would be his last flight into space.

Yet now, two years after STS-26, the first scientific Spacelab mission in the post-Challenger era was at risk. Early on 3 December, the IPS started to experience difficulty 'locking' onto its guide stars. An alternative plan was quickly worked out to help the astronauts to manually point the telescopes and track targets on the HUT's television camera using a hand paddle. This worked reasonably well and allowed them to aim the telescopes with an accuracy of around three arc-seconds ... but it was not a positive start. "The mood is one of concern," said Flight Director Bob Castle. "We'd certainly like the system to work perfectly, but there is no panic. People are working to solve the problem and we have confidence we will solve [it] in a fairly short time."

Worse, though, was to come. Late the previous evening, Vance Brand had picked up the scent of warm electrical insulation, which was traced to an overheated DDU. It was bad news, for the twin display units were critical in enabling the astronauts to control the IPS and the ASTRO-1 telescopes. By the afternoon of the 3rd, Bob Castle admitted to journalists that the mission's observation timeline had been thrown at least six hours behind schedule. By 7:00 am on the 4th, when scientific work resumed, Flight Director Al Pennington – one of a team of flight directors which also included Castle, Wayne Hale and Gary Coen – described the outlook as somewhat brighter, although no fewer than two dozen astronomical targets for that day had been lost. "What we have to do," reflected ASTRO-1 Mission Scientist Ted Gull, "is make sure we reallocate what is left to the higher-priority objects that have been lost." Indeed, by the end of the day, the pace had quickened and the telescopes scrutinised the bright galaxy NGC 4151, which was thought to contain a massive black hole at its core, whilst the BBXRT – whose pointing system was independent of ASTRO-1 and was thus largely unaffected – successfully acquired X-ray spectra of the Crab Nebula. Still, the Shuttle and her crew were well behind schedule.

With its computer woes finally resolved, WUPPE came fully to life late on the evening of 4 December and was directed to observe a variable binary system known as HR-1099 and HD-22468. Such systems consist of two stars, in close proximity to each other, in which one 'stirs up' its companion and causes massive blobs of material to spiral away from it. The Wisconsin experiment was later employed to study a rapidly rotating star, 21 Velpecula, although the history of problems meant that it was being handled with kid gloves. As the telescopes found their feet, other efforts were underway to obtain full operational capability from the IPS' Optical Sensor Package, whose star trackers offered one means of locking onto celestial targets. Support from MSFC in Huntsville and JSC in Houston enabled successive refinements to be made to its pointing geometry and Sam Durrance was able to accomplish the first operational identification of a desired target: a distant white dwarf.

On the whole, as STS-35 ended its third day, recovery seemed to be on the horizon and Mission Manager Jack Jones reported that the payload appeared to be healthy. Ted Gull agreed, admitting early on the 5th that ASTRO-1 was "really coming alive" and that he could smile at long last. After extensive troubleshooting, a string of more refined calibrations of the star tracker optics had been uplinked to Columbia and the observatory was able to perform automatic target acquisitions. Hoffman and Durrance were able to utilise this capability by acquiring one target, followed immediately by another, with no need for recalibration of the instruments between each sighting. The pointing stability, though, was lower than it should have been, primarily because control of the Spacelab pallets, the IPS and the telescopes were being conducted using a single DDU on the aft flight deck. It was hoped that both situations would improve, leading to an increase in the quality of the ultraviolet data, but such hopes seemed dashed early on 6 December when the second DDU overheated and failed. Initial efforts to restart it were unsuccessful and Mission Control asked the astronauts to remove its lower panel, look at the internal components and check the air-intake filters for the presence of lint.

Hoffman vacuumed out a small quantity, but it did nothing to revive the unit. At this point, ASTRO-1 had completed 70 of its planned 250 observations; it had achieved a measure of remarkable success, but the second DDU failure was a devastating blow for the rest of the flight. (So mixed were the emotions that William Blair, the assistant project scientist for HUT, admitted that he did not know whether to cry or grin from ear to ear.) Years later, Bob Parker recalled that the prognosis for the flight was grim. "Suddenly, we couldn't point," he told the NASA oral historian. "The ground came up with a scheme where they could control the telescopes ... but what they *couldn't* do in near-real-time was guide the telescopes. Everybody put a brave face on it, but it was a far cry of what we had intended it to be." In fact, for a few hours on the 6th, the crew had little to do but enjoy the glorious view of Earth, drifting serenely 'below' them. This view, however, was somewhat fleeting on STS-35, depending on the observation schedule. "We weren't pointed at the Earth very much," admitted Mike Lounge, "which was a little unusual, because the other flights [had their] payload bay down at the Earth all the time. We were pointed at the stars, so we could be, depending on the target, any attitude."

Once more, the remarkable ground specialists managed to bounce back from the DDU failures and by 7 December they were able to command all but the final motions of the telescopes. It was then left to the astronauts to fine tune them for each observation run. "We've had a lot of setbacks," said Art Davidsen, "but success is at hand." Ted Gull agreed, although he cautioned that making the mission work would require an enormous team effort. Revised procedures involved operating the telescopes sequentially – first the UIT, which had the observatory's largest field of view, followed by the HUT and lastly the WUPPE – and even the BBXRT was indirectly affected, since it had to cease its own work whenever Columbia entered a 'safe' attitude. In one observation, Alternate Payload Specialist Ken Nordsieck guided Durrance to acquire Supernova 1987A with the UIT. As Durrance controlled the telescope manually with a joystick, Nordsieck provided him with second-by-second pointing instructions. A total of six minutes of high-quality ultraviolet spectra were collected. More was to follow with the observation of a radio-quiet, extremely luminous quasar, Q1821+643.

The crew, too, had established themselves into a routine, working effectively and efficiently with their colleagues on the ground to maximise scientific output. Typical clipped exchanges between Durrance and Nordsieck on the space-to-ground communications link were crisp and terse: "Sam, you're within an arc-minute ... Okay, give me a 'mark' when you're happy ... The data's looking real good, Sam ... We're seeing lots of photons down here." Although Nordsieck would never fly into space, his involvement with ASTRO-1 was equally important; as a long-time professor of astronomy at the University of Wisconsin at Madison, he worked extensively on the development of WUPPE and was a member of its science analysis team. For the pilots, Brand, Gardner and Lounge, it was simply fascinating to watch the research being undertaken by the four professional astronomers. In total, ASTRO-1 recovered from its troubles to make 231 observations of 130 different celestial objects and ran for 143 hours to accomplish 70 percent of its pre-mission goals. Despite being affected by the problems with the observatory, the BBXRT also returned an enormous volume of data. One of its most important targets was Markarian 335, a bright, compact object, 325 million light years from Earth and thought to have a black hole at its heart. The X-ray imagery enabled astronomers to 'see' emissions from material being sucked into the hole. Yet even the BBXRT suffered its own trials with the TAPS support structure: its original observation timeline had to be dropped in favour of a less efficient, day-by-day method. An improperly compensated gyro drift rate meant that no stable pointings could be accomplished for the first 60 hours of the flight and TAPS itself – whilst able to *point* the telescope – was rather sluggish in its ability to acquire targets.

1-10: Jeff Hoffman changes a roll of camera film aboard Columbia's flight deck, with Earth visible through the aft and overhead windows.

Still, the mission proved an outstanding example of triumph in the face of adversity and achieved far more than might have been expected after so many problems. In their final ultraviolet science briefing, Ted Gull asked Art Davidsen what the final results were in the 'match' between 'The Huntsville ASTROs' and 'The Universal Secrets'; Davidsen could only reply, with a grin and a twinkle in his eye, that the ASTROs had won by a mile!

Although the main thrust of the flight had been the observatory in the payload bay, the astronauts participated in a number of other activities. One of the most noteworthy was the 'Space Classroom', which gave Jeff Hoffman the dubious honour of becoming the first person to wear a tie in orbit. "All the male teachers wore ties," he told Mission Control. "I know nobody has ever worn a tie in space, so I thought I'd give it a try and see what it looks like." Then, with the tie secured in place by a piece of Velcro, he introduced the first school lesson ever transmitted from space. Years later, in his NASA oral history, Hoffman provided some additional detail on the tie story. "My uncle was a New York lawyer," he recalled, "and he represented a lot of foreign firms. Among them was the Hermès Corporation, which makes fancy silk scarves and silk ties."

According to Hoffman, one of Hermès' publicity agents gave NASA the idea that the corporation was using a visit to the space agency for commercial purposes. NASA abruptly cancelled the visit, refusing to allow the use of its government-furnished facilities for advertising. Fortunately, Hoffman's uncle put the agent in touch with Hoffman to patch things up. "It was very nice," he said later, "because we ended up getting invited to a lot of the Hermès parties." When Hoffman mentioned that he wanted to take a tie with him into space, the corporation was only too happy to provide one. But there was a problem. Hermès' ties were all silk and NASA fire-safety guidelines demanded cotton. Without hesitation, they made him a cotton tie; a beautiful patterned one with an image of an astronaut emblazoned upon it.

The classroom experiment was part of a project entitled 'Assignment: The Stars' and was intended to encourage students' awareness of science, mathematics and engineering. On 7 December, the four astronomers gathered on Columbia's flight deck and spoke to students in Huntsville and Greenbelt by means of a two-way televised link. The main emphasis was upon the electromagnetic spectrum, which Durrance likened to a musical symphony. To press the analogy, he played two taped versions of the same piece; the first, he explained, was unrecognisable because its high and low notes had been removed, whereas the second turned out to be the theme from *Star Wars*. "You need to hear *all* the notes," Durrance told the students, "to appreciate the sound." Likewise, he concluded, the Universe was playing its own symphony of sorts, across the electromagnetic spectrum, some parts of which were harder to detect than others.

As the mission entered its homestretch, it became clear that the gremlins were not yet finished with STS-35. Originally scheduled to fly for almost ten days, managers were forced to confront the reality that Columbia might need to return to Earth some 24 hours early. Weather conditions at Edwards Air Force Base – the primary end-of-mission landing site – were expected to be unfavourable between 12-14 December and the decision was taken to bring the crew home a day earlier than intended, overnight on the 10th/11th. This forced astronomers to hurriedly revise their priorities for the final observations of the flight. ASTRO-1 was finally deactivated and stowed about 12 hours before landing, followed, six hours later, by the BBXRT. A cold front on the California coastline was being tracked and rain showers and heavy cloud were predicted in the vicinity of Edwards.

In addition to speaking to schools, the astronauts also made ham radio contacts, thanks to licenced operator Ron Parise, and a couple of days before landing an important visitor arrived at Mission Control in Houston. Eduard Shevardnadze, then-foreign minister of the Soviet Union, was planning to speak to the STS-35 crew and Vance Brand – who spoke Russian as a result of his involvement in the Apollo-Soyuz mission – had put together a short speech. (The nature of dialogue between the United States and the Soviet Union had reached such a level of cordiality by this time that, had Columbia launched as planned, in May 1990, her crew was scheduled to perform a ship-to-ship voice link-up with cosmonauts Anatoli Solovyov and Alexander Balandin aboard the Mir space station; the first time that such contact had ever been made.) Unfortunately, in the case of Shevardnadze, the news that Columbia would be returning a day earlier than planned meant that the astronauts' sleep patterns changed. It turned out that Brand would be asleep at the time of the Shevardnadze 'comm pass'. At first, NASA decided to simply cancel the event, because it was unfair to wake the commander in the midst of sleep. Then the agency changed its mind. Hoffman, Lounge and Durrance were awake, on the 'graveyard' shift, when the call came up that they *would* be speaking to Shevardnadze ... on their *next* orbital pass!

The three men exchanged anxious glances. One of them poked his head downstairs into the darkened middeck. Brand was sleeping soundly. They knew that he would be responsible for landing the orbiter and did not want to wake him. It was Hoffman who came to the rescue. "I had studied a little bit of Russian in college," he said, "but that was a *long* time ago." When the time for the comm pass arrived, he burrowed into his mental Russian language vault and produced a couple of sentences – worded to the effect of 'Greetings from space from the Shuttle Columbia' – and thought no more about it. One person who was listening intently, though, was JSC Director Aaron Cohen, who bubbled with excitement over Hoffman's performance. Shortly after landing, Queen Elizabeth II and the Duke of Edinburgh visited JSC and Hoffman, whose wife, Barbara, was British, delicately asked Cohen if there was any chance that they could be allowed to see the royal couple. Cohen did better than that. He was so impressed by the astronaut's diplomatic skills that he asked the Hoffmans to serve as the Queen and Prince Philip's guides. If it turned into a memorable experience for Barbara Hoffman, then it certainly also paid dividends for her father, back home in England, who got free drinks at the pub for months afterwards.

Shortly before 9:00 pm PST (midnight EST) on 10 December 1990, Vance Brand and Guy Gardner fired Columbia's OMS engines for the 231-second de-orbit burn and an hour later, at 9:54 pm PST on the 10th (12:54 am EST on the 11th), the vehicle swept through the darkness to alight on Edwards' concrete Runway 22. "We're home," radioed an elated Brand as Columbia rolled to a halt, "and glad to be back." Although he had three previous missions to his credit, this was his first opportunity to land in the hours of darkness and the differences were quite profound. "If you're landing at Edwards in the daytime," he later explained, "out of the corner of your eye you can see the sagebrush and it's giving you an impression of how high you are above the ground ... but at night you're really relying on light patterns that you see." From the flight engineer's seat, Mike Lounge also remembered re-entering the atmosphere in darkness was particularly colourful, as waves of superheated plasma rippled across the vehicle's surfaces in a rainbow of colours: reds, oranges, yellows and pinks.

Aside from the problems faced by ASTRO-1, the mission had gone well. In fact, just a few months later, NASA announced that a reflight of the three ultraviolet telescopes would be scheduled for late 1994. To Jeff Hoffman, though, it illustrated a fundamental flaw in the original mandate of the Shuttle to be able to fly frequently. "ASTRO was a payload that was developed at the time when people were saying that the Shuttle would be able to fly these payloads once or twice a year," he said. "It was a very expensive payload to develop and in the end it only flew twice. It would have been much more cost-effective had the Shuttle been capable of the sort of flight rates that were originally anticipated."

There were a few niggling glitches elsewhere, however, included a troublesome text and graphics machine, which jammed time and time again, and a blocked waste water line to the toilet, which meant that the urinal was unusable. Mike Lounge had the unenviable task of tending to the latter. "It was messy," he told the NASA oral historian, "and we had to stow the waste from the waste tank into plastic bags and seal them up. It was stinky. It was *not* glamorous space flight!" For Jeff Hoffman, it also highlighted the somewhat ludicrous effort to prepare for each mission. Before launch, the astronauts participated in a bench review, inspecting each and every item to be carried aboard. Most of the items were for contingencies and almost certainly would never be used ... but the all-male crew noticed a whole boxful of *female* urine collectors!

"There's *seven men* on this flight," they protested. "Why are we carrying a box of female urine contingency devices?"

"Well," came the reply, "some flights *have* women, and some don't, but the *paperwork* that would be involved to take this thing on and off, depending on whether you had women on the flight, would be so onerous that it's easier just to leave it on every flight!" During the mission, as they worked through the limited supply of male urine collectors, these tended to leak tiny yellow bubbles. Fortunately, the men's sense of smell was very much depressed, partly due to fluid shifting in microgravity, although Hoffman remembered working with Lounge to mop up urine deposits with old socks. "It was probably just as well," he concluded, "that there *weren't* any women on that flight. It was pretty gross!" He pitied the technicians who had to clamber into Columbia's middeck after landing to begin the process of cleaning everything up ...

HOW TO SAVE A 'GREAT OBSERVATORY'

Almost a century ago, a young scientist named Arthur Holly Compton was working at Washington University in St Louis, Missouri, when he made a significant discovery in the field of X-ray and gamma ray physics; a discovery whose consequences would earn him lasting renown and a Nobel Prize. Compton came from an academic family – his father had been a university dean and Arthur and his brothers Karl and Wilson all earned PhDs from Princeton – and today he is best remembered for his identification of the 'Compton effect', which would be central to the gamma ray detectors aboard a spacecraft that would one day launch into orbit, bearing his name.

Gamma rays represent some of the most energetic events in the known Universe and typically arise from astrophysical processes which involve the production of very high-energy electrons. In fact, gamma rays usually have energies between 10 keV and 10 MeV. Life on Earth is protected from them by our planet's relatively thick atmosphere and, since the dawn of the Space Age, this has meant that the majority of gamma ray research from an astrophysical perspective has been undertaken by orbiting satellites. In 1923, Arthur Holly Compton's work on the interaction of high-energy radiation and matter demonstrated the wave/particle duality of nature and played a fundamental part in the development of modern physics. His subsequent work in the observation of cosmic radiation proved ground-breaking in that it laid the cornerstone to our present understanding of how gamma radiation is created.

The 'Compton effect' and 'Compton scattering' jointly demonstrated that light could not be explained as purely a wave phenomenon and won Compton the Nobel Prize for Physics in 1927.

More than half a century later, in September 1991, the name of Arthur Holly Compton returned to the fore when it was bestowed upon a spacecraft whose own contributions to gamma ray astrophysics went on to prove crucial. Originally known as the 'Gamma Ray Observatory' (GRO), the spacecraft was part of a family of four 'Great Observatories', developed by NASA to provide new insights into the origin and evolution of the Universe across a broad segment of the electromagnetic spectrum. Firstly, the Hubble Space Telescope (named in honour of the US astronomer Edwin Hubble) would observe at visible and part of the ultraviolet wavelengths, after which GRO would examine gamma radiation. Subsequent missions included the Advanced X-ray Astrophysics Facility (AXAF, later renamed for the Indian astrophysicist and Nobel laureate Subrahmanyan Chandrasekhar) and the Space Infrared Telescope Facility (SIRTF, later renamed for the US theoretical physicist Lyman Spitzer). The Hubble Space Telescope was placed into orbit by the crew of Space Shuttle Discovery in April 1990 and GRO would follow a year later in April 1991. However, the project had been significantly delayed – and changed – by the loss of Challenger.

Yet the observatory's basic premise remained the same: to answer a variety of questions relevant to our understanding of the Universe, such as the formation of the elements, the structure and dynamics of the Milky Way Galaxy, the nature of mysterious pulsars in the cosmos, the existence of black holes and anti-matter and the physical processes responsible for the evolution of the Sun, other stellar objects and dying stars. Four major scientific instruments, overlapping energy ranges from 10 keV to 30 GeV, would be employed to undertake the research, developed jointly by NASA and ESA, together with contributions from Germany (originally West Germany in the days before reunification), the Netherlands and the United Kingdom. Planning for the mission encompassed an initial 15-month Phase 1, in which an all-sky survey and high-priority studies would be conducted, after which a year-long Phase 2 would hand over 30 percent of observing time to guest investigators and a year-long Phase 3 would expand this still further to 50 percent. Further extensions were expected and GRO was scheduled to remain operational for at least five or six years. As circumstances transpired, it spent almost a decade in orbit, performing high-quality science.

Physically, GRO was a three-axis-stabilised spacecraft, with a gross weight of 15,876 kg, more than a third of which was taken up by scientific payload. At the time of its launch, this made GRO the heaviest astronomy satellite ever placed into orbit. It reminded astronaut Jerry Ross of a diesel locomotive in appearance. "The thing was *huge*," he told the NASA oral historian. "Everything on it was real bulky, real thick, real heavy, and it was just very impressive of the stoutness of the satellite. Most times you go up to a satellite and you're almost afraid to *breathe* on it, because it may fall apart on you." Not so the GRO. Its massive internal beams, which formed the central backbone of the spacecraft, were essential to supporting the large scientific instruments.

It was designed to be capable of pointing at celestial targets for periods of days or weeks at a time, with an accuracy of just 0.5 degrees, and its hydrazine supply was to be used not only for station-keeping, but also to execute a controlled re-entry at the end of the mission. (This would prove critical in view of the nature of that re-entry in June 2000.) Built by TRW in Redondo Beach, California, GRO was equipped with a pair of solar arrays and a set of nickel-cadmium batteries to provide up to 4,000 watts of electrical power. It was designed to operate at an altitude of 450 km, high enough to avoid excessive atmospheric drag and low enough to avoid the Van Allen radiation belts, which might compromise its observations.

Aboard GRO, the Burst and Transient Source Experiment (BATSE), built by NASA's Marshall Space Flight Center, consisted of eight detectors to search for 20-600 keV gamma ray bursts and to perform all-sky surveys. It had the capability of alerting the other three instruments in the event of a major burst. The Naval Research Laboratory's Oriented Scintillation Spectrometer Experiment (OSSE) covered gamma rays at the 0.05-10 MeV range, enabling it to study both supernovae and high-energy novae, which were believed to be the locations of the formation of heavy elements. Meanwhile, the Imaging Compton Telescope (COMPTEL),

1-11: The huge Gamma Ray Observatory – later named in honour of physicist Arthur Holly Compton, a few months after its launch – is exposed to space as Atlantis' payload bay doors are opened. The doors contain radiators on their inside faces and must be opened within hours of liftoff, to begin the process of removing excess heat from the vehicle. One of the OMS pods, the vertical stabiliser tail fin and, at bottom-right, the RMS mechanical arm, are clearly visible.

developed by the Munich-based Max Planck Institute, together with the University of New Hampshire, the Netherlands Institute for Space Research and ESA, was tuned to 0.75-30 MeV and its primary purpose was to produce a detailed all-sky map as seen in 'moderate' gamma rays. Finally, the Energetic Gamma Ray Experiment Telescope (EGRET) focused on the high-energy 20 MeV-30 GeV range and was created jointly by NASA's Goddard Space Flight Center, the Max Planck Institute and Stanford University in California. In addition to categorising and characterising no fewer than four pulsars (revealing them to be particularly strong sources of gamma radiation), EGRET would ultimately facilitate the mapping of the entire gamma ray sky and observe 271 astronomical sources, 170 of which were previously unknown. Perhaps more so than any of the other instruments, EGRET enabled scientists to reaffirm many of the long-standing theories about the origins of gamma radiation. Having said this, the contributions of the others were valuable: BATSE averaged one gamma ray burst, every day, throughout its operational lifetime, OSSE extensively surveyed the centre of the Milky Way and COMPTEL made an all-sky map of the unstable radioactive isotope of aluminium, known as 26Al, which is produced in supernovae and distributed through interstellar space.

In the bulletproof days before Challenger, GRO was to be launched in May 1988, loaded with 1,800 kg of hydrazine station-keeping propellant and the plan was for a Shuttle mission to refuel the observatory in 1990. Contracts to develop a coupling mechanism for a hydrazine transfer unit were awarded by NASA in December 1984, for completion and delivery just 15 months later. There *were* safety mechanisms in place – with *triple* redundancy – but it was with a great sigh of relief that the destruction of Challenger and its aftermath terminated all such plans. Six months after the Shuttle returned to flight operations, in April 1989, NASA announced the names of five astronauts to deploy GRO on STS-37. They were scheduled to fly in April 1990, although hardware and other delays pushed their launch considerably to the right ... and contributed to the addition of both a *planned* and an *unplanned* spacewalk, properly termed an Extravehicular Activity (EVA).

In command of the mission was Steven Ray Nagel, an astronaut selected as a pilot by NASA in January 1978, but whose route to space took a rather different turn. Typically, Shuttle pilots flew at least one mission in the pilot's seat, before upgrading to become a commander, but in a few cases – including that of Nagel – it was decided to fly some of them as mission specialists in the first instance, before upgrading to pilot and finally commander. Much has been written over the years about why certain pilots were flown in this fashion; some have argued that it demonstrated favouritism of Air Force astronauts (like Nagel) over their Navy counterparts, whilst others have suggested that there were simply not enough pilot slots and flying first as a mission specialist enabled them to at least gain space experience at an earlier date. Either way, Nagel's career panned out well: he served as a mission specialist on Mission 51G in June 1985 and served as the flight engineer, before being upgraded to the pilot's seat on Mission 61A, just four months later.

Privately, at first, Nagel questioned his own abilities: was NASA telling him that he was not good enough to be a pilot? "Nothing against mission specialists," he told the oral historian. "I'd trade my pilot's slot to go be a mission specialist and do an EVA." The most likely rationale seems to be that the 1978 astronaut class was particularly large and there were many more mission specialist places available than pilots and, in order to fly sooner rather than later, NASA management decided to front-load several pilots onto flights as mission specialists. "The *other* way of looking at it," said Nagel's friend, fellow astronaut John Fabian, "was that they were doing Steve a favour. Better to give him a flight, flying the middle seat as a flight engineer, which would mean that he was learning the procedures necessary to fly the ascent and the entry, rather than to keep him sitting on the ground." As the flight engineer, Nagel trained on the same systems as the pilots and, years later, he admitted that it served him well. If nothing else, he gained mission-specific EVA training out of the assignment. His uncertainty was also calmed by words from George Abbey, the head of Flight Crew Operations: NASA *would* fly him as a pilot soon ... and he kept his word. In February 1984, Nagel was assigned as pilot of a joint Spacelab flight with West Germany, then scheduled for launch in September of the following year. At this point, Nagel's *first* mission was expected to fly in October 1984 and that would have meant that his two flights would be spaced about a year apart. "The spacing was nice," he said. Problems arose when his first flight slipped into the spring – then the summer – of 1985, whilst his second flight *didn't move*. As a result, Nagel *did* fly as a pilot, in quick succession, but far quicker than he could have anticipated.

Like so many of his contemporaries, Nagel grew up with a love of aviation. He was born in Canton, Illinois, on 27 October 1946. His father was not a certified instructor, but owned a small Piper Cub and took his son flying from a very early age – as an *infant*, in fact – and after taking lessons the young Nagel soloed on his 16th birthday. By this time, the first two groups of NASA astronauts had been selected and, although he aspired to such a career, Nagel spoke little about it, preferring instead to enter the military and fly jets. He applied to the Air Force Academy, but was placed on an alternate list and eventually entered the University of Illinois at Urbana-Champaign, in his home state, to study aeronautical and astronautical engineering. Whilst there, he enrolled in the Reserve Officer Training Corps and upon graduation in 1969 entered the Air Force. With the war in Vietnam at its height, he said, "the people pipeline was wide open, so the classes were big in pilot training".

Nagel completed his flight instruction at Laredo Air Force Base in Texas, then commenced training in the F-100 Super Sabre at Luke Air Force Base in Arizona and later served in a tactical fighter squadron in Louisiana. By this time, in the late summer of 1971, the war was winding down. Nagel served for a year in Thailand as a T-28 Trojan instructor pilot, then returned to the United States as an instructor and flight examiner for the A-7D Corsair II. His dream of NASA had matured over the years and Nagel wrote to the agency, asking what sorts of aircraft he should fly. "That was a *dumb* question," he admitted to the oral historian. "They couldn't have told me! Nobody answered me anyway." The path which would eventually lead to space took him next to Edwards Air Force Base in California, and test pilot school.

On graduating in December 1975 he remained to conduct test work on the A-7D and the F-4 Phantom. At one stage during this period, the Approach and Landing Tests of Enterprise were being conducted at Edwards and Nagel found himself providing A-7D flight instruction to astronauts Joe Engle, Dick Truly, Fred Haise, Gordon Fullerton and Vance Brand. In a close-knit community of test pilots, Nagel found that when NASA came calling for astronaut applicants, virtually *everyone* wanted to be considered. As each group came back from interview, they had their own stories: some thought the process was a piece of cake, others were mortified that the panel asked them about current affairs. Some of the pilots even went out and bought copies of *Time* magazine to keep themselves updated with world events. Nagel himself was summoned to Houston in September 1977 and at the time of his selection in January of the following year he was in the process of completing a master's degree in mechanical engineering at California State University.

More than a decade later, and with two Shuttle flights under his belt, Nagel was ready to command his first mission. It was quite different to his previous missions because, he admitted, "if it goes well you take the pats on the back, but if it goes poorly you take the blame". His responsibility encompassed not only getting himself and his crewmates ready to fly, but also interacting with flight directors and mission management. Yet working with fellow astronauts meant that he was working with a team which, despite their space flight experience or lack thereof, was both highly focused and incredibly motivated. Nagel had no input in the selection of the STS-37 crew, but he *was* responsible for dividing up the duties of his team. When it came to contingency EVAs, the one man who stood out as an obvious candidate was the only other veteran member of the crew, Jerry Lynn Ross.

Ross' involvement with EVA is so significant that, today, he stands in third place on the list of the world's most experienced spacewalkers, having completed no fewer than *nine* excursions over an impressive span of seven Shuttle missions. It was a remarkable achievement for a highly talented mechanical engineer ... who could not *swim*. "I got assigned to EVA because I was the class *rock*," Ross told the oral historian. "When we did our swimming training and our scuba training, I was the guy that sank to the bottom. We had a couple of ... people in the astronaut office who had Red Cross swimming training ... try to teach me how to swim." One day, fellow astronaut Mary Cleave showed him a few strokes in the swimming pool. "Either I sank to the *bottom*," Ross remembered, "or I went *backwards!*" In the end, Cleave gave up. Training underwater had long been standard practice for EVAs and Ross eventually learned to scuba dive and remained qualified throughout his time as an astronaut.

After qualifying in the space suit he began to focus on satellite servicing procedures, and worked extensively on the Manned Manoeuvring Unit. "Once I got into the EVA area and worked EVA," Ross said, "I *never* let it go. Even if I wasn't officially assigned to it, I continued to work in it and to do whatever I could to have opportunities to get in the [water] tank [to] do development or testing work."

Ross was born on 20 January 1948 in Crown Point, Indiana, and represents one of only a handful of individuals to have dedicated virtually his adult working life to the astronaut business. His EVA accomplishments have already been noted, but in

April 2002 he became the first human to chalk up as many as *seven* missions into space. It is a record which was tied a few months later by fellow astronaut Franklin Chang-Díaz and which both men jointly hold to this day. Ross grew up at a time when the Cold War was at its peak and the idea of rockets, whether for carrying explosives or men, was steadily entering the popular consciousness as something more than a facet of science fiction. As a child, he watched television shows about space stations, read articles in *Life* magazine, created scrapbooks of space-related events and watched in awestruck astonishment when the Soviets launched Sputnik and America responded with Explorer-1. Even at this young age, he was introduced to the word 'engineer'. "I truly didn't fully understand what an engineer *was*," Ross told the NASA oral historian, "but I knew that they had to use a lot of math and science. I *liked* math and science, so I thought that's what I wanted to do. I wanted to become an engineer." By his own confession, this gave him a one-track mind, working on farms to earn money for a bank account which would someday pay his way through the prestigious Purdue University.

His three-step plan was relatively straightforward: (1) Be an engineer, (2) Go to Purdue and (3) Get into the space programme. In Ross' mind, it was as cut-and-dried as that. Unlike so many others, he stuck doggedly to his plan and not only achieved it, but surpassed it. After completing high school in 1966, he entered Purdue to study mechanical engineering and received his bachelor's degree in 1970 and a master's in 1972. He entered active duty with the Air Force and worked on computer-aided design of ramjet engines and captives tests of supersonic ramjet missiles at Wright-Patterson Air Force Base in Ohio. Ross graduated at the top of his class from the Air Force Test Pilot School's flight test engineer course in 1976 and later served as project engineer for the flying qualities of the B-1 Lancer bomber. His role included both the training and supervision of all B-1 flight test engineers and also mission planning for the bomber's offensive avionics. "The B-1 at the time was the Air Force's highest-priority programme," he remembered, "and I was given the opportunity to come on-board as a B-1 flight test engineer and to work in the stability and flight controls areas of the B-1." It was shortly after the B-1 effort that Ross learned about NASA's plans to hire astronauts; he was one of thousands of hopefuls who submitted applications in 1977 and, though summoned to Houston for interview, was not successful. Still, he persevered and George Abbey offered him a position as a payload officer at JSC, working on the integration of military payloads into the Shuttle, with the hint that it might stand him in good stead for possible future selection. Two years later, in May 1980, after reviewing six thousand applications and interviewing 120 of those, NASA selected 19 new astronauts ... including Ross!

By the time of his assignment to STS-37, Ross had flown twice, including a pair of EVAs on Mission 61B in November 1985 and a classified Department of Defense assignment. His third voyage was not originally supposed to include a spacewalk – it was to spend five days in orbit, deploying GRO on the third day – but the situation changed markedly as 1989 wore into 1990. "*Everybody* wants to do an EVA," said Nagel, "and I just used my own best judgement on that and tried to give people what they want or have an aptitude for." One man with limitless reserves of 'aptitude' was

Jerome Apt III – nicknamed 'Jay' in the astronaut office – who came from Massachusetts, where he was born on 28 April 1949. He received his early education in Pittsburgh, Pennsylvania, and entered Harvard University to study physics. After gaining his degree in 1971, Apt moved to the Massachusetts Institute of Technology and gained a PhD in laser spectroscopy in 1976. He then served as a staff member at the Center for Earth and Planetary Physics at Harvard, working on NASA's Pioneer Venus Orbiter mission, and was assistant director of Harvard's Division of Applied Sciences. He applied unsuccessfully for selection as an astronaut in 1978 and 1980. His NASA career began in 1980 as a planetary scientist at the Jet Propulsion Laboratory in Pasadena, California, and later as payload flight controller. It was third time lucky for Apt when NASA selected him into its astronaut corps in June 1985. In his memoir, *Skywalking*, astronaut Tom Jones – who later flew with Apt – described him as a consummate professional, highly experienced in a technical sense and a good friend and colleague, but admitted that he was "a man who had perfect confidence only in himself".

The two other members of the crew included the STS-37 pilot, Ken Cameron, a lieutenant-colonel in the Marine Corps, and civilian physicist Linda Maxine Godwin, whom Steve Nagel later married. (In fact, at their wedding in 1996, Jerry Ross acted as best man.) Godwin came from Cape Girardeau, Missouri, born on 2 July 1952, but spent her formative years in Jackson and earned a degree in mathematics and physics from Southeast Missouri State University. She then completed a master's degree and a doctorate at the University of Missouri in 1976 and 1980, respectively, with a research focus on low-temperature solid state physics, including electron tunnelling and the vibrational modes of absorbed molecular species on metallic substrates at liquid helium temperatures. Like Jay Apt, she worked for NASA long before her astronaut days, beginning her career in the Payload Operations Division of the Mission Operations Directorate to work on Shuttle payload integration. A keen saxophonist and clarinet player, Godwin entered the astronaut corps, alongside Apt, in June 1985.

As with many astronauts, Godwin lavished glowing praise on her parents and teachers for their inspiration – in fact, one of her science teachers would attend *every one* of her four Shuttle launches – although she admitted in a pre-flight interview before her final mission, in 2001, that she was not particularly *drawn* to the astronaut business. "I grew up watching a lot of the coverage of the early US space programme," she said, "so that made me interested in NASA, but I never thought it was something I could do." That changed about two-thirds of the way through her PhD, in 1978, when the space agency hired its first group of female astronauts. Two attempts to enter the astronaut corps in 1980 and 1984 were unsuccessful, but Godwin – like Apt – found that her career took a third-time-lucky turn.

Jay Apt had worked extensively on GRO issues during his first few years as an astronaut, but it was Godwin who would be the primary operator of the Shuttle's RMS arm which would be used to lift the observatory out of the payload bay and deploy it into space. The $100 million arm was Canada's contribution to the Shuttle programme – a contribution that dated back to 1974, when Spar Space Robotics Corporation was contracted by the country's National Research Council to build a device for deploying and retrieving satellites from orbit and, ultimately, assembling the components of a space station. In May 1979, the first contracts were signed with NASA for the production of three RMS units and their supporting hardware and software. The challenges involved in building an arm of such complexity and dexterity were enormous: it needed to operate both autonomously and under manual control and meet strict weight and safety requirements. Moreover, nothing quite like it had ever been built or used in space before, which made Spar's task yet more difficult. Although a functional floor rig was built to test its joints, the first real demonstration did not come until it was actually uncradled on the STS-2 mission in November 1981. Measuring 15.2 m long in order to be able to reach the far end of the payload bay, it consisted – just like a human arm – of shoulder, elbow and wrist joints, linked by two graphite epoxy booms. Other components were constructed from titanium and stainless steel. To protect it from thermal extremes in space, the arm was covered in white insulation and fitted with heaters to maintain its temperature within required limits. Without a payload attached, it could move at up to 60 cm/min, but this was reduced to a tenth of that speed when fully loaded.

Ingeniously, the means by which the arm 'picked up' and 'put down' objects was achieved by the so-called 'end effector' – essentially a hand that employed a kind of three-wire snare to capture a prong-like grapple fixture attached to a deployable or retrievable payload. The Hubble Space Telescope, at that time scheduled for launch in the mid-1980s, had an in-built grapple fixture that would enable it to be deployed, retrieved and serviced by future Shuttle crews. During 'operational' missions, astronauts would use two television cameras on the arm's wrist and elbow to guide the end effector over a target's grapple fixture, before commanding the three metal ties of the snare to close around it at precisely the right instant. When this was done, it would impart a force of 500 kg onto the grapple fixture, establishing a grip sufficient for the RMS to move the target. Although the arm was controlled by the Shuttle's General Purpose Computers, its movements were directed by an astronaut using a joystick on the aft flight deck. As the astronauts issued each instruction, the computers examined it and determined which joints needed moving, their direction and their speed and angle. Meanwhile, the computers also looked at each joint at 80-millisecond intervals and, in the event of a failure, automatically applied a series of brakes and notified the crew.

Although GRO would be deployed by the RMS, a complex series of steps were required whilst in the Shuttle's payload bay before it could be released. The gigantic craft had been transported by flatbed truck from TRW's Redondo Beach facility to Los Angeles International Airport in early February 1990 and airlifted to Cape Canaveral Air Force Station a few days later. Since then it had undergone extensive preparations for launch. Aboard Atlantis, it was held in place by one keel trunnion fixture and four attachment points along the port and starboard sills of the bay. During ascent, it was unpowered – except for provisions to keep its star tracker shutters in a closed position – and its Airborne Electrical Support Equipment (AESE) in the bay was to be powered up by the astronauts within 90 minutes of reaching space. In the early stages of the mission, this would provide GRO with heater power and limited alarm telemetry for monitoring its critical system functions.

1-12: Grappled by the Canadian-built RMS mechanical arm, the Gamma Ray Observatory – one of the largest payloads ever deployed by the Shuttle – is raised into position, ahead of its deployment. Its electricity-generating solar arrays are clearly visible, as are the sparkling bluish domes of the COMPTEL and EGRET instruments. The spacecraft remained in orbit until mid-2000.

Twenty-one hours into the STS-37 mission, an in-bay checkout of the observatory was to begin, running through command and telemetry systems, control systems and communications. This would serve as a partial rehearsal for the actual deployment on the third day of the flight. At length, Godwin would use the RMS to grapple GRO and raise it high above the payload bay, whereupon the solar arrays would deploy, dual telemetry links with the Shuttle and the geostationary Tracking and Data Relay Satellite (TDRS) network would be established and the observatory's batteries would undergo a charging cycle. When each of these steps had been satisfactorily completed, Godwin would release GRO and Nagel and Cameron would manoeuvre the Shuttle away from the payload. Next, the observatory's systems would be powered up within a strict six-hour time period and, within five days, it would be ready to begin scientific operations.

Whilst on the end of the RMS, the potential for something to go awry was vast; the solar arrays might not unfurl correctly, for example, or the communications antennas might not open, and it was the responsibility of Jerry Ross and Jay Apt to be in a position to perform a contingency EVA to rectify such problems. In fact, Apt had worked extensively on GRO during his time as a payload officer and one of his achievements was helping to implement contingency EVA capabilities – including astronaut-friendly handholds – on the spacecraft. For much of 1989, the two men trained extensively in the Weightless Environment Training Facility (WET-F), a huge water tank at JSC, but did not anticipate that a real spacewalk would come their way. However, as part of its drive to prepare for the EVA-heavy workload involved in building Space Station Freedom, in June 1989 NASA began development of an experimental Crew and Equipment Translation Aid (CETA), a kind of hand-propelled 'cart' that would travel along rails to quickly deliver astronauts and their gear along the station's exterior. By the time that CETA received approval, the final payload reviews for STS-37 had already taken place, but the interest of the crew in performing the test led it to be added to their mission in March 1990. Word of an EVA on STS-37 had long preceded the official NASA announcement; certainly, *Flight International* noted that planning was underway in December 1989.

By this time, the launch had slipped from April to June to November 1990, and would ultimately slip into the late spring of 1991, giving the astronauts plenty of time to get ready for the CETA trials. Early plans envisaged CETA as something like an oversized golfing caddy, which engineer Charles 'Ed' Whitsett described as "overkill ... like taking a bus when all you need to do is go out to the back field on your motorcycle". At the opposite extreme, astronauts performing a hand-over-hand translation along the station's exterior was expected to take time and impose unwanted duress on both their space suits and the structure itself, as well as making it difficult to carry large pieces of hardware from place to place. CETA's beauty was in its simplicity – like a railroad handcar of the Old West – and it would run along a set of rails built into Freedom's backbone-like truss. For the STS-37 test, Ross and Apt would use a cart on an extendible track in the Shuttle's payload bay, some 14 m long, and would translate along it in prone positions, hand-over-hand, or angled slightly 'upwards', using pedals. Each technique – the manual, mechanical and electrical – incorporated brakes and provisions to move in reverse and the task of the spacewalkers was to assess the amount of energy required to perform each movement, as well as comfort, security, control and visibility. The six-hour EVA would also see them performing additional tests by using the Crew Loads Instrument Pallet (CLIP) to measure forces and torques imparted whilst using tools.

Not surprisingly, both astronauts were excited, particularly Ross, who had performed the *last* EVA of the Shuttle programme, just a few weeks before the Challenger accident. However, he stressed that the astronaut office was anxious to build up its pool of EVA experience, since the construction of Freedom would involve a 'wall' of EVAs, far more demanding than anything ever attempted in history. In fact, only a handful of astronauts with recent EVA experience remained on active status. That had to change ... and quickly. Ross later worked on orbiter-based EVAs, whilst Apt had worked on Freedom-based EVAs, and both men pushed very hard for a series of 'stand-alone' EVAs to build up experience. The paucity of experience in the corps actually convinced Ross, at first, that Steve Nagel would assign Apt and Godwin to perform the EVA.

Nagel thought differently. "I understand what you're saying," he told Ross, "and

I think that's a good thing to consider, but I'd really look stupid if we had to do some kind of contingency EVA on a primary payload and you *weren't* one of the two guys that was outside!"

Ross laughed. "Okay. Sounds good to me." He was under no illusion that a contingency EVA would realistically be needed to help GRO on its way.

The hydrogen leaks which grounded Columbia and Atlantis during the summer of 1990 conspired to push other missions further downstream and STS-37 was eventually rescheduled for launch early in April 1991. As a direct result of these delays, Jerry Ross had gotten himself an amateur radio licence, courtesy of Ken Cameron. Several earlier missions had carried the Shuttle Amateur Radio Experiment (SAREX) and Cameron, Nagel and Apt were particularly enthusiastic about making contact with ground-based hams during the course of their mission. Every so often, during training, Cameron would pass Ross' office and dump a pile of application paperwork on his desk. Ross was not particularly enamoured with amateur radio (he was busy enough with the GRO work, the EVA work, a pair of middeck experiments and his duties as STS-37's flight engineer), but he finally told Cameron that if the launch was delayed beyond a certain date, he would complete the requirements, including studying Morse code, fill out the application forms and get his licence. As STS-37 slipped to April 1991, Ross caved in and agreed. Cameron took him to George Bush Intercontinental Airport in Houston for the exam and, a few months later, in orbit, Ross (with the callsign of N5SCW) became the first amateur radio operator to talk to other hams only from space. "I've *never* used one, physically, on the ground, to talk to anybody else," he said.

Still, STS-37 marked the first Shuttle mission on which *all* members of the crew were licenced amateur radio operators and all five astronauts participated. SAREX's development was jointly sponsored by NASA, the Amateur Radio Relay League and the Radio Amateur Satellite Corporation. Today superseded by the Amateur Radio on the International Space Station, the concept was pioneered by astronaut Owen Garriott, who carried a hand-held radio into orbit in November 1983 and convinced NASA of its viability in getting students interested in space. Although all of the STS-37 astronauts used SAREX during their time aloft, Ken Cameron had specifically scheduled sessions in his pre-sleep and post-sleep periods, each day, and the entire crew participated whenever possible. Yet their workload was heavy. In addition to the GRO and EVA commitments, a number of experiments were housed in Atlantis' middeck. The Department of Defense's Radiation Monitoring Experiment – previously flown on STS-41 – was aboard, as was a materials dispersion investigation co-funded by BioServe Space Technologies and Instrumentation Technology Associates, Inc., and a protein crystal growth study which utilised bovine insulin. In the payload bay, the Ascent Particle Monitor measured levels of contaminants during launch and ascent.

In the spring of 1991, STS-37 was to come hard on the heels of another flight, STS-39, during which a seven-man crew aboard Discovery would operate a complex payload for the Department of Defense. Unfortunately, in late February, shortly before the STS-39 launch, cracks were found in four hinges associated with the two umbilical door drive mechanisms between the External Tank and the orbiter. "The cracks are not in the door hinges," explained a NASA Headquarters news release on 28 February, "but rather in metal that supports the mounts for electric mechanisms that open and close the doors." It all boiled down to the fact that the doors were critical to the success of a flight; they *had* to close properly following the jettison of the External Tank in order to protect Discovery from the extreme heat of re-entry into the atmosphere. Although documented events were suspected of overly stressing the doors and possibly initiating the cracks, no conclusive evidence could be found to pin down their origin and NASA management opted, conservatively, to investigate further.

The decision was made as Discovery underwent final preparations at Pad 39A and the scheduled 9 March launch was cancelled and rescheduled for late April or early May. Space Shuttle Program Director Bob Crippen announced that the delay would adversely affect the 1991 manifest; original plans called for up to seven missions – STS-39 and STS-37, followed by a life sciences flight aboard Columbia in May, the deployments of a Tracking and Data Relay Satellite, a Defense Support Program satellite and NASA's Upper Atmosphere Research Satellite in July, August and November and, finally, an international Spacelab dedicated to microgravity science in December. The delay would mean that the last of these missions would slip into the spring of 1992.

In the meantime, the STS-39 stack was returned to the VAB and the Shuttle removed and rolled into the OPF for repairs on 15 March. Hinges from Columbia – themselves exhibiting minor cracks – were removed, reinforced and installed onto Discovery. With STS-37 now scheduled to begin on 5 April, NASA decided to give it priority. On 1 April, Discovery was returned to Pad 39A, with a new launch date for STS-39 of the 23rd. By now, Atlantis had sat patiently on Pad 39B for several weeks and Steve Nagel and his crew arrived at KSC in their fleet of T-38 jets.

Yet cracks had also been found in Atlantis' disconnect doors. Admittedly, they were smaller than those of Discovery, but still sufficient to raise concern and a potential two-week launch delay. As circumstances transpired, liftoff of STS-37 was delayed only by about five minutes on the 5th, due to low cloud. This carried the potential to threaten the maximum allowable cloud ceiling if Nagel and Cameron were forced to perform a Return to Launch Site abort in the first few minutes of ascent, as well as having implications for wind effects on blast propagation. Both conditions were found to be acceptable and Atlantis speared for orbit at 9:22 am EST. After insertion into orbit – at 450 km, one of the highest yet attained by the Shuttle – the process to prepare GRO for deployment ran like a well-orchestrated symphony. Early on 7 April, GRO hung above Atlantis' payload bay, backdropped by the blue and white Earth, still in the firm grasp of the RMS. To preserve the opportunity for Ross and Apt to go outside on a contingency EVA, the cabin had already been depressurised from its normal 101.3 kPal to around 70.3 kPal to prepare them for the 29.6 kPal pure oxygen of the space suits, as well as clearing nitrogen from their bloodstreams to avoid an attack of the bends.

Ross had earlier checked out the two suits in Atlantis' airlock, as Apt assisted Godwin on the flight deck on GRO pre-deployment tasks. They had also donned their biomedical body sensors and gotten into their liquid-cooled long underwear to enable a rapid response if an EVA was deemed necessary. The underwear contained a threaded network of 91.5 m of plastic tubes and took the form of a single-piece, zip-up suit, officially known as the Liquid Cooling and Ventilation Garment. During an EVA, it would pump coolant water, supplied by the life-support backpack, through the tubes to control the astronauts' body heat, exhaled gases and perspiration. Anti-fog compound was applied to the insides of their helmet visors. With Ken Cameron on hand to assist them with donning the space suits, Ross and Apt were ready to go.

So was GRO itself ... or so it seemed. The electricity-generating solar arrays unfolded perfectly and without incident. Apt looked over at Ross and offered a word of encouragement: "Well, I guess everything's downhill from here!" Nagel also felt that they were out of the woods. After all, the two arrays were very mechanically complicated, unfolding one at a time, like a pair of giant accordions, to reach a total wingspan of more than 20 m. The false sense of confidence would quickly overtake them. The next step was the deployment of the observatory's critical high-gain antenna and commands were transmitted from the ground to extend it out on its 3 m boom.

It did not move.

The crew exchanged anxious glances. Attempt after attempt – six in total – were made and Nagel and Cameron even tried shaking it loose with a burst from Atlantis' manoeuvring thrusters, but to no avail. Privately, Nagel never had much confidence in the latter procedure, "because the [RMS] is so 'spongy' with the satellite on the end that you fire a jet and it doesn't do *anything* ... all the motion is damped out by the time it gets to the satellite". Godwin tried moving the RMS sharply, and stopping abruptly, but *sharpness* and *abruptness* were misnomers with the mechanical arm, which moved relatively slowly, and these efforts had no effect. Jerry Ross was by now stationed in the pilot's seat, monitoring data, and glanced over at Steve Nagel in the commander's seat. Both men knew that an EVA was the only realistic option. Ross removed his wedding ring and handed it to Nagel.

"Steve," he said, "I'm going downstairs to get ready."

Nagel nodded. No more than a minute later, a call came up from Mission Control, asking them to do just that. Assisting the two spacewalkers was Kenneth Donald Cameron, the crew-cutted Marine Corps aviator who would go on to become NASA's first head of operations in Star City, near Moscow, during the early days of Shuttle-Mir in the mid-1990s. Cameron was born in Cleveland, Ohio, on 29 November 1949, and enlisted in the Marines at Parris Island, South Carolina. He completed Officer Candidate School at Marine Corps Base Quantico in Virginia in 1970 and later graduated from the Infantry Officer's Course and Vietnamese Language School and served as a platoon commander in the latter days of the bitter and highly divisive conflict in south-east Asia. After this year-long tour of duty, Cameron was also part of a company of Marine Security Guards at the US embassy in Saigon. In 1972, he reported to Naval Air Station Pensacola in Florida for flight training and became a naval aviator the following year, flying A-4M Skyhawks.

Selected to attend Massachusetts Institute of Technology, Cameron commenced undergraduate studies in aeronautics and astronautics and received a bachelor's degree in 1978 and a master's degree in 1979. After graduation, he flew for a year out of Marine Corps Air Station Iwakuni in Japan and worked at the Pacific Missile Test Center, prior to admission into the Naval Test Pilot School at Patuxent River, Maryland, in 1982. He was then assigned as project officer and test pilot for the new F/A-18 Hornet fighter, a position he held at the time of his selection as an astronaut by NASA in May 1984.

Seven years later, on his first space mission, Cameron now found himself scurrying around Atlantis' middeck to prepare a pair of Extravehicular Mobility Units (EMUs, the formal name for the Shuttle space suit) for Ross and Apt. Before actually clambering into the two-piece space suits, electrical harnesses were attached to their 'hard' upper torsos to provide biomedical and communications links through the backpack. A wrist mirror and spiral-bound, 27-page checklist were placed on each suit's left arm, followed by the insertion of a small fruit and nut food bar and water-filled drink bag. The next step was the connection of a soft black-and-white communications hat – famously nicknamed the 'Snoopy cap' since Apollo days – to the top of the torso. Physically, the suits were nothing less than $2.5 million miniature spacecraft in their own right, consisting of 'upper' (above-waist) and 'lower' (below-waist) segments, together with helmet, gloves and backpack, known

1-13: Grinning after a job well done, and with the Gamma Ray Observatory looming over his shoulder, Jerry Ross had good cause to be proud of his work to set the second of NASA's Great Observatories on the road to astronomical discovery.

as the Portable Life Support System. The suits had been developed under a series of contracts with Hamilton Standard of Connecticut. Ross and Apt firstly pulled themselves into the lower torso, which featured joints at its hips, knees and ankles and a metal body seal closure for connecting to a ring on the upper torso. It also included a large bearing at its waist for greater mobility and allowed the astronauts to twist whilst their feet were held firmly in restraints.

After donning the trousers of the suit, their next step was to plug the airlock's service and cooling umbilical into a display and control panel on the front of the upper torso. This would provide coolant water, oxygen and electrical power from the Shuttle until shortly before they were scheduled to go outside, thereby conserving the limited consumables available in their backpacks. The two men finally entered the airlock, where the upper torsos 'hung' on opposing walls and, through a half-diving, half-squirming motion, manoeuvred themselves into the top halves of their suits. With arms outstretched, and Cameron nearby to assist, they slipped into the upper torsos and their waist rings were brought together, connecting the coolant water tubing and ventilation ducting of the long underwear and the biomedical sensors to their backpacks. Cameron helped them to lock the body seal closure rings at their waists. The hard upper torso was essentially a fibreglass shell under several fabric layers of a thermal and micrometeoroid garment. On its back, it held the life-support system and on its chest the display and control unit by which the spacewalker would manage oxygen, coolant and other consumables; in fact, due to the difficulties in seeing 'down' to read labels on the unit, the mirrors on the suits' left wrists would help immeasurably. For additional ease, the labels were written backwards!

Next step: the gloves. Snapped into place on the wrist rings of the upper torso, these had silicone rubber fingertips to provide a measure of tactile sensitivity when handling tools in Atlantis' cavernous payload bay. Finally, the enormous polycarbonate bubble helmets were lifted over the astronauts' heads and clicked into place on the neck rings of their upper torsos. Over the top of each helmet was an assembly containing manually adjustable visors to shield their eyes from solar glare, together with two EVA lamps to illuminate work areas out of range of the Sun or the Shuttle's own payload bay floodlights. Mobility in the neck rings was unnecessary, because the helmets were easily big enough to allow the astronauts to move their heads around. Unlike previous Apollo space suits, the modularised Shuttle ensemble, with its waist closure ring, eliminated the need for pressure-sealing zips and therefore had a much longer shelf life. Additionally, the use of newer, stronger and more durable fabrics enabled space suit engineers to design joints with better mobility, resulting in lower weight and a reduction in overall cost. Jerry Ross and Jay Apt, by now floating motionless in Atlantis' tiny airlock, were, in effect, small spacecraft in their own right. However, they were not yet 'self-contained', as their oxygen, electricity and coolant water were still being provided by the Shuttle's systems; not until shortly before the two men ventured outside would they transfer to their suits' life-support consumables.

At length, a go-ahead came from Mission Control and Ross commenced the final depressurisation of the airlock and pushed open the outer hatch into the payload bay. Sunlight flooded into both of their faces and, beyond, the enormous bulk of the

GRO hovered on the end of the RMS. It was a nervous time. In fact, on his previous EVAs, Ross had never gotten as excited or anxious as he did on the spacewalk to save GRO. "I didn't know if we could fix it or not," he told the NASA oral historian, "and here we are, on the spot to try to go out and fix this thing ... and if we *can't*, then we've got this great big lead weight. What are we going to do with it? We may not be able to bring it home, because the solar arrays [have] already been deployed and the antenna's partly released. Oh, man!" And Nagel agreed. Without a functional high-gain antenna, GRO *could* have completed its mission, but it would have been very cumbersome to relay scientific data to the ground. Coming on the heels of the embarrassing start to the first Great Observatory mission – the Hubble Space Telescope – it was imperative that GRO be deployed in full working order.

Quickly, the two spacewalkers split up: Apt moved to the port side of the payload bay to set up tools, whilst Ross, on the starboard side, immediately attended to the task at hand. Godwin moved the RMS slightly to tilt the GRO towards him. "The antenna," Ross said later, "was on the back side, facing the aft bulkhead of the orbiter, and the guys in the cockpit couldn't see it from the aft windows." By the time he reached the location of the antenna, Atlantis was passing out of direct communication with Mission Control – a Loss of Signal period – and Ross used the quiet time to position himself in such a way that he could shake it open. He knew that the antenna was relatively close to GRO's hydrazine tanks, and he definitely did not want to ding *those* and risk having a highly-toxic leakage on his hands. By now, Nagel, Cameron and Godwin were getting views from the aft payload bay cameras on their monitors and were able to offer Ross some additional guidance. He gave the antenna a couple of lateral shakes. It still felt solid; immovable. A few more tries achieved a measure of success, as it started to loosen up a little.

"I was probably putting in 45 to 50 pounds of force, is my recollection," Ross said later, "and I could tell it was starting to walk out. Finally, it came *free* and swung out about 30 to 40 degrees from the stowed position." A thermal blanket had become 'hung up' on a bolt and Ross' intervention had taken just 17 minutes. He let out "a war whoop" and returned to the port side of the bay to join Apt. The two men gathered their tools and Ross returned to GRO to begin the process of manually locking the antenna's boom into its deployed position. "And *that* was a pretty good feeling," he reflected. "I felt that I'd probably earned my keep for that day!" At one stage, Ross took a breather, moving close to Atlantis' aft flight deck windows to grin at his crewmates inside the cockpit. Nagel snapped his photograph. It was a photograph which would return to haunt him – literally – a couple of years later ...

At 5:36 pm EST on 7 April, some three orbits and more than 4.5 hours later than planned, Linda Godwin finally released a perfectly functional GRO from the grasp of the mechanical arm. The Great Observatory, with all the grace of a space-age diesel locomotive, drifted serenely away into the inky blackness to commence its mission. Ross and Apt had a ringside seat for much of this activity, for they were able to remain outside whilst the final checks on the observatory were performed. "During that period of time," Ross told the oral historian, "Jay and I were allowed to stay outside on the spacewalk and to do a series of force measurements." Using the CLIP, they evaluated three force torque sensor plates, a soft stowage assembly and a foot restraint. The intention was to understand the kind of loads an astronaut might experience whilst in and out of a foot restraint and Ross performed "a whole series of manoeuvres of turning wrenches, turning handles, manoeuvring myself ... a whole series of things [and] we were recording the data so that we could get more information". Years later, even the powerfully-built Ross remembered working up a vigorous sweat, which streamed into his eyes at one stage, such was the level of exertion and exhaustion.

The spacewalkers were supposed to be back inside Atlantis' airlock in time for the actual GRO release. It did not entirely work out that way. "I think the only thing that *was* still in the airlock when we released the satellite," said Ross, "were our *toenails*! Jay and I were pretty well outside of the hatch ... and then they fired the jets to move the orbiter away. *That* was really cool. We were over North Africa at the time that we released it and we were above the satellite, looking down. That was a pretty awesome sight."

To make up for their ringside seat, Ross and Apt prepared dinner for their crewmates, later that evening, to celebrate a quite remarkable day.

The fourth day of the flight was the day originally timelined for the EVA to perform the CETA tests. Having spent three hours and 30 minutes outside on the 7th for the contingency activity, the two astronauts would increase that still further on the 8th with an excursion of five hours and 47 minutes. Designated as the 'Extravehicular Activity Development Flight Experiment' (EDFE), it consisted of the trials of the CETA and rail track, the CLIP activity and a translation evaluation. With GRO gone, Ross and Apt unpacked the track segments – stored in two 7 m halves – and extended it along the payload bay sill. Half of the track, in the forward end of the bay, was already in position, whilst the other half, in the aft end, could not be assembled until after GRO was gone. For each test, the three CETA carts were mounted on a red 'truck', attached to the rails. Of the three, the men agreed that manually propelling it, hand over hand, whilst held in place by foot restraints, worked best for them. "That's the one we ultimately went with" for the International Space Station, said Ross, "because it's just easy to do and it's the least amount of overhead and the simplest." As for the electrical cart, Ross had his own reservations. During trials in the WET-F, before launch, safety concerns meant that the training version was not equipped with a real electrical generator. Instead, the WET-F divers drove him up and down the track, as Ross went through the motions of turning the crank on the 'generator'. In orbit, for real, he started cranking the generator ... and nothing happened.

"Man, these engineers," he muttered to himself. "I *told* them this thing wouldn't work. It's a piece of junk."

Still, he persevered, cranking the generator handle as hard as he could. After the mission, Linda Godwin told him that she could *feel* the motion rippling through Atlantis' structure. After a while, Ross tired himself out and paused for a break. Glancing down, he quickly realised why he seemed to be making so little progress after so much exertion: the *brake* was still firmly in place! Elsewhere, Jay Apt used a fish scale to measure the loads that the RMS could handle. Overall, it was a highly successful demonstration of the techniques which would later be employed – on two missions, by Ross himself – to construct the International Space Station.

Re-entering Atlantis' cabin, the astronauts were in the process of removing their gloves when Apt noticed a big bloody spot on his right hand. At first he thought it was a blister which had burst. Only when the crew returned to Earth and the space suits were inspected did the full story emerge. His glove's 'palm bar' – a C-shaped mechanism which kept the palm configuration fixed – had twisted and one of its tips had punctured the pressure bladder, causing it to rub against his skin. The resultant leak was very small, far too insignificant to trigger Apt's secondary oxygen supply, and the astronaut's hand and the dried *blood* helped to re-seal it.

Landing was scheduled for Edwards Air Force Base on the 10th, but was postponed by 24 hours due to unacceptable weather conditions, both in California *and* at the Kennedy Space Center in Florida. The following morning, the situation at Edwards had improved and at 4:55 am PDT (9:55 am in Florida), Atlantis swept into the desolate base at the cusp of daybreak. Due to an incorrect call on high-altitude winds, Nagel brought the orbiter down a couple of hundred metres short of the runway, though, fortunately, this did not pose a serious problem, since it was on a dry lakebed and not obvious to most spectators. However, if it had been in Florida, whose SLF runway was fringed by swamps, the 'low-energy' landing would have been more obvious.

Years later, Nagel – who went on to become an instructor in Grumman's Gulfstream Shuttle Training Aircraft after his astronaut career – would blame himself for the shortfall. "It wasn't a real great day to land at Edwards," he told the oral historian. "There were high-altitude winds aloft and a big wind shear." As he prepared to perform a 270-degree turn to reach the runway, he allowed Atlantis to depart from the turn and "was not really aggressive about correcting back". He smoothly corrected back and rolled out onto his final approach, a little low, but the presence of the wind shear caused him to lose a lot of airspeed. "If I'd been real aggressive in how I flew," Nagel continued, "really slowed it down and stretched it a little bit, I could've been back on the glide path okay, but I wasn't that aggressive with it." Due to the nature of the dry lakebed, he was not even aware of how short he *did* land. Not much attention was paid to the event in the newspapers, although *Aviation Week & Space Technology* mentioned it and, certainly, in the astronaut office it was definitely analysed. Ultimately, Nagel totally accepted his responsibility, but admitted that the wind shear information and its negative effects factored into his decision-making on the day.

After touchdown, with all six wheels on the lakebed, Atlantis required less than 2,000 m to come to a complete halt. "It was a *real* short landing rollout for the Shuttle," he said, "because it was so darn windy. But that was that. I licked my wounds after that. They reassigned me to another flight. It didn't hurt my career." Quite the opposite, in fact, for later in 1991 Nagel was chosen to serve as acting chief of the astronaut office for Dan Brandenstein, who had entered active training to command STS-49. More than a decade after the event, as he taught new astronaut pilots to land the Shuttle, Nagel often reminded them of his low-energy final approach on STS-37. Flying the Gulfstream, he would set up similar approach profiles for his students and explain to them: "If you *ever* get here, don't mess around. Get back on the glide path, right away."

MILKSHAKES AND DOGGIE BISCUITS

The loss of Challenger on 28 January 1986 brought with it a healthy dose of reality for many NASA astronauts. Some had already announced plans to retire, either outraged by the agency's cavalier attitude towards the most sophisticated and dangerous flying machine in the world or in search of other challenges. Others simply did not wish to wait three or four years for another flight. In those dismal days, such thoughts certainly passed through the mind of the first African-American spacefarer, Guion Stewart Bluford Jr. For all of his life, Bluford had pushed the boundaries in a society in which skin colour was perceived to be a disadvantage. When he was chosen as an astronaut in January 1978, he was one of only two military members of the class to hold a PhD and also have an impressive career in the Air Force.

Born in Philadelphia, Pennsylvania, on 22 November 1942, the son of a mechanical engineer father and a special education teacher mother, Bluford attended Overbrook High School and Pennsylvania State University, where he studied aerospace engineering on an Air Force officers' training programme. Graduation in September 1964 was followed by a commission as a second lieutenant, flight training at Williams Air Force Base in Phoenix, Arizona, and the award of his pilot's wings in February 1966. With many fellow Air Force pilots going to Vietnam, Bluford was no exception; after completing survival school and several months of radar and intercept training, he ultimately flew 144 combat missions – half of which were directly over the communist North – in F-4C Phantom jets. "These missions included combat air patrol, close air-to-ground support and air superiority flights," he told the NASA oral historian, "throughout North and South Vietnam, as well as Laos." By the summer of 1967, back in the United States, he had been designated as a T-38A Talon instructor pilot at Sheppard Air Force Base, near Wichita Falls in Texas, and later served as a standardisation and evaluation officer and an assistant flight commander. Bluford attended Squadron Officers School in 1971 and returned as an executive support office to the deputy commander of operations.

Whilst working at Sheppard, he began to explore opportunities for becoming an aerospace engineer. His parent service was not particularly enamoured by the idea – "the Air Force was critically short of pilots at that time," Bluford explained, "and thus needed my skills as an instructor pilot, versus as an engineer" – but it was prepared to support him on a master's degree course. He completed the course in aerospace engineering at the Air Force Institute of Technology in 1974 and, whilst studying, one of Bluford's professors advised him to continue towards a doctorate. "I applied and got accepted ... whilst still completing my master's degree requirements," he explained. "I dovetailed some of the PhD coursework among my master's degree courses, so that I could complete the coursework for *both* programmes in *two and a quarter years!*" In March 1974, after completing his PhD coursework, Bluford was assigned to the Flight Dynamics Laboratory at Wright-Patterson Air Force Base in Dayton, Ohio, as a staff development engineer and, later, as deputy for advanced concepts in the Aeromechanics Division and as branch chief of the Aerodynamics and Airframe Branch. Whilst there, he completed his

Milkshakes and doggie biscuits   87

1-14: After suffering a herniated disk in his back, STS-39 astronaut Guy Bluford had to stand at the far right, close to a chair, for the official crew portrait. Joining him in this striking group shot are (from the left) Lacy Veach, Don McMonagle, Greg Harbaugh, Mike Coats, Blaine Hammond and Rick Hieb.

doctoral thesis. "It was a great opportunity," he said, "for me to use both my technical skills and my flying experience in developing advanced technologies for future aircraft. I led an organisation of 45 to 50 engineers, who were doing basic aerodynamic research, in such areas as forward swept wings, supercritical airfoils, advanced analytical aircraft techniques, inlets, axisymmetric nozzles and computational fluid dynamics."

By 1977, the Air Force was pressuring Bluford to return to active flying, as an instructor pilot for the T-37 Tweet training aircraft, and during this period a notice from NASA, calling for Shuttle astronaut candidates, caught his attention. In Bluford's mind, it was the perfect opportunity to fulfil his flying requirements for the military, whilst also putting his technical skills to good use and expanding his knowledge. "I could do it *all* as a NASA astronaut," he exclaimed. "What a deal!" With more than the minimum mandated 1,000 hours of pilot-in-command jet time, Bluford was able to apply for both the pilot *and* mission specialist categories – although he was not a test pilot – and was ultimately summoned to Houston in early November. It was the ninth and second-to-last group to be interviewed for the 1978 selection. Bluford heard the news of the astronaut selection one day in January 1978 and assumed that he had been unsuccessful ... only to arrive at work one Monday morning and receive a call from George Abbey, the head of Flight Crew Operations, informing him of his selection. "I later discovered that NASA had called *all* 200 finalists that morning," Bluford recalled, "and told them of their decision."

Interestingly, several of the new candidates – Bluford included – were feverishly working to complete their doctoral dissertations at the time of selection. "I had given myself until the end of the year to complete the document," he told the oral historian. "NASA wanted me in Houston in July and thus I had to expedite the writing! I later learned that both Sally Ride and Kathy Sullivan were in the *same* situation with their PhD dissertations. I defended my research and completed my dissertation in June 1978, just before I left for my new assignment as a NASA astronaut." In fact, the Blufords – Guy, his wife and their two children – were in the process of moving house from Dayton to Philadelphia, early that month, and he stayed behind to finish the dissertation. "I eventually completed the document," Bluford concluded, "made six or seven copies of it, dropped it off on my dissertation advisor's desk one Sunday evening and left for Philadelphia to pick up the family." Years later, he would consider getting his PhD from the Air Force Institute of Technology as his crowning achievement.

By the beginning of 1986, Bluford had just returned from his second Shuttle mission and, as the astronaut most current in Spacelab operations, he was assigned to follow payload safety changes in the wake of Challenger. The bleakness of the future with regard to flying again made him take stock of where his life was heading. He took preparatory courses at the University of Houston at Clear Lake for a master's in business administration, which he received in 1987, but his astronaut career was far from over.

The Department of Defense had long harboured a great interest in the Shuttle; as has been seen, the Air Force provided significant political lobbying for the reusable spacecraft in the early 1970s and in the pre-Challenger era expected to devote one orbiter to a series of classified missions out of Vandenberg Air Force Base in California. Many payloads were scheduled to be flown, including radar-imaging satellites, reconnaissance and intelligence satellites, military communications and relay satellites, early-warning satellites, global-positioning satellites and several 'mixed-cargo' research missions. When the Shuttle returned to normal operations in September 1988, the Department of Defense took priority over many NASA science payloads and five major military satellites were placed into orbit within the first two years. At the same time, other efforts were underway to conduct research sponsored not only by the Department of Defense and the Air Force, but also by the Strategic Defense Initiative (SDI), the 'Star Wars' organisation, created by President Ronald Reagan as a means of transporting advanced military systems and technologies into space. One such mission, 'Starlab', involved a dedicated Spacelab payload, with a long module and pallet, to conduct laser impingement and tracking experiments in conjunction with instruments on Wake Island in the central Pacific Ocean. For this mission, a pair of payload specialists – an Air Force officer named Dennis Boesen and a civilian SDI physicist, Ken Bechis – were selected for training in August 1987, but ultimately Starlab never flew. The Shuttle's hydrogen leaks in the summer of 1990 caused great concern for Starlab's sponsors (by that time, it had already been postponed until January 1992 at the earliest) and it was eventually cancelled in favour of experiments aboard an unmanned free-flying satellite. Certainly, by the time NASA released its December 1990 Shuttle manifest, Starlab had disappeared.

One other research mission which did reach fruition, though, was the mixed-cargo flight which eventually became STS-39. Shortly after the resumption of mission operations, Bluford and fellow astronaut Charles Lacy Veach were teamed to work with the Air Force in developing their manned payloads for the Shuttle. The pair worked at the Air Force Space Systems Division and the Aerospace Corporation in El Segundo, California. Both men were ideal candidates for the task, since both had extensive Air Force backgrounds, in addition to their astronaut training. Veach came from Chicago, Illinois, where he was born on 18 September 1944, although much of his education was in Hawaii. The notion of becoming an astronaut came early for Veach. When he was only six years old, he took a comic book to his mother, pointed to a picture of a man standing on the Moon with what looked like a goldfish bowl on his head, and declared: *"That's* what I'm going to be!" He completed Punahou School in Honolulu in 1962 and entered the Air Force Academy, receiving a degree in engineering management. His military career commenced with receipt of his pilot's wings from Moody Air Force Base in Georgia in 1967 and Veach undertook fighter gunnery training at Luke Air Force Base in Arizona. He spent the next 14 years as an active-duty fighter pilot, flying the F-100 Super Sabre, the F-111 Aardvark and the F-105 Thunderchief in the United States, Europe and – significantly – Vietnam, where he performed no fewer than 275 combat missions, was shot down twice and received a Purple Heart. Veach was a member of the famed 'Thunderbirds', the Air Force's aerial demonstration squadron, and although he left active military service in 1981 he continued to fly the F-16 Fighting Falcon as a pilot with the Texas Air National Guard. His NASA career began in January 1982 as an engineer and research pilot at JSC in Houston, instructing astronauts to fly the Gulfstream Shuttle Training Aircraft, and Veach was himself selected as one of them in May 1984.

STS-39 was perhaps the most complex of all the military Shuttle flights. When Bluford and Veach were assigned, along with fellow astronaut Rick Hieb, in May 1989, their principal payload consisted of Air Force Project-675 (AFP-675), assembled and integrated by Lockheed, and the SDI's Infrared Background Signature Survey (IBSS). Unlike Bluford and Veach, who were both Air Force pilots with an extensive military heritage, Richard James Hieb – of Russian-German ancestry – was a civilian aerospace engineer. Born on 21 September 1955 in Jamestown, North Dakota, his mother was a teacher in the elementary school in his home town. Hieb earned a degree in mathematics and physics from Northwest Nazarene College in 1977 and a master's credential in aerospace engineering from the University of Colorado at Boulder, two years later. Immediately after graduation, he joined NASA to work on the development of Shuttle crew procedures and mission activity planning, working in Mission Control in Houston. An unsuccessful application to enter the astronaut corps in May 1984 was followed by success in June of the following year and during his time as an astronaut Hieb would worked extensively on rendezvous and proximity operations. This experience would serve him well on STS-39, whose rendezvous commitment was perhaps heavier than any previous Shuttle flight.

The first payload, explained Guy Bluford, "was a collection of experiments designed to measure background radiation and ultraviolet emissions, identify contamination in the orbiter environment and demonstrate X-ray imaging". The five AFP-675 experiments were the Cryogenic Infrared Radiance Instrument for Shuttle (CIRRIS), the Far Ultraviolet Camera (FARUV), the Uniformly Redundant Array (URA), the Horizon Ultraviolet Program (HUP) and the Quadropole Ion Neutral Mass Spectrometer (QINMS). All were mounted on a cross-bay Experiment Support Structure. Primary targets included Earth's upper atmosphere, its aurorae and astronomical objects, as well as analysing the spectrum of gases released from or around the Shuttle itself, as part of efforts to better understand the difficulties of identifying spacecraft with remote sensors and distinguishing them from natural phenomena. In addition to the Air Force and the SDI, sponsorship for the AFP-675 experiments came from the Phillips Laboratory's Geophysics Directorate, the Naval Research Laboratory and the Los Alamos National Laboratory.

Of the experiments, CIRRIS and HUP had flown before. As its name implies, the former operated at infrared wavelengths and was intended to obtain simultaneous spectral and spatial measurements of atmospheric 'airglow' and emissions from terrestrial aurorae; in fact, STS-39's launch was precisely timed to ensure proper visibility of auroral displays. On its first flight in June 1982, CIRRIS was unable to operate properly, due to a jammed lens cap, but on STS-39 it was expected to answer fundamental questions about the optimum atmospheric 'windows' for detecting cold-body targets, together with background radiance levels in various regions, the spatial structure of the background and the variability of Earth's limb emissions during auroral events. The results of these experiments were anticipated to contribute substantially to the development of more advanced surveillance systems. Such was the sensitivity of CIRRIS that it demanded a so-called 'gravity gradient' attitude, in which the Shuttle flew in a tail-to-Earth orientation, thereby minimising its use of thrusters which might produce unwanted contaminates. Adjoining CIRRIS was the Air Force Geophysics Laboratory's HUP, which would monitor the spectral characteristics of Earth's horizon at ultraviolet wavelengths and measure contamination of the orbiter's payload bay.

The remaining AFP-675 experiments were devoted to ultraviolet and X-ray research. Of these, the Naval Research Laboratory's FARUV observed both natural and man-made emissions in near-Earth space, including atmospheric airglow, diffuse northern and southern aurorae and chemicals and particulates in the environment around the Shuttle. In the latter case, RCS and OMS firings and 'surface glow' effects were of significant interest. At the same time, FARUV was to be directed towards astronomical objects, including diffuse nebulae, the Milky Way 'background', neighbouring galaxies and various star fields. Finally, Los Alamos' URA experiment was an X-ray detector, to be employed for studies of astronomical sources, whilst the Phillips Laboratory's QINMS would explore the levels of contamination in the payload bay – particularly hydrogen, oxygen and water vapour – and gather data whilst Discovery passed through the auroral zones and polar latitudes.

The second major payload, IBSS, was mounted atop the second Shuttle Pallet Satellite (SPAS-II), which had earlier flown aboard STS-7 in June 1983. Designed and built by the West German aerospace firm Messerschmitt-Bolkow-Blohm (MBB), under a June 1981 agreement with NASA, the SPAS was designed to accommodate scientific and technical experiments provided by fee-paying customers. Roughly triangular in shape, it measured 4.2 m across, 0.7 wide and 1.5 m high and weighed 1,500 kg when fully laden. During missions, it could operate in the payload bay – secured by one keel and two longeron trunnions – or be deployed for up to 40 hours in autonomous free flight. For STS-39, the $13 million platform was dominated by the IBSS experiment, with its barrel-shaped infrared cryostat, which sought to obtain scientific data for use in the development of missile defence sensors. Flying freely from the Shuttle at various ranges, IBSS-SPAS would make spectral, spatial and temporal radiometric observations of the orbiter's exhaust plumes and replications of booster firings. Interactions of these plumes with the ionosphere would be analysed, as would the region around the engine nozzles.

A crucial element of this activity was the Chemical Release Observation (CRO), which involved the deployment of three tiny subsatellites from the Shuttle's payload bay at rates of about a metre per second, enabling them to separate until they trailed IBSS-SPAS by 50-200 km. Signals transmitted from Vandenberg Air Force Base would expel chemicals – nitrogen tetroxide, unsymmetrical dimethyl hydrazine and monomethyl hydrazine, all typically used as station-keeping propellants – from the satellites, which would quickly vaporise into clouds, to be observed, simultaneously, by IBSS-SPAS, by sensors on the ground in California and by sensors aboard a tracking aircraft. It was hoped that this combined data would enable characterisation of the interactions of the chemicals with the upper atmosphere.

Elsewhere, the Critical Ionisation Velocity (CIV) experiment involved the ejection of pressurised xenon, neon, carbon dioxide and nitric oxide gases from four canisters in the payload bay, at different angles to the orbiter's own velocity, in order to enhance ionisation with the thin upper atmosphere. This was expected to enable researchers to examine a theory which held that gases could be 'ionised' if passed through a magnetised plasma and their kinetic energy caused to exceed their ionisation potential. "Ions so created," noted NASA's STS-39 press kit, "would then flow along the local magnetic lines of force and generate emissions which can be detected by space-borne sensors, thereby permitting tracking of the vehicle releasing the gases."

The sensors aboard IBSS-SPAS would acquire spectral data on these chemical releases, as well as Earth's surface under various conditions of light and darkness, hard earth and water and clouds and cloudlessness. Additional measurements of the so-called 'orbiter glow' – the phenomenon in which rarefied atmospheric gases struck the Shuttle's surfaces, particularly its tail, causing visible and infrared radiance – would be pursued, as part of ongoing efforts to understand their cause and precise nature. During two days of free flight, IBSS-SPAS would observe more than 60 orbiter manoeuvres, making STS-39 not only the most complex Department of Defense mission, but also one of the most complicated missions ever undertaken by NASA.

At the time of the assignment of Bluford, Veach and Hieb, the flight was scheduled for launch in July 1990, but found itself repeatedly delayed. In a way, this was beneficial, because training was incredibly complex. Major Rob Crombie, an Air Force officer who had been selected earlier in the 1980s as a manned spaceflight engineer, worked with them to develop flight procedures, including planning to handle malfunctions with the AFP-675 and IBSS-SPAS payloads. Every few weeks in the summer of 1989, Bluford and Veach would fly out to Lockheed Martin's Space & Missile facility in San Jose, California, to work on these procedures in a full Shuttle flight deck simulator, equipped with payload controls and displays. They also worked independently on other areas, to ensure a better grasp of the material. As the mission slipped further towards the end of 1990, there was also a brief respite to do other things. "Lacy [Veach] had to take three to four months off," recalled Bluford in his NASA oral history, "to go to McConnell Air Force Base in Wichita, Kansas, for training in the F-16. We adjusted his training flow to take that into account and Lacy never missed a beat when he returned." Ever supportive, Rob Crombie acted as their crew representative and Bluford would laud his performance as "critical" in the success of STS-39.

Several months passed before the remainder of the crew was announced by NASA. Unlike previous Department of Defense flights, which typically carried a crew of five, STS-39 would feature no fewer than seven astronauts – and, for the first time in the Shuttle era, would carry as many as *five* mission specialists. Certainly, *Flight International* reported that the additional two NASA crew members took the places of two Air Force manned spaceflight engineers, part of a cadre which had effectively been stood down in the wake of Challenger. (The identities of these two engineers remain unidentified by NASA, although Guy Bluford mentioned Rob Crombie as having played a large role in STS-39 training and in May 1988 *Flight International* suggested the name of another Air Force officer, Scott Yeakel, as being associated with the IBSS payload.)

At the end of September 1989, veteran astronaut Mike Coats – then serving a temporary duty as acting chief of the astronaut office – was named to command STS-39, with rookie Blaine Hammond as pilot and two mission specialists, Greg Harbaugh and Don McMonagle. By this time, launch had shifted to no earlier than November 1990 and would follow hard on the heels of the Ulysses mission. Herein lay STS-39's first obstacle. So critical was the need to meet the narrow Ulysses window, and so packed was the Shuttle schedule at the end of 1990 that NASA decided to remove STS-39 and postpone it by a full year. In its January 1990 manifest, the agency announced that AFP-675 and IBSS-SPAS would fly under a new designation, STS-51, aboard Atlantis, no earlier than January 1992. The rationale was that this move would protect the Ulysses launch window and provide additional flexibility for KSC's Shuttle workforce. By the end of March 1990, however, the schedule had changed again. STS-39 returned to its original position, but was pushed back slightly to January 1991. For Mike Coats, who intended to retire after STS-39, it must have been a relief.

Milkshakes and doggie biscuits 93

1-15: The barrel-shaped infrared cryostat of IBSS is prominent atop the Shuttle Pallet Satellite in this image, captured shortly after the payload was deployed from the RMS mechanical arm. The end effector of the arm itself is visible at bottom right.

In spite of the complexities of the flight, only Coats and Bluford had flown before. (In fact, they would later have a humorous crew photograph taken, with the five rookies wearing Houston Rockets basketball kits, Bluford dressed as the team referee and Coats wearing a coach's uniform.) Michael Lloyd Coats was born in Sacramento, California, on 16 January 1946, and attended high school at Riverside. He entered the Naval Academy, received his degree in 1968 and was designated a naval aviator in September of the following year. Coats initially trained on the A-7E Corsair II attack aircraft and served two years aboard the USS *Kitty Hawk*, flying more than 300 combat missions over Vietnam. Upon his return to the United States, he became an A-7E instructor pilot in California and was selected for test pilot school in 1973. Graduation the following year led to duties as project officer for the A-7 and for the A-4 Skyhawk attack aircraft. Coats' final flying assignment before selection into NASA's astronaut corps was as an instructor at the Naval Test Pilot School in Patuxent River, Maryland, from April 1976 until June 1977. At the same time, he completed a master's degree in the administration of science and technology from George Washington University. He was attending the Naval Postgraduate School in Monterey, California, studying for another master's in aeronautical engineering when he was summoned to Houston in August 1977 as one of the *first* group of candidates. In Mike Mullane's mind, at least, Coats was quite different from the stereotypical military aviator: a quiet family man, devoted husband and father of two.

By now, Coats had flown twice into space – once as pilot and once, on STS-29, as commander – and confidently expected that to be his lot. It was not that he did not enjoy space flight. In fact, like many of his comrades, he was enthralled by the view from Earth. "The feeling that is still so vivid in my mind," he once told an interviewer, "is looking out the windows when you get into orbit and looking at the Earth for the first time from space and seeing this *living* planet. The *colours* – the blues of the oceans, the greens, the browns – it's just really obvious this is a living planet ... and this living planet is going through this big, black void. *That's* the perfect word, because there is *nothing* out there. The stars don't twinkle out there; they're just harsh points of light. The blackness of space and the bright, living beauty of the Earth just amaze me and I think they have the same effect on anyone who has been up there."

Rather, Coats' decision to leave NASA was entirely based around his family, who had endured the relentless training grind for more than a decade. The loss of Challenger had profoundly affected his wife, Diane, and she made him promise that STS-29 would be his final mission. Timing changed all of those plans. When STS-29 flew in March 1989, it was the first Shuttle mission of George H.W. Bush's administration and Diane Coats suggested flying something for the new president, "and he'll invite us to the White House". This fell flat when it became apparent that several items had been flown for Bush on earlier missions, during his tenure as Vice President under Ronald Reagan, so Diane suggested carrying something to present to First Lady Barbara Bush instead. At the last minute, it was decided to fly a little gold Shuttle charm. During the mission, the astronauts spoke to Bush and told him about the charm. The president was excited – nobody had ever flown anything into space for his wife before – and, lo and behold, came an invite to the White House.

Shortly after STS-29 landed, Mike and Diane Coats flew to Washington, DC, where Barbara Bush personally gave them a guided tour of the presidential mansion. Whilst there, Diane gave a Shuttle-shaped biscuit to the presidential dog, Millie, and Mike Coats proudly stood for photographs in the Oval Office. A couple of weeks later, at his desk in the astronaut office in Houston, Coats received an unusual message, asking him to call the White House. They had so impressed the Bushes with the biscuit for Millie that they were invited to a state dinner, where Mike and Diane found themselves in the reception line with Audrey Hepburn and Bob Hope.

During the dinner, President Bush turned to Coats and pointedly asked him: "Well, when are you going to fly again?"

Coats thought for a moment. "Well, I promised my wife I would just fly that second flight."

"Do you want to fly again?"

"Well, sure."

"Let me see what I can do." With that, Bush got up from the table and approached Diane, who was seated between the White House's ambassador for protocol and the wife of Vice President Dan Quayle. Coats watched as his wife frowned, then gradually smiled. A few minutes later, Bush returned.

"I think it's okay if you fly one more time."

Coats was stunned. "What in the world did you tell my wife to make me expendable?"

Bush grinned. "That's between me and her."

After dinner, Coats asked Diane what had made her change her mind. The answer: Bush had promised them both a second invitation to the White House for dinner! The moral of the story, Coats later explained, was to always *listen* when one's wife came up with a crazy idea. Yet his third mission on STS-39 would most definitely be his last. In December 1989, fellow astronaut Mike Mullane went to see Coats to announce his own impending retirement from NASA. In his capacity of acting chief of the astronaut office, Coats accepted Mullane's words and shared a few of his own. "As he continually flipped his pen," Mullane wrote in his memoir, *Riding Rockets*, "he confided to me that he had already made the same decision." In the aftermath of Challenger, both men accepted that the Shuttle had the potential to kill them ... and destroy their families in the process.

So it was that Coats' final mission would be the most complex of his astronaut career. Seated alongside him on Discovery's flight deck for STS-39 would be an Air Force test pilot named Lloyd Blaine Hammond Jr. Exactly six years younger than Coats, he was born in Savannah, Georgia, on 16 January 1952, the son of a General Electric sales manager. He received much of his schooling in Missouri and grew up watching F-4 Phantom jets flying into Lambert Field in St Louis and dreaming of someday becoming a pilot. Hammond received his degree from the Air Force Academy in 1973 and a master's in engineering science and mechanics from Georgia Institute of Technology in 1974. A year later, after flight training, he received his pilot's wings at Reese Air Force Base in Texas and flew F-4s in Germany, before becoming an instructor in the F-5B. Hammond attended the Empire Test Pilot School in the United Kingdom in 1981 and returned to the United States to work at Edwards Air Force Base as a trainer and lecturer in spin/stall theory. He was selected as an astronaut by NASA in May 1984.

The complexity of STS-39, and its rendezvous commitment, required the flight to involve 24-hour operations, with the crew split into 'red' and 'blue' shifts to work around the clock. With overall responsibility for mission safety and success, Mike Coats was not assigned to either team, but tended to anchor his schedule with the red shift. He was joined by Blaine Hammond, with Lacy Veach responsible for the AFP-675 and IBSS-SPAS payloads and Rick Hieb operating the RMS. On the blue team, rookie astronaut Don McMonagle was placed in charge of the flight deck, whilst Guy Bluford supervised the payloads and Greg Harbaugh operated the mechanical arm. "We had to do rendezvous, multiple translation manoeuvres, extended station-keeping and deployment and retrieval of the SPAS with the RMS," recalled Bluford.

"This involved precision orbiter manoeuvring, IBSS-SPAS commanding, CIRRIS and IBSS observation sequences and multi-body management in a very intensive timeline. A lot of co-ordination was required on the flight deck, synchronising orbiter and SPAS manoeuvres and documenting key events. There was approximately 36 hours planned for rendezvous and proximity operations." On each shift, the payloads, RMS and orbiter often had to be managed simultaneously.

With this in mind, it is unsurprising that STS-39 demanded a seven-man, all-NASA crew, and the presence of Donald Ray McMonagle, selected as a pilot in June 1987, in the role of a mission specialist on his first flight added to this base of expertise. McMonagle was born in Flint, Michigan, on 14 May 1952 and entered the Air Force Academy to study astronautical engineering. His military career began immediately after graduation, when he completed flight instruction at Columbus Air Force Base in Mississippi, then undertook training in the F-4 at Homestead Air Force Base in Florida. He flew for a time in South Korea and served as an instructor pilot in the F-15 Eagle, before admission into test pilot school at Edwards Air Force Base in 1981. He completed his studies there as the outstanding graduate of his class (earning the coveted Liethen-Tittle Award) and was the project test pilot for the Advanced Fighter Technology Integration F-16, serving as an operations officer at Edwards until his selection into the astronaut corps. By this time, he had also gained a master's degree in mechanical engineering from California State University at Fresno. McMonagle was a rising star at NASA from the outset, serving on an engineering 'tiger team' to investigate the damage to Atlantis' thermal protection system on STS-27, and would become the first pilot in his class to command a Shuttle mission. In his later years with the space agency, he would establish a new EVA Project Office and serve as manager of launch integration at KSC, with the enormous responsibility for the final preparations before each Shuttle launch. The final – and youngest – member of the crew was Gregory Jordan Harbaugh, born on 15 April 1956 in Cleveland, Ohio. After high school, he gained a degree in aeronautical and astronautical engineering from Purdue University in 1978 and was immediately employed by NASA as an engineer at JSC in Houston. During the pre-Challenger era, Harbaugh supported Shuttle flight operations in Mission Control as a data processing systems officer and earned his master's degree in physical science from the University of Houston at Clear Lake in 1986. A year later, in June 1987, he was selected as an astronaut candidate.

Operating around the clock required the astronauts to shift their sleeping patterns in the final weeks before launch. In fact, a new sleep regime had been implemented by NASA prior to the dual-shift STS-35 mission, whose crew spent the final week before launch awake all night in an all-white room, under harsh fluorescent lighting, to adjust their irregular sleep cycles. "Their exposure to bright light will be timed so that sunlight will not disrupt their bodies' shifted circadian rhythm," noted *Flight International* in the days before STS-39, "when they emerge from the crew quarters during the day for training. The astronauts' body clocks will be 'reset' in such a way that the Sun will be interpreted as evening light, rather than morning light, because the body's circadian rhythm is apparently synchronised with sunrise and sunset."

By the autumn of 1990, the hydrogen leaks had effectively pushed STS-39 into the spring of the following year, with launch expected in late February. Then, in November 1990, Bluford suffered a herniated disk in his back. Physical therapy worked well for a while, but when he continued to experience problems during one of his training sessions it was considered prudent to take a closer look. "I started sensing numbness in my right shin," Bluford remembered, "and found that I couldn't stand for any longer than 30 minutes, before my leg started aching." (When the crew had their official photograph taken, Bluford was positioned at the far right side, so that he could stand close to a chair.) The only option was an operation and there was talk about removing Bluford from STS-39, but judicious rescheduling of crew tasks by Mike Coats allowed for his crewmate to undergo the operation and recovery and rejoin the mission. In total, Bluford was out of work for barely two weeks and, within a month, flight surgeons verified that he was fit and healthy. Years later, he believed that Rob Crombie – the Air Force manned spaceflight engineer, who had trained extensively with them – would "probably" have flown in his stead if the result of the operation had been different. "He would have done an excellent job," Bluford said, "if the circumstances dictated."

If the hydrogen problems which plagued Columbia and Atlantis in 1990 left Discovery virtually unscathed, it was only a matter of time, with such a complex machine, before a technical issue arose. Two leaking thrusters pushed the STS-39 launch back from 26 February to no earlier than 3:49 am EST on 9 March, at the opening of a three-hour 'window', precisely timed to achieve maximum auroral observations. The flight would last a little over eight days, touching down at Edwards Air Force Base in California on the 17th. It was during the final checks of the vehicle on Pad 39A that technicians discovered cracks on all four lug hinges on the two External Tank umbilical door drive mechanisms, necessitating a rollback to the VAB on 7 March and replacement. On the 21st, NASA announced that STS-37 would fly first, with STS-39 rescheduled for early May. Discovery returned to the pad at the beginning of April and on the 15th the space agency formally announced a new launch attempt on the 23rd. This came to nothing when a transducer on one of the main engines' high-pressure oxidiser turbopumps exhibited out-of-specification readings. The transducer was replaced and retested.

Following a slight delay, Discovery roared into space at 7:33 am EDT on the 28th and the red team of Coats, Hammond, Veach and Hieb took charge of operations to activate and check the expansive STS-39 payload. For AFP-675, this proved somewhat problematic, when two tape recorders stopped working after just four hours of operation, requiring a complicated bypass repair, in which the astronauts rerouted wires and attached a splice to the Ku-band antenna to permit direct downlink to ground stations. Their blue team counterparts, meanwhile, bedded down for their first sleep period. In addition to AFP-675 and IBSS-SPAS, the red team activated another series of experiments, called the Space Test Payload (STP), which was mounted on a cross-bay bridge structure, known as a 'Hitchhiker'. Managed by NASA's Goddard Space Flight Center, the Hitchhiker provided a mounting location for five military payloads: a Naval Research Laboratory ultraviolet limb imager to observe atmospheric airglow at altitudes of 50-100 km, an advanced liquid feed demonstration of electronic pressure regulators and ultrasonic propellant sensors for future spacecraft propulsion systems, a space kinetic infrared test to examine the infrared signature of the Shuttle glow phenomenon and enable identification of the responsible chemical species, an experimental data processing system and a particle monitor to measure particulate debris emerging from the payload bay during flight.

To conserve electrical power, the red team placed Discovery's systems into a 'Group B Power-Down' mode, preparatory to the IBSS-SPAS deployment, scheduled for the third day of the mission. This deployment was postponed by 24 hours, to enable AFP-675 to acquire more data before its cryogenic coolant was depleted, but IBSS-SPAS was finally positioned high above the payload bay, at the end of the RMS, early on 1 May. At 4:18 am EDT, it was released into free flight, kicking off an intensive two-day period of activities, during which time Discovery's OMS engines would be fired on 16 occasions and her RCS thrusters an impressive 41 times. Due to their complexity, all of the OMS burns were performed at shift changeovers, when all seven men were awake. During this time, IBSS-SPAS observed the bursts, as well as the gas releases from the CIV canisters in the payload bay and the deployments of the three CRO subsatellites.

Immediately after deployment, Coats and Hammond fired Discovery's RCS thrusters to raise their orbit slightly above IBSS-SPAS, causing the Shuttle to drift about 10 km 'behind' the free-flying satellite and enabling the so-called 'far-field' observations to begin. During this time, it observed Discovery performing a particularly rapid sequence of manoeuvres which the crew had nicknamed 'The Malarkey Milkshake', in honour of the head of the guidance team, Rockwell International engineer John Malarkey, who led its development. (Malarkey served as STS-39's Orbit 2 Rendezvous officer.) The milkshake involved Discovery rotating out of plane: firing one of her OMS engines for 20 seconds to propel the orbiter 'northwards', off its previous ground track, without altering its altitude. Immediately after this burn, the crew performed a 'fast-flip' yaw manoeuvre, using the RCS to turn Discovery's nose 180 degrees to the 'south'. A single-engine OMS braking burn then returned the orbiter to its original ground track.

With the completion of the far-field observations, Coats and Hammond performed a retrograde RCS firing to slightly lower their orbit, causing them to move to a position 2 km behind IBSS-SPAS. At length, late on 2 May, the red team extended the RMS to grapple IBSS-SPAS and berth it back in the payload bay at 7:15 pm EDT. A second phase of operations, involving the unberthing of the payload by the RMS, but no physical deployment, began at around midday on the 3rd and lasted for approximately 23 hours.

Meanwhile, in Discovery's middeck, the hand-held Radiation Monitoring Experiment continued observations of gamma rays, electrons, neutrons and protons in the cabin and the CLOUDS experiment (the Cloud Logic to Optimise the Use of Defense Systems) served to quantify variations in apparent cloud cover as part of efforts to develop new meteorological observation models. One other activity which was far quieter was the deployment of STS-39's only classified payload, ejected by Bluford from a Multi-Purpose Experiment Canister (MPEC) on 5 May. Years later, he joked that his crewmates disappeared downstairs to the middeck and 'pretended'

1-16: Like a shimmering snake, Earth's aurorae are beautifully juxtaposed with a thruster firing which silhouettes Discovery's OMS pods and vertical stabiliser tail fin. Manoeuvres, engine burns and auroral observations formed a central tenet of the STS-39 mission.

not to notice as he remained on the flight deck to oversee the deployment. In truth, only Bluford and Coats were privy to the classified nature of the payload and other than the name of its stated sponsor (the Air Force Space Systems Division) its purpose remains unclear to this day.

The mission of STS-39 had one additional challenge in store, however, when the Edwards landing was called off, due to unacceptably high crosswinds. As a result, for the second time in less than six months, a returning orbiter was directed to KSC in Florida and Coats and Hammond guided their ship onto Runway 15 at 2:55 pm EDT on 6 May. The landing did not go well. Touching down at 387 km/h – the second-fastest Shuttle landing, eclipsed only by STS-3 – the right-hand main gear hit the runway 60 m before the left-hand gear, causing one of the right-side tyres to shred. Unfortunately, the men's wives were out in California and would not meet their husbands again until they returned to Houston.

THREE WOMEN OR THREE DOCTORS?

With the safe landing of Discovery, her sister ship Columbia was well on the way to flying the second dedicated scientific voyage of the post-Challenger era, later that same month, although STS-40 would be devoted not to astronomy, but to medical research. NASA already had plans to send the queen of the Shuttle fleet to prime contractor Rockwell International's Palmdale facility in California, later in 1991, for several months of modification and refurbishment to enable her to fly longer missions of up to 16 days. The delays enforced by Columbia's hydrogen leaks had already meant that two of her flights – the first International Microgravity Laboratory (IML-1) and the first Atmospheric Laboratory for Applications and Science (ATLAS-1) – had already been shifted onto Discovery and Atlantis. This had left only STS-40, which had been waiting for many years to fly. In fact, the 'science' crew had been assigned early in 1984 and included two mission specialists and four research scientists to fill two payload specialist positions.

The payload specialist candidates had been selected by NASA in January 1984. They consisted of cardiologist Drew Gaffney, biochemist Millie Hughes-Fulford, physiologist Bob Phillips and biophysicist Bill Williams. A few weeks later, on 2 February, the space agency announced the selection of physicians Rhea Seddon and Jim Bagian – together with a third astronaut, John Fabian – as mission specialists. Designated 'Mission 61D', their seven-day flight aboard Columbia was scheduled for January 1986 and its payload was known as 'Spacelab-4'. In the aftermath of the Challenger accident, it was renamed as the first Spacelab Life Sciences mission (SLS-1). By this time, Fabian had retired from NASA.

It is not surprising that Margaret Rhea Seddon was assigned primary responsibility for 61D's medical research, for in addition to her qualifications as a physician and a surgeon, she had already been following Spacelab-4 for several years. Shortly after her selection as an astronaut in January 1978, she approached one of the senior members of the corps, physician Joe Kerwin, to discuss future opportunities on the Shuttle in the life sciences arena. In her NASA oral history, Seddon recalled that she "probably" expressed her formal interest in Spacelab-4 to George Abbey, the head of Flight Crew Operations, although she was assigned to another mission, 41F, at the time. With 41F scheduled for launch in August 1984 and Spacelab-4 not due until the spring of 1986, Abbey told her that it might not be a problem. It wasn't. Within weeks, she received the assignment. 'Formal' training did not begin for many months, however, since Seddon's first mission was repeatedly delayed and did not fly until April 1985. "I did what I could," she said, "but it was all kind of a scramble, because I didn't have much time to devote to it."

The situation improved at the end of January 1985, when astronauts Vance Brand and Dave Griggs were assigned as the mission commander and pilot. By the time of the Challenger disaster, however, Spacelab-4 had slipped until no earlier than March 1987, partly due to Shuttle delays and partly due to the sheer complexity of the scientific payload. By this time, Bill Williams had resigned from the mission (in February 1985), citing a desire to spend more time with his family, and three payload specialist candidates – Phillips, Gaffney and Hughes-Fulford – remained. In April 1985, Phillips and Gaffney were selected to fly. "It was very over-subscribed," Seddon recalled of Spacelab-4. "There was too much to do, too many experiments, too much crew time required." She had only recently flown, and Jim Bagian had not flown at all, and neither were skilled in the development of complex timelines. There were problems with the experiments, too, including animal holding facilities which had not performed particularly well on earlier flights. With the loss of Challenger, the payload specialists returned to their research institutions and Seddon and Bagian found that the mission gained relatively low priority after the resumption of Shuttle operations. Tracking and Data Relay Satellites to permit near-continuous space-to-ground communications needed to be launched, as did a handful of time-critical planetary missions and sensitive Department of Defense payloads. When Seddon and Bagian were formally named to STS-40 in February 1989, their launch was anticipated no earlier than June of the following year.

Around this time, in March 1989, Seddon gave birth to her second son, Dann. She was married to Shuttle pilot Robert 'Hoot' Gibson and in July 1982 their first son, Paul, had become the first child born to parents who were both US astronauts. Seddon came from Murfreesboro, Tennessee, where she was born on 8 November 1947, the daughter of an attorney. She developed a keen love for the sciences at an early age, graduating from Central High School and entering the University of California at Berkeley to study physiology. At Berkeley, she first encountered the Free Speech Movement and became aware for the first time that careers previously barred to women – medicine and aviation, for example – were within reach. Seddon received her degree in 1970 and was accepted into the University of Tennessee College of Medicine. "I was pretty sure when I started medical school that I wanted to be a surgeon of some sort," she told the NASA oral historian. In 1973, having gained her medical degree, she completed a surgical internship at the University of Tennessee and became the only woman on their three-year general surgery residency programme. "I think I was probably the *second* woman they had *ever* accepted," she recalled, "so that was interesting!" During this period, her interests expanded to cover the nutrition of surgical patients – the techniques and technologies needed to feed intravenously were still in their infancy – and it was this mix of skills that Seddon believes attracted her to NASA. She did not believe that she stood a chance of making the cut. "They were hiring people for July 1st 1978," she said, "and I was finishing my residency June 30th." If it did not work out, she told herself, she would return to Plan B: a surgical subspecialty or a PhD in nutrition. As events unfolded, Plan A succeeded and she was selected by NASA in January 1978. She continued working part-time in hospitals in the Houston area and married Gibson in May 1981.

"It was time for us to get back together and nail everything down," Seddon said of the first few months after assignment to STS-40, "the timeline, the equipment, what we could do, what we could sign up to do." By this time, Bill Williams had long since resigned from the mission and the prime payload specialist slots were assigned to Gaffney and Phillips, with Hughes-Fulford serving as a backup. Francis Andrew Gaffney came from Carlsbad, New Mexico, and was born on 9 June 1946. After high school, he entered the University of California at Berkeley and graduated with a degree in psychology in 1968; interestingly, Seddon was at the same institution, a couple of years behind Gaffney, studying physiology. After earning his degree, Gaffney attended the University of New Mexico and received his medical doctorate in 1972. He served his internship and residency in internal medicine at Cleveland Metropolitan General Hospital in 1975, before taking a fellowship in cardiology at the University of Texas Southwestern Medical Center, where he was later a professor. Like many astronauts, Gaffney served a spell in the military, working as a flight surgeon with the Texas Air National Guard and, as an Air Force reservist, rising to the rank of colonel. From June 1986 until his announcement as a member of the STS-40 crew, Gaffney served as a visiting senior scientist in the Life Sciences Division at NASA Headquarters.

The second payload specialist, Bob Phillips, had also earned himself distinction in physiology and veterinary medicine, receiving doctorates in both fields. Phillips spent seven years of his life working on SLS-1, before a medical problem unexpectedly caused him to be grounded in 1989. Yet his knowledge of the payload was immense and he elected to remain as an alternate payload specialist, supporting the mission from NASA's Marshall Space Flight Center in Huntsville, Alabama. "He really wasn't a backup," said Rhea Seddon, "but he trained with us; he knew everything we knew. We just thought the world of Bob, so we were pleased that he was going to stay on with us." Many would have understood if Phillips had thrown in the towel and returned to his previous position as a professor of physiology at Colorado State University, but he remained in a support capacity. His place was taken by Millie Elizabeth Hughes-Fulford, who came from Mineral Wells, Texas, born on 21 December 1945. After completing high school in her home town, she entered Tarleton State University and earned a degree in chemistry and biology in 1968. That same year, she moved to Texas Women's University as a graduate fellow, studying plasma chemistry and would later earn her PhD in biochemistry in 1972. Her post-doctoral career began at the faculty of the Southwestern Medical School at the University of Texas at Dallas, focusing on cholesterol metabolism, and she wrote extensively on the regulation of bone and cancer growth. Like Drew Gaffney, she served a spell in the military, reaching the rank of major in the Army Reserve Medical Corps. She applied, unsuccessfully, for admission into NASA's first Shuttle astronaut class as a mission specialist candidate in January 1978.

Hughes-Fulford's addition to STS-40 meant that the mission would earn a new record as the first space flight to include as many as *three* women. In April 1989, NASA had announced the names of commander Bryan O'Connor, pilot John Blaha and flight engineer Tammy Jernigan to fill the remaining crew positions. For Tamara Elizabeth Jernigan, who would go on to record five Shuttle missions by the time she was 40, the assignment brought another added accolade; for when she eventually launched in June 1991 she was the youngest American woman ever to fly into space, pipping the previous record-holder, Sally Ride, by a matter of days. It is a record that Jernigan still holds for the United States to this very day. Yet Jernigan would exhibit a no-nonsense competence. "In meetings," wrote astronaut Tom Jones, who later flew with her, "she would quickly get to the crux of any issue on the table." (In her later career with NASA, she would serve as deputy chief of the astronaut office.) She came from Chattanooga, Tennessee, and was born on 7 May 1959, but much of her early education took place on the West Coast, in California. Aged only ten, she vividly recalled stepping out of her front door to look at the Moon, one summer evening in 1969, knowing that a pair of astronauts – Neil Armstrong and Buzz Aldrin – had just landed there. For Jernigan, *that* was her first memory of aspiring to become an astronaut herself, someday. The dream took on a more serious turn after high school in Santa Fe, when she moved to Stanford University to study physics. "I was a physics major," she told a NASA interviewer in the weeks before her final Shuttle mission. "In 1978, they started taking astronauts from the scientific community in earnest and also taking women. It was in 1978 that I thought I might have a chance to be selected."

A keen intercollegiate athlete and varsity volleyball team member whilst at Stanford, Jernigan earned her degree in 1981, followed by a master's credential in engineering science, two years later. Her career with NASA began immediately after graduation, as a research scientist in the Theoretical Studies Branch of the agency's Ames Research Center in Mountain View, California. A second master's degree, in astronomy, from the University of California at Berkeley, followed in 1985. In June of that same year, she was selected as an astronaut, aged only 26. Alongside the demanding programme of astronaut candidate training, Jernigan pursued her PhD in space physics and astronomy and achieved her doctorate in 1988. Aged only 29 when she was assigned to STS-40, Jernigan would serve as the mission's flight engineer, seated behind and between O'Connor and Blaha to offer a crucial additional set of eyes to monitor the instruments and displays and respond to emergency situations.

When Hughes-Fulford came aboard, Bryan O'Connor was once asked by an interviewer if he was 'afraid' of having three women on the flight. "No," the straight-laced Marine Corps colonel replied, "I'm more worried about having three *doctors* on the flight!" O'Connor's sly humour, though, could not mask his very real concern, in the early days of training, about the compatibility of members of his crew; in particular, his payload specialists. After the medical disqualification of Bob Phillips, the arrival and involvement of Hughes-Fulford was far greater than it had been previously. This caused friction with Drew Gaffney. "Sometimes," O'Connor told the NASA oral historian, "I've thought that Millie and Drew were like oil and water and it was a pleasant surprise for me when, seeing how they operated or *didn't* operate together in the office or after the training's done or whatever, to where they would take *all* that baggage, old concerns with one another's performance, disagreements about the science, and we'd get into the simulator and there was *none* of that. These two trained like two professionals." Yet when the pair left the simulator, O'Connor compared their relationship to the 19th century feud between the Hatfields and the McCoys. In her NASA oral history, Rhea Seddon agreed that there was "some stress, just some frictions" in the crew relationship.

From his position as the mission's commander, O'Connor found this worrisome. He was concerned that, should some of their "underlying issues" affect work in space, during a critical time, it might impair their ability to get the job done. As a Marine, he had done much of his training and flying in single-seat aircraft and compared his fears to those of a bomber pilot or transport pilot, trying to get crews to work effectively together. O'Connor concluded that the seven of them should sit down and do a personality assessment, very similar to the Myers-Briggs psychometric test. This revealed that, unlike the military, in which two or three of around half a dozen personality types tend to dominate the rest, the Shuttle crew featured a much broader range and demanded a broader understanding. "You've got to be a little more forgiving of certain things," O'Connor reflected, "and be more sensitive to other things to communicate properly and to operate as a crew."

The personality assessment enabled them to smooth the road to a much better working relationship. "I think everybody just assumed that everybody would get along," remembered Seddon. "Sometimes that's hard to do, especially when you've been training together in close quarters for a long time. Everybody has their little quirks." It underlined not so much a need for the astronauts to attempt to alter their personalities, but to adjust their approaches to how they 'led' in certain areas and how in other areas they could 'follow' and 'interact' more harmoniously. By the time STS-40 closed in on its launch in June 1991, O'Connor was more than satisfied that his crew would work like an oiled machine, with "no bad baggage". He was not trying to get Gaffney and Hughes-Fulford to be best buddies – "after the mission was over, they probably never spoke to each other again" – but to complete the best work in orbit. It worked and the payload specialists' performance was nothing short of outstanding, with a full plate of intense medical investigations completed.

Joy and sorrow can be close bedfellows and, certainly, in the case of STS-40, happiness and sadness pervaded their first few weeks together as a crew. The assignment of Bryan O'Connor to the commander's seat was particularly welcomed and Seddon lauded his hard work and in-depth knowledge and praised him overall as a "good fellow". It was no accident that O'Connor became the first pilot from NASA's 1980 astronaut class to receive a flight assignment *and* the first to receive a command, although as the schedule writhed and contorted, both before and after Challenger, he ended up flying later in line. He originally applied for the first group of Shuttle astronauts, selected in January 1978, but was rejected, along with four other pilots, due to NASA's requirement to take more women astronauts. Two years later, he made the cut.

Bryan Daniel O'Connor was born in Orange, California, on 6 September 1946. He grew up in a military family – his father was a career Marine Corps pilot – and spent weekends looking at aircraft on the flight line. After high school, he entered the Naval Academy and received a degree in engineering in 1968. Two years later he got a master's credential in aeronautical systems from the University of West Florida. At first, he had mixed feelings about where he wanted his own military career to take him. "Maybe I would get into submarines or ships," he said, "but the aviation bug was too deep in there and so when it came time to select a service and an occupation, I went for Marine Corps aviation." After Basic School he received his naval aviator's wings in June 1970 and flew as an attack pilot in the A-4 Skyhawk and the AV-8A Harrier. "*That's* when I got bit by the test pilot bug," he said. In 1975, he was accepted at Naval Test Pilot School and later served within the Strike Test Directorate, participating in the evaluation of various vertical takeoff and landing aircraft. Going on to graduate from the Naval Safety School, he was in charge of Harrier flight testing when he applied for NASA's 1978 astronaut selection. Rejected for what he considered to be his lack of experience, he spent 1979 gathering more flight test time in the Harrier. "I don't think I would have flown the Shuttle any sooner had I been picked up in that first class," he said later. "They were going to hire 40 and they cut five pilots out, because they didn't need them, right at the last minute. I found out much later that I was one of those five. The reason they didn't need them is because in that time frame, Columbia was having problems. All the tiles had fallen off on a flight across country. One of the main engines had blown up on a test stand and the whole Shuttle programme was slipping, so if they had picked me up in 1978 and I had joined the astronaut office, I probably would have eventually flown the same time anyway as I did."

A decade later, O'Connor had flown as pilot on Mission 61B in late 1985 and would have flown again in the summer of 1986, had Challenger not brought the Shuttle programme to its knees. Now he found himself seated alongside Air Force colonel John Blaha, one of his comrades from the 1980 astronaut class, and both would be making their second Shuttle flights. (Blaha had flown as pilot on STS-29 in March 1989.) Within ten weeks, however, it all changed. Fellow astronaut Dave Griggs was killed in a vintage aircraft crash on 17 June 1989, only five months before he was due to fly as pilot on STS-33. Since Blaha had recent flight experience, it was decided to assign him to fill Griggs' shoes and a new pilot, an Air Force major named Sid Gutierrez, was named to STS-40.

O'Connor and Gutierrez had worked together before. In 1986, following the Challenger disaster, O'Connor had been asked to lead an action centre at NASA Headquarters in Washington, DC, to help plan the agency's response to the recommendations of the Rogers Commission. Every so often, other astronauts would rotate in and out of O'Connor's office. Among them was Sidney McNeill Gutierrez, the first astronaut of Hispanic descent to fly as a Shuttle pilot. Born in Albuquerque, New Mexico, on 27 June 1951, he completed high school in his home town and entered the Air Force Academy to study aeronautical engineering. Whilst at the academy, Gutierrez was a member of the National Collegiate Championship Parachute Team, accomplishing more than 550 jumps and attaining the rating of Master Parachutist. After graduation in 1973, he began pilot training at Laughlin Air Force Base in Del Rio, Texas, and remained for several years thereafter as an instructor in the T-38 Talon jet. In 1977, he completed a master's degree in management from Webster University and continued his Air Force career, flying the F-15 Eagle fighter at Holloman Air Force Base in New Mexico. Graduation as a test pilot in 1981 led to several years of work on the F-16 Falcon Combined Test Force, working on the new fighter's airframe and propulsion systems. Intensely proud of his Hispanic heritage, Gutierrez would be honoured many times after his selection as a pilot by NASA in May 1984. Even before his first Shuttle mission, he received the Congressional Hispanic Caucus Award and, after commanding STS-59 in April 1994, was named by the magazine *Hispanic Business* as one of the 100 Most Influential Hispanics. Today, with the Shuttle now a significant footnote in the history books, Gutierrez remains the only Hispanic ever to have piloted and commanded the reusable spacecraft.

The final member of the STS-40 crew, James Philip Bagian, had worked on the mission for as long as Seddon and the payload specialists, having been assigned to the flight in its original incarnation of 'Spacelab-4', back in February 1984. Of Armenian ancestry, Bagian was born in Philadelphia on 22 February 1952 and entered Drexel University to study mechanical engineering. He received his degree in 1973 and a doctorate in medicine from Thomas Jefferson University in 1977. Whilst working on his medical studies, Bagian was a process engineer for the 3M Company and, later, a mechanical engineer for the Naval Air Test Center at Patuxent River, Maryland. By 1978, he had secured a post as a NASA flight surgeon and research medical officer and applied, unsuccessfully, for admission into that year's astronaut class. (Incidentally, Bagian was invited to Houston in the same interview group as his future crewmate, Rhea Seddon.) Two years later, in May 1980, alongside Bryan O'Connor and John Blaha, he was finally selected and his early years as an astronaut were amply demonstrative of his amazing breadth of talent. He qualified as a free-fall parachutist and private pilot, was active in the mountain rescue community from 1981 onwards, served on the Denali Medical Research Project on Mount McKinley in Alaska, was a snow and ice rescue instructor on Mount Hood in northern Oregon, received Professional Engineer Certification in 1986, was board-certified by the American College of Preventive Medicine in 1987 *and* reached the rank of colonel in the Air Force Reserve. At the time of the Challenger accident, Bagian was not only assigned to the Spacelab-4 mission, but also to another flight, scheduled for September 1986 to retrieve NASA's Long Duration Exposure Facility from orbit. In March 1989, he made his first Shuttle flight. Rhea Seddon summed up Bagian's credentials, with a hint of understatement, as "very outstanding".

STS-40 would be one of the most complex research flights ever undertaken by the Shuttle and centred on a bus-sized laboratory, housed in Columbia's payload bay. The origin of 'Spacelab' can be traced back to the first few months after Neil Armstrong and Buzz Aldrin landed on the Moon. In late 1969, NASA outlined a number of key directions for the US space programme after Apollo. President John Kennedy's challenge to land a man on the lunar surface before the end of the decade had been met, but it was not until this time that serious consideration was given to what would come next. The establishment of some kind of permanent, or at least 'frequent', human presence in space was of major importance – hence the Shuttle – and in December 1972, the European Space Research Organisation (ESRO) – which, in 1974, merged with the European Launcher Development Organisation (ELDO) to establish today's European Space Agency (ESA) – agreed at a ministerial conference in Brussels to develop a modular, multi-purpose laboratory to fly in the payload bay of the winged orbiter. Originally described as a 'sortie module', it was later renamed 'Spacelab'. To NASA, the word 'sortie' merely reflected the low-cost, short-duration nature of the research facility ... but in Europe, and particularly France, the similarity of the word with the verb 'to leave' – *sortir* – generated some distaste. Indeed, the French word *sortie* is equivalent to the English word *exit*. In the United States, even the name 'Spacelab' was initially accepted with some hesitation, for the agency had only recently concluded its 'Skylab' series of missions and was worried that the two might become confused. "However," wrote Douglas Lord, "despite NASA's objections, once the Europeans had committed to the programme, they unilaterally decided to use the name 'Spacelab', and 'Spacelab' it became." The deal called for Europe to develop the facility in exchange for flying its own astronauts on specific missions. Spacelab, it seemed, would permit NASA to neatly sidestep the biggest obstacle in its financial battles with Congress: how to have both the Shuttle *and* a temporary space station in a decade of decreasing budgets. Unfortunately, the reality proved an unhappy prelude to the project's eventual success.

For Europe, involvement in Spacelab initially proved the consolation prize, after two other proposals were rejected by NASA and the Department of Defense. During 1971, ESRO and ELDO had invested $20 million in a series of studies to develop one of three components for the Shuttle: its payload bay doors, the sortie module or a reusable 'space tug'. Consensus favoured the latter, but it was turned down by NASA in June 1972, apparently because the Pentagon – envisaged to be the tug's main user – was reluctant to have a 'foreign entity' building a booster to place its top-secret payloads into orbit. (In April of that year, the Air Force had cited concerns over "national security issues" and told NASA that it would not guarantee buying a foreign-built tug if that did not meet its required specifications.) Constructing the clamshell-like payload bay doors was also promptly rejected because, said NASA, the Europeans lacked the expertise or organisation to make such a vital contribution to the orbiter's structure. The sortie module, on the other hand, required less sophisticated technology, had a well-defined interface with the Shuttle and schedule slips or budgetary overruns would not hamper the reusable spacecraft's ability to fly. When NASA offered this option to Europe, it received a lukewarm reception. Many ESRO member states hesitated to participate, questioning what they had to gain from an effort that demanded a $250 million investment and yet would be used principally by the United States. These lingering doubts were overruled, said political scientist John Logsdon, by a desire to "pursue co-operation on almost any terms, no matter how one-sided". In truth, it provided western Europe with its best chance of advancing its ambitions in space and sending its own astronauts aloft.

Indecision gave way to approval of involvement with the sortie module in December 1972; moreover, ministers agreed to form a single European space organisation. In January 1973, ESRO's member states voted in favour of building the sortie module and, from May, began working with NASA on the details of a formal Memorandum of Understanding. On 24 September, the contracts between the two organisations were finally signed in Washington, DC, by NASA Administrator Jim Fletcher and ESRO Director-General Alexander Hocker and the wheels of the largest international collaborative venture in space so far were set in motion. The majority of the funding for Spacelab (54.1 percent) came from West Germany, with Italy ranking second at 18 percent. In the terms of the contract, ESRO assumed responsibility for the "definition, design, development, manufacturing, qualification, acceptance testing and delivery" of an engineering model and spaceworthy flight unit to NASA by 1978. For its part, NASA would operate Spacelab, fly 'payload specialists' on selected missions and possibly procure additional hardware. Project management in the United States went to MSFC in Huntsville, Alabama, whose increasing involvement in selecting and training scientists for 'payload specialist' positions on Spacelab missions led to an expression of concern from JSC. Director Chris Kraft was reluctant to permit MSFC to choose and train payload specialists, arguing that they should be "selected from the present corps residing in Houston". Unfortunately for Kraft, the decision had already been made by NASA Headquarters, but it highlighted the different way in which Spacelab would operate, compared to other flights. Unlike pilots and mission specialists, who were selected formally as 'career' astronauts by the agency, payload specialists would be picked by an Investigators Working Group (IWG), whose panel included principal scientists responsible for Spacelab experiments. For the first time, a centre outside Houston was infringing on JSC's territory by providing mission-specific training. In all other areas, however, the two centres remained separate: JSC having responsibility for Shuttle operations and MSFC for Spacelab payloads.

Difficulties with the European partnership also arose when it became clear that NASA would not purchase more than one additional flight unit; in its 1973 plan for Shuttle utilisation, the agency expected to buy five Spacelabs to support two dozen research missions annually. However, a multitude of technical problems pushed the reusable spacecraft's first flight beyond 1980, only two units had been commissioned and, as ESA's budget declined, it became worryingly clear that the Europeans would have little funding available to even use Spacelab for their own experiments. The $250 million facility, they feared, would inevitably slip under American control. Perhaps the most important defining phase came in March 1978, when NASA and ESA initiated a nine-month critical design review, which provided a final opportunity to incorporate significant technical changes. Concerns remained, however. The first three Spacelab missions would all exceed the Shuttle's cargo-carrying limit as a result of orbiter-supplied equipment. MSFC opted to upgrade the orbiter's landing capability as a possible solution, although NASA Headquarters felt this would set a bad precedent. Nor was reducing Spacelab's payload weight desirable. Ultimately, ways were found to absorb the weight excess without seriously impacting each of the missions. Ongoing budgetary problems caused further woes. The European member states initially pledged up to 120 percent of their individual financial commitments to accommodate cost overruns, but even those had been consumed by September 1979. After protracted deliberations, ESA was obliged to propose increasing its members' funding to 140 percent. Only Italy refused to accept the new cost ceiling and by the end of the decade both the Europeans and Americans had reason for disappointment: the former resented paying more than expected, while the latter was disappointed that ESA was unprepared to take risks or bear full responsibility for Spacelab's early missions.

Inevitably, these worries impacted the flight schedule. When ESA delivered a pair of Spacelab modules to the United States – one under the terms of the original Memorandum of Understanding, the second as a follow-on procurement – NASA's budget was slashed in the first year of Ronald Reagan's presidency. This imposed one-year delays to several Spacelab missions. "Over the past four years," lamented James Harrington, head of the Spacelab effort at NASA Headquarters, "the Spacelab-1 launch has slipped three years! Additionally, the manifest of Spacelab flights has been reduced from four or five flights per year to the current two flights per year through 1986." As the problems of this increasingly unequal partnership were being thrashed out, the miniature space laboratory gradually took shape throughout the mid to late 1970s. Its bus-sized pressurised module comprised two components: a 'core' segment, which housed data-processing equipment, a workbench and a set of air-conditioned research racks lining its walls, and an 'experiment' segment, providing additional room for scientific operations. Although the core could be flown on its own, this configuration was ultimately never used and all module flights employed both segments joined together, with a pressurised volume of 75 m^3, to form a 'long module'. When one considers the dimensions of the short module, it is clear why NASA opted to use the longer version for a total of 16 missions between November 1983 and May 1998: the long module was over 7 m long and virtually doubled the amount of 'rack' space in which to carry experiments.

Racks were refrigerator-sized facilities which could be 'rolled' into the module's cylindrical shell and the facility also offered a central aisle of floor space, onto which experiments could be affixed, and provided a pair of ceiling openings for viewing windows or scientific airlocks. "The racks were pretty much standard," remembered Gene Rice, a former head of space life sciences with NASA. "You either had a drawer in a rack or you had a whole rack or ... a double rack, depending on the magnitude or size of the experiment. We would help [experiment customers] through the process of designing their experiment, integrating it into a Spacelab rack, doing the testing that they needed to do [and] getting it to a [NASA] centre. They would have to show that they met the safety requirements to put it into the Spacelab and to fly it." The racks contained air ducts to cool experiments and power-switching panels. On the first dedicated Spacelab mission, STS-9, the module contained 12 racks, of which the pair nearest the entrance to the module were devoted to control subsystems. The ceiling of the core section provided a 0.3 m wide opening for a high-optical-quality Scientific Window Adaptor Assembly (SWAA), through which Earth observation cameras could be directed, and that of the experiment section provided a Scientific Airlock (SAL), into which samples requiring exposure to space could be inserted and retrieved. As with other payload bay hardware, the exterior of Spacelab was covered with a layer of passive thermal protection material to protect them from the extremes of sunlight or frigid orbital darkness. Closed at each end by a pair of truncated cones, the module was held in place by three longeron fittings on the payload bay walls and one in its floor at the midpoint of the bay to avoid violating the Shuttle's centre-of-gravity constraints during ascent and re-entry. It was linked to the crew cabin by a 5.8 m long tunnel, built by McDonnell Douglas. Since the Spacelab hatch was 1.5 m 'higher' than that of the middeck airlock, a 'joggle' section was included in the tunnel to provide this vertical offset. Of course, astronauts might still need to perform contingency EVAs – perhaps to close the payload bay doors – and with this mind a 'mini-airlock' was built into the roof of the tunnel and spacewalkers could close off the whole section without having to depressurise either the crew cabin or the Spacelab. (Nevertheless, mission rules prohibited the module from being occupied during an EVA.) Although an EVA was never actually performed during a module flight, towards the end of STS-40 such action was briefly considered when a problem with some thermal insulation threatened to interfere with the closure of the payload bay doors. For the first Spacelab Life Sciences mission (SLS-1), the long module would be packed with biomedical apparatus for 18 major investigations.

Although the 'science' team – Bagian, Seddon, Gaffney and Hughes-Fulford – would be largely responsible for the research work during the nine-day mission, the 'orbiter' team of O'Connor, Gutierrez and Jernigan showed early willingness to participate as 'guinea pigs' for blood draws and other tests. "Tammy Jernigan had already signed up for everything," O'Connor recalled. "She was a scientist herself and certainly was interested and very engaged in the science training." For the pilots, on the other hand, it was more problematic. Much of their training revolved around flying the Shuttle itself, and landing it after a hypersonic descent through the atmosphere, which meant that their physical and mental health was paramount. As a result, they objected to only the experiments which might pose a risk to their ability to fly the orbiter, such as those focused on the behaviour of the eyes or the vestibular system.

FIRST MEDICAL RESEARCH FLIGHT

When the crew began training in the late summer of 1989, they confidently expected to fly in June of the following year, but delays pushed their mission to September and, with the hydrogen leaks experienced by Columbia and Atlantis, eventually into the middle of 1991. By the end of March, the Spacelab module had been installed into the Shuttle's payload bay and a few days later, on 3 April, the connecting tunnel to link it with the middeck airlock was fitted. Problems with the Solid Rocket Boosters presented some concern, when gauges on their aft skirts began to give spurious readings; a decision was made that their hold-down posts were probably misaligned and both were destacked, checked and restacked. Early in May 1991, the STS-40 vehicle rolled out to Pad 39B and the seven astronauts were actually in Florida, performing their countdown demonstration test, as Discovery touched down a few kilometres away to conclude STS-39. A couple of weeks later, on the 19th, O'Connor and his crew returned to Florida for their own launch, scheduled for the 22nd. "We're all ready to go," exulted Tammy Jernigan. "Light 'em!" For her part, Rhea Seddon wished for clear skies and smooth sailing.

Clear skies did not appear to be a problem, with forecasters predicting a 90 percent chance that the weather would co-operate. Unfortunately, on this occasion, the Shuttle would not. On 20 May, NASA decided to postpone the launch, when technicians discovered a leaking liquid hydrogen transducer in Columbia's aft compartment. This had been removed and replaced during the extensive hydrogen leakage investigation, but now it failed an analysis and exhibited weld defects. If one of these welds were to fail during the highly dynamic climb to orbit, there existed a very real danger that one of the nine transducers projecting into the main engines' fuel and oxidiser lines might crack, break off and be ingested into the high-speed turbopump.

Such an event would prove catastrophic. In September 1990, a leaking transducer was removed from Columbia, following one of the STS-35 delays, and was returned by Rockwell International for attention ... but to the *wrong* contractor. Inexplicably, noted *Flight International* in its summary of the events, it was three months before

First medical research flight 111

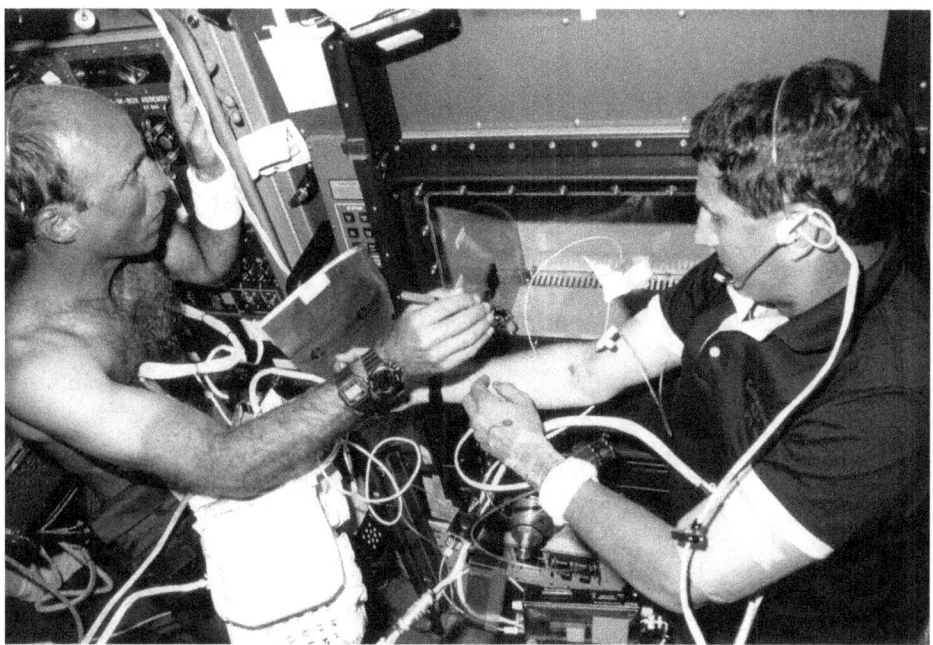

1-17: Jim Bagian (left) removes a catheter from Drew Gaffney's arm, shortly after Columbia reached orbit. The catheter formed part of an experiment to measure blood pressure in the 'great veins', close to the heart.

the error was caught and the transducer was despatched to the correct contractor, RFD. Not until April 1991 was a report finally submitted to Rockwell, indicating that the transducer exhibited serious structural cracking and could easily have caused a disaster on STS-35. By this time, of course, STS-35 had *already* flown and had thankfully dodged a bullet. However, it was evident that it was only a matter of time before the tip of the transducer would break off and become ingested into the main engine turbopump. Not until further Rockwell investigations got underway in mid-May 1991, only days before the first STS-40 launch attempt, was the decision made to remove the transducers from all orbiters. Of these, a total of ten – from both Columbia and her sister, Discovery – were found to be badly designed and six of them revealed cracks around their steel welds.

O'Connor was not alone in expressing his disappointment that this problem emerged so soon before launch, but accepted that it had to be rectified. So too did a failed General Purpose Computer on the Shuttle's flight deck, which was quickly replaced and tested, *and* a multiplexer-demultiplexer (MDM), part of the Shuttle's ordnance. With all faulty components replaced, launch was rescheduled for 9:00 am EDT on 1 June. *That* attempt proceeded right down to T-20 minutes, when it was scrubbed following a problem with the calibration of one of Columbia's inertial measurement units, a key piece of the navigation hardware. A new attempt was set for the morning of the 5th.

The crew arose early and began suiting up for launch; as part of one of the medical experiments, a catheter was inserted into Drew Gaffney's arm to monitor his cardiovascular changes and fluid shifts during ascent. Weather posed something of an obstacle, keeping Columbia on the ground for an hour and a half, before the Shuttle finally roared aloft at 9:24 am EDT. "Thanks for the great ride," O'Connor told Mission Control after he and Gutierrez established themselves in an orbit of 250 km, inclined 39 degrees to the equator. Normally, the opening hours of a Shuttle mission were a flurry of activity and STS-40 was no exception, particularly since – unlike most Spacelab flights – this would operate on a single-shift system, rather than having two teams working around the clock. Having said this, the astronauts ended up averaging 14-hour working days during the flight, their circadian rhythms closely monitored to provide for a more uniform set of biomedical data points.

SLS-1 consisted of 18 experiments to investigate the fundamental problems affecting the biology of humans, animals and fish in the microgravity environment of space. Ten of these investigations required the astronauts as direct subjects, whilst seven utilised a complement of 28 rats and one used almost 2,500 jellyfish. During the mission, the crew explored how the heart, blood vessels, lungs, kidneys and hormone-secreting glands responded to the new conditions, studied the causes and effects of space sickness in greater depth and examined minute changes to muscle and bone structure whilst in orbit. Scientists were particularly interested in their physiological response in the first few hours of flight. In support of this objective, Bagian, Seddon, Gaffney and Hughes-Fulford underwent an extensive battery of demanding tests before launch, and after landing, with a full 24 hours of 'head-down' bed rest to evaluate whether this ground-based analogue for space flight produced similar cardiovascular responses. Additionally, a sequence of vestibular tests were conducted to assess their sensitivity to linear acceleration.

The mission also had a pronounced international flavour, with researchers from France, Russia, Germany and Canada participating in a biospecimen-sharing project. This kind of co-operation would feature with increasing prominence and regularity in subsequent Spacelab flights, laying the groundwork for today's International Space Station. Like ASTRO-1, the SLS-1 effort got underway in 1978, when NASA issued an Announcement of Opportunity for medical and biological experiments to be flown on future Shuttle missions. Testing of the equipment began on Spacelab-3 in April 1985 and highlighted a number of operational flaws, notably contamination, leaks and odour problems from animal-holding facilities in the pressurised module. Late in 1985, the *Newsletter of the American Society for Gravitational and Space Biology* revealed that these problems needed resolution before the dedicated medical research flight on Spacelab-4, then planned for early 1987. Before Challenger, NASA hoped to fly at least one Research Animal Holding Facility (RAHF) on Spacelab-4, housing as many as 24 rats, and transfer them, in space, to a new unit called a General Purpose Workstation (GPWS). This made the adequate containment of particulate debris even more critical.

In the wake of Challenger, a three-year down time was used by NASA's Ames Research Center to modify the RAHF and a 12-day 'biocompatibility' test was undertaken in August 1988 to verify that the improvements worked. Its ability to contain debris – particularly food bar crumbs and faeces – and deal with odours and micro-organisms were identified as key problem areas. A single-pass auxiliary fan was fitted to aid the environmental controls and testing in March 1989 satisfied project management that the issues had been resolved. When it flew on SLS-1, the hardware performed admirably; it could indeed capture crumbs, flecks of the rats' hair and faeces and emitted no noticeable odours. Moreover, at one stage in the flight, when the science crew moved rats from the RAHF to the GPWS, it marked the first time that any rodents had floated freely in space, outside their cages. "He didn't want to get stuck out in the middle of nowhere with nothing to hold on to," Seddon recalled of one rat's experience outside the cage. "He would just hold onto your hand and then once we got him turned loose, we didn't want to throw him; we just wanted to get him off our hand. He floated around until he could grab onto the cage." The rat was very docile, she continued, and did not try to bite her.

The 28 rats (*Rattus norvegicus*) aboard Columbia were actually part of a much larger group of 74, of which 45 were kept on the ground as 'control' specimens and another had been dropped from the flight, just hours before launch, when a clogged water line failed in his cage. ("I imagine he's having a happy life now," NASA Test Director Mike Leinbach joked at the time.) The rats weighed an average of 275 g and were about nine weeks old; the 20 (soon to become 19) flight specimens were loaded into the RAHF a few days before launch. The device required an entire Spacelab module rack and its cages provided the rats with food, water and waste-management facilities, as well as controlling humidity, temperature and ventilation. The rats' movements were monitored and recorded by infrared light sources and sensors mounted inside each cage. During the mission, the RAHF was extensively tended by the science crew, who evaluated its capabilities. Generally, it proved favourable and maintained high levels of particulate containment.

Its environment was so good, in fact, that it was even possible for several of the rats to be moved to the GPWS in a portable transfer unit. (On the SLS-2 mission in October 1993, scientists would oversee the first dissections of rats in space for studies of their vestibular mechanisms.) Nine other rats were carried in a pair of Animal Enclosure Modules in Columbia's middeck. Like the RAHF, these provided food, water, waste-management, air and lighting, but did not enable the astronauts to actually handle the rats. These nine rats flew in the middeck because they were loaded aboard so late in the countdown – only 15 hours before launch – that technicians could not gain easy access to the Spacelab module. At first, in orbit, the rats clung to the sides of their cages, but upon realising that they would not fall, they became more relaxed and floated freely around.

One fundamental thrust of the SLS-1 rodent research were studies of the effect of microgravity on their muscles and bones, particularly through measurements of changes in circulating levels of calcium-metabolising hormones and the uptake and release of calcium in their bodies. These changes posed a considerable concern, because they appeared to be similar to those observed in human patients with osteoporosis and a clear understanding of the mechanisms behind them was seen as vital in the planning and execution of future long-duration space missions. Bone growth in the rats' legs, spines and jaws was observed and the loss of calcium and phosphorus was measured, revealing decreased skeletal growth and reduced leg-bone breaking strength and spinal mass. Other experiments explored decreases in the strength and endurance of muscles. In general, the rats returned from STS-40 much more lethargic than they had before launch, with reduced muscle tone, and were found to use their tails much less frequently as an aid for balance. Additionally, their red and white blood cell quantities decreased, although they were in far better overall shape than anticipated.

The astronauts participated in the calcium studies, too, by carefully monitoring their daily intakes of food, fluids and medication and weighing themselves frequently. In particular, the science crew took blood samples to better determine the role of calcium-regulatory hormones on the observed changes in their calcium balance. For Bagian, Seddon, Gaffney and Hughes-Fulford, to say that SLS-1 *was* their 'home' for most of the nine-day mission would not be an exaggeration, for as well as working in the Spacelab module they also *slept* there, finding it much more comfortable than the cramped middeck. "They all thought it was a great place to sleep," Bryan O'Connor told Mission Control on one occasion. "It was nice and dark and quiet back there," agreed Rhea Seddon. "We were doing single shifts, so the lab was essentially buttoned up for the night. It was dark and we could cool it down, so we just hung our hammocks back there." Every so often, the quiet would be disturbed – the module was situated near the end of the payload bay, so the science crew could hear the *boom boom boom* of the orbiter's manoeuvring thrusters going off, as well as the mice in their cages and the refrigerators switching on and off – but Seddon considered it a far more peaceful place to relax than the flight deck or middeck.

Moreover, all seven astronauts devoted everything to the science, typically beginning their duties ahead of time, working through meals and continuing well past bedtime. The result was a substantial increase in the overall scientific yield from SLS-1, whose effects would be amply illustrated in the analysis of results. In some quarters, the work in the Spacelab module was described as "slavish", owing to its unending intensity, but the reality was rationalised by Seddon. "Jim and I had each flown a flight," she told the NASA oral historian, "but the things that happened on those flights were pretty much stand-alone chunks of time: deploy the satellite on this day, film the toys on this day, do this on that day. They *didn't* interact. And suddenly, on *this* one, I don't think we completely understood how things could go awry if one piece of your equipment didn't work and they had to re-plan the day." In fact, so delicate was the need to keep to the timeline that it had even been thrown into some disarray by the launch delay of 85 minutes or so on 5 June, which meant the astronauts had to follow an 'off-nominal timeline' for Day One of the mission. As a consequence, they were frequently kept up late, troubleshooting or working an experiment which had earlier experienced difficulties. Every so often, the voice of Bryan O'Connor would be heard over the module's intercom, calling them back up to the flight deck for lunch.

One particularly significant experiment focused on the heart, lungs and blood vessels and was slightly delayed on 5 June when the Gas Analyser Mass Spectrometer developed difficulties and needed additional time to stabilise itself for operations. Detailed observations of the cause of the light-headedness reported by many spacefarers were made and one experiment investigated the theory that it might arise when the 'normal' reflex system responsible for regulating blood pressure behaves differently after adapting to a weightless state. Crew members wore a special neck cuff, a little like a whiplash collar, which recorded blood pressure levels. Other experiments evaluated how quickly the astronauts adapted to the new environment, through prolonged expiration and 'rebreathing' – inhaling previously exhaled gases – whilst at rest or pedalling a stationary bicycle ergometer. This produced surprising new data on the amount of blood being pumped out by the heart, together with oxygen usage and carbon dioxide production rates. Still other tasks involved differing gaseous mixtures and examined the influence of terrestrial gravity, and its absence, on lung function. Measurements of blood pressure in the 'great veins', close to the heart, were conducted using the catheter in Drew Gaffney's arm. This experiment was part of a cardiovascular adaptation study, developed by Dr Gunnar Blomqvist of the University of Texas Southwestern Medical Center in Dallas.

The catheter study had an interesting history in itself. In pre-Challenger days, still known at that time as Spacelab-4, concerns had been raised about the safety aspect of having a catheter threaded into Gaffney's arm towards his heart, but it became far riskier in the months and years after the accident, when the Shuttle programme did away with the practice of sending astronauts into space in lightweight flight suits and introduced a new partial-pressure suit, made by the David Clark Company. This pumpkin-orange ensemble was bulky and cumbersome and, although it provided hyperbaric and cold-water immersion protection for its wearer, it caused a headache for the catheter experiment. "Once you put it under a pressure suit," Seddon recalled, "where you can't get to the catheter that's threaded up at the crook of the elbow, you had to go through *another* round of safety assessments." It placed Gaffney in the difficult situation of potentially having to bail out of the Shuttle, in an emergency situation, with the catheter embedded in his arm. "How do you sit in your chair during ascent," asked O'Connor, rhetorically, "with a catheter to your heart, hooked up to some electronics that goes into the orbiter's recording systems and *then* be able to bail out or egress from the cabin on the launch pad if you have to go very quickly?" It was a difficult question to answer and one which consumed many months of effort and training and planning and modifications to the flight hardware.

Notably, disconnects had to be configured to ensure that Gaffney would not bleed to death if the catheter pulled loose in an emergency. The catheter had been inserted on 4 June, a full day before launch, and was removed by Jim Bagian about four hours after reaching orbit. Its data indicated the degree of body fluid redistribution and the rate at which this redistribution occurred. It revealed, unsurprisingly, that his blood pressure rose whilst on the launch pad, rose sharply and peaked during ascent and steadily returned to normal after a few minutes in space. This appeared to refute earlier suggestions that a rise in blood pressure resulted from fluid shifting into the upper body as a result of weightlessness.

A substantial number of blood-related experiments were performed to investigate the mechanisms responsible for decreasing numbers of circulating red blood cells – known as 'erythrocytes' – in space and a subsequent reduction in the oxygen-

1-18: The SLS-1 research module, together with part of the tunnel linking it to Columbia's cabin, is pictured in the payload bay during STS-40.

carrying capability of the blood. Samples were taken from each crew member before, during and after the flight and their relative volume to plasma was measured to check the rate of production and destruction of blood under 'normal' terrestrial and microgravity conditions. Although the SLS-1 data did not provide conclusive answers, it did indicate that a drop in red blood cell production *was* a contributory factor. Televised downlinks from the Spacelab module revealed what appeared to be a modern torture chamber, with all of the blood work in progress. One time, Gaffney calmly offered his arm to Hughes-Fulford for a needle and submitted for *another*, wielded by Bagian, an hour later. The blood work was frustrating at times, Rhea Seddon reported, as samples took longer to acquire in microgravity "and veins are not co-operating", but progress was otherwise exceptionally smooth.

Another set of experiments focused on the brain, central nervous system, eyes and inner ear and included studies from a joint US-Canadian project to investigate the impact of space sickness on the performance of the astronauts. On STS-40, members of the science crew placed their heads inside a rotating dome, which induced a sense of half-rotation in the direction opposite to that of the dome's own rotation. The astronaut subject then used a joystick to indicate his or her perception of self-motion.

Space sickness – properly termed 'Space Adaptation Syndrome' – is now known to affect around half of all astronauts and cosmonauts. Research over the last five decades has concluded that it is a nauseous malaise akin to motion sickness, which typically lasts no more than two or three days. It was first noted by Soviet cosmonaut Gherman Titov in August 1961, although this did not become known in the West until many years later. The first occurrence in an American spacecraft was suffered by Frank Borman on Apollo 8 in December 1968. The malaise usually manifests itself in sensations of disorientation and discomfort, coupled with dizziness and recurrent headaches. In some cases, it also induces fits of vomiting. Even today, explanations and countermeasures for it remain imprecise. It appears to be aggravated by a subject's ability to move around freely in weightlessness, and seems to be more prevalent in 'larger' spacecraft, as indicated by the fact that 60 percent of Shuttle astronauts report the complaint. Modern thinking postulates that the influence of weightlessness on the human vestibular system – the workings of the inner ear, which control balance – could offer a possible root cause. This disorientation arises when sensations from the eyes and other sensory organs conflict with those from the vestibular apparatus and with information in the brain derived from a lifetime of 'normal' gravity on Earth. A 'repatterning' of the central memory network occurs over the first few days, such that the unfamiliar sensations from eyes and ears begin to be correctly interpreted. Today, motion sickness medicines have been shown to help, but are rarely used, because most space fliers prefer to adapt naturally in orbit, rather than risk starting their missions in a drowsy state.

"Scopolamine and Dexedrine worked real well here on the ground," said Rhea Seddon, "but we found in space it solved the problem during the timeframe that you were taking it, but then there was an adaptation after you *quit* taking it and you got sick then." On his STS-29 mission in March 1989, Jim Bagian tried intramuscular Phenergan – an anti-nausea medication normally prescribed for post-operative or chemotherapy patients – and found that it worked well. Fifty milligrams of *that* could easily knock someone out on Earth, but as a countermeasure for space sickness it worked like a charm.

Awareness of position in space is important, particularly during re-entry and landing, when astronauts need to be able to reach levers and switches. The general results of the SLS-1 motion experiments pointed to a loss of sense of orientation and limb position in the absence of visual cues. On several occasions, the members of the science crew were also blindfolded and asked to describe the positions of their limbs in reference to their torsos and point towards familiar structures inside the Spacelab module. Other work investigated changes in the inner ear, which has long been known to be highly sensitive to gravity and responsible for inducing disorientation in space.

The measurements of the nervous system involved the first flight of jellyfish on the Shuttle; 2,478 Moon jellyfish (*Aurelia aurita*), to be precise, which were housed in a pair of containers in one of Columbia's middeck lockers. They were chosen because they are one of the simplest organisms known to possess a nervous system, employing structures known as 'rhopalia' to maintain a correct orientation in water, very similar to mammalian otoliths. The main aim of the jellyfish experiment was to determine their reproductive abilities and the impact of the new environment on their gravity-sensing organs and swimming behaviour. They were videotaped throughout the flight and finally 'fixed' on 12 June to preserve them for their return to Earth. Overall, the jellyfish polyps, which developed into sexually reproductive ephyrae in space, proved 'normal' in most respects, despite hormonal changes and abnormalities in their swimming behaviour after landing.

However, the problem-free appearance of the flight was deceptive. Scientifically, SLS-1 was proving to be a grand success – "The mission has exceeded our expectations," gushed Arnauld Nicogossian, head of NASA's Space Life Sciences Division – but a potentially catastrophic problem with Columbia's thermal insulation cropped up shortly after arrival in orbit. Within minutes of opening the payload bay doors on 5 June, a camera revealed that a number of thermal blankets on the aft bulkhead had become detached. Moreover, part of the payload bay door seal strip was displaced. The danger, of course, was that the seal and insulation could hamper the successful closure of the doors prior to re-entry. Lead Flight Director Randy Stone assured the press that there was no cause for alarm. "We don't believe this to be any issue with respect to safety or mission duration," he said. "The latches on these doors are very strong and we believe that even if the seal *was* in the way, we could collapse the seal and close the doors safely with no problem."

Nevertheless, it was decided to ship a section of seal to JSC in Houston, where astronaut Kathy Sullivan simulated an EVA to manually remove it. Encased in a bulky space suit in the waters of the WET-F tank, she verified managers' belief that the doors could close properly, despite the loose seal, and a contingency spacewalk would be unnecessary. If Jim Bagian and Tammy Jernigan did need to venture outside, it was considered relatively straightforward to either cut off the broken seal or push it back into its retainer. By 8 June, Mission Control told the crew that they did not believe an EVA to be necessary and were surprised by a rare note of disagreement from Bryan O'Connor. He was worried that the seal might snarl on a mechanical 'fork' which assisted in the closure of the doors. After several discussions, and an uplinked explanation of the procedures, he acquiesced that "you've answered our big questions".

In the meantime, the mission was running smoothly as it entered its homestretch. On the evening of 9 June, Bagian, an amateur magician, decided to play a trick on fellow astronaut Marsha Ivins, sitting at the capcom's desk in Mission Control, alongside Flight Director Al Pennington. With the assistance of an apprentice, Sid Gutierrez, he asked Pennington to pick a card from a new deck. He also had a deck of cards in orbit and said before the flight that he had selected a card and turned it opposite to the other cards. Bagian predicted that the card he had selected before launch would match the one Pennington chose on the ground. The flight director picked a four of spades, as did Bagian. "Truly incredible," was all the astonished Ivins could say, as she succumbed to the same trick.

As the days wore on, another issue arose over the issue of taco sauce. Astronauts had long been aware that their sense of taste became dulled in the weightless environment, due to fluids shifting in the body. To spice up their meals the STS-40 crew asked for plenty of shrimp cocktails and spaghetti and meatballs, and condiments such as taco sauce, to be included in Columbia's pantry. "Despite these efforts, some of the food was still pretty bland," admitted Sid Gutierrez. "Although I didn't do it myself, I observed crewmates putting taco sauce on Rice Krispies in the morning!" Late in the mission, the taco sauce quantities had begun to run low and Bryan O'Connor decided to place an embargo on them, taking control of the stocks and sharing them out equally to each crew member. "Thereafter," concluded Gutierrez, "taco sauce became the medium for exchange. If it was your turn to clean the latrine, you could pay someone a taco sauce or two to do it for you."

Aside from the joys of magic and taco sauce, none of the astronauts tired of the astonishing view of Earth through the windows. On her final evening in orbit, Millie Hughes-Fulford floated alone in the darkened Spacelab module, working out to the haunting echo of Enya from her Walkman. At length, tired and hot, she stopped and paused by the window to look out. "The Earth was in darkness," she recalled, "and I saw great lightning storms below. The storm cells were almost a hundred miles across! When the lightning struck, the *entire* cloud canopy would light up underneath, like giant hot air balloons." Startled, she looked around for a camera to capture this amazing scene. There were none within reach and Hughes-Fulford knew that by the time she had returned to the flight deck to get one, the view would be gone ... to be replaced by another. It summed up the thoughts of another astronaut, Bob Cabana, who had made his first flight a few months earlier. In Cabana's mind, it was important to realise that the opportunity to journey into space was a privileged one and something that very few individuals have been able to experience. On her last night in orbit, Hughes-Fulford put her nose up to the window and planted a memory of what she saw in her brain, where it has remained. "And don't take a picture of it," Cabana cautioned later, in a Smithsonian interview, "because you would be disappointed when you get home. You'll have it with you always and nobody can take it from you."

At length, on 13 June, the orbiter crew set to work readying Columbia's systems for the impending return to Earth. The moment of truth for the payload bay doors came early the following morning, but to play things safe it was decided to orient the troublesome seal towards the Sun to 'thermally condition' it. In the meantime, Bagian entered the Spacelab module for the final time, partly to store a few bags of blood samples, but also to videotape the closure of the doors. The port-side door closed and latched without incident, followed by the starboard door, and Jernigan, floating on the aft flight deck, confirmed that they were both properly sealed shut. An hour later, at 7:39 am PDT, O'Connor and Gutierrez guided their ship to a safe landing on concrete Runway 22 at Edwards Air Force Base.

More than 7,500 people had gathered at the California site to witness the touchdown, but would not actually see the astronauts disembark from Columbia. On this occasion, a specially designed airport-style 'people mover' – the Crew Transport Vehicle (CTV) – had been commissioned to whisk them away for medical checks. "It's like a big trailer house on high that can be jacked up," explained former space suit technician Jean Alexander. "They [the astronauts] come inside and take the suits off and kind of get their land legs back and the doctors check them out. If there's any medical experiments that have to be done after landing, there's gurneys in there." The rats, too, were quickly taken away, but to a somewhat different fate, for 15 of them (and their ground-based 'controls') were killed and dissected a few days later for analysis of their inner ear mechanisms.

For Millie Hughes-Fulford, the landing after nine days in weightlessness produced peculiar sensations. The position of the Shuttle's nose meant that the entire cabin was angled slightly downwards and she found it hard to get out of her seat and stand up. After removing her suit in the CTV, she finally had the chance to walk down the steps to see her husband and daughter. "Because my equilibrium was totally gone," she told an interviewer from the Smithsonian, years later, "I was holding onto the rail, trying to move as quickly as possible and not walk like a little old lady." When her daughter arrived, Hughes-Fulford gave her a hug and whispered in her ear: *"Help hold me up!"*

"EVEN THE GIRLS ..."

Whenever John Elmer Blaha went into space, in some small way, he always changed. It was a purely psychological thing – nothing physical – but he was convinced that each person who experienced the peculiar condition of 'weightlessness' and beheld the Home Planet from such a unique vantage point turned, somehow, into a different kind of human being. Late in November 1989, he had just returned from his second Shuttle flight and was finishing up his medical debriefing with one of the NASA flight surgeons.

"Who else gives you a debrief like me?" Blaha asked.

The flight surgeon grinned. "Shannon Lucid."

"Is that right?"

"Yeah."

Blaha knew Lucid well. They had been selected as astronauts, just two years apart, their children were approximately the same age and they were good friends. In late May 1990, both were overjoyed when NASA announced that they would be flying together on STS-43, planned as a five-day mission in May of the following year. When he heard the news, Blaha went straight to Lucid's office with a request: after a day or two in orbit, they would exchange views on the extent to which each other had 'changed' whilst off the planet. Blaha knew that he changed; he loved space exploration with every fibre of his being and after two Shuttle missions as pilot, he was getting ready to command his first flight on STS-43.

When he was selected as an astronaut in May 1980, Blaha was jointly the senior ranking member of his class, alongside Roy Bridges; both held the rank of a lieutenant-colonel in the Air Force. Blaha came from San Antonio, Texas, where he was born on 26 August 1942, the son of an Air Force pilot, although he received his high school education in Virginia. Blaha entered the Air Force Academy and received a degree in engineering science in 1965 and a master's in astronautical engineering from Purdue University in the following year. He earned his pilot's wings at Williams Air Force Base in Arizona in 1967 and undertook 361 combat missions in Vietnam, flying the F-4 Phantom, the F-102 Delta Dagger, the F-106 Delta Dart and the A-37 Dragonfly. By the time he returned from Vietnam and was accepted into the famed test pilot school at Edwards Air Force Base, Blaha had his sights on becoming an astronaut. "I had read the biographies of some of the early astronauts," he explained, "and realised many flew different types of airplanes and attended the test pilot school." Apollo 11 veteran Edwin 'Buzz' Aldrin happened to be commandant of the school at this time and Blaha got to know him and shared his goal. Aldrin recommended that he stay at Edwards and teach in the NF-104 Starfighter research aircraft, which, with rocket augmentation, achieved altitudes of over 100,000 feet. After test pilot school, Blaha followed Aldrin's advice and flew the NF-104 to an altitude of more than 104,400 feet – around *thirty kilometres*. Blaha next served as an F-104 instructor and taught his students to fly low lift-to-drag approaches, stability and control and spin flight test techniques. In 1973, he was assigned as an exchange pilot with the Royal Air Force at Boscombe Down in Wiltshire and flew a variety of aircraft, including the Jaguar, the Hawk, the Harrier and the Buccaneer. "Our time in England," he said, "was the best three years of our lives. We could eat breakfast together as a family and then I went to work at nine o'clock in the morning. By five o'clock in the evening, I was home." Graduation from the Air Command and Staff College at Maxwell Air Force Base in Alabama followed in 1976 and Blaha was assistant chief of staff for studies and analysis at the Pentagon when he answered NASA's first call for astronaut candidates. He was unsuccessful on his first attempt, but was selected two years later. In his book, *Dragonfly*, Bryan Burrough summarised many perceptions of Blaha from engineers, managers and fellow astronauts. Blaha often presented himself as "a dolt", who needed things repeated to him. "But John's far from stupid," engineer Mark Bowman said of Blaha. "He just wants to make sure he *gets it*. He wants to start from scratch and go over and over and over."

As for Shannon Matilda Wells Lucid, the story of her life is one of triumph over terrible odds. She had been born in war-torn China on 14 January 1943 and her experiences during her formative years make it unsurprising that she grew up with a "zest for life, steely determination and resourcefulness", according to writer Peggy Mihelich. Her parents, Oscar and Myrtle Wells, were Baptist missionaries and they, together with Shannon, her younger brother, Joe, her aunts and an uncle and her grandparents were taken captive by the Japanese army and held in Shanghai's Chapei Civil Assembly Centre prison camp. She learned to walk in early 1944, whilst aboard the Swedish ship *Gripsholm*, which returned her family to the United States as part of an exchange of non-combatant citizens of the warring nations. It was a long and arduous voyage and, during a stopover at Johannesburg in South Africa, she received her first pair of shoes. After the war, the family returned to China – living at times in Shanghai, Nanking and Anking – and Shannon found herself the centre of attention, due to her blonde hair and blue eyes. Her fierce desire to learn to read prompted her parents to place her in a Chinese elementary school. Aged five,

122 **Rise from the ashes**

1-19: Central and southern Florida, together with the Bahamas, loom 'behind' Atlantis during STS-43. The doughnut-shaped 'tilt table' was earlier used to support the deployment of the TDRS-E payload.

she took her first flight. As the DC-3 flew over mountainous terrain and landed on a gravel runway, the young girl was convinced that *flying* was the most remarkable thing for a human being to do ... and steeled herself to do the same when she grew up. Her family was expelled from China in 1949, after the Communist Revolution, and the young girl received her schooling in Bethany, Oklahoma.

She entered the University of Oklahoma to study chemistry and received her degree in 1963. By now, the first teams of astronauts had already been selected and Lucid was astonished that *all* of them were *male*; in fact, she had written to *Time* magazine in 1960, criticising NASA for choosing only men. Space exploration had fascinated her, ever since she read about the rocket experiments of Robert Goddard

... but there was another motive. "The Baptists wouldn't let women preach," she once said, "so I *had* to become an astronaut to get closer to God than my father!" During her undergraduate studies, she took flying lessons, gained her licence and encountered another cruel and harsh reality of life. One day, in her final year of study, she sat down with her professor to discuss her options for getting a job. The professor looked at her blankly. "A job?" he asked. "You plan on *working*? But you're a *girl*!" It underlined the reality that women were not taken seriously in many professional careers. Despite having a private licence, her efforts to become a commercial pilot led nowhere, for the same reasons. Fortunately, the Kennedy and Johnson administrations, with their incessant civil rights campaigning, smoothed the road over the next few years and she found work in academia as a teaching assistant and research chemist, firstly at her *alma mater*, then at the Oklahoma Medical Research Foundation in Oklahoma City and finally at the Kerr-McGee oil and gas corporation. By now married to Michael Lucid, she returned to study at the University of Oklahoma, earning a master's degree in biochemistry in 1970 and a PhD in 1973. With her doctorate, Lucid gained a job as a research associate with the Oklahoma Medical Research Foundation and remained in this position until NASA called for astronaut applicants. Lucid "scrambled" to complete and submit her application. In late August 1977, she was invited to Houston as part of the third group of finalists to be interviewed ... a 20-strong group which included a subset of individuals whose presence, a decade earlier, would have been inconceivable: *eight women*.

Three of those eight women – Lucid, Anna Fisher and Rhea Seddon – would form half of the female component of the astronaut class announced in January 1978. "It's a remarkable story," fellow astronaut John Fabian said of Lucid's life. "It's a story of the human spirit and I love to tell it ... because kids don't realise what opportunities really lie ahead of them. Some are very quick to worry about the disadvantages that they have in their own lives, or as they perceive in their own lives, and I think the Shannon Lucid story is just a great story about overcoming obstacles and blasting through ceilings and knocking down doors and never letting anything get in the way of doing the things that you believe are right."

Bryan Burrough has cited comments made about John Blaha that he relied too heavily on others, and particularly upon Lucid, during their space missions together. Yet others, including the legendary flight director Gene Kranz, whilst admitting that his seniority and experience never turned him into a 'rising star' within NASA, Blaha was, nonetheless, "steady and dependable". On STS-43, he would lead a mission to deploy the fourth Tracking and Data Relay Satellite (TDRS-E), part of a network of communications platforms in geostationary orbit, some 35,000 km above Earth. Since the beginning of the manned space programme, astronauts had relied heavily upon ground stations and tracking ships for communications and it was only possible to maintain contact for around 20 percent of each 90-minute orbit. Moreover, during re-entry, super-heated plasma around the vehicle caused a period of radio 'blackout'. With the arrival of TDRS, it would be possible to communicate with astronauts during 85-98 percent of each orbit *and* throughout re-entry.

As the Shuttle gained momentum in the mid-1970s, it was envisaged that a pair of these powerful satellites – one stationed over the equator, just off the north-eastern corner of Brazil, known as 'TDRS-East', and a second over the central Pacific Ocean, near the Phoenix Islands, known as 'TDRS-West', would fill this urgent communications and tracking need. TDRS-A was launched in April 1983, but was almost lost when its Boeing-built Inertial Upper Stage (IUS) failed to insert it into its proper orbit. Only by using the satellite's own hydrazine thrusters were controllers able to gradually manoeuvre it into its final location, although the result was that its operational lifetime was shortened. Ongoing problems with the IUS meant that it was almost three years before the 'West' satellite, TDRS-B, could be launched ... and *that* was the primary payload aboard the ill-fated Challenger on 28 January 1986. Two more TDRS satellites (C and D) were launched in September 1988 and March 1989, the former replacing the doddery TDRS-A in the west (slightly south of Hawaii) and the latter taking up position in the east, near Brazil. Unfortunately, TDRS-C also succumbed to anomalies which affected its Ku-band relay capability. The fourth satellite, TDRS-E, would therefore be positioned at 175 degrees West longitude to become the primary provider of communications services over the Pacific from October 1991. TDRS-A and TDRS-C, meanwhile, would be relegated to the status of on-orbit 'spares'.

Yet TDRS was no miracle worker. It could not process or adjust communications traffic in either direction. Rather, it operated as a 'bent pipe' repeater, relaying signals and data between its Earth-circling users and the highly automated ground terminal. Signals processing, therefore, was done on the ground and the satellite's sophistication was devoted to its very high throughput. Located in the New Mexico desert, White Sands provided a clear line of sight with both satellites and its limited amount of annual rainfall meant that weather conditions would not interfere with their Ku-band uplink or downlink channels.

By the summer of 1991, NASA prepared to launch its fourth TDRS into orbit on STS-43. The hydrogen leaks had already pushed it back from its original target date and problems with Discovery, earlier in the year, had caused the mission to be shifted onto her sister ship, Atlantis. Discovery, meanwhile, had already been assigned to deploy the Upper Atmosphere Research Satellite (UARS). In March, NASA announced that this switch "preserved the agency's capability to fly Discovery" on the UARS mission during a critical 'science window' between September and November 1991.

Launch of STS-43 was scheduled for 23 July, but was postponed when a faulty controller on one of the orbiter's main engines required replacement. A second attempt on 1 August was held at the T-9 minute point when a vent valve failed to produce a 'closed' status indication, as it should, after the completion of the cabin pressure check. Although the vent valve was cycled several times, it still failed to display its closed indication. At length, it became clear that the valve *had* closed properly – it was nothing more than an instrumentation glitch – and the countdown continued ... but by this time, the weather at KSC had begun to deteriorate and the launch attempt was scrubbed. Finally, Atlantis and her crew of five thundered into space the next morning at 11:02 am EDT.

Preparations to deploy TDRS-E got underway almost immediately, with the release of the payload scheduled for a little over six hours into the mission. Since these deployments had been 'baselined' on previous flights, TDRS was considered a relatively 'ho-hum' assignment by some astronauts – "we even let the *girls* deploy TDRSes", joked the notoriously chauvinistic Mike Mullane in his memoir, *Riding Rockets* – but that took nothing away from the awesome size and impressive credentials of these satellites. Like Ulysses, the TDRS was mounted atop a gigantic IUS booster and, when accommodated in a doughnut-shaped Airborne Support Equipment (ASE), the combination filled three-quarters of the Shuttle's 15 m payload bay. The ASE served as a kind of 'tilt table' and was used to raise the TDRS-IUS 'stack', firstly to an angle of 29 degrees, where telemetry checks were performed and the booster transferred to its internal batteries, and finally to the deployment angle of 58 degrees. At that point, the IUS' ordnance separation device, fitted with compressed springs, would physically eject the payload from Atlantis at a rate of 12 cm/sec.

Deployment occurred some six hours after launch and the stack swept silently into the inky blackness, passing directly over the Shuttle's flight deck windows. Twenty minutes later, John Blaha and his pilot, Michael Allen Baker, fired both OMS engines to create a safe separation distance before the ignition of the IUS. For Baker, the only rookie member of the crew, it was the beginning of an impressive four-flight career as an astronaut. He came from Memphis, Tennessee, where he was born on 27 October 1953. Baker received much of his later schooling in California and earned his degree in aerospace engineering from the University of Texas in 1975. He then entered the Navy, completed initial flight training and gained his gold wings as an aviator at Naval Air Station Chase Field in Beeville, Texas. Baker served as an attack pilot, flying the A-7E Corsair II from the USS *Midway*, homeported in Yokosuka, Japan, and later worked as an airwing signal officer. In 1981, he entered the Naval Test Pilot School and, after graduation, undertook carrier suitability structural tests and catapult and arresting gear trials for the Navy's fleet of A-7s. In the two years before his selection by NASA in June 1985, Baker worked as an instructor, both at the Naval Test Pilot School and at the Empire Test Pilots School in Boscombe Down in the United Kingdom. After completing astronaut candidate training in July 1986, six months after the loss of Challenger, he was intimately involved in the critical redesign and modification of the Shuttle's landing gear, including its nosewheel steering mechanism, brakes, tyres and a new 'drag chute' to be introduced into the fleet in 1992.

Within days of deployment, TDRS-E was fully functional and ready to commence several weeks of rigorous testing. By the beginning of October, the newly renumbered 'TDRS-5' resembled a colossal 'windmill', measuring 17.4 m across its twin solar panels, which extended from a hexagonal spacecraft 'bus'. The panels generated 1,800 watts of electrical power, supplemented by on-board nickel-cadmium batteries when in Earth's shadow. Inside the bus, the communications payload was capable of transmitting in a single second the entire contents of a 20-volume encyclopaedia. Mounted atop the bus were the satellite's antennas, capable of receiving transmissions from White Sands, amplifying them and re-transmitting them to the 'user' spacecraft and vice versa. The main space-to-ground link was a circular, 2 m antenna, which operated across the Ku-band frequency, while data from other spacecraft was routed through one of two umbrella-shaped, 4.9 m dual-feed S-band/Ku-band single access parabolic dishes. Constructed from gold-clad molybdenum wire mesh, both had transmission rates in the order of 300 megabits per second, capable of handling heavy traffic from Hubble, the Shuttle and, in particular, the 1982-launched Landsat-4 Earth resources platform. For multiple access service, an S-band 'phased' array of 30 helical antennas was mounted directly onto the body. This incorporated a forward link, which transmitted command data to the 'user' spacecraft, and a return link to relay signal outputs directly to White Sands. Upon receipt, the ground terminal 'de-multiplexed' the signal and distributed it to 20 sets of beam-forming equipment, which discriminated among the 30 signals to select the unique signatures of individual users.

With TDRS-E successfully deployed, the crew turned its attention towards a battery of experiments in the middeck and aboard Atlantis' payload bay. Originally, STS-43 was baselined only as a five-day mission, but in late 1990 was almost doubled to nine days in order to focus on an extensive programme of biomedical and technological research. This scientific focus was also highlighted by the unusual shape of the crew's mission patch – resembling an Erlenmeyer laboratory flask, it was flat-bottomed, with a conical body and a cylindrical 'neck' – which also bore a close resemblance to the shape of the Mercury capsule which Al Shepard had flown, 40 years earlier. In the Shuttle's cabin, investigations into protein crystal growth, the processing of polymer membranes, combustion science studies and liquid-to-liquid diffusion in microgravity conditions were undertaken, as were observations of terrestrial aurorae and measurements of the effect of acceleration on delicate experiments. Additionally, feasibility experiments were carried out to evaluate the usefulness of fibre optic technology for video and audio communications between the payload bay and the crew cabin. In the payload bay was the Shuttle Solar Backscatter Ultraviolet (SSBUV) instrument, whilst Atlantis was herself used as part of the Air Force's Maui Optical System (AMOS) experiment, in which her orbital progress was tracked using a series of high-precision electro-optical sensors based on the Hawaiian island of Maui.

All five crew members participated in these experiments, including the final two mission specialists, George David Low and James Craig Adamson. For his part, Low had been one of the youngest astronauts ever chosen by NASA when he joined the tenth class in May 1984, aged only 28. He was born in Cleveland, Ohio, on 19 February 1956, and grew up with the space programme all around him, since his father had worked for both NASA and its predecessor, the National Advisory Committee for Aeronautics. The elder Low had been intimately involved in the planning of Projects Mercury, Gemini and Apollo and later headed the Apollo Spacecraft Program Office in Houston, forming part of the team which committed Apollo 8 to the audacious goal of orbiting the Moon. He later served as Deputy Administrator and Acting Administrator of the agency in 1969-76 and saw his son admitted into the astronaut corps, only to die just two months later, in July 1984. (Sadly, father and son would both succumb to cancer.) As a child, David Low was fascinated by science and declared his intent to become an astronaut, aged only nine years old. He entered Washington & Lee University to study physics and engineering and graduated in 1978. He undertook a master's degree in physics and engineering at Cornell University in 1980 and a *second* master's, this time in aeronautics and astronautics, from Stanford in 1983. During this period, he worked in the Spacecraft Systems Engineering Section of the Jet Propulsion Laboratory, involved in the systems engineering design of the Galileo spacecraft and the Mars Geoscience/Climatology Orbiter (later the Mars Observer). One of his contemporaries, Frank Culbertson, labelled him as "more academic than the rest of us", but admitted that Low was a good operator and a skilled mechanic, who worked on cars, but understood the physics behind them and communicated this understanding well. Described by United Press International as "an intense young astronaut" and "a man not given to frivolity", Low admitted that it was the influence of his father that provided a yardstick by which he measured his own life and how he treated others.

Army Lieutenant-Colonel Adamson came from Warsaw, New York, where he was born on 3 March 1946. He received a degree in engineering from the Military Academy in 1969 and was commissioned into the Army as a second lieutenant. During his career, he undertook pilot and paratrooper training, Arctic water and mountain survival training, nuclear weapons training and graduated from the Navy's test pilot school at Patuxent River, Maryland. Adamson flew as a scout pilot and air mission commander over Cambodia during the Vietnam conflict and later earned a master's degree in aerospace engineering at Princeton University. He subsequently taught aerodynamics – including fluid mechanics and aircraft performance – as an assistant professor at West Point. He arrived at JSC in 1981 as a research pilot and aerodynamicist and was selected as an astronaut three years later.

Atlantis broke new ground at the end of her STS-43 mission, by becoming the first post-Challenger Shuttle flight in which the Kennedy Space Center was scheduled as her primary landing site. With the successful returns of STS-38 and STS-39 to the Shuttle Landing Facility in November 1990 and May 1991, NASA's improved sense of confidence prompted the agency to place Edwards Air Force Base in California on 'reserve' status for the first time in more than five years. Landing in Florida eliminated the $1 million cost and the additional week of processing time needed to fly the orbiter across the continent from California and was seen by NASA as a key step in moving from the post-Challenger mindset of over-conservatism to one of full Shuttle operations.

In the early summer of 1991, the decision by Bill Lenoir, a former astronaut and now the agency's Associate Administrator for Space Flight, to resume landings at KSC aroused great criticism, with many engineers and managers arguing that Shuttles should continue to land at Edwards until "tougher tyres" had been fitted and tested. The damage to STS-39's tyres after touching down in a crosswind raised further concern. Even Shuttle Program Manager Bob Crippen insisted that KSC landings would only be approved if strict rules, including those associated with crosswinds, were met, and in the weeks before STS-43 he announced publicly that it was "likely" that Atlantis would be directed instead to Edwards. Ultimately, conditions in Florida were perfect and on 11 August John Blaha and Mike Baker guided their ship smoothly onto Runway 15, touching down at 8:23 am EDT. The *next* mission, STS-48 in September, would go a step further by attempting to land at KSC *in darkness*, but Crippen remained cautious. "We're still going to land at Edwards," he told journalists. "The weather is going to end up dictating that. I'm budgeting for about 60 percent of the flights landing at Edwards and 40 percent at KSC." By the end of the Shuttle era, in July 2011, Crippen' 60-40 prediction had proven accurate ... but fell in favour of KSC, rather than Edwards: out of the 133 Shuttle missions which successfully landed, a total of 78 touched down in Florida, 54 at Edwards and a single flight at White Sands in New Mexico.

OZONE WATCHER

Late in September 2011, the skies above the Pacific Ocean were illuminated by an astonishing – though not unexpected – fire show. NASA's 5,900 kg Upper Atmosphere Research Satellite (UARS), launched exactly two decades earlier, in September 1991, returned to Earth in a blaze of glowing debris, with the remnants splashing down in a remote region of the Pacific. Originally expected to function for a couple of years, the UARS mission was extended several times and even when budget cuts forced it to be decommissioned in June 2005 no fewer than *six* of its ten scientific instruments were still fully functional. Its orbit was slightly lowered by flight controllers in December 2005, in anticipation of an eventual destructive re-entry, and in October 2010 the crew of the International Space Station was obliged to perform a debris avoidance manoeuvre in response to a conjunction with the aging satellite. Its eventual descent to Earth on 24 September of the following year brought a rather high-profile closure to a mission which had proven instrumental in changing our perception of the Home Planet.

Built by General Dynamics, under an initial $145 million contract with NASA which soon expanded into a billion-dollar programme, UARS could trace its heritage to well before the loss of Challenger. It was originally scheduled to be launched by the Shuttle in October 1989 as a long-duration observatory to monitor gas concentrations and pressures, the effects of solar irradiance and ozone concentrations in Earth's fragile atmosphere. From its 600 km vantage point, inclined 57 degrees to the equator, the satellite would have the capability to observe up to 80 degrees in latitude, thus affording it essentially global coverage of Earth's stratosphere and mesosphere. This, in turn, would enable UARS to make measurements over the full range of local time zones in all major geographical locations, every 36 days. The instruments aboard the satellite were expected to produce the most complete data on solar energy inputs, terrestrial winds and atmospheric composition ever gathered. These explored the composition and distribution of nitrogen and chlorine compounds, together with ozone, water vapour and methane, measured thermal emissions to create vertical abundance profiles of atmosphere gases, determined temperatures and pressure characteristics of high-altitude winds and analysed the effect of solar ultraviolet radiation.

Significantly, UARS' ten instruments were designed to operate as a single experiment, capable of providing atmospheric scientists with an opportunity to make simultaneous measurements of all factors affecting ozone depletion. Four of those instruments measured the concentrations and distributions of gases known to play an integral role in the depletion process. The Cryogenic Limb Array Etalon Spectrometer examined concentrations of nitrogen and chlorine, ozone, water vapour and methane, whilst the Improved Stratospheric and Mesospheric Sounder also monitored carbon dioxide, nitrous oxide, nitric acid and carbon monoxide. Meanwhile, the Microwave Limb Sounder provided the first 'global' data set on chlorine monoxide (the key intermediate compound in the ozone-destruction cycle) and the Halogen Occultation Experiment observed vertical distributions of hydrofluoric acid and members of the nitrogen family.

Two other instruments – the High-Resolution Doppler Imager and the Wind Imaging Interferometer – provided the first direct measurements, on a global scale, of horizontal winds responsible for dispersing chemicals and aerosols in the upper atmosphere. Elsewhere, the Solar Ultraviolet Spectral Irradiance Monitor, the Solar Stellar Irradiance Comparison Experiment and the Particle Environment Monitor studied the effect of incoming solar energy on the atmosphere. Additionally, the Active Cavity Radiometer Irradiance Monitor offered accurate determination of total solar activity over the long term to aid climatic studies.

When the STS-48 crew was announced in December 1990, they confidently anticipated launch in November of the following year, but Shuttle delays in the spring and summer of 1991 actually conspired to bring their flight *forward*. Original plans for Discovery saw her flying the STS-39 mission for the Department of Defense in March, followed by the deployment of TDRS-E on STS-43 in July and finally UARS in November. When Discovery found herself grounded for more than six weeks, due to cracks on the lug hinges of her External Tank's umbilical door drive mechanism, her STS-43 flight was pushed onto sister ship Atlantis and the UARS mission was brought forward to October and, finally, September. With a scheduled operational lifetime of 20 months, an autumn launch was highly desirable in that it would enable UARS researchers to observe at least two winters in the northern hemisphere and at least one season examining the much-publicised 'hole' in the ozone layer above Antarctica. For STS-48 commander John Oliver Creighton – nicknamed 'JO' – it made a pleasant change to fly somewhat *earlier* than planned.

Creighton came from Orange, Texas, where he was born on 28 April 1943, although his family moved to Seattle, where he received much of his schooling. "I've been interested in flying since I can remember," he told the NASA oral historian. "I can remember going and watching the Blue Angels fly … during the hydroplane races and it was just something that I've always wanted to do." After a year at the University of Washington, he entered the Naval Academy to study for a degree in aerospace engineering, graduated in 1966 and began flight training. He received his wings in October 1967 and undertook two combat deployments to Vietnam, flying F-4 Phantoms off the USS *Ranger*. Upon returning to the United States, in June 1970 he attended Naval Test Pilot School at Patuxent River, Maryland, and the following year took up a position as a project test pilot at the school. His duties as Propulsion Project Manager focused on the development of engines for the new F-14

Tomcat fighter. "I got in on the ground floor of the F-14 programme," he said, "and got an opportunity to be one of the first Navy pilots to fly the F-14." In July 1973, he began a four-year assignment as a member of the first operational F-14 squadron and completed two deployments in the Western Pacific aboard the USS *Enterprise*. This cruise was meant to be a peacetime exercise, since the United States had virtually ended its offensive operations in Vietnam. At its end, the *Enterprise* dropped anchor in Manila, in the Philippines, but after barely *half an hour* was sent back to the coast of South Vietnam for 30 days and Creighton and his squadron found themselves providing F-14 fighter cover over the evacuation of the US Embassy in Saigon on the night that the city fell to the communist North. "I saw that, first-hand," he said of the events of 30 April 1975, "from about 10,000 feet. We were just there to make sure that no MiGs came down and tried to harass our helicopters that were evacuating the personnel out of there." Creighton returned to the United States in July 1977 and served as operations office and programme manager for the F-14 at the Naval Air Test Center's Strike Directorate, and later that same year was called to Houston for astronaut screening. Most of the testing, he recalled, was physiological: they tested his eyes and heart, looked in every orifice of his body and took X-rays "until you glow in the dark". Years later, Creighton admitted that the selection panel was *not* something that he could realistically have trained himself to handle. "They don't ask you to derive any differential equations," he said. "They just want to talk to you and, I think, get a sense of the kind of person you are. They'll throw you some oddball questions, just to see how you think on your feet."

By the time he was named to STS-48, Creighton had already flown two Shuttle missions, the most recent of which – STS-36 in February 1990 – had been delayed on a number of occasions, most notably when he came down with a cold. After his return from STS-36, he worked for a time in the astronaut office's Mission Development Branch and by his own admission was surprised when he received his next flight assignment so early. With launch almost a year away, Creighton and his crewmates – his pilot, Ken Reightler, and mission specialists Charles 'Sam' Gemar, Jim Buchli and Mark Brown – were given low priority in the simulators for the first few weeks. That changed with Discovery's delays. "All of a sudden," Creighton recalled, "we found out we were going to leapfrog a couple of other flights and be sooner, rather than later, so then we really had to scramble. That was a tough nine or ten months of very intense training."

In addition to its high orbital inclination of 57 degrees to the equator, UARS also demanded one of the highest altitudes ever reached by the Shuttle: some 600 km, slightly less than the Hubble Space Telescope. "We set a world altitude record for a winged vehicle" at a 57-degree inclination, Creighton proudly said of STS-48 and to this day he retains the commemorative plaque to prove it. His launch was finally set for 12 September and after a brief, 14-minute delay, caused by noisy interference on the air-to-ground communications link, Discovery roared into space at 7:11 pm EDT.

Most of the STS-48 crew had flown before – Jim Buchli, in fact, was only the second person to fly as many as four times on the Shuttle and had served a stint as deputy chief of the astronaut office – but for Kenneth Stanley Reightler Jr, the pilot, it was his first mission ... and it surpassed all of his expectations. Years later, he told a Smithsonian interviewer, with more than a hint of humour, that the task of the pilot was "to start the Auxiliary Power Units, to lower the landing gear and to make the mission commander look good". That said, Reightler's role in an emergency situation would be quite different: to instinctively respond to any contingency, to hold a great deal of knowledge about the Shuttle's systems and hundreds of switches and controls and to potentially exercise that knowledge, in a timely manner, to save his own life and those of his crewmates. Jim Buchli had likened a Shuttle launch to being strapped onto the front of a freight train at full speed. Not surprisingly, on STS-48, Reightler was anxious to perform at his peak. One concern was his bulky orange partial-pressure suit, which he felt was 'pulling' on his body and might make

1-20: The STS-48 stack rolls to the launch pad, atop the Mobile Launch Platform and 'crawler'. Typically, the journey from the Vehicle Assembly Building to the pad surface – a distance of 5.6 km – required around six hours.

it difficult for him to reach switches. "To bolster my confidence, right after liftoff, I started systematically reaching around the cockpit as the G-forces started to build," he related. At length, glancing periodically over at his rookie pilot, Creighton politely asked *"Would you knock that off?"* Reightler did just that, "and the rest of the ride to orbit was a dream come true".

For a naval aviator who rose to the military rank of captain and flew two Shuttle missions in his NASA career, it is perhaps appropriate that Reightler was born in Patuxent River, Maryland, the home of the Naval Test Pilot School, on 24 March 1951. (His father-in-law, Commander William H. McHenry, was also a retired naval officer.) Reightler was educated in Virginia, received a degree in aerospace engineering from the Naval Academy in 1973 and was designated a naval aviator in August 1974. His early flight training was in the P-3C Orion and he served as a mission commander and patrol commander in Jacksonville, Florida, later participating in overseas deployments to Iceland and Sicily. Reightler transitioned to jets, graduated from the Naval Test Pilot School in 1978 and worked as a project officer for numerous test programmes, involving the P-3 and other aircraft, as well as a flight test instructor. Aboard the USS *Dwight D. Eisenhower*, he made two deployments to the Mediterranean Sea, after which he was selected by the Naval Postgraduate School for a master's degree in aeronautical engineering. Reightler also pursued master's studies in systems management at the University of Southern California. He applied unsuccessfully for admission into NASA's astronaut corps in 1985, but was finally selected in the twelfth class, nicknamed 'the GAFFers', in June 1987.

Reightler's arrival in space brought with it the excitement faced by every first-time astronaut. Following orbital insertion, his time came to unbuckle from his seat in Discovery's forward flight deck and float downstairs to the middeck to doff his bulky suit ... and as soon as he did so, the *view* through the overhead windows arrested him. "The orbiter was inverted, flying over Asia," he recalled, "so I was looking *straight down* at Earth. I was totally unprepared for the colours, textures, detail and vastness of the scene. It literally took my breath away. No amount of training or looking at slides and movies can prepare you for that moment." It was a moment his four crewmates could well understand.

James Frederick Buchli was born in New Rockford, North Dakota, on 20 June 1945, and was making his fourth Shuttle mission on STS-48. After high school in Fargo, he went to the Naval Academy and, upon graduation in aeronautical engineering in 1967, was commissioned into the Marine Corps. Buchli underwent basic infantry officer training and was despatched to Vietnam as a platoon commander, executive officer and company commander. Returning to the United States in 1969, he completed flight training in Pensacola, Florida, earned his naval aviator's wings and spent two years at Marine Corps air stations in Hawaii and Japan. A tour of duty in Thailand was followed by completion of his master's degree in aeronautical engineering systems from the University of West Florida in 1975 and graduation from the test pilot school in 1977. Buchli was selected as part of NASA's first class of Shuttle astronauts in January 1978. The fourth member of the STS-48 crew, Mark Neil Brown, had been attached to the space agency for almost as long as Buchli, albeit in a different capacity. Brown hailed from Valparaiso, Indiana, born on 18 November

1951 and received his degree in aeronautical and astronautical engineering from Purdue. Upon graduation in 1973, he entered the Air Force, received his wings and flew the T-33 Shooting Star and the F-106 Delta Dart as a fighter-interceptor pilot at Sawyer Air Force Base in Michigan. He was selected to attend the Air Force Institute of Technology in 1979 and earned a master's degree in astronautical engineering the following year. Shortly afterwards, Brown was transferred by the Air Force to NASA's Johnson Space Center in Houston as a flight dynamics engineer, participating in the development of contingency procedures for the Shuttle. Admission into NASA's tenth group of astronauts followed in May 1984 and STS-48 marked his second space mission. Rounding out the crew was STS-38 veteran Sam Gemar. In anticipation of any problems, he and Buchli were trained to perform a contingency EVA and much of the second day of the mission was spent preparing space suits and lowering Discovery's cabin pressure accordingly.

Deployment of UARS was accomplished using Discovery's RMS mechanical arm, deftly operated by Mark Brown, at 12:23 am EDT on 16 September. Its release came an orbit later than planned, due to the need to attend to a minor communications issue, but inaugurated a spectacular mission, whose results continue to reverberate to this day. "One of the things they saw," John Creighton told the oral historian, "is that the chloroflurocarbons that come from the industrialised northern hemisphere were migrating down over the Antarctic and that's what was a direct correlation to what was causing the destruction of the ozone layer in the Antarctic spring. It would release all of these things that were trapped in the lower atmosphere and then it would spiral up, because of the circular wind patterns, up into the ozone and then create the 'hole'. That was kind of exciting to see that what people had long suspected was *proven*." For a mission devoted to atmospheric science, it proven somewhat serendipitous that Mount Pinatubo – a stratovolcano in the Cabusilan Mountains of the island of Luzon in the Phillippines – erupted in mid-June 1991, producing the 20th century's second-largest terrestrial eruption.

The events surrounding the event were complicated by Typhoon Yunga, which added a lethal mixture of ash and rain, and although predictions led to the successful evacuation of tens of thousands of inhabitants from the disaster zone, Pinatubo stands as one of the most devastating environmental calamities in history. Earlier volcanic deposits were remobilised, infrastructure was destroyed wholesale and river systems were significantly altered by the effects of hundreds of billions of kilograms of magma, as well as volcanic dust which created an aerosol-rich layer of sulphuric acid haze in the stratosphere that spread around the world and lowered global temperatures by around half a degree Celsius. In the first few weeks after its deployment from Discovery, UARS' instruments were directed to measure carbon dioxide emissions from the Pinatubo eruption. "And lo and behold," said Creighton, "there was destruction of the ozone layer, right around the equator, because of the eruption of the volcano." Over the next decade, UARS contributions to atmospheric and solar science were so significant that plans were even afoot for a Shuttle mission, involving a single EVA, to either retrieve it or install new instruments. These plans never reached fruition and by September 2001 had been shelved by NASA.

STS-48 remained in orbit for five days and the astronauts were fully occupied by a battery of scientific, medical and technical experiments. Investigations in protein crystal growth, muscular atrophy in rats and polymer membranes were conducted, measurements of cosmic and gamma radiation were taken and an intriguing demonstration of a model truss structure for Space Station Freedom was assembled in the middeck. This MIT-funded study, known as the Middeck Zero-Gravity Dynamics Experiment, or 'MODE', incorporated accelerometers and strain gauges to examine the behaviour of truss members and the sloshing of fluids (represented in the test by water and silicone oil) and assess the performance of 'rotary joints' to steer the station's massive solar arrays. For the first time, an Electronic Still Camera – a modified Nikon F-4 35 mm camera, fitted with a charge-coupled device for digital image storage – was carried into space aboard the Shuttle. Today, it is easy to take the real-time downlink capabilities of electronic still cameras for granted, but on STS-48 this was demonstrated for the first time. "The ability," noted NASA's pre-mission press kit, "is expected to greatly improve photo-documentation capabilities in Earth observations and on-board activity on the Space Shuttle, as well as future long-duration flights, such as Space Station Freedom or a human mission to Mars."

For John Creighton, the MODE tests were particularly memorable. "It was almost being a kid with a Tinker Toy set," he recalled, "putting all those things back together again. We had a little shaker with a bunch of strain gauges on it to vibrate this structure to see how it would react in space. What they were trying to do is verify the computer models down on the ground to make sure that the vibration and the characteristics of this truss would react in space the way they thought they were predicting they would in computers ... and *that* was successful." NASA had been flying spacecraft for several decades, of course, but the behaviour of fluids in partially-empty propellant tanks under microgravity conditions was still imperfectly understood. MODE's clear Plexiglas vials of water and silicone oil enabled the first detailed analyses to take place.

The mission briefly entered the news, early on 17 September, when it became the first Shuttle flight to take evasive action to avoid a piece of space junk. Four days after launch, Mission Control advised the astronauts that their trajectory would carry them uncomfortably close to the Soviet Union's car-sized Cosmos 955 satellite; perhaps as close as just 1.2 km. "With the planned space station," Mark Brown told journalists after the flight, "which will be less manoeuvrable than a Shuttle, NASA will need to study the potential problem of collisions with debris much more carefully." (These proved ironic words, considering the fact that in October 2010 an International Space Station crew performed an avoidance manoeuvre associated with the very same satellite, UARS, that Brown had deployed.) A short, seven-second burn of Discovery's RCS thrusters established a 16 km wide berth between the crew and the defunct Cosmos 955, which had been in orbit for 14 years. "We were all busy conducting experiments," Ken Reightler recalled, years later, "so no one gave this operation much time or attention. I didn't even take a look out the window to see if I could see the intruder go by. I just wanted to make sure we *didn't* make the news back home!"

For Reightler, his big event at the end of STS-48 was lowering the Shuttle's landing gear, a few seconds prior to touchdown ... and *that* was something his wife, Maureen, constantly reminded him to do. Discovery's 13th return from space was unusual in two ways: firstly, it was only the second post-Challenger mission to have the Kennedy Space Center as its *scheduled* landing site, and secondly, it would touch down, for the first time in Florida, in darkness. As a result, much of the astronauts' re-entry training had been in darkness, crossing the darkened United States and alighting on a swamp-fringed runway. The actual event turned out to be somewhat different, when rain clouds in Florida forced a one-orbit extension to the mission and an eventual decision to land at Edwards Air Force Base in California. Creighton brought Discovery perfectly onto concrete Runway 22, in darkness, at 12:38 am PDT on 18 September, concluding a mission of just over five days.

As the orbiter's wheels kissed the runway, the astronaut careers of three of the STS-48 crew came to an end. Mark Brown would work extensively on the early transition of the ailing Space Station Freedom into the International Space Station, partnered with post-Soviet Russia, before retiring from NASA, whilst Creighton and Buchli would both depart the astronaut corps in mid-1992. "I could have stayed on and flown once or twice more," Creighton told the oral historian, "but I wanted to get back up to the Northwest sooner or later." His wife was completing her medical residency and the couple moved to Seattle, where Creighton took a position with Boeing as a production test pilot and instructor. Buchli, too, entered Boeing, but remained in space circles, as manager of Space Station Systems Operations and Requirements within the corporation's Defense and Space Group in Huntsville, Alabama. Ken Reightler and Sam Gemar would each fly one more Shuttle mission and, for the latter, the weeks before his final launch would be unfortunately smeared by ugly revelations of an extra-marital affair.

INFRARED EYES

With the return of Discovery, only one more Shuttle mission remained on NASA's revised schedule for 1991. STS-44 would be the penultimate flight to be totally dedicated to Department of Defense objectives and, although it was for the most part declassified, its primary payload – a Defense Support Program (DSP) early-warning satellite – had long since proven its capabilities as an integral part of the United States' surveillance arsenal. When the crew for the mission was named in May 1991, their launch was originally scheduled for March of the following year, but the hydrogen leaks enforced an unavoidable delay until August ... and eventually November, when Atlantis was retasked to fly Discovery's STS-43 mission. In addition to flying many months later than planned, STS-44 had picked up a unique crew member, the first and only military payload specialist to fly aboard the Shuttle in the post-Challenger era, and had also *lost* its commander.

That commander was originally Dave Walker, a two-flight Shuttle veteran and a man renowned as one of the astronaut office's most competent pilots. A combination of misfortune and cruel circumstance had led to his removal from the STS-44 crew roster. In May 1989, only days after Walker returned from his second mission, STS-30, he had been flying from Houston to Washington, DC, aboard one of NASA's fleet of T-38 jets, when he inadvertently passed within a few tens of metres of a Pan Am commercial aircraft. The near-miss was later partially attributed to air traffic controller error, but the episode and other 'infractions of NASA flying rules' prompted chief astronaut Dan Brandenstein to temporarily ground Walker from T-38 flight status. The 60-day period of grounding came into force in early July 1990, just weeks after Walker's assignment to STS-44, and since all Shuttle pilots and commanders needed to be of 'current' flight status in the T-38 jet trainers it cost him the command of the mission. On the 9th, NASA announced that another veteran astronaut, Fred Gregory, would lead STS-44 in Walker's stead.

Frederick Drew Gregory had already established his credentials as the first African-American commander of a space mission, when he led the crew of STS-33 in November 1989. As one of only three African-American candidates to be chosen for astronaut training in January 1978, the gruff Air Force colonel would go on to serve as NASA's deputy administrator. Gregory came from Washington, DC, where he was born on 7 January 1941. His father was an engineer and took every opportunity to expose his son to new experiences, frequently visiting Andrews Air Force Base in Maryland. "In the late '40s and early '50s, they had sports car racing at Andrews," Gregory told the NASA oral historian. "They would use the taxiway and the runways for these car races. He would always position himself and me across from a hangar and there would always be airplanes ... and though the object was to watch the sports car racing, you couldn't avoid seeing the airplanes in the background." Such sights were undoubtedly influential in the young boy's maturing mind, as were several of his father's friends, who had been members of the Tuskegee Airmen, the first all-black aviation unit in the Second World War. By the time he was in his mid-teens, Gregory was aware of the link between aviation and the military and decided that it would form the basis of his future career. When he met Barbara Archer (later to become his wife of more than four decades), Gregory took her on a first date ... to Andrews Air Force Base to watch an air show. "She was either very patient," he said with a laugh, "or in fact had those same kinds of motivations."

Gregory entered the Air Force Academy to study for his science degree, becoming one of only a handful of African-Americans ever to be admitted to the prestigious institution. (Three others, Charles Bush, Isaac Payne and Roger Sims, were admitted shortly prior to this.) "These were high-quality people," he reflected. "These were *not* tokens. They weren't brought in just to change the colour of the Academy; they were brought in because they were absolutely equal to the other members of the class." Gregory noted that, with segregation of black and white still widespread across the United States, the *military* actually seemed more open to full racial integration than many other sectors of society, having taken serious steps towards this end as early as 1947, the year the Air Force became an independent service. He received his bachelor's degree in 1964 and began training as a helicopter pilot at Stead Air Force Base in Nevada, receiving his wings the following year. In June 1966, he undertook his "very fulfilling" first deployment to North Vietnam, as an H-43 combat rescue pilot, and upon returning to the United States he flew the UH-1F helicopter as a missile support pilot at Whiteman Air Force Base in Missouri. Then, in January 1968, his aviation experience changed direction, when the opportunity presented itself to transfer to fixed-wing aircraft, firstly on the T-38 Talon and later the F-4 Phantom. In the meantime, he had also been accepted into Naval Test Pilot School and after graduation in June 1971 he was despatched to Wright-Patterson Air Force Base in Ohio, where he tested both fighters and helicopters. His next assignment, in June 1974, was a detail to NASA's Langley Research Center as a research pilot. NASA specifically wanted a test pilot with experience in rotary aircraft *and* fighters and, years later, Gregory would admit that had he *not* chosen helicopter training earlier in his career, "I probably would not be where I am right now, because I would have been just like any other test pilot with a single capability". By 1977, his interest in remaining a research pilot was waning and it was Nichelle Nichols – the African-American actress who had inspired women and blacks to apply for the astronaut programme – who suggested that Gregory apply. His worry was his relative paucity of fighter experience and, briefly, he considered resigning from the Air Force and entering the astronaut corps as a civilian. By this time, he had earned a master's degree in information systems from George Washington University, and in November he was invited to Houston for interview.

As an astronaut, Gregory's experience as a former helicopter pilot made rides with him aboard the T-38 a thrilling experience. "Apparently," wrote Mike Mullane, "helicopter pilots believed they would get a nosebleed if they ever flew above a few feet altitude, or at least I got that impression from flying with Fred ... We would pass over the tops of windmills with just yards of clearance. The only thing that protected us from running into buzzards and hawks was that *they* had sense enough to cruise at higher altitudes." When a power line loomed, Gregory would hop the jet over it. On one occasion, they swooped down into the yawning Rio Grande River Gorge in New Mexico and found themselves looking *up* to see the canyon's rim! "In what is truly a remarkable irony," Mullane concluded, "many years later, Fred was appointed NASA's Associate Administrator for Safety. I guess we *all* eventually grow brains!"

For Gregory, the prospect of flying a third time was rendered even more exciting by the fact that it was aboard a third different orbiter, Atlantis. (His two earlier flights were aboard Challenger and Discovery.) "Each of these ladies," he said of the three vehicles, "had slightly different personalities. You could get inside and hear things on one that you wouldn't hear on the other, but Discovery and Atlantis were very close to each other. Challenger was a little earlier and so it had some slightly different characteristics." By the time of his assignment to STS-44, Gregory was already thinking about how to close his astronaut career and one option was to fly each of the five orbiters, including Columbia and the new, Challenger-replacement vehicle, Endeavour. "*That* would be a reason to stop flying," he told the oral historian, "so this one just kinda fit right into the scheme. I was pretty excited about it." As events transpired, STS-44 would be Gregory's final mission.

His 'core' crew consisted of representatives of each of the major military services, with a fellow Air Force officer, Terence 'Tom' Henricks, as the pilot, and a trio of mission specialists: Army flight test engineer Jim Voss, civilian physician and former enlisted Marine Corps aviator Story Musgrave and Navy oceanographer and meteorologist Mario Runco. All of them, save Musgrave, had yet to experience the exhilaration of space flight ... and yet Musgrave's credentials in the astronaut business incomparably surpassed them all. Scientist, physician, engineer, pilot, mechanic, poet and literary critic, Franklin Story Musgrave had amassed extensive flying time in the Marine Corps and by 1987 had secured no fewer than *six* academic degrees. He was never called by his first name – not even by his parents – and his middle name, 'Story', honoured an old family surname from several generations back. "I got into this business to be on the intellectual and physical frontier," he explained of his decision to pursue a career with NASA. "I wanted a *transcendental* experience – an existential reaction to the environment. I'm not talking about an illusion, or seeing something that wasn't there, but a magical, emotional reaction to the environment. *That's* what I've been after all my life: to experience and feel new sensations."

That life, certainly, began badly for Musgrave, and in *NASA's Scientist-Astronauts*, historians Dave Shayler and Colin Burgess would characterise his formative years as "a childhood filled with despair". Born in Stockbridge, Massachusetts, on 19 August 1935, his parents were alcoholics – his mother meek and acquiescent, his father malicious and brutal – and the family's isolation on a farm meant that visitors were rare. Two things 'saved' him: one was that his ancestry on his father's side had boasted *nine* straight generations of doctors and the other was the ability to escape into the natural environment. "The unhappy situation," wrote Shayler and Burgess, "would often cause young Story to flee his home by night, making his way into the embrace of a nearby forest, where he would lie on his back, look up and marvel at the stars. He recalls doing this when aged *only three*, but the darkened forest held no fears for him." Even at this age, Musgrave considered nature to be his home – "my solace," he said, "a place where there was beauty, in which there was order" – and he was soon building rafts and becoming more self-reliant.

His salvation came in 1945, when his abused mother finally decided that she could no longer bear a violent existence with her husband and fled, taking Story with her. Yet tragedy was never far away: both of his parents *and* his younger brother would eventually commit suicide, whilst his older brother died in an aircraft accident. Psychologically, these calamities helped to improve Musgrave's self-reliance and mould him into the man that he would become. Surprisingly, in view of his later academic accomplishments, he did not shine at high school. "He hated school and all that was associated with it," wrote his biographer, Anne Lenehan, on Musgrave's website, www.storymusgrave.com. "He was constantly in trouble with the school authorities and was subjected to almost continual disciplinary action." A motorcycle accident caused him to miss out on his final exams, but Musgrave's perspective was that home life and school life offered a "fantastically narrow" window on the world. "I felt the urge to expand my horizons," he told Lenehan, "and to see other worlds." He joined the Marines in 1953 and trained as an aviation and instrument technician, completing active-duty assignments in Korea, Japan and Hawaii and serving aboard the aircraft carrier USS *Wasp* in the Far East. These years also rekindled an earlier interest in aviation and he resumed his studies to gain his pilot's licence.

Clutching a National Defense Service Medal and an Outstanding Unit Citation from his Marine Corps squadron, Musgrave left the military and enrolled at Syracuse University in New York to study mathematics and statistics. Shortly before gaining his bachelor's degree, he was employed as a mathematician and operations analyst by the Eastman Kodak Company in Rochester, New York. He then followed up with a master's in business administration from the University of California in 1959, a bachelor's credential in chemistry from Marietta College in 1960 and a doctorate in medicine from Columbia University in 1964. With his medical degree under his belt, Musgrave began his own research into the human nervous system, with a one-year surgical residency at the University of Kentucky Medical Center in Lexington. His achievements enabled him to win post-doctoral fellowships from both the Air Force and the National Heart Institute. During this period, Musgrave's interests broadened to encompass aerospace physiology, temperature regulation and clinical surgery ... and *another* master's degree (this time in physiology and biophysics) in 1966. When NASA announced its intention to select a second group of scientist-astronauts, Musgrave was convinced that his experiences had led him to the door of space. At first, NASA considered him to be *over-qualified* – five degrees so far, an active laboratory, a surgical practice and a licenced commercial pilot, flight instructor and accomplished parachutist (he would ultimately log over 500 jumps) – but his potential was noticed and he was selected in August 1967.

Musgrave's advantage over the other members of his class of scientist-astronauts was that he was alone amongst them in being a qualified pilot ... although even he had never flown high-performance jets. Jungle survival training and flight instruction were therefore mandatory for all group members. Musgrave went to Reece Air Force Base in Lubbock, Texas, completing his 53 weeks of training with the highest scores ever recorded at the base and a commendation. In April 1969, he was detailed to the Apollo Applications Program, subsequently renamed 'Skylab'. Less than two years later, his name was announced as the backup science pilot for the first Skylab mission. At the time, plans existed for a second space station – 'Skylab B' – and Shayler and Burgess have speculated that had this actually flown, in around 1975-77, Musgrave would have been a primary candidate for a seat on one of its missions. Sadly, budget cuts and the emphasis on getting the Shuttle ready to fly ultimately sounded the death knell for Skylab B. In 1974, he was assigned to the life sciences branch of the astronaut office and in October of that year participated in a medical development test of Spacelab with Dennis Morrison of JSC's Bioscience Payloads Office. The two men spent seven days inside a Spacelab mockup and conducted a series of biomedical demonstrations in order to perfect operational procedures for real missions. Musgrave participated in a second such test for five days in January 1976, together with nuclear chemist Robert Clark and cardiopulmonary physiologist Charles Sawin; on this occasion, they lived and worked in a full-scale mockup of not only the Spacelab facility, but also the middeck and flight deck of the orbiter itself.

During this period, Musgrave continued clinical and scientific training as a part-time surgeon at Denver General Hospital and as a part-time professor of physiology and biophysics at the University of Kentucky Medical Center. In his NASA capacity, he participated in the development of the Shuttle space suit – together with

1-21: The power differential between the twin SRBs (80 percent) and the Shuttle's three main engines (20 percent) is nowhere better illustrated than in a launch view like this one. STS-44 in November 1991 lifted off at night, presenting a stunning contrast between the golden plumes of the boosters and the ethereal blue-white exhaust and 'Mach diamonds' from the cluster of main engines.

the airlock, life-support systems and a new Manned Manoeuvring Unit – as well as working in the Shuttle Avionics Integration Laboratory. Moreover, for an astronaut who was selected with *no* jet experience, Musgrave would eventually amass 17,700 hours of experience; more than a third of this time was in NASA's T-38 Talon jets, but in all he flew over 160 different types of aircraft. By comparison, no other astronaut has come close to this total, with most former chiefs of the office averaging 7,000 hours. In fact, even the most flight-experienced chief astronaut, John Young, logged 15,200 hours in his career. "From the beginning of his NASA days," wrote Anne Lenehan, "Story flew just about every day, sometimes *twice* a day, and around a hundred hours a month ... an extraordinary amount, given his other commitments." Nor was Musgrave a cautious aviator. Fellow astronaut Mike Mullane recalled one stomach-churning episode aboard a T-38, shortly before STS-6. In *Riding Rockets*, Mullane wrote that Musgrave "asked [Air Traffic Control] for a block of altitude and then went into a series of spiralling rolls and violent manoeuvres that alternately had me slammed into my seat at 4 Gs and lifted from it in negative Gs. My head snapped back and forth like a palm tree in a hurricane. Within a minute I was ready to blow my last meal ... and had to plead with him to stop." Musgrave's remarkable attention became legendary. He was nicknamed 'Dr Details' by his fellow astronauts. He exuded self-confidence both in the missions he flew and in the tasks he fulfilled. "It was sheer play for me," said the man whose life had begun under such a cloud of menace, "to be able to so completely interact with my environment."

STS-44 was Musgrave's fourth Shuttle flight; he and Gregory had also previously flown together on STS-33 and were perhaps the ideal pair to guide three rookies through training for a complex mission. Gregory's approach as their commander was to identify each crew member's talents and fill the gaps through training. "The commander is responsible for the success of the mission," he explained. "That's a given. A secondary role for the commander is to ensure that the crew has had *fun*." Like a family – though "*not* a dysfunctional family," Gregory cautioned – a Shuttle crew learned over time to accept its own strengths and weaknesses, in which no single member became overly headstrong and the skills of each astronaut worked harmoniously with the rest. In précis, he concluded, "a good crew is like a ballet ... it's musical in some ways and it's very co-ordinated and beautiful". By November 1991, as their training drew to a close, Gregory could finally stand back and gaze with pride upon the near-perfect oil painting into which STS-44 had turned.

In time, Terence Thomas Henricks would go on to make four space flights and become the first Shuttle pilot and commander to log more than a thousand hours in orbit. He was born in Bryan, Ohio, on 5 July 1952 and earned his degree in civil engineering from the Air Force Academy. After graduation in 1974, he underwent flight instruction at Craig Air Force Base in Alabama and trained on the F-4 Phantom. Henricks flew in fighter squadrons in England and Iceland and was selected to attend test pilot school at Edwards Air Force Base in 1983. He subsequently remained in California as an F-16C Falcon test pilot and headed the 57th Fighter Weapons Wing Operating Location until his selection by NASA in June 1985. In addition to his flying accomplishments, Henricks was a master parachutist, with 749 jumps, and earned a master's credential in administration from Golden Gate University.

In the case of James Shelton Voss, a childhood love of science fiction eventually guided him to space ... though, in his youth, the career of 'astronaut' seemed closed to him. Born in Cordova, Alabama, on 3 March 1949, Voss became a keen wrestler and footballer in his formative years and studied aerospace engineering at Auburn University. After graduation in 1972, he entered the Army, but was permitted to delay his entrance into active duty until he had earned a master's degree in aerospace engineering sciences from the University of Colorado. Voss' military career began well: he was the Distinguished Graduate of the Army's Infantry Basic Course and received the Honor Graduate and Leadership Award upon completion of the Ranger Course. (The latter has been variously described as "the toughest combat course in the world" and "the most physically and mentally demanding leadership school the Army has to offer".) Voss' accomplishments, even at this early stage in his Army career, give an impression of his character.

Yet an astronaut career eluded him, for virtually all of the astronauts then serving with NASA were test pilots. "And though I was in the military," he told a NASA interviewer before his final space mission, "I was an engineer and I didn't have vision that was good enough to be in the programme then." Everything changed in January 1978, when the first selection of Shuttle astronauts – including a range of military and civilian personnel, engineers and scientists, pilots and physicians – was made. Voss tendered his first unsuccessful application to NASA in that very year and it would take no fewer than *six* attempts before he was finally accepted into the astronaut corps in June 1987.

Having said this, Voss was not wholly consumed by NASA. His Army work was something he pursued with a passion – "The things that I did in my military career ... were things I *wanted* to do," he said – and he spent time as a platoon leader, intelligence staff officer and company commander in West Germany, attended the Infantry Officer Advanced Course in 1979 (making the Commandant's List) and taught mechanics as a professor at the Military Academy at West Point, receiving the William P. Clements Jr Award for Excellence in Education. Graduation from Naval Test Pilot School in 1983 brought with it the Outstanding Student Award, after which he entered the Armed Forces Staff College and served as a flight test engineer with the Army's Aviation Engineering Flight Activity. He began working for NASA in November 1984 as a vehicle integration test engineer, seconded from the Army, and supported four Shuttle missions, including the final mission of Challenger. In the wake of that disaster, Voss participated in the Rogers investigation. Selection by NASA in the summer of 1987 was unquestionably one of the high points of his life, but, fundamentally, he was driven by a love of aviation ... and space flight was simply a logical extension of aviation. Years later, he would pay tribute to a number of individuals for guiding his steps: his grandparents, Jim and Millie Wright, who raised him, together with Major Jack Damewood, his ROTC instructor at Auburn University, whose model of a good soldier and officer steeled Voss to follow his example. Others included his wresting and football coaches, Bobby Barrett Ray Campbell and Swede Umbach.

The final NASA member of the STS-44 crew was Mario Runco Jr, whose quite remarkable career path had seen duty as a police state trooper, then a Navy oceanographer and meteorologist and ultimately an astronaut. Born in the Bronx, New York, on 26 January 1952, of Italian-American parentage, Runco's close physical resemblance to Mister Spock – Leonard Nimoy's character in *Star Trek* – proved the root of many jokes whilst at NASA. He studied Earth and planetary sciences at the City College of New York and, whilst there, played intercollegiate ice hockey. After graduation in 1974, he moved to Rutgers University for his master's degree in atmospheric physics, again played ice hockey, and spent two years as a research hydrologist, carrying out groundwater surveys for the US Geological Survey on Long Island. In 1977, Runco joined the New Jersey State Police and spent a year as a patrol trooper, then entered the Navy in June of the following year. His early military duties, not surprisingly, took full advantage of his academic and work training: he served as a research meteorologist at the Navy's Oceanographic and Atmospheric Research Lab in Monterey, California, then as a meteorological officer aboard the USS *Nassau* from 1981-83. During this latter assignment, Runco was designated a Naval Surface Warfare Officer. He then worked as a laboratory instructor at the Naval Postgraduate School and performed hydrographic and oceanographic surveys of the Java Sea and Indian Ocean, aboard the naval survey vessel *Chauvenet*. Selected by NASA, alongside Jim Voss, in June 1987, Runco's police background led to the inevitable nickname of 'Trooper'.

"The word had gotten around," Runco told a Smithsonian interviewer, "that our crew was a little playful and somewhat irreverent – not what one would expect from a crew with only two veterans and four rookies." A few months before launch, a memo had circulated in the astronaut office and came to be nicknamed 'The Hair Memo'. Its content surrounded an incident in which astronaut Marsha Ivins had allowed her long hair to float freely on an earlier mission; NASA management feared that this could prove dangerous and had directed that hair was to be tied up. Thankfully, the crew included bald Story Musgrave and *another* idea entered Fred Gregory's mind: to walk out for their terminal countdown demonstration test, *with no hair*! Runco and Voss bought some rubberised latex head coverings and five pairs of aviator-style sunglasses, like those worn by Musgrave, together with a handful of *Story Musgrave* name tags for their partial-pressure suits. When they walked out of the Operations and Checkout Building, it seemed to the gathered journalists that *six* Story Musgraves were emerging ...

As NASA prepared to launch STS-44, late in November 1991, the nature of its DSP payload, mounted atop an IUS, was well-known and, even in pre-Internet days, information about it was freely available. After this mission, only one more dedicated flight for the Department of Defense would be flown by the Shuttle; for all intents and purposes, the Pentagon had long since lost interest in the reusable orbiters. In the wake of the Challenger disaster, only payloads which could not easily be reconfigured for launch aboard expendable rockets had been kept on the Shuttle. When the last classified mission returned to Earth in December 1992, more than two decades of military involvement and domination of the Shuttle came to an end.

The fate of Challenger, of course, had much to do with this. Losing a vehicle and a human crew imparted unwelcome public and political scrutiny on manned launches and eventually drew the Department of Defense back to expendable rockets. Edward 'Pete' Aldridge was Undersecretary of the Air Force from 1981-86 – and chair of the National Reconnaissance Office during the same period – and very quickly saw the writing on the wall. "I believe Jimmy Carter wrote a presidential directive [in 1978] that the Space Shuttle ... would meet *all* the demands of *all* the users," he told NASA's oral historian. The capabilities of the winged vehicle were expected to render all other boosters then used by the military – the Atlas, the Titan and the Delta – unnecessary and the expectation was that they would be steadily phased out between 1978-86. The Shuttle would take centre stage as the 'National' Space Transportation System.

Aldridge's worries were twofold. Firstly, when the first Shuttle mission got underway in April 1981, it was abundantly clear that promises of weekly flights could not be delivered upon. Turnaround times were far longer than anticipated, the vehicle itself was complex and difficult to service ... and *two* of the orbiters were too heavy for effective use by the Department of Defense. Significant cost savings, which NASA promised would be a third of the nominal fee for an expendable rocket, never materialised, and in terms of the flight schedule the Shuttle was incapable of flying more than 12-18 times per year. This abject failure to deliver prompted Aldridge to approach Secretary of Defense Caspar Weinberger in 1983 to advise against the termination of expendable rocket production. Weinberger agreed, and so did President Ronald Reagan, and proposals were put in place to continue at least the Titan booster production for another five years. "NASA got *very* upset about it," added Aldridge. "Administrator Jim Beggs [who headed the agency from 1981-85] saw that as a ploy of the Air Force to remove itself, ultimately, from the Shuttle."

For his part, Beggs said that he had no concern about the Air Force opting for expendables, as a "backup" capability, but stressed that those rockets were equally as vulnerable as the Shuttle to failure. (In fact, *two* Titans would be lost in separate accidents, a few months either side of Challenger.) At length, after discussing the issue with Weinberger, a compromise was reached between Beggs and Aldridge, in which the Department of Defense would use a third of each year's Shuttle missions for its purposes. Yet the seeds of doubt had already been sown. When the STS-44 crew was announced in May 1990, the manifest already showed theirs to be one of the last military Shuttle flights.

Yet the crew had picked up a rather unique payload specialist: Thomas John Hennen, an Army chief warrant officer and a recognised expert in the fields of intelligence and surveillance. Hennen was born on 17 August 1952 in Albany, Georgia, the son of an Air Force chief master sergeant, and a career in the armed forces beckoned from an early age. After high school, Hennen attended Urbana College from 1970-72, on an academic and athletic scholarship, and entered the Army. His introduction to the intelligence began at Ford Hood, Texas, in 1973, during which time he was involved in the planning, development and conduct of operational and force development tests in support of Army combat activities. In particular, Hennen worked on firing methodologies for the AH-1 Cobra attack helicopter, testing new camouflage netting, uniforms and painted equipment, and the development of the Army's first remotely-piloted vehicles.

He later served in a variety of roles in military intelligence and imagery analysis and from 1981-86 was attached to the Army Intelligence Center at Fort Huachuca in Arizona, as project officer for the Imagery Intelligence Tactical Exploitation of National Space Capabilities Program. Hennen's work led to the development of training programmes and he authored a number of works in military intelligence, including radar training plans. Selected as a payload specialist candidate, alongside fellow Warrant Officer John Hawker and Sergeant Mike Belt, in September 1988, Hennen was tasked with helping to develop and operate a battery of experiments to assess the capabilities of a professional imagery analyst to observe terrestrial targets from the Shuttle.

His payload, known as 'Terra Scout', included the Spaceborne Direct View Optical System (SPADVOS) and the Military Man in Space (M88-1). The former consisted of a series of viewing site 'packets', including selected maps and photographs, which Hennen was trained to observe using a combination of small-aperture, long-focal-length optics and a charge-coupled device (CCD) camera to produce high-resolution digital imagery. The nature of these sites, which included Army, Navy and Air Force units, varied from airfields to ports to ships at sea and the overall intention of the experiment was to assess the military benefit of having a trained observer in orbit. Adjoining this work, M88-1 assessed both visual and communications capabilities, with near-real-time information relay and quantification of Hennen's visual resolution limits.

In the months after Challenger, other missions were also planned, featuring Air Force meteorologists, Army geologists and geographers and physicists from the Strategic Defense Initiative (SDI). These plans formed part of the Department of Defense's Space Test Program and encompassed not only Terra Scout, but also 'Terra View', using cameras and binoculars to observe ground-based targets, and 'Terra Geode', employing specialised equipment to assess terrain suitability for tactical movements of troops. In fact, some of the Terra View and Terra Geode work was actually carried out – the former by Army astronaut Jim Adamson during the classified STS-28 mission and the latter by geologist-astronaut Kathy Sullivan during the Hubble deployment flight, STS-31 – although plans to fly dedicated Army specialists aboard the Shuttle for these two payloads never bore fruit. Similarly, a military Spacelab mission (known as 'Starlab') retained a place on the Shuttle manifest until 1990, before budgetary support dried up and it was cancelled.

For Fred Gregory, having Hennen aboard was something totally new. "We had never flown a military non-officer before and it was actually pretty exciting," Gregory told the NASA oral historian, "because he was from the photo interpretation field. I know many had talked about what you can actually see from space, suggesting that there may be a battlefield advantage of being in space and so, with Tom on board, he was the one who was going to come in and use whatever small optical devices he had to determine whether it was possible to do a good amount." By the time STS-44 launched, at 6:44 pm EST on 24 November 1991, the crew had been moulded into such a close-knit team that Gregory considered the four

146 Rise from the ashes

1-22: Its infrared telescope primed to begin a mission of military early-warning, the DSP satellite is raised to its deployment angle in Atlantis' payload bay during STS-44. The attached IUS booster and the tilt table can also be barely discerned in this image.

first-time astronauts as "rookie by name, but very, very experienced when they got on-board". In fact, Jim Voss was placed in charge of the deployment of Atlantis' primary payload, the 2,380 kg DSP satellite, several hours into the mission.

Launch had already been postponed five days from the 19th, due to the need to replace a malfunctioning inertial measurement unit on the DSP's booster and even the second attempt on the 24th met with 13 minutes of delay to allow an orbiting satellite to pass out of trajectory range and to permit the replenishment of the External Tank with liquid oxygen after minor repairs. Despite the lack of experience, the conversation in Atlantis' cockpit in the minutes before launch was filled with much humour and kidding – "the nervous sort of banter," recalled Voss, "you get on a sports team, before it goes out to play" – until Tom Henricks noticed that Story Musgrave had fallen quiet.

"Story," Henricks asked, "how come you're so quiet over there?"

Musgrave, who had been an astronaut for almost a quarter of a century and had flown three previous Shuttle missions, replied, in all seriousness: "Because I'm scared to death." At that moment, Voss realised that if *this* guy, with all of his experience, was nervous, then it was time for the rookies to get serious. After insertion into orbit, Voss remembered, the Shuttle was enveloped in darkness and it was difficult to see Earth. He unstrapped from his seat and floated downstairs to the middeck to doff his partial-pressure suit, then helped Runco and Hennen with theirs. After an hour or so, Voss finally got chance to return to the flight deck ... and his attention was instantly arrested by the grandeur of Earth, literally filling every one of the windows. "It was like someone grabbed me," he told the Smithsonian interviewer. "I just latched onto the window and whatever it was that I was going to do went completely out of my mind. The view out the window was so spectacular that I had to think – *gosh* – I just *have* to stop here and look for a minute."

With the scheduled deployment of the DSP satellite and its IUS booster looming ahead of them, there was little time to ponder their new surroundings. Under the direction of Jim Voss, the mandatory checks were carried out and around six hours into the mission the forward payload restraints were released and the aft frame of the airborne support equipment tilted the gigantic combination to a 29-degree angle. Gregory and Henricks manoeuvred Atlantis into the deployment attitude and the satellite's electrical power source was transferred from the orbiter to its own internal supply.

At 1:03 am EST on the 25th, a little more than six hours after reaching orbit, Voss commanded the IUS tilt table to raise the payload to a deployment angle of 58 degrees and released the $400 million DSP into space at a rate of 10.6 cm/sec. The tilt table was then lowered to its stowed configuration and Atlantis retreated to a safe distance, in preparation for the IUS first-stage solid-rocket engine firing. Forty-five minutes after deployment, the pyrotechnic inhibitors were removed and the engine roared to life soon afterwards, burning for 146 seconds. Approximately five hours later, at the peak of the geosynchronous transfer orbit, the second stage motor ignited – thrusting for a further 108 seconds – to establish the DSP in its operational orbit, whereupon the solar array paddles and communications antennas were unfurled.

The satellite was the latest in a long line of geostationary early-warning platforms, with a heritage dating back to November 1970. Measuring 10 m long and 7 m wide when fully active in orbit, the TRW-built DSP was dominated by a tube-shaped, wide-angle Schmidt infrared telescope, and was spin-stabilised about its Earth-facing axis, rotating six times per minute. Twin, paddle-like solar arrays provided upwards of 1,400 watts of electrical power. Detection of infrared sources – such as missile plumes or ground-based rocket tests – was accomplished by the telescope, which was aimed off the DSP's axis, utilising an array of 'staring' photoelectric cells. As the detector passed over an infrared source of interest, it developed an electronic signal for subsequent transmission to Air Force ground stations. Earlier in 1991, other

DSPs detected Iraqi Scud missile launches and have been used for other, civilian tasks, including the observation of forest fires and volcanic eruptions.

Having thus accomplished their primary objective in spectacular fashion, the astronauts settled down for what was expected to be ten days of military experiments. In addition to SPADVOS and M88-1, a series of radiation and contamination monitors – many of which had flown before – were located in Atlantis' cabin and payload bay. The orbiter itself was used again as a tracking target for the Air Force's electro-optical sensors on the Hawaiian island of Maui and a hand-held binocular-type experiment was used to assess changes in the astronauts' vision in microgravity. Shuttle thruster firings were also observed by an ultraviolet plume-detection instrument aboard the Naval Research Laboratory's Low-Power Atmospheric Compensation Experiment (LACE), launched in February 1990. Since STS-44 was planned for ten days, it would be among the longest Shuttle missions to date, and thus a suitable flight on which to demonstrate medical investigations in support of the forthcoming Extended Duration Orbiter (EDO). Specifically, the astronauts used a lower-body negative pressure garment – which drew fluids into the legs as a countermeasure for the punishing effects of adaptation to terrestrial gravity – and drank heavily-salted water in the expectation that it would increase their ability to stand upright after spending so long in the weightless environment.

With the emphasis of much of the mission being on Earth observations, and coming only months after the devastating Mount Pinatubo eruption, the astronauts of STS-44 were able to make comments about significant changes in the atmosphere. Certainly, Gregory and Musgrave were vocal in their conviction that atmospheric pollution and thickening layers of haze had worsened in the two years since their last flight together. Cloud cover had also hampered a number of the SPADVOS and M88-1 experiments, performed by Hennen and Runco, although the crew were able to make useful observations, successfully counting the number of ships in the Guantanamo Bay naval base in Cuba and even spotting *crates* aboard an oil tanker. From his perspective, Runco would later remark that the observations were of only marginal use for military surveillance purposes and even Tom Hennen, whose expertise in the field spanned more than 20 years, only managed to observe around a dozen of his 30 targets, partly due to bad weather and partly due to an event outside of anyone's control. Late on the morning of 30 November, IMU-2 – one of Atlantis' three inertial measurement units, an integral part of the Shuttle's navigational hardware – failed. The crew attempted to cycle power to the device, in the hope of reviving it, but to no avail. Flight rules dictated that with one IMU down, a Minimum Duration Flight had to be declared and Atlantis should return to Earth at the next available opportunity.

Less than a week into their planned ten-day mission, STS-44 was coming home.

Later that afternoon, Gregory, Henricks and Musgrave performed a 'hot-fire' test of the orbiter's RCS thrusters and flight controls, to ensure their preparedness to support re-entry into the atmosphere. In the meantime, the other crew members continued to gather as much data from their experiments as possible, in anticipation of a landing at Edwards Air Force Base in the mid-afternoon of the following day. Original plans targeted STS-44 towards the Kennedy Space Center, but the failure of the IMU led mission managers to call up Edwards, whose large dry lakebed runways offered additional margins of safety for an incoming orbiter with a degraded navigational capability. The payload bay doors were closed and latched at 1:46 pm EST on 1 December and the 183-second de-orbit burn commenced at 4:28 pm. Thirty-five minutes later, Atlantis encountered the first traces of the upper atmosphere and after a sweeping hypersonic descent across the Pacific Ocean and into California, she touched down on the dry lakebed runway at 2:34 pm PST (5:34 pm EST). Her mission had lasted just 70 minutes shy of seven full days... and three days shorter than it should have been.

Moreover, the switch of landing site from Florida to California also meant that none of the crew's families were in attendance. "It would have been spectacular to watch," Tom Henricks told a Smithsonian interviewer, "because we landed on lakebed Runway 5, which meant we came right over the top of the buildings. Someone in the control tower could have looked in the Shuttle windows as we went by and if you had been on the ramp, where NASA keeps its planes, you practically could have jumped up and *touched* our wheels!" STS-44 was not only the very first Shuttle landing on Runway 5, but it was the *last* ... and also the last ever to touchdown on a dry lakebed. Despite the disappointment at losing the last few days, Fred Gregory remained philosophical. "We had completed our primary mission," he reflected, "which was to deploy [the DSP]." Many of the scientific and medical experiments were successfully completed and with a flight log rapidly approaching 7,000 hours he decided it was time to hang up his space helmet.

Six months later, on 30 May 1992, Gregory flew a T-38 from Houston to Whiteman Air Force Base in Missouri, with fellow astronaut Kathy Sullivan in his backseat. It was his last flight. The pair stopped, had lunch and returned to Houston. Sullivan had recently flown her third Shuttle mission and had already decided to retire and accept a position with the National Oceanic and Atmospheric Administration; in fact, she was due to start her duties as NOAA's Chief Scientist in mid-August 1992 and was confirmed by the Senate the following March. That flight with Sullivan pushed Gregory through 7,000 hours on his aviation log and he decided to end his flying career there. "I left everything in the airplane," he said, "helmet, helmet bag, checklist, gloves, everything. I got out of the airplane and walked away. It was a clean break." Gregory would remain with NASA in a managerial role, eventually rising to become the agency's Deputy Administrator in 2002, but his departure was one of many during this period, as members of the 'Old Guard' – the original Shuttle astronauts, selected a decade or more earlier – moved on to pastures new and passed the baton to a newer, younger breed of spacefarer. In 1992, which had been designated 'International Space Year', in order to commemorate the 500th anniversary of Christopher Columbus' discovery of the Americas, the Shuttle would fly more times than it had ever done since the pre-Challenger era. That year would truly represent a new dawn for the reusable fleet of orbiters, as they transitioned from a post-Challenger mindset of cautious recovery to one of fully operational missions which would culminate in longer flights, a dramatic repair of the Hubble Space Telescope, a rocky love affair with Russia and a permanent space station.

2

The last Soviet citizen

A NEW START

On 8 September 1989, two cosmonauts aboard the Soyuz TM-8 spacecraft caught their first glimpse of their quarry: the Mir orbital station. It glinted like a star on the horizon. Alexander Viktorenko and Alexander Serebrov should have launched earlier in the year, but their planned five-month mission had been repeatedly delayed. Mir was always intended for continuous occupancy and from February 1987 until the end of April 1989, rotating teams of men had accomplished precisely that objective. However, delays in preparing a pair of 20,000 kg research modules for the station – one equipped with a dedicated airlock to support spacewalks, the other loaded with microgravity science experiments – had prompted managers to leave Mir unmanned for several months. In February 1989, the Soviet Union announced that the launch of the first module, 'Kvant-2', would be delayed primarily because the production of the second, 'Kristall', had itself slipped. (As the year wore on, it became clear that the Kvant-2 launch – originally scheduled for March – had also been affected by a faulty batch of microcircuits for its rendezvous hardware.) To understand the reasons behind this decision, it is important to understand the background of Mir itself.

Starting with the name, 'Mir' is roughly translatable as either 'World' or 'Peace' and seems to have formed part of a propagandist Soviet effort to present their space programme as one of 'Star Peace', diametrically opposed to US President Ronald Reagan's belligerent 'Star Wars'. This theme was reinforced by former cosmonaut Alexei Leonov, speaking shortly after Mir's February 1986 launch, that "by naming the space station in this way, we want to emphasise once again that the Soviet programme for space research and for the use of outer space is intended solely for peaceful purposes". This notion was supported by Mir's first commander, Leonid Kizim, who responded to a question about whether the new station would be used for military purposes. "The programme for our work," he replied, "does *not* contain any experiments for military purposes. As for the statements by US officials, it seems to us that they are being made in order to justify their *own* plans for transferring the arms race to space." Relations between the two nations were improving, it is true, and the attempts of General Secretary Mikhail Gorbachev – elected in March 1985 – to reform the Soviet Union were bearing fruit, but the intrinsic mistrust between East and West remained strong.

The roles and capabilities of the Mir space station were at least an order of magnitude higher than anything previously attempted. Unlike the 'monolithic' Salyut stations, launched between 1971 and 1982, Mir represented something entirely novel: a first effort to create a 'modular' complex in space. After the insertion into low-Earth orbit of its 'core' module, equipped with a unique, multiple-port docking adaptor, the plan was for it to be expanded over several years into a fully-fledged research facility, with the addition of large laboratory modules, each weighing around 20,000 kg, devoted to astrophysics, remote sensing and the life and microgravity sciences. For a decade and a half, counting from the launch of its core in February 1986 to the end of its life in March 2001, Mir would prove itself to be one of the most successful and influential space stations ever launched. Record after record would be secured by its cosmonauts, many of which still stand to this day, including records for the longest time spent in orbit, and it would be occupied, uninterrupted, for more than ten years. In total, more than a hundred individuals from a dozen nations – Russia, Syria, Bulgaria, Afghanistan, France, Japan, the United Kingdom, Austria, Germany, the United States, Canada and Slovakia, to say nothing of the many Soviet and Russian cosmonauts who traced their racial and cultural ancestry to Azerbaijan, Kazakhstan, Kyrgyzstan and elsewhere – would work aboard Mir, would gaze at the Home Planet through its windows, would share meals at its dinner table, would exercise on its treadmill and would share in its successes, trials and near-disasters. From its high vantage point, around 250 km above Earth, Mir would not only eclipse the achievements of every space station which preceded it, but would also lay the critical cornerstones and smooth the way for today's International Space Station.

Mir's story began in February 1976, when a Soviet decree identified it as quite distinct from the other Salyuts: it would possess docking ports at each end of its core module, just like them, but its spherical multiple adaptor would carry two radial ports for additional research laboratories. By the summer of 1978, this plan had changed to a final configuration of a single aft port and no fewer than *five* ports on the multiple adaptor: one at the front and another four radial ports, spaced at 90-degree intervals. These radial ports would accommodate small laboratories, weighing in the order of 7,000 kg apiece, and probably derived from the Soyuz spacecraft but with the orbital and descent modules replaced by a pressurised research compartment. Plans shifted yet further in the spring of 1979, when a government resolution called for the project to be merged with a cancelled military space station project, known as 'Almaz', and Mir's ports were reinforced to handle larger modules based on the TKS (*Transportniy Korabl Snabzheniya*, known for a time in the West as the 'Star' and 'Heavy Cosmos') spacecraft in size and shape, each weighing around 20,000 kg.

Mir would benefit from digital flight computers and gyrodyne flywheels for attitude control, an automated rendezvous system, a dedicated satellite communications network and improved oxygen generators and carbon dioxide scrubbers. Unfortunately, financial resources during this period were very much directed into the Soviet Union's Buran shuttle programme and it was not until the spring of 1984 that Valentin Glushko – a legendary rocket engine designer and head of the Kaliningrad-based Energia design bureau – was ordered to place the new station into orbit by February-March 1986, to coincide with the 27th Communist Party Congress. (It is probably not entirely coincidental that Glushko received his order only weeks after Ronald Reagan announced plans for the United States to build its own space station.) It would be difficult to get Mir ready for launch in such a short span; in fact, original plans had not envisaged the core module being placed into orbit before late 1986. Static and dynamic testing of the core was completed in December 1984, but the remainder of the processing was far from complete.

There were other headaches, too. The weight of electrical cabling pushed the core module a couple of thousand kilograms outside its Proton launch vehicle's performance envelope. In response, a sizeable chunk of the experiment hardware had to be removed and remanifested onto subsequent Progress supply missions, but the weight overspill was still unacceptably high. At length, in January 1985, the planned 65-degree orbital inclination for Mir was relaxed to 51.6 degrees, like the Salyuts, although this would reduce the photographic coverage of the Soviet Union. Difficulties with the new Strela ('Arrow') flight computers forced engineers to install older-specification units, as a short-term measure, in order to meet the politically-mandated launch target. In April 1985 it was decided to ship Mir's core module to the launch site in what is today the independent nation of Kazakhstan to complete systems testing and integration. When it arrived at the desolate launch site in early May, more than *eleven hundred* wires – almost half of the total – required substantial rework, but by October it had completed its final clean room inspections and the focus changed to communications checks (using the Cosmos 1700 satellite which had recently been inserted into geostationary orbit).

Mir left Earth from one of the most desolate locations on the planet. 'Tyuratam', as it is properly known, is a railway junction, some 200 km east of the Aral Sea, and in the local Kazakh tongue its name is translated as the gravesite of Tyura, beloved son of the great Mongol conqueror Genghis Khan, whose medieval empire spanned much of Asia. According to some sources, it began as an ancient cattle-rearing settlement on the north bank of the Syr Darya River, although at least one Soviet-era journalist gave it a more modern origin, hinting at its foundation in 1901 as an outpost to replenish steam engines passing between Orenburg and Tashkent. Its importance over the last half a century, though, cannot be disputed. It was from this sparsely inhabited expanse of steppe, five decades ago, that the first strides of a journey far more audacious, much longer and considerably harder than any the Great Khan could have foreseen were taken. It was from this place that Yuri Gagarin, the first man in space, began his flight in April 1961, changing our perception of the Universe forever. For all of its historic attributes, this remote corner of old Soviet Central Asia – an area swarming with scorpions, snakes and poisonous spiders, whose climate produces vicious dust storms, soaring summertime highs of 50 degrees Celsius and plunging wintertime lows of -25 degrees Celsius – was one of the most secretive, mysterious and closeted places in the world. In fact, even its *name* was kept strictly under wraps, as part of a deliberate effort to mislead and confuse prying Westerners. Today, it is still variously called 'Tyuratam', after the tiny railhead, or, more often, 'Baikonur', which covers a broader and different geographical location to the north-east.

The first launch attempt for Mir's core module came on 16 February 1986, less than three weeks after America had reeled from the loss of Space Shuttle Challenger. It was scrubbed when Mir failed a communications test, but the launch occurred successfully at 12:28 am Moscow Time on the 20th, with the Proton booster inserting the 20,400 kg infant station perfectly into a 170 × 300 km orbit, inclined 51.6 degrees to the equator. By 6 March, the very day on which the 27th Communist Party Congress ended, Mir was established in its operational orbit. Physically, the core measured 13.1 m long and 4.2 m in diameter and, with its twin solar arrays fully deployed, had a wingspan of 20.7 m. These arrays each covered 76 m^2 and utilised high-performance gallium arsenide cells, with an initial electrical power yield of around nine kilowatts. (A third, 'dorsal' solar panel, presumably omitted at launch, due to weight considerations, would finally be installed during a spacewalk in June 1987.) Two liquid-fuelled main engines, with an individual thrust of 300 kg, were situated in an unpressurised compartment at the rear of the core module, but could not be used after the spring of 1987, due to the arrival and permanent docking of the first scientific module, the astrophysics laboratory 'Kvant-1' ('Quantum').

From 'front' to 'back', Mir's most visible feature was the spherical multiple adaptor, which measured 2.2 m in diameter and 2.8 m long. "It serves as a lobby for spacecraft docking," noted a 1987 US Senate report on Soviet space activities, "and houses the airlock for egress into space." It was intended that arriving modules would firstly dock at the 'forward' port, after which an arm-like manipulator device would robotically separate them and manoeuvre them around to one of the radial ports. A pair of 'sockets' for these devices were situated on the multiple adaptor. Moving out of the adaptor, the cosmonaut entered Mir's cylindrical working and living compartment, 7.7 m in length and comprising two sections, which measured 2.9 m and 4.2 m in diameter, connected by a short conical frustum. Although this dual-section compartment took the form of a stepped cylinder, it provided a single internal volume. At the rear was a service propulsion area and transfer adaptor, the same diameter as the main compartment and measuring 2.3 m long, which carried communications and rendezvous equipment, and four propellant tanks for the 32 thrusters of the attitude control system. Lastly, it housed a pair of 300 kg main engines.

Inside the core, Mir differed greatly from the earlier Salyuts. With the exception of the station's control consoles, it was largely devoid of experimental apparatus. It had instead been designed as a 'habitat' or living area, which the cosmonauts came to affectionately nickname 'the base block', and could comfortably accommodate a crew of two. Each man was provided with his own private cabin, equipped with a window, a sleeping bag and a desk. It also housed a toilet, an entertainment system for movies and music, exercise equipment, medical supplies, and a galley and fold-down dinner table with food warmers and movable 'chairs'. "The crews cook on a hot plate," noted the US Senate report, "and are allowed to select their own food, as long as they consume the required number of calories." *Spatial awareness* was provided by a dark green 'carpet' on the 'floor', light green 'walls' and a white 'ceiling' on which were installed fluorescent lamps, and the installation of equipment was deliberately arranged to mimic this Earth-like orientation. Communications were to be provided via a network of command and control satellites, known as 'Luch' ('Ray'), which could trace their genesis to the very same February 1976 decree which had brought Mir into existence, although at that time there was only one satellite in operation. Similar in function to NASA's Tracking and Data Relay Satellites, the Luch network provided a high-rate space-to-ground link for the station and its visiting Soyuz spacecraft, together with communications support for the Soviet Navy. In total, a trio of three-axis-stabilised Luch satellites were launched between October 1985 and December 1989, under the cover names of Cosmos 1700, Cosmos 1897 and Cosmos 2054. Although much of the hardware destined for Mir would be launched by Progress or aboard subsequent modules, the core did arrive in orbit with some of its research facilities in place: crystal growth apparatus, melting and solidification facilities, electrophoresis equipment, photometers and spectrometers, metallurgy hardware and cameras.

A little over a year after the launch of Mir's core, in April 1987, the first additional module, Kvant-1, arrived and was successfully docked onto the rear port. The next pair of laboratories, Kvant-2 and Kristall, were to be positioned on opposing radial ports of the multiple docking adaptor and it was considered desirable that they should be installed in rapid succession to preserve the 'balanced' configuration of the complex. Their arrival was expected early in 1989 and Soyuz TM-7 cosmonauts Alexander Volkov and Sergei Krikalev – launched in November 1988 – had trained extensively to commission the new modules and perform a series of EVAs from Kvant-2's airlock, at least one of which was to demonstrate a Soviet equivalent of America's Manned Manoeuvring Unit (MMU) jet backpack.

Certainly, crew assignments indicate that the new modules were expected at around this time. Space analyst Jim Oberg speculated in mid-1988 that veteran cosmonaut Alexander Serebrov's lengthy 'disappearance' from either a prime or backup crew slot in this period was indicative of his being 'rested' in preparation for a dedicated mission involving one or more of the new modules. Instead of launching aboard Soyuz TM-8 in April 1989 to replace Volkov, Krikalev and physician-cosmonaut Valeri Polyakov and continue the permanent occupancy of Mir, for the second time in its orbital life Mir was left uncrewed. However, when Serebrov and Soyuz TM-8 commander Alexander Viktorenko finally boarded the station on 8 September 1989, they would switch on the station's lights, figuratively and literally, for almost a full decade of uninterrupted operations. Not until the dying months of the 20th century would the lights aboard Mir be dimmed and in those ten years, from the beginning of September 1989 until the very end of August 1999, there would always be at least two humans circling Earth. It is a remarkable accomplishment which cannot, and *should not*, be underestimated, for one should bear in mind that, even by the turn of the millennium, the pioneering flight of Yuri Gagarin remained very much in living memory.

2-01: Pictured during the third Shuttle-Mir docking mission in March 1996, this view shows the station's base block at centre, with the Kvant-1 astrophysics module 'above' and a Soyuz-TM spacecraft at the top of the frame.

In that decade, Mir would seesaw violently between triumph and near-disaster: it would expand, just as the Soviets originally intended, into an impressive 'dragonfly' (as Bryan Burrough called it) with modules, arrays, cranes and antennas sprouting in a myriad directions, with no fewer than *five* research laboratories, and would host dozens of crews from a dozen nations. On the other side of the coin, however, it would fall victim to the kind of funding issues which were an integral feature of post-Soviet Russia and from 1989 onwards the country's space budget established itself on a slippery slope which would see it plummet by 80 percent. That lack of funding would delay modules, scrub missions and oblige cosmonauts to live aboard an outpost which was no longer fit for purpose. As early as April 1989, half of Mir's equipment was effectively inoperable and the lack of additional modules made the interior cramped, as Krikalev observed. Batteries would not properly hold charge, creating chronic power supply difficulties.

PERESTROIKA AND *GLASNOST*

A wind of change grasped the Soviet Union in the mid-1980s. It originated in the socialist states of Eastern Europe and gusted into a whirlwind towards the end of the decade as failed economic policies, flawed Five-Year Plans, detestable political surveillance and religious persecution and the actions of a string of unelected dictators intensified a popular cry for reform and democratic change. The Berlin Wall, once uniformly grey, now had virtually no grey left on its western face, so ubiquitous was the colourful, angry graffiti which covered it. West and East cried out for reunification. The wind intensified in March 1985, when the elderly Konstantin Chernenko died and the 54-year-old Mikhail Gorbachev – a long-time member of the Politburo and a lawyer by profession – was elected as General Secretary of the Communist Party. However, if Western leaders had reacted to the arrivals of Chernenko and Yuri Andropov with concern, then their response to Gorbachev was far warmer; for here was a man who was considerably younger than his predecessors, a man who had travelled widely on Soviet business, a man who had seen how the West had prospered through freedom of speech, political transparency and the respect of human rights, and a man who had seen the policies of the Communist Party bring his once-proud nation to an economic standstill.

Within weeks of entering office, Gorbachev became the first Soviet leader to admit that the economy was virtually stagnant and reorganisation was crucial; a "vague programme of reform" was acutely needed to bring about rapid technological modernisation, increased industrial and agricultural productivity, and a resharpened, more efficient and less corrupt bureaucracy. His efforts to streamline the economy, maintain quality control, battle inferior manufacturing and combat rampant alcoholism necessitated broader reforms on the political level, but Gorbachev remained a staunch Communist. He was far from willing to surrender the Soviet notion of a centrally-planned economy in favour of a free market. Speaking that summer to the economic secretaries of the Eastern Bloc states, he scoffed at the idea that 'the market' might prove to be their saviour. "Comrades," he told them, "you should not think about lifesavers, but about the *ship* ... and the ship is *socialism!*"

Yet the fact remained that the ship was immovable in some areas and downright leaky in others. *Perestroika* (restructuring) was a term which had become entrenched in Gorbachev's political ideology by the late spring of 1986. At face value, it called for the creation of more effective and dependable mechanisms "for accelerating economic and social progress", encouraging the initiative of the individual – a *real* Soviet first – and balancing a need for order and discipline with opportunities for criticism and self-criticism. By early 1987, the Central Committee proposed multi-candidate elections and the appointment of non-Party members to government posts, opponents of Stalin were rehabilitated and in 1988 *glasnost* (openness) brought a measure of real freedom to the Soviet people for the first time. Thousands of political prisoners and dissidents were freed, private ownership of businesses was permitted and in March 1989 the Soviets cast their votes for the Congress of People's Deputies, marking the first open elections in the Union for more than seven decades. Today, many Western analysts see *perestroika* and *glasnost* as the two principal nails in the coffin of the Soviet Union, although that outcome could hardly have been further from Gorbachev's mind in 1985. Rather, he wanted to see socialism work more effectively for its people, with less 'cronyism' and corruption and greater transparency into government affairs.

As circumstances transpired, this snowballed much further than he could have anticipated: relaxed censorship on the state media led to the exposure of severe social and economic problems – previously denied or concealed – as well as food shortages, the terrible price paid to the effects of alcoholism, pollution and environmental destruction, dramatic mortality rates and the first evidence of the 'purges' imposed by Joseph Stalin on his own people, half a century earlier. Under Gorbachev, steps were taken to end a bloody war of attrition in the mountains of Afghanistan and, as time passed, his weakening of the authority of the Communist Party and the removal of its power to control the media emboldened and strengthened nationalists in the Eastern Bloc. Simmering discontent in Estonia, Latvia and Lithuania, illegally annexed by Stalin, would boil over, whilst other republics, including Ukraine, Azerbaijan and Georgia, would rise in nationalistic revolt. In the final year of the 1980s – the 'Year of Revolutions' – the Eastern Bloc would collapse in dramatic and spectacular fashion. The Berlin Wall would be toppled, one dictator would be gunned down by firing squad and republic after republic would throw off the Communist yoke, announce elections and set foot on the thorny path to new nationhood. By the time Gorbachev was obliged to hand over the reins of power in Russia to Boris Yeltsin, late in 1991, the Soviet Union as a political entity would be over ... and years of economic turbulence, civil strife, crime and outright military conflict would ensue.

NEW ARRIVALS, NEW CHALLENGES

Difficulties in space translated to more everyday issues on Earth; pressing and crucial issues, such as people's livelihoods. Across the vast expanse of the largest nation on Earth, from St Petersburg to Vladivostok, from Pevek and Anadyr in the east to Derbent and Baltiysk in the west, millions of citizens of the old Soviet Union – and, from January 1992, the 'Commonwealth of Independent States' – would wait weeks, or even months, to receive even the most meagre pay. Lights in Star City would be dimmed to save electricity, flight controllers would moonlight as taxi drivers to make ends meet and the West and its allies would be increasingly courted: a Japanese television station would pay millions of dollars to fly one of its journalists to Mir in December 1990 and visitors would fork out $200 for a day exploring flight simulators and trying out space suits in places still shrouded in military secrecy only a few years earlier. If the mid-1980s and Gorbachev offered a glimmer of hope for political reform, then the 1990s yielded a glimpse of the harsh reality of transitioning from single-party authoritarianism to multi-party pluralism; "the demons of democracy", as Burrough put it. What happened to the old Soviet Union during those times makes it easier to understand the attraction of Communism – in which the price of having a secure job and healthcare and a pension was paid by a lack of real freedom of expression.

With such severe austerity and profound economic strife on Earth, there was little to justify massive expenditure on a space station whose relevance to the average Russian in the street no longer carried the same sense of propagandist pride enjoyed in the days of Nikita Khrushchev or Leonid Brezhnev. Nor were the cosmonauts themselves any longer revered as the 'heroes' of yesteryear; rather, they were 'workers', earning the equivalent of a thousand dollars for an EVA, and Mir's commanders would worry about mission success for more than just the obvious reasons... not only from a professional perspective or a sense of national pride, but with a wary eye focused on the bonuses (or lack thereof) that he could expect to receive in pay and perks upon his return to Earth. Such perks and professional accolades and a sense of national pride must certainly have been on Alexander Volkov's mind on the morning of 27 April 1989, when he separated Soyuz TM-7 from Mir and brought the capsule and his comrades, Krikalev and Polyakov, back to Earth at 5:59 am Moscow Time. Polyakov had experienced a unique mission: observing the physiological adaptation of two long-duration crews of cosmonauts, whilst undertaking a long-duration mission of his own.

For Volkov and Krikalev, on the other hand, their *next* mission together, late in 1991, would come at one of the most divisive and tumultuous times in Russian history. Their country would have morphed from a vast Union, covering a sixth of the world's land area, spanning 10,000 km from east to west and 7,000 km from north to south, and a dozen time zones, into a loose arrangement known initially as the Commonwealth of Independent States. In some cases, the 'independence' of its former satellite republics was hotly contested and in the 1990s many of the old Soviet states were torn by political and ethnic conflict. Nowhere was 'The End' of the Soviet Union more clearly delineated than in space, when Sergei Krikalev launched to Mir in May 1991 as a 'Soviet' cosmonaut... and returned to Earth, ten months later, as a citizen of an entirely new and very different nation. A new era had begun and its twists and turns would affect the future progress of our species in ways which could hardly be imagined.

That era, and that change, was already underway when cosmonauts Viktorenko and Serebrov blasted off from the darkened Tyuratam aboard Soyuz TM-8 at 12:38 am Moscow Time on 6 September 1989 to commence their two-day 'chase' and rendezvous with Mir. Both men had flown into space before; Viktorenko once and Serebrov twice. Alexander Stepanovich Viktorenko was born in Olginka, in northern Kazakhstan, on 29 March 1947 and after a stellar career in the Soviet Air Force was selected as a cosmonaut candidate in May 1978. Together with fellow selectee Nikolai Grekov, they joined a larger detachment of civilian engineers and physicians the following December to commence training in earnest. However, whereas the remainder of the group graduated in October 1980, Viktorenko was delayed by a serious training accident and did not complete his final examinations until February 1982. Nevertheless, he quickly established a reputation for himself as a dedicated professional. Of the eight members of his 16-strong group who went on to actually fly in space, Viktorenko was fifth.

In time, he would undertake four space missions, spend a cumulative 489 days in orbit and would be in command of Mir in March 1995 when the first American visitor, NASA astronaut Norm Thagard, arrived. "A pleasant fellow, but also extremely competent", was how Thagard later described him. "Sasha Viktorenko was one of those who was probably more like you would expect to find in an American commander. Although we've got some American commanders who are fairly authoritarian ... I think if you looked at the scale of things, you'd find the Russians ... a little bit further down towards the less authoritarian end." After completing his high school education, Viktorenko entered the Advanced School of Military Aviation in Orenburg, graduated in 1969 and entered the Soviet Air Force, where he rose rapidly through the ranks. In 1985-86, he served as backup commander for both the Soyuz T-14 and T-15 missions, then flew his first space flight aboard the Soviet-Syrian Soyuz TM-3 in July 1987.

Joining Viktorenko as flight engineer was the gregarious Alexander Alexandrovich Serebrov. In his book *Dragonfly*, Bryan Burrough described Serebrov as "garrulous" and "headline-loving", with a penchant for fast cars. Serebrov was born in Moscow, on 15 February 1944, and graduated from the capital's Institute of Physics and Technology. He was chosen as one of seven civilian cosmonaut engineers in December 1978 and flew aboard Soyuz T-7 in August 1982 and Soyuz T-8 in April 1983; the latter was intended as a long-duration mission to the Salyut 7 station, but was unable to dock and was brought back to Earth after just two days. Serebrov continued his preparations for a lengthy mission and might have flown the 11-month Soyuz TM-2 mission to Mir with Vladimir Titov in February-December 1987 had he not apparently failed a medical check a few weeks before launch. Both prime crew members were replaced by their backups, Yuri Romanenko and Alexander Laveykin.

Soyuz TM-8 represented the latest in a long line of piloted spacecraft flown by the Soviets since the mid-1960s. Soyuz was the brainchild of Chief Designer Sergei Korolev, with the original intention of supporting both Earth-circling missions and lunar ventures to rival the United States' Apollo programme. As early as 1964, the design and definition of Soyuz was well underway and technical documentation and a mockup revealed it as a craft capable of lofting two or even three cosmonauts. Even its *name* was no accident: for the Soviet Union's official moniker – *Soyuz Sovietskikh Sotsialisticheskikh Respublik*, the Union of Soviet Socialist Republics – was often popularly known amongst its citizenry as 'Soyuz' (the 'Union'). Therefore, the name of the spacecraft not only reflected its role in supporting rendezvous and orbital stations, but was also a highly symbolic and political statement. In *Challenge to Apollo*, historian Asif Siddiqi noted that when Korolev first saw the mockup, he proudly declared that Soyuz was "the machine of the future". Indeed, Soyuz has had the most long-lasting impact on the world. Since its first manned flight in April 1967, Soyuz and a modified version of its original R-7 launch vehicle continue to be used operationally today; a fitting legacy to an enduring talent.

By the end of the decade, seven manned Soyuz spacecraft had rocketed into orbit. They employed a pair of large rectangular solar panels, mounted on the instrument module, to generate electrical power. The total surface area of these wing-like appendages was 14 m^2, with each wing measuring 3.6 m long and 1.9 m wide. The remainder of the craft's design comprised the spheroidal orbital module, 2.65 m long and 2.25 m wide, the bell-shaped descent module, itself 2.2 m long and 2.3 m wide at the base, and the instrument module, a cylinder 2.3 m long and 2.3 m wide. This shape emerged at the end of almost a decade of planning, theoretical work and aerodynamic modelling. As early as 1958, Mikhail Tikhonravov and Konstantin Feoktistov, both engineers at Korolev's OKB-1 design bureau, envisaged a multi-purpose craft capable of both Earth-orbiting and circumlunar missions. Space historians Rex Hall and Dave Shayler noted that the shape of the descent module was decided at least partly by a desire to touch down on land, rather than in water, and several designs were sketched out. The first utilised aerodynamic surfaces, facilitating an aircraft-like return to a runway, whilst the second adopted a 'missile principle', entering space in a ballistic manner and descending beneath parachutes. By 1961, concerns about mass and the need for adequate thermal protection during re-entry had eliminated the winged design from consideration. The missile principle, though, needed further work to man-rate it: a ballistic descent would impose significant duress on the vehicle and its occupants and Tikhonravov and Feoktistov moved instead toward the concept of a 'glancing' re-entry to reduce stress.

In 1962, the shape of the craft had evolved into something approximating a contemporary car's headlamp, which aerodynamic simulations predicted would avoid the high deceleration and thermal loads of a ballistic descent and have sufficient lift to be able to steer towards a given landing site. A plethora of proposals also surrounded the means of landing, with helicopter-like rotors, fan-jet or liquid-propelled engines, controlled parachutes, ejection seats and shock-absorbing inflatable balloons all being considered. By 1963, however, Korolev had approved the design which remains in use today: a combination of braking parachutes and a soft-landing apparatus of solid-fuelled rockets. Even as the descent module was taking shape, the appearance of the spacecraft remained somewhat fluid and early designs for a space station ferry and a lunar-going concept both utilised a descent module for the crew, attached to an instrument module for propulsion and power. Already, the design was expanding further to encompass a habitable orbital module and there was disagreement about where this should be located. In some initial drawings it appeared *between* the instrument module and the descent module and in others it was *above* the descent module. The idea of placing the orbital module below the descent module was soon rejected, since it would require cutting a hatch into the descent module's base, potentially compromising its heat shield. The final layout, with the descent module in the middle, was in place by the end of 1962. By this time, it had also received the name of 'Soyuz'.

In spite of Korolev's assertion that it was the machine of the future, Soyuz became mired for some years in technical and bureaucratic problems, to such an extent that by 1964 its development was virtually paralysed by the Soviet drive for the Moon. Early plans called for it to carry one or two cosmonauts, but by December 1963 the

162   **The last Soviet citizen**

2-02: First conceived under the genius of Chief Designer Sergei Korolev, Soyuz is the world's longest-serving piloted spacecraft. Since its first manned mission in April 1967, variants of the craft have carried human passengers on more than 100 discrete flights.

basic design of the Earth-circling version, known as 'Soyuz-7K-OK' ('Orbitalny Korabl' or 'Orbital Ship'), had grown to accommodate a three-man crew. Its purpose was to support automated rendezvous and docking, spacewalking, manoeuvring and scientific research, thereby fulfilling the key requirements for a space station ferry. During 1964, Korolev directed a small group under Boris Chertok, one of his deputies at OKB-1, to explore other uses for the basic 7K-OK craft. One proposal called for docking two Soyuz together in orbit to demonstrate their rendezvous capabilities and having a cosmonaut spacewalk from one ship to the other. Not only would this ambitious plan offer valuable engineering experience, but it also supported early ideas for a Soyuz-based Moon mission in which a cosmonaut would transfer from the command ship to the landing craft in lunar orbit by 'extravehicular activity' (EVA). In February 1965, Korolev presented this 'new' version of Soyuz to the Scientific-Technical Council of the State Committee for Defence Technology and was told to proceed.

Beginning at its base, the instrument module, also known as the 'service module', carried chemical batteries and two large solar panels to charge them, together with a thermo-regulation radiator and an integrated propulsion and attitude-control system. The latter, designated 'KTDU-35', comprised a pair of engines, one primary and one backup, sharing the same oxidiser and fuel supply. The primary engine had a thrust of 417 kg and was capable of a change in velocity of some 2,750 m/sec, equivalent to a specific impulse of around 280 seconds. On the basis of early reports, which speculated that this engine could boost Soyuz to an altitude of 1,300 km, Phillip Clark suggested that the spacecraft required a propellant capacity of 755 kg. Propellants took the form of unsymmetrical dimethyl hydrazine and an oxidiser of nitric acid, housed in spherical tanks within the instrument module. Attitude control came from 22 primary and eight backup hydrogen peroxide thrusters. Guidance, rendezvous, communications and environmental gear filled the remainder of the cylindrical compartment. The descent module sat directly above the instrument module and housed the crew during ascent and re-entry. It had a habitable volume of some 2.5 m^3. The commander's seat was located in the centre, flanked by positions for a flight engineer and a research cosmonaut or 'test' engineer. Many of Soyuz' flight regimes were pre-programmed from the ground. Consequently, the main instrument panel presented the crew with readouts and visual displays of the performance of on-board systems, together with a monitor for the external television camera, an optical periscope called 'Vzor' ('Visor') for attitude manoeuvres and the 'Globus' ('Globe') device to show the spacecraft's position above Earth. In the event of a failure of the automatic systems, and to facilitate rendezvous and docking, it was expected that the commander could assume manual control. As a result, two hand controllers (one for velocity, the other for attitude) were located directly underneath the instrument panel.

Rendezvous and docking were supported by the Vzor, together with a system of gyroscopes, attitude-control sensors and thrusters and the 'Igla' ('Needle') radar. The latter would automatically navigate the spacecraft to its target and draw to a halt at a range of 200-300 m, after which the crew would take charge and accomplish the final approach and docking. The systems to facilitate physical contact had undergone extensive development since 1962. At first, OKB-1 engineers Viktor Legostayev and Vladimir Syromiatnikov advocated a 'pin-cone' device to allow two vehicles to dock. At this stage, however, there was no provision for an internal transfer of cosmonauts from one craft to the other and, sometime in 1965, Korolev's proposal to change this was rejected by Feoktistov on the basis that a significant amount of work had already been done and additional revisions would put the development further behind schedule.

The descent module would be the only component capable of surviving the intense heat of atmospheric re-entry and bringing the cosmonauts back to Earth. At the end of a mission, the instrument and orbital modules would be jettisoned and the descent module would employ half a dozen hydrogen peroxide engines, each producing a thrust of 10 kg, to provide roll, pitch and yaw controllability during the early stages of re-entry. To protect its occupants, it was coated with a heat-resistant ablator, together with a thermal shield at its base that would detach shortly before touchdown to expose the four solid-propellant landing rockets. A 14 m^2 drogue parachute would deploy 9.5 km above the ground in order to stabilise the craft, prior to deploying the main canopy. If a problem occurred, a secondary canopy could be deployed. Seconds before touchdown, an altimeter would command the landing rockets to fire to cushion the impact. Atop the descent module in space, the spheroidal orbital module held a bunk, a cupboard for food and water, life-support gear, a latrine, controls for experiments, cameras and a variety of other equipment appropriate to each individual mission.

Difficulties aside, Soyuz promised to be one of the safest manned spacecraft ever built, possessing as it did the Soviets' first 'true' launch escape system. This consisted of a tower atop the R-7's payload shroud and a multiple-nozzle, solid-fuelled rocket engine. In the event of an emergency from 20 minutes before launch until 160 seconds into the ascent, the shroud would split at the base of the descent module and the escape tower's engine would lift the descent and orbital modules to safety. At the top of the arc, the descent module would be released to parachute back to Earth, landing a couple of kilometres from the pad, in the case of a pad abort. Early predictions estimated that the crew could be exposed to gravitational loads as high as 10 G during such a scenario.

Launching Soyuz, which was considerably heavier and more complex than the earlier Vostok craft, demanded further improvement of Korolev's original R-7 rocket. The basic design of the missile, physically, remained the same: a two-stage behemoth, fed by liquid oxygen and a refined form of kerosene known as 'Rocket Propellant-1' (RP-1). Strapped around its lower stage were four tapering boosters, each 19.6 m long. With the escape tower in place, the upgraded R-7 stood 49.3 m tall and was rolled to the launch pad horizontally on a railcar; a method still used today. Four cradling arms, known as the 'tulip', supported the booster and a pair of towering gantries provided pre-launch access.

Originally designed for a crew of three, Soyuz had effectively become a two-person vehicle after the deaths of cosmonauts Georgi Dobrovolski, Vladislav Volkov and Viktor Patsayev in June 1971. During the spacecraft's development, Korolev felt that wearing a pressurised space suit would be just as uncomfortable and impractical as wearing a wetsuit inside a submarine and opted to do away with them. It was a fateful decision, for during the re-entry of Soyuz 11 a pressure valve inadvertently opened and Dobrovolski, Volkov and Patsayev died when their air leaked out of the cabin. In the wake of the tragedy, the valves were modified and it was decreed that, in future, space suits would be worn for *all phases* of a mission in which depressurisation was a possibility. In response to this requirement, a 'Sokol-K' ('Space Falcon') suit would be tailored for each cosmonaut and was compatible with the seat liners aboard Soyuz. A prototype was completed within weeks of the disaster and by the spring of 1972 had been fully tested and signed off as flight-ready. Since the Soyuz 12 mission in September 1973, the suit and its descendents have been worn by every cosmonaut during launch, docking, undocking, re-entry and landing. "In the event of decompression," wrote Hall and Shayler, "the [Sokol] is automatically isolated from the cabin environment and supplied directly with either pure oxygen or an oxygen-rich mixture from a supply in the cabin or from self-contained systems." It included a soft helmet which could be pushed back over the head when not in use, a removable, white-topped 'skull-cap' for communications headgear and pressure-sealed gloves. The Sokol could also be used in the emergency transfer of cosmonauts from one spacecraft to another, with the aid of small hoses connected to the spacecraft's life-support system or through a portable backpack, although this has never been done. Testimony to its success is that, since 1971, no cosmonaut has lost his or her life through the decompression of their spacecraft; indeed, the hardware has proven so reliable that there have been no other instances of depressurisation, *at all*, aboard a Soyuz. However, in order for the suits to be properly accommodated in the cramped confines of the descent module, the third crew seat – that of the research cosmonaut or test engineer – was eliminated and its place taken by a system which could automatically pump air into the cabin in the event of decompression. Not until November 1980 and the arrival of an upgraded version of Soyuz would another three-man crew venture aloft.

This upgraded vehicle was known as 'Soyuz-T' ('Soyuz-Transport') and, although it looked much the same as its predecessors, its capabilities were greatly improved. It could survive, independently, for up to 14 days, *without* the full powering-down of its systems, and could be kept in 'orbital storage' for six full months. Its orbital module could also be left accessible during long-duration missions docked to a Salyut to provide a few extra cubic metres of storage space *and* could be jettisoned *prior* to the de-orbit burn, thereby allowing a ten-percent reduction in propellant to around 250 kg at the end of each mission. This allowed for the inclusion of a third (fully-suited) crew member or two cosmonauts and a hundred kilograms of cargo. As for the descent module itself, this component included improved window covers, capable of being jettisoned after re-entry to allow better views of parachute deployment, and the cosmonauts themselves would benefit from an improved Sokol suit, weighing only eight kilograms, which was lighter and considerably more flexible than earlier models. The escape tower atop the R-7 booster was also modified. It could be jettisoned 123 seconds after liftoff, rather than the previous 160 seconds, and the upgraded solid-fuel rocket meant that in an abort situation the orbital and descent modules would be pulled to a higher altitude, thereby enabling the *main* parachute – rather than the less reliable backup canopy – to deploy and bring the craft safely down to the ground. Six soft-landing rockets in the base of the Soyuz-T descent module replaced four in the previous model and, internally, the 'Chaika' ('Seagull') control systems incorporated a digital computer called 'Argon', cathode ray tube displays and lightweight circuitry. The new instrument module had been designed in a similar manner to that of the Progress cargo ship, with smaller attitude-control thrusters incorporated into the main propulsion system, so that both could draw their nitrogen tetroxide and unsymmetrical dimethyl hydrazine propellants from the same supply. Finally, two wing-like solar arrays spanned 10.6 m (slightly smaller than those of the original Soyuz) and generated a little over half a kilowatt of power.

Soyuz-TM ('Transport Modified'), aboard which Alexander Viktorenko and Alexander Serebrov rode to Mir, was thus the fourth generation in this spacecraft family. An unmanned version of the craft had flown in mid-1986, ahead of the first piloted mission, with Romanenko and Laveykin aboard Soyuz TM-2 in February 1987. Development of Soyuz-TM got underway with the signing of draft plans in April 1981 and most of its engineering blueprints were complete by the spring of the following year. Physically, it closely resembled its predecessor, with an independent lifetime in orbit of 14 days and an 'orbital storage' duration of about six months, but its systems were much improved. So too was its launch vehicle, which was equipped with lighter escape motor to enable an increased payload capacity in the region of 200 × 250 km into a 51.6-degree orbit. Rex Hall and Dave Shayler remarked in *Soyuz: A Universal Spacecraft* that, had Mir flown, as intended, into a 65-degree orbit, Soyuz-TM might have been unable to transport as many as three cosmonauts to the higher inclination. The changes made in 1984-85 effectively made Soyuz-TM "more flexible", Hall and Shayler wrote, "in executing its assigned mission, should the real-time situation require a deviation from the flight plan". Soyuz-TM's orbital module included an additional forward-facing window and rendezvous controls, whilst the landing system of the descent module featured improved parachutes with new synthetic fibres woven into shroud lines and lighter canopies.

Another of Soyuz-TM's capabilities, from which Viktorenko and Serebrov benefitted during their final approach to the space station early on 8 September 1989, was the new 'Kurs' ('Course') rendezvous mechanism. This accomplished radar contact with Mir from a distance of 200 km, rather than the standard 30 km achievable with the older Igla device. Soyuz TM-8 then acquired a final radar 'lock' on Mir at a distance of 20-30 km. The Kurs antenna was omni-directional, which offered additional flexibility, and the final approach could be completed by the spacecraft's on-board computer. Unfortunately, as the ship drew within 4 m of the station, the Kurs malfunctioned and Viktorenko was obliged to assume manual control. He withdrew to a distance of 20 m and proceeded to complete a flawless docking.

It was the start of an ambitious expedition, lasting more than five months, which included the arrival of the Kvant-2 module, the conduct of six spacewalks and a demonstration of an EVA aid, nicknamed 'the space motorbike'. In anticipation of Kvant-2, whose launch was scheduled for mid-October, Viktorenko and Serebrov spent their first few weeks installing equipment in Mir's docking mechanism to accept the new module. Then, on 10 October, the Soviets revealed that problems with a batch of faulty microcircuits from PO Electronica in Voronezh had forced a six-week postponement of the Kvant-2 launch. It would now fly no sooner than late November. In the meantime, the cosmonauts were hard at work unpacking the first of a new type of unmanned resupply craft, known as 'Progress-M'. The development of the Progress concept had been in the works since 1973 and the first draft plans were laid in February of the following year. Early Salyut space stations had to be launched with almost all of the cosmonauts' supplies aboard and, over time, they would also require periodic re-boosts to maintain their altitude against orbital decay ... re-boosts which demanded many hundreds of kilograms of propellant expenditure *every year*. "To maintain a 250 km orbit," wrote Phillip Clark, "would require [4,300 kg] of propellant. At 350 km, this requirement dropped to about 600 kg. Of course, the Soviets allowed the Salyuts to decay to lower orbits and then raised them again in large manoeuvres; this was more economical in terms of fuel expenditure. All in all, to maintain a fully-functioning Salyut in orbit for two years with a crew permanently on-board required about [18,000 kg] of consumables. It was ... impractical to launch all these supplies with the station – even if a Soviet booster had the necessary lifting power, which none had at the time." Progress was the obvious solution. It was modelled closely on Soyuz, with the orbital and 'descent' modules redesigned to house up to 1,300 kg of foodstuffs, water, experiments – including biological payloads which had to be operated within a short period of time – and up to 1,000 kg of propellant. For ease of removal, the lattice-like framework of cargo racks within the Progress could be unfastened with simple half-turn bolts.

The Progress was capable of docking automatically with the space station. Its orbital module carried two television cameras to permit flight controllers on Earth to observe the station as the procedure was executed. The descent module was replaced by a framework with four large tanks to carry several hundred kilograms of propellants for the station. These would be fed into the storage tanks in Salyut 6's aft compartment using an ingenious system of pipes which mated at the exterior of the docking collar. "The refuelling can be conducted either by the crew," explained Clark, "or automatically under the control of the ground. Once Progress has docked at the back of Salyut, the propellant lines and connections are checked for integrity. This done, a compressor slowly reduces the nitrogen pressure in the propellant tanks. The nitrogen is pumped back into its storage bottles ready for the refuelling operation itself. The fuel and oxidiser are transferred at different times for safety reasons." Firstly, the unsymmetrical dimethyl hydrazine would be fed through a system of pipes into Salyut, the lines would be purged with high-pressure nitrogen to prevent contamination or spillage, because the propellants are 'hypergolic', meaning that they ignite upon coming into contact. Next, the nitrogen tetroxide oxidiser would be transferred and the lines purged. In practical terms, as time would tell, the Soviets were able to proudly boast that from the first Progress mission in January 1978 the propellant transfer system never once failed. In their history of Soyuz, Rex Hall and Dave Shayler speculated that the name 'Progress' may have come from the implication that it suggested significant progress in space station operations, although its precise heritage is unclear.

By the summer of 1989, Progress had flown several dozen times to the Salyut 6 and 7 stations, as well as to Mir itself, but the launch of Progress M-1 (a 'Modernised' version of the craft) on 23 August marked the start of something quite different. For a start, the spacecraft docked with Mir's *forward* port, instead of the rear one, which implied that the complex possessed two sets of propellant-transfer mechanisms. "This produced far greater redundancy than on either Salyut 6 or Salyut 7," wrote Hall and Shayler, "and was a feature that was to become very useful throughout the long life of the station." The new vehicle also benefitted from a 100-

200 kg increase in cargo capacity to around 2,400 kg, whilst twin solar arrays – with a total span of 10.6 m from tip to tip – increased its independent flight time to 30 days and enabled it to remain docked with Mir for up to six months. Progress M-2 followed in December 1989, carrying among its payload a protein crystal growth experiment from a private US company, Payload Systems, Inc. Much speculation existed in the Western press about other capabilities of the 'new' Progress, with *Flight International* remarking in October 1989 that "a recoverable capsule", capable of returning up to 150 kg of materials to Earth, may form part of a future payload. This 'Raduga' ('Rainbow') capsule would make its first flight with Progress M-5, launched in September 1990. According to Rex Hall and Dave Shayler, the Raduga concept was approved in July 1988 and one of its primary aims was to ignite future commercial opportunities in space. It was carried aboard the Progress' dry cargo compartment and was typically used to return exposed camera films, magnetic data tapes and samples from materials processing furnaces. It was intended to be flown two or three times per annum. The capsule measured 1.4 m long and 78 cm in diameter and weighed 350 kg, including its return payload. It would be attached to the docking system of the cargo module and ejected from the de-orbiting Progress-M at an altitude of around 120 km to re-enter the atmosphere autonomously and parachute to a touchdown on Soviet soil.

By now, the Kvant-2 module had also docked, giving the station an unusual L-shaped configuration. It was launched from Tyuratam atop a heavy-lift Proton booster on 26 November and was scheduled to arrive at Mir on 2 December, but problems were experienced unfurling one of its electricity-generating solar arrays. By rolling Kvant-2 and rotating the 'flopping' array, simultaneously, ground controllers succeeded in fully deploying and locking it into position. Then, as the module closed to within 20 m of Mir, its Kurs rendezvous system sensed that it was approaching its quarry too rapidly and called an abort. Nor were the difficulties peculiar to Kvant-2; aboard the older Kvant-1 module, attached to the aft end of Mir, a computer error had caused the attitude-control gyrodynes to shut down. At length, the glitches were corrected and Kvant-2 successfully docked at the forward axial port at 3:21 pm Moscow Time on 6 December, whereupon Mir's mechanically-driven manipulator arm – the short and stubby 'Ljappa' – grappled it, pivoted it by 90 degrees and repositioned it on the 8th to its permanent location on the 'dorsal' port of the multiple adaptor. The procedure took about an hour to complete. "The arm," wrote Tim Furniss in *Flight International* in April 1989, "is attached to the module and rotates and grasps an attachment ... The module then undocks, moves away a very short distance and is rotated around to a side port for hard docking – not by power from the arm, but by very careful thrusting by Mir to pitch the module correctly."

Four days later, on 12 December, Viktorenko and Serebrov boarded Soyuz TM-8 and relocated it from the aft port of the Kvant-1 module to the front port of Mir's multiple docking adaptor, thus freeing up the former to accept future Progress visitors. They also began working inside Kvant-2, previously known as 'the D-module', for *Dushnashcheniye*, or 'augmentation'. Measuring 13.7 m in length, with three pressurised sections and a total habitable volume of 60 m^3, Kvant-2 was described by the Soviets as a "utility module" and perhaps its most notable element was the large EVA airlock at the far end, with a one-metre-diameter specialised hatch that hinged outward. A set of star sensors and six gyrodynes were designed to reduce propellant expenditure for the orientation of Mir. It also ferried the new 'Salyut-5B' computer to replace Mir's original 'Argon-16'. One of the new module's key aims was to improve living conditions aboard Mir, with the Elektron electrolysis system to provide oxygen from recycled water. It also incorporated the 'Vozdukh' apparatus to scrub carbon dioxide from the cabin's air, together with a pair of water-regeneration devices and sanitation and shower facilities. It also ferried a large amount of equipment into orbit: East Germany's MKF-6MA Earth resources camera, capable of imaging across six spectral bands, the KAP-350 topographic camera and the ARIZ X-ray spectrometer, together with a handful of spectrometers, television cameras, cosmic dust detectors, a quail-egg incubator ... and a peculiar EVA contraption known as the *Sredstvo Peredvizheniy Kosmonavtov* (SPK, the 'Cosmonaut Manoeuvring Equipment'). This was of similar purpose to NASA's Manned Manoeuvring Unit, a jet-propelled space suit backpack used on three Shuttle missions in 1984, and in addition to its testing Viktorenko and Serebrov were assigned an intricate series of five EVAs to fully integrate Kvant-2 into the Mir complex and prepare it for the arrival of the next module, named 'Kristall' ('Crystal').

Preparations for the first EVA got underway in late December and on 8 January 1990 the two cosmonauts poked their helmeted heads into the vacuum of space, during orbital darkness, kicking off an excursion which would run for two hours and 56 minutes. Their suits were known as 'Orlan' ('Sea Eagle') and marked the third generation of pressurised ensembles, first trialled outside the Salyut 6 station in December 1977. Built by the Zvezda design bureau, the Orlan consisted of flexible limbs and a one-piece, rigid body-helmet unit. The cosmonaut entered the suit through a hatch in the rear of the torso and, for Salyut 6 operations, it had a maximum operating time of around three hours. For Mir, this increased to nine hours. In its earliest guise, the Orlan's integrated design meant that it did not need external oxygen hoses and its operating pressure required a relatively short pre-breathing period of only 30 minutes. Electrical power came from umbilicals connected to the station and the cosmonauts controlled the function of the suit from a chest panel. The distance to which a cosmonaut could venture beyond the hatch was therefore limited by the length of his umbilical. The Orlan received significant modification over the years. By the Salyut 7 era, in the early 1980s, its 'DM' configuration featured bright lights at the headset temples for illuminating suit controls and benefitted from a generally sturdier construction, with rubberised fabric shoulder belts, and afforded greater flexion for the wearer. The 105 kg 'DMA' version, used aboard Mir, was first trialled by Soyuz TM-4 cosmonauts Vladimir Titov and Musa Manarov in October 1988. Its improvements included lighter, more flexible and tougher composite fabrics in the arms and legs – which could be removed from the suit for repair or replacement – as well as having more durable electrical motors and more dexterous gloves. The Orlan-DMA also carried an inflatable forearm cuff, which used air from the suit's backup oxygen tank to

170 The last Soviet citizen

2-03: Cosmonauts at work outside Mir's Kvant-1 module, clad in Orlan space suits.

respond to a glove puncture; in essence, this sealed off the glove until the cosmonaut could return to the airlock. Its integral backpack measured 1.2 m long and 48 cm wide and operated at pressures of 26.2-40 kPal. Like its predecessors, the Orlan-D of the Salyut 6 era and the Orlan-DM of the Salyut 7 era, the DMA possessed dual polyurethane rubber pressure bladders, an integrated 'Korona' voice communication system with dual microphones and earphones and primary and backup transceivers and amplifiers. The Korona antenna was embedded within the DMA's outer fabric layer.

Viktorenko and Serebrov's primary objective on 8 January 1990 was to install a pair of 80 kg star trackers onto Kvant-1. Although this was successfully accomplished, their pressurised garments presented their own minor irritations: a broken wire in Viktorenko's suit prevented water temperature monitoring, whilst Serebrov suffered a leak in his suit's coolant loop. The two men also retrieved a number of materials specimens from the exterior of the station. Three days later, on the 11th, they were back outside for almost three hours, using Mir's transfer compartment as an airlock for what would turn out to be the last time until 1995 and using the Orlan-DMA suit with a power umbilical for the final time. They retrieved an experiment which had been left outside the station by a French astronaut in

December 1988, removed the supports for the French-built ERA assembly and installed a number of exposure 'cassettes' laden with non-metallic material samples, along with a device to measure Earth's ionosphere and magnetosphere. Finally, the cosmonauts returned to the depressurised transfer compartment and moved the Konus-2 drogue from the +Y port (where Kvant-2 was docked) to the opposing -Y port, in readiness for Kristall. After the arrival of the new module, and in the wake of its 'hard' docking and activation of connecting latches, Konus-2 would be removed, again in preparation for the arrival of the third module. (The Konus-1 drogue remained at Mir's forward -X port).

Viktorenko and Serebrov's third EVA, on 26 January, was devoted to testing the Orlan-DMA space suit and preparing Mir for the SPK backpack evaluations. The men again spent three hours outside and saw the cosmonauts demonstrating an add-on package to supply power, telemetry and cooling to their suits, thus rendering obsolete the umbilicals used on previous Soviet excursions. This was perhaps the DMA's most significant improvement, since it provided a degree of suit autonomy for the first time. The men were linked to Mir only by safety tethers and, for the first time, departed the station not through the multiple docking adaptor, but through Kvant-2's new airlock, whose large-diameter hatch provided passage into space for the SPK backpack. Looking like an overstuffed and overly padded armchair, this device was named in honour of the legendary Daedalus' foolhardy son, Icarus, whose wings of feathers and wax melted as he drew too close to the Sun. Fortunately, nothing of the sort transpired to affect Alexander Serebrov during the mission's fourth EVA, on 1 February 1990, as he put Icarus' spacefaring namesake through its inaugural paces.

The 218 kg SPK was, like the Orlan suit itself, built by the Zvezda design bureau. It was propelled by 32 compressed-air thrusters, half of them 'primary' and half 'redundant'. Prior to its first test, the cosmonauts had installed a recovery winch and mooring post for the manoeuvring unit outside Kvant-2, removed the new module's obstructing Kurs antenna and installed a pair of exposure cassettes and a camera onto the hull. Serebrov remained tethered to Mir by a 60 m tether for the SPK tests, because the station could not manoeuvre to recover him in the event of a failure. He started by gingerly making a trio of short flights to a distance of around 5 m and then out to 33 m, before returning. During his final exercises, he realised that he was approaching the Kvant-2 'dock' somewhat off-course and, although he was able to correct the problem, he noticed that the tether caused him to flip backwards and rock "like a pendulum". Four days later, on 5 February, a second SPK run was completed in a dramatic EVA which ran to almost four hours. After installing the Spin-6000 device to the front of the SPK's belly band to measure the radiation outside Mir, Viktorenko drove the manoeuvring unit to a distance of 45 m and completed an 'aerobatic' roll, covering a total of about 200 m. Viktorenko needed helped from Serebrov to redock, because the Spin-6000 blocked his field of view. Although the SPK remained attached to the external Kvant-2 dock for several years, and was kept ready to perform external inspections of Mir, it was never used again.

The quintet of EVAs, and the SPK trials, together with the arrival of commissioning of Kvant-2, had proven the highlight of Viktorenko and Serebrov's residency aboard the station, but the third new module, Kristall, was expected to arrive in mid-1990, during the six-month mission of the Soyuz TM-9 crew. The arrival of the module would transform Mir's shape again from an 'L' into a 'T', with Kvant-2 and Kristall on opposite sides of the multiple docking adaptor at the forward end of the station. The timing of Kristall's arrival was also critical and in July 1989 Chief Designer Yuri Semenov of the Energia bureau was quoted by *Flight International* as noting that a gap of a few months was the maximum period of time allowable for the 'asymmetry' of the station.

In command of Soyuz TM-9 was a man who had waited for more than a decade for his first chance to venture into space; a man who would go on to record five missions and 651 days in orbit; a man who would narrowly, and by his own hand, miss out on commanding the first crew of the International Space Station; and a man who, even today, has spent more time spacewalking – over 82 hours in 16 excursions – than any other human being. That man was Anatoli Yakovlevich Solovyov, born on 16 January 1948 in Riga, today's capital of Latvia. He completed the Lenin Komsomol Chernigov Higher Military Aviation School in 1972 and served until 1976 as a senior pilot and group commander in the Far Eastern Military District. Selected as a cosmonaut candidate in August 1976, Solovyov commenced more than two years of evaluation and training and served as backup commander for the Soyuz TM-3 Syrian mission. His military experience and upbringing had imbued him with a staunch support for Communism – he had been a party member since 1971 – and, during the Shuttle-Mir effort with the United States in the 1990s, it has been remarked that he was "not especially friendly" towards the Americans and gave NASA reason to suspect that he was "a stern critic of the two countries' collaboration". Bryan Burrough described him as "the Chuck Yeager of the Russian programme ... there is a bit of the Old Soviet in Solovyov". If Alexander Viktorenko was one of the most 'Americanised' of the Soviet commanders, in terms of his personable nature, then Solovyov was the reverse. "He gave *orders*," wrote Burrough, "harsh orders, often shouted." Having said this, NASA astronaut Dave Wolf, who flew aboard Mir with him in 1997-98, praised Solovyov's stance and felt that a more 'patronising' tone would not have made him feel like an integrated member of the team. Yet Solovyov's credentials stand for themselves and do not just extend to EVA: at the time of writing, in early 2013, he holds sixth place on the list of most experienced spacefarers in history, is one of only half a dozen cosmonauts to have completed five missions and is one of only two cosmonauts to have commanded all five of his flights.

For Solovyov's civilian flight engineer, Alexander Nikolayevich Balandin, Soyuz TM-9 would be his first and only space mission. Born in Fryazino, a 'science town' (or *Naukograd*), near Moscow, on 20 July 1953, he completed high school with an engineering diploma and was selected into the cosmonaut corps – alongside Alexander Viktorenko and Alexander Serebrov – in December 1978. During his early career, Balandin worked on the development of the Soviet Union's Buran shuttle programme and at one stage formed part of the backup team for the reusable craft's first two-man flight, originally scheduled for late 1991. Despite Soviet assertions that Buran, which undertook an unmanned inaugural voyage in November 1988, would fly with a crew to Mir, such plans never reached fruition and were ultimately cancelled, due to budget cuts. Balandin shifted over to Soyuz-TM training and, with Solovyov, fulfilled backup duties for Viktorenko and Serebrov, ahead of their own launch.

Perhaps indicative of the steady thaw in relations between the United States and the Soviet Union, a group of NASA astronauts – including Dan Brandenstein, then-chief of the corps, and Jerry Ross – travelled to the Tyuratam launch site to watch Soyuz TM-9 spear for the heavens at 9:16 am Moscow Time on 11 February 1990. The rousing ascent was soon followed by disturbing news when Solovyov and Balandin reached orbit, for three of their descent module's eight thermal insulation blankets had come loose at the end abutting the heat shield. They appeared to be 'flapping' as the Soyuz commenced its first manoeuvres in space ... and *this* carried the risk of them obstructing optical and infrared navigation sensors which needed to align the craft for re-entry. Moreover, the loose blankets could cause the Soyuz to either overheat or cool excessively and any resulting condensation could trigger an electrical short. As Soyuz TM-9 approached Mir, engineers on the ground worked feverishly to lay out procedures to accommodate the problem and repair it. In the meantime, Solovyov and Balandin docked successfully at the rear Kvant-1 port on 13 February and were greeted with bear hugs from the resident cosmonauts, Viktorenko and Serebrov.

The issue with Soyuz TM-9's thermal blankets raised further eyebrows in the days and weeks to come. It was feared that explosive bolts binding the instrument module to the descent module might fail after long-term exposure to the harsh space environment or that the heat shield – vital to ensuring Solovyov and Balandin's survival during their fiery return to Earth, six months hence – might itself have been compromised. An EVA to tend to the problem might cause further damage. Soyuz TM-10 was scheduled to launch in early August and for a time consideration was given to launching that craft with just a single cosmonaut, reserving the additional two seats to 'rescue' Solovyov and Balandin if their own ship was declared unsafe. In the meantime, on 19 February, Viktorenko and Serebrov boarded Soyuz TM-8 in order to return to Earth. After undocking, they performed a fly-around inspection of Solovyov and Balandin's craft and reported that one of the loose blankets projected at a 90-degree angle and the others at about 60 degrees. Viktorenko likened them to "petals of a flower". A few hours later, Soyuz TM-8 touched down safely at 7:36 am Moscow Time after a 166-day mission, leaving the new crew alone for six months.

Early analysis suggested that the damage to the thermal blankets probably occurred around 165 seconds into Soyuz TM-9's ascent, as the aerodynamic shroud separated from the rocket at an altitude of some 80 km. The photographs taken by Viktorenko and Serebrov enabled engineers to determine that an EVA might be able to either reattach the blankets or, in the worst-case situation, cut them off. Since the tools to complete the work had to be fabricated and tested on the ground, and procedures developed by fellow cosmonauts in the 'hydrolab' at the Star City training centre, near Moscow, it seemed likely that an EVA would occur late in the mission. The equipment and training videotapes for the cosmonauts – who *had* received EVA training, but not for a spacewalk of this complexity – would be launched inside the Kristall module. In the meantime, Mir was manoeuvred into an attitude which alternately placed Soyuz TM-9 in or out of direct sunlight in order to maintain an adequate thermal environment.

Solovyov and Balandin's expedition featured the arrival of two Progress freighters – one of the newer 'M' series, launched in late February, and the last of the older-generation series, launched in early May. (Interestingly, the latter demonstrated an ejection seat for the Buran shuttle, drop-testing it from a compartment installed in place of the rocket's escape tower engine.) Progress deliveries aside, the major new component for Mir which arrived during the Soyuz TM-9 residency was the Kristall module, which blasted off from Tyuratam atop a Proton booster at 1:33 pm Moscow Time on 31 May 1990. The launch came after lengthy delays, having been originally scheduled for December 1989, then late March 1990, and then mid-April, before being postponed a further six weeks for further efforts to turn over control of Mir from the old Argon-16 computers to the new Salyut-5B hardware. Docking with Mir was scheduled for six days later, but was delayed due to a problem with one of the new module's orientation engines; Kristall was finally secured to the front port of the multiple docking adaptor at 3:47 pm on 10 June. Next day, it was moved by the Ljappa arm to its new position on the radial port, opposite Kvant-2, and Solovyov and Balandin began work inside the module on the 15th. It was expected to be a significant step forward in the steady Soviet commercialisation of Mir. According to *Flight International*, the 100 kg of space-manufactured products from Kristall during the five-month Soyuz TM-9 expedition was expected to yield a profit of 25 million roubles (around $42 million); covering more than half of the cost of Solovyov and Balandin's entire mission. In November 1990, Deputy Flight Director Viktor Blagov would be quoted as stating a possible 97-million-rouble revenue stream from Soyuz TM-9, although it was stressed that this remained a hypothetical figure as none of the Kristall-produced photography or materials samples had been sold on the international market.

Like Kvant-2, the new Kristall module weighed a little under 20,000 kg and measured approximately 11.9 m in length. Before launch, it was nicknamed 'the T-module', for *Teknologia*, in reference to its scientific aims. Its purpose was described by the Soviets as a research facility for materials processing and biological experiments and its two compartments offered a total pressurised volume of around 60 m^3. The instrument-payload compartment housed food containers, a 'hot-house' for growing radishes and leaf lettuce, the Aniur electrophoresis device and the Krater-3, Optizon-1, Zona-2 and Zona-3 semiconductor processing equipment. A 0.8 m hatch at the far end led to a junction-docking section with its own spherical docking mechanism. The latter was equipped with two androgynous docking assemblies, one mounted axially and the other radially. The axial unit had an androgynous docking system and was intended for Buran. (It would later be used by the US Shuttle in the late 1990s, but the radial port was never employed.) The docking section also contained the Priroda-5 Earth-resources camera. Kristall's twin solar arrays spanned 36 m and produced a total power output of up to 8.4 kilowatts. Unlike the station's own arrays, the 500 kg Kristall panels could be folded or

2-04: The cramped and cluttered interior of Mir, as seen in 1996, during one of the Shuttle rendezvous and docking missions.

unfolded to suit requirements and one of them would be relocated to Kvant-1 in the spring of 1995, ahead of the first Shuttle-Mir docking mission.

By mid-July 1990, less than a month before the scheduled return of Solovyov and Balandin to Earth, plans finally crystallised for an EVA to attempt to repair the damaged thermal blanketing on the exterior of Soyuz TM-9. The situation had already prompted the Western media to speculate that the cosmonauts were 'marooned' in space; an assertion rightly denied by the Soviets, who possessed the means to dispatch a 'rescue' Soyuz if needed. At length, on the 17th, the cosmonauts were ready to depart the Kvant-2 airlock, laden with ladders and tools. They had trained for the repair by videotape sent up from the ground aboard Progress and also by watching practice sessions from the Star City hydrolab. To make Soyuz TM-9 itself more easily reachable for the repairs, on 3 July they transferred it from the rear of Mir to the front port of the multiple docking adaptor. During the process to open the outer hatch, it would appear that they violated one of their egress procedures. They should have turned a handwheel until a slit, just 1-2 mm wide, opened around the lip of the hatch to allow residual air to escape. Furthermore, they should have ensured that the airlock was in vacuum with a hand-held measuring device *before* turning the handwheel any further to release retaining hooks and allow the hatch to swing open. It appears that they turned the handwheel too far, prematurely releasing the hooks. The air pressure in the chamber was 5 kPal and it caused the hatch to spring back against its hinges with a force of around 400 kg. "The implications of this error," wrote Rex Hall and Dave Shayler, "were not immediately apparent."

However, the need to attend to it would lead to *another* contingency EVA, later in the mission.

Departing Kvant-2, Balandin traversed the entire 13.7 m length of the module, hand over hand, with clamps and long and short tethers, which required almost 90 minutes. (After the mission, the duo quipped that they had *plenty* of hand holds, but needed a few more *street signs*!) As Mir entered orbital darkness, they rested briefly, and it was not until three hours into the EVA that they finally managed to erect the 7 m ladder across the gap between Kvant-2 and the Soyuz TM-9 descent module at the forward port of Mir's multiple docking adaptor, spanning the orbital module on which there were no hand holds. A 'curved' ladder was also installed to enable them to reach the Soyuz' heat shield. Television camera cables were too short to reach the work site, which meant that tense mission controllers could not see the cosmonauts. Still, the reports from Solovyov and Balandin were encouraging: there was no obvious damage to the explosive separation bolts or to the heat shield itself, which spelled good news for the impending re-entry. However, because the flapping thermal blankets had *shrunk* and could not be reattached to the locking ring, the cosmonauts were obliged to fold two of them in half and leave the third – the most severely torn – alone. By now, they were well behind schedule and hastened back to the Kvant-2 airlock in darkness by the inexorable ticking of the clock. The EVA had now reached the five-hour mark and although the Orlan-DMA could sustain a lengthy excursion, it was preferred for safety reasons not to use it for much beyond six hours. When the cosmonauts were back in the airlock, the six-hour safety limit had passed ... and now they hit another obstacle: the hatch *would not close*. Time was running out and they were forced to leave the outer hatch slightly ajar and use Kvant-2's inner compartment as a contingency airlock. This left the airlock proper exposed to vacuum. By the time the spacewalk ended, Solovyov and Balandin had been outside for seven hours and 16 minutes, securing a record for the longest Soviet or Russian EVA to date.

If the need to repair Soyuz TM-9 could hardly have been predicted in the days and weeks before the cosmonauts' 11 February launch, then neither could their next challenge, which was to attempt to repair the damaged Kvant-2 airlock hatch and retrieve the ladders and tools so that they would not block the docking port when the spacecraft departed. The sheer duration of the first EVA, and the anticipated demands of the second, led a Soviet State Commission to convene in order to specifically authorise Solovyov and Balandin to remain outside for up to nine hours, if needed. On 25 July, the two men set off outside for what would fortunately require only 3.5 hours, but which quickly highlighted the severity of the problem. They televised images of the Kvant-2 damage and clearly showed mission controllers that one hinge was deformed. Despite difficulty in gaining sufficient leverage, they managed to force the hatch closed, then repressurised the airlock and the instrument compartment and sealed themselves inside the latter. After a day or so, the external hatch showed no indications of a leak and the cosmonauts were given permission to reopen the inner hatch to the airlock.

A full repair was still needed, but not during Solovyov and Balandin's mission. A week after their second dramatic EVA, the Soyuz TM-10 spacecraft arrived at the aft port of the Kvant-1 module, bearing the next resident crew, Gennadi Manakov and Gennadi Strekalov. They had launched from Tyuratam at 12:32 pm Moscow Time on 1 August and arrived at the station two days later. *Flight International* suggested in July that the core crew "may be joined by a cosmonaut from Lithuania", citing the name of Rimantas Stankyavichus, although this never transpired. Stankyavichus was a Buran shuttle pilot trainee and was widely expected to fly the reusable vehicle's first piloted mission, but he was tragically killed in an aircraft accident just a few weeks later in September 1990. Original plans called for the Buran pilots to fly 'space familiarisation' missions aboard Soyuz in the first instance, as Igor Volk had done to Salyut 7 in July 1984 and Anatoli Levchenko to Mir in December 1987. Other possible 'third-seaters' for Soyuz TM-10 included a Soviet journalist, for which two candidates – Svetlana Omelchenko of *Air Transport* and Pavel Mukhortov of *Soviet Youth* – had been selected for training in May 1990. As circumstances transpired, no third-seat option was taken on Soyuz TM-10 and neither the Lithuanian cosmonaut, nor the Soviet journalist, ever flew.

In spite of the disappointment over the Kvant-2 hatch damage, cosmonauts and mission controllers were exultant that Soyuz TM-9 was safe to return to Earth and the scientific and commercial accomplishments of the six-month mission had been notable, too, with a protein crystal growth experiment described as generating a profit of 25 million roubles. Finally, on 9 August, after 179 days aloft, the cosmonauts prepared for re-entry. The orbital and service modules were jettisoned simultaneously, in an effort to minimise the chances of the thermal blankets becoming snagged, but the return to Earth proved entirely nominal and the Soyuz TM-9 descent module thumped onto the desolate steppe of Kazakhstan, 72 km north-east of the town of Arkalyk, at 10:35 am Moscow Time.

Aboard Mir, Gennadi Mikhailovich Strekalov was the 'old man', both in terms of age and prior experience, aboard the new Soyuz TM-10 mission. He came from Mytishchi, a major industrial hub, situated to the north-east of Moscow, where he was born on 26 October 1940, and judging from the nature of his birthplace it is perhaps not surprising that he forged a career for himself in science and engineering. His father was killed in 1945, during the Red Army's liberation of Poland, only weeks before the end of the bloody conflict with Nazi Germany. The young Strekalov completed his schooling and became an apprentice coppersmith, before enrolling at the prestigious Bauman Moscow Higher Technical School. He received his engineer's diploma in 1965 and moved directly to work for the Moscow-based organisation which evolved from OKB-1 into TsKBEM and eventually Energia, helping with the design of Soyuz. Strekalov was chosen as a civilian cosmonaut in March 1973 and within months began formal training. His first crew assignment was as backup flight engineer on the Soyuz 22 mission in September 1976. Two years later, he recommenced flight training and at the end of 1979 was teamed with Vasili Lazarev for the 13-day Soyuz T-3 mission to Salyut 6. Six months later, following changes to that mission, he was reassigned as backup flight engineer to Konstantin Feoktistov. When Feoktistov was grounded in October 1980, Strekalov found himself back on the prime crew. In the wake of this first flight, he trained for a long-duration Salyut 7 expedition, but was twice thwarted: in April 1983, aboard Soyuz T-8, his crew unsuccessfully attempted to dock with the space station, and the following September he and comrade Vladimir Titov narrowly escaped death when their rocket exploded on the launch pad. Surprisingly, within months of this near-disaster, in April 1984, Strekalov flew into space a third time. After a decade of waiting, his four-month mission to Mir in August-December 1990 – and his fourth space voyage overall – would yield his first taste of long-duration space flight.

Joining Strekalov aboard Soyuz TM-10 was Gennadi Mikhailovich Manakov, who was born in Yefimovka in the Orenburg Oblast on 1 June 1950. A lengthy career in the Soviet Air Force, together with test-piloting credentials, led Manakov to selection as a member of the Buran shuttle group in September 1985, alongside Viktor Afanasyev and Anatoli Artsebarski, who would follow him into orbit as the commanders of Soyuz TM-11 and TM-12. Manakov transferred to the Star City cosmonaut detachment in January 1988 and served with Strekalov as the backup crew for Soyuz TM-9.

Manakov and Strekalov were not the only 'crew' members of the new Mir mission, for they had carried with them into orbit four live Japanese quails, which were transferred to the Kvant-2 incubator shortly after docking. In fact, one of the quails laid an egg *en-route* to Mir. The expedition of Soyuz TM-10 was relatively quiet and uneventful and many sources agree that the cosmonauts' experiment programme was relatively modest, though somewhat varied in scope, with work in geophysics and astrophysics, biology and biotechnology and materials investigations undertaken. Manakov and Strekalov completed an emergency drill to handle situations which might arise aboard the enlarged space station, connected Kristall's attitude-control systems to the rest of Mir and received a pair of Progress visitors. Of the latter, one carried equipment for the forthcoming visit of a Japanese journalist aboard Soyuz TM-11 and the other carried the first Raduga sample-return capsule, capable of bringing up to 150 kg of payload back to Earth. Unfortunately, when the Raduga commenced its return home on 28 November, heading for a touchdown in Kazakhstan, its beacon signal was lost during re-entry and it presumably was destroyed in the atmosphere.

An EVA to examine and tend to the damaged Kvant-2 hatch was scheduled for 19 October, but was slightly delayed when Strekalov caught a cold. Eventually, on the 29th, the cosmonauts floated into open space and utilised a special tool to remove insulation from the outside of the hatch ... only to reveal that the buckled hinge was beyond their ability to repair. However, it was later noted by *Flight International* that mission managers did not regard the hatch problem as serious, "since it involves damaged hinges which merely make it difficult to close completely". Manakov and Strekalov installed a latch to ensure adequate closure and returned inside Mir. Another EVA by the two men to prepare for the transfer of solar arrays from Kristall to Kvant-1, where they would be shaded less frequently, was postponed until after the replacement of the Kvant-2 hinge. At the age of 50, Strekalov became the oldest Soviet spacewalker, an achievement he shared with Frenchman Jean-Loup Chrétien; the pair would jointly hold the record until Story Musgrave performed three EVAs in December 1993 at the age of 58.

As the year drew to a close, Mir had returned to a regular pattern of crews rotating in and out of the complex at approximately six-monthly intervals and at 11:13 am Moscow Time on 2 December 1990 Soyuz TM-11 roared into orbit, carrying two Soviet cosmonauts and a Japanese journalist, Toyohiro Akiyama, who became the first citizen of his nation to enter space. He was born on 22 July 1942 in Setagaya, one of the 23 wards of the Tokyo Prefecture, and earned his degree from the International Christian University in Mitaka. Akiyama joined Tokyo Broadcasting System (TBS) in 1966 as a journalist and later worked for the BBC World Service, before returning to TBS as a foreign news correspondent and as the chief correspondent in Washington, DC. In the late 1980s, TBS decided to fund a mission of one of its journalists to Mir and Soyuz TM-11 became the first commercial flight to the Soviet station; the television network reportedly paid a $28 million fee. During the mission – variously dubbed 'Kosmoreporter' or 'Space Journalist' – the Japanese astronaut would make a ten-minute televised broadcast and a pair of 20-minute radio broadcasts each day. Much of the electrical power demands of the video and television equipment, which weighed a total of 170 kg, were incompatible with Mir and obliged TBS to make extensive use of convertors. In August 1989, Akiyama and a young TBS camera operator, Ryoko Kikuchi, were selected to attend the cosmonauts' training centre in Star City for a year of preparations.

Britain's first astronaut, Helen Sharman, worked closely with both Akiyama and Kikuchi during their training together and remembered them both well. In February 1990, she performed her first parachute jump, immediately after Kikuchi. "I had met Toyohiro in Star City," Sharman wrote in her autobiography, *Seize the Moment*, "and he told me that before starting the training he was completely unfit: he was a two-packs-a-day smoker and the heaviest thing he had ever lifted was a pen. He had gone through the training ... and at the end of it he emerged as a fully-capable cosmonaut, fit as a fiddle, part of a professional team."

Whilst it was hardly peculiar for a Japanese to enter space, it came as something of a surprise that Akiyama represented a news corporation and was not a 'professional' astronaut. Until October 2003, Japan's space programme was nominally the responsibility of three separate government organisations – the National Space Development Agency (NASDA), the Institute of Space and Astronautical Science (ISAS) and the National Aerospace Laboratory (NAL) – and these had contributed enormously to the island nation's position in the space exploration field since the 1960s. Under their auspices, the first Japanese satellite, Osumi, had been launched in February 1970, making them only the fourth discrete nation to launch its own satellite on its own rocket. More recently, the joint US-Japanese Spacelab-J Shuttle mission had been planned and early work on the expansive Japanese Experiment Module (JEM) for today's International Space Station had gotten underway. In 1985, Mamoru Mohri, Takao Doi and Chiaki Naito had been selected as astronaut candidates to support Spacelab-J, originally scheduled for mid-1988, but extensively delayed by the Challenger tragedy. The mission eventually took place in September 1992, offering the Soviets a neat coup of flying the first Japanese citizen into orbit.

Shoulder to shoulder with Akiyama aboard Soyuz TM-11 were Viktor Afanasyev, the commander, making his first flight, and Musa Manarov, the flight engineer, making his second. Viktor Mikhailovich Afanasyev was born on 31 December 1948 in Bryansk, some 400 km south-west of Moscow, and entered the Soviet Air Force shortly after graduating from Kachynskoye Military Pilot School. He spent six years as a pilot, senior pilot and aircraft flight commander and in 1976 attended the Test Pilot Training Centre, subsequently serving as a test pilot and senior test pilot and receiving a Class 1 military test pilot certification. In 1980, he graduated from Moscow's Ordzhonikidze Aviation Institute. Several years later, Afanasyev began basic cosmonaut training at Star City, proceeded to advanced training in 1988 and by 1989 had been assigned as backup commander for Soyuz TM-10. In his book *Dragonfly*, Bryan Burrough originated an informal cosmonaut tradition with Afanasyev: the checking of all household appliances, before launch. "The joke," wrote Burrough, "and there is some truth to it, is the moment a cosmonaut blasts into space, his family's apartment begins falling into disrepair." Before each of his launches, Afanasyev checked the refrigerator, the stove, the stereo, the lamp bulbs, the radiators and the television. It was a tradition followed by several subsequent Mir crews.

Musa Khiramanovich Manarov – the spelling of his first name has appeared, variously, as 'Musakhi', 'Musat' or 'Musachi' – had previously been part of the Soyuz TM-4 crew which completed the world's first year-long manned space mission between December 1987 and December 1988. Born on the southern shore of the Absheron Peninsula, in the industrial hub of Baku, it is hardly surprising that Manarov grew up with a keen engineering mind, a voracious interest in aviation and an ambitious outlook on the world around him; for Baku stands as the political, scientific, cultural and industrial centre of today's independent Azerbaijan. (At the time of writing, this impressive capital city had failed in its bid to host the 2020 Summer Olympics, but was expected to advance a new bid for the 2024 event.) Manarov was born here on 22 March 1951, at a time when Azerbaijan was a republic of the Soviet Union. He earned an engineering diploma from Moscow Aviation Institute in 1974, entered the Soviet Air Force – eventually rising to the rank of colonel in the reserves – and was selected as a cosmonaut candidate in December 1978. It would appear that Manarov was originally teamed with Romanenko on the Soyuz TM-2 backup crew in September 1985, but was temporarily removed from flight status due to an undisclosed medical issue and replaced by Laveykin. By March 1987, Manarov was apparently back on active status and was paired with Vladimir Titov for Soyuz TM-4. At the end of his second mission, in late May 1991, Manarov became the world's most experienced spacefarer, with more than 540 cumulative days under his belt. It was a record that he would hold for more than three years.

Arriving at Mir on the morning of 4 December 1990, Viktor Afanasyev guided Soyuz TM-11 smoothly to a docking with the aft port of the station's Kvant-1 module, thus kicking off a week of joint work with Gennadi Manakov and Gennadi Strekalov and a flight with a distinctly Japanese flavour. During the course of that week, Akiyama's custom-moulded couch was transferred from the launch craft over to Soyuz TM-10, in which he would return to Earth with Manakov and Strekalov on

10 December. The cosmonauts also began loading the descent module with equipment and camera films. At length, in the early hours of landing day, Soyuz TM-10 undocked from Mir and touched down some 68 km north-west of Arkalyk at 9:08 am Moscow Time, concluding an eight-day voyage for Akiyama and a little more than 130 days for Manakov and Strekalov. TBS had insisted on placing a television camera inside the descent module, to capture the cosmonauts' reactions upon landing. It would seem that the Japanese journalist suffered a particularly unpleasant bout of space sickness, remaining strapped into his couch during the journey to Mir and never really gaining his 'space legs' whilst in orbit. As a result, Akiyama's reaction upon landing back on *terra firma* must have been quiet relief.

Alone for the next five months, Afanasyev and Manarov settled down to an extensive, and expansive, programme of scientific and technological work aboard the T-shaped Mir complex. They also performed no fewer than four EVAs, the first of which – on 7 January 1991, the day of Russian Orthodox Christmas – successfully replaced the damaged hinge on the Kvant-2 airlock hatch. The cosmonauts spent almost five and a half hours outside the station and their work was described as "very complex and very delicate", since the hinge was never intended to be exchanged in orbit; nor was it suitable for manipulation by a repairman clad in a bulky suit and thick gloves. Nevertheless, a little more than four hours into the EVA, Afanasyev and Manarov entered Kvant-2's airlock and closed and sealed the hatch to check their work, then reopened it and ventured outside to carry out additional tasks. They moved equipment and parts for the forthcoming solar array transfer, removed a space exposure cassette and took down a camera for the Gemma-2 environmental monitoring unit for repair. Two weeks later, on 23 January, the duo returned outside for five and a half hours. They installed a 45 kg telescoping boom, known as 'Strela' ('Arrow'), onto Mir's base block, as part of efforts to move the 500 kg array from Kristall to Kvant-1. To test the new boom, Manarov rode at its tip, whilst Afanasyev operated the cranking mechanism. At one stage in the spacewalk, Mir passed over the war-torn Persian Gulf region, and the cosmonauts were able to observe fires, smoke plumes and an oil spill triggered by Saddam Hussein's retreating forces. Finally, on the 26th, they spent six hours and 20 minutes installing a pair of solar array supports for the arrays onto either side of Kvant-1. The work brought them particularly close to the module's Kurs rendezvous antenna, which would bite them later in their mission.

In the meantime, Afanasyev and Manarov pressed on with their materials science experiments in the Kristall module and extended an instrumented pole into space as part of the 'Diagramma' programme to characterise the magnetic environment outside Mir. Then, in the third week of March, the time came for the arrival of a new resupply freighter, Progress M-7. The spacecraft was launched from Tyuratam on 19 March and followed an entirely nominal two-day approach and rendezvous profile, arriving at Mir on the 21st. In addition to food, water and oxygen for the crew, it also carried scientific equipment and the second Raduga return capsule. As it manoeuvred to line up with the aft port of Kvant-1, things began to go awry. Progress M-7 ceased approaching the station at a distance of 500 m and a second attempt on 23 March was called off due to a "catastrophic error", in which it would

182 The last Soviet citizen

2-05: The interior of Mir's multiple docking adaptor, which provided access to various additional modules at the station, including the base block.

appear that the freighter came within 5-7 m of a collision. Mission controllers commanded the Progress into dormant mode as the effort to investigate the issue got underway. A backup Progress was readied for launch, to deliver supplies in the event that its sister failed to dock. On 26 March, Afanasyev and Manarov undocked Soyuz TM-11 from Mir's front port and flew around the station to dock at the Kvant-1 rear port. It was during this operation that it became clear that the problem was with the station itself, for they encountered the same difficulty and had to dock manually. This left the front port of the multiple adaptor free for Progress M-7, which docked without incident on the 28th. The problem was traced to Kvant-1's Kurs rendezvous antenna, which was situated near where the cosmonauts had worked during their third EVA in late January.

At around this time, the joint Soviet-British Soyuz TM-12 mission was postponed from its original 12 May launch date to sometime between 16-19 May and a spacewalk was planned for the end of April to inspect the Kurs hardware. Accordingly, on the 23rd Afanasyev and Manarov ventured outside for the fourth time. Their departure was delayed slightly by the need to reconnect a cable in the Kvant-2 airlock, but the cosmonauts quickly made up for lost time, with Afanasyev remaining near the airlock and Manarov clambering some 30 m along Mir's exterior. (This actually violated the Soviet 'buddy' policy of always having a pair of cosmonauts work as a team and mission managers would rebuke Manarov after his return to Earth.) When he reached the rear of Kvant-1, Manarov videotaped images of Kurs for engineers; it was obvious that one of the 23 cm parabolic dish antennas was *missing*; it had likely been kicked loose during the 26 January EVA. A replacement antenna would be installed in late June by the next Mir crew, Soyuz TM-12's Anatoli Artsebarski and Sergei Krikalev.

"NO EXPERIENCE NECESSARY"

Few British space enthusiasts can possibly forget the famous plethora of posters, newspaper advertisements and radio and television announcements which materialised, nationwide, throughout the summer of 1989. *Astronaut Wanted*, they crowed, *No Experience Necessary*. It was typically indicative of the understated humour of an island kingdom whose imperial history, wealth and technological prowess in the 19th and 20th centuries had given way to a surprising dearth of activity in the human space flight sphere. Although the United Kingdom had become the sixth nation to launch its own homegrown satellite into the heavens – tiny 'Prospero', atop a Black Arrow rocket from Woomera in South Australia in October 1971 – its aspirations in space exploration almost exclusively focused upon unmanned research, with particular emphasis upon the Earth sciences. During the course of the 1980s, despite the formation of the British National Space Centre (BNSC), the Conservative government of Margaret Thatcher virtually gutted any chance of a human space programme.

Nonetheless, efforts were afoot in the pre-Challenger era to send at least one Briton into orbit aboard the Shuttle. Squadron Leader Nigel Wood of the Royal Air Force was slated to fly in June 1986 as a payload specialist to observe the deployment of the Ministry of Defence's Skynet 4A communications satellite. Planning for this mission had been formally announced more than two years earlier, in January 1984, when the MoD revealed that it would launch Skynet 4A and 4B on the Shuttle. However, it reserved the right to fly its third satellite – Skynet 4C – aboard Ariane, thereby demonstrating visible support for the European booster. "The £60 million or so that it will cost to fly two Skynet 4s on Shuttle is probably slightly cheaper than flying on Ariane," *Flight International* reported at the time, "and no doubt was a factor in selecting the American launcher." Rumours of shortlisted candidates from the Royal Air Force and Royal Navy proved accurate, when, in February 1984, four finalists were selected. In addition to Wood, they were civilian physicist Christopher Holmes, who was the deputy manager of the Skynet 4 project, together with two military officers, Peter Longhurst of the Navy and Tony Boyle of the Royal Signals. (Boyle was replaced by fellow Signals officer Richard Farrimond in July.) Wood's selection as the prime payload specialist was revealed in May 1985 and he trained to conduct a series of six British experiments, focusing on the effects of cosmic radiation, changes in head-eye co-ordination and adaptation to weightlessness, studies of adhesive bonding, the ability to estimate mass in microgravity, motor skills – including postural control – and an ergonomics experiment. The loss of Challenger in January 1986 sadly ended Wood's chances of ever reaching space.

Then came 'Project Juno', the effort to send a Briton into space aboard a Soviet Soyuz-TM spacecraft for a week-long mission to Mir. And for Helen Patricia Sharman the radio announcement of *Astronaut Wanted, No Experience Necessary* came in June 1989, whilst she was sitting in a traffic jam during the drive home from her employer, the Mars confectionery company, in Slough, Berkshire, where she worked as a research chemist, investigating the flavourant properties of chocolate. The 26-year-old Sharman was heading back to her flat in Surbiton, south-west London, and in her autobiography, *Seize the Moment*, co-authored with Christopher Priest, she described the events of that evening as "the rarest of moments", which hindsight would convince her was "the crucial, pivotal moment in my life". From that moment onwards, she was an integral part of the selection process for Britain's first astronaut ... and less than two years later, in May 1991, she became not only the first Briton ever to enter space, but also made the United Kingdom the first of only three nations to date to have a woman as its first astronaut. (Sharman has since been joined, at the time of writing, by Iran's Anousheh Ansari in 2006 and by South Korea's Yi So-yeon in 2008.)

Sharman was born in Grenoside, a suburb of the industrial city of Sheffield in Yorkshire, on 30 May 1963, the daughter of a college lecturer father and a nurse mother, and received much of her early education in her home town. In her autobiography, she described her upbringing as decidedly unremarkable and eventually entered the University of Sheffield to study chemistry, graduating in 1984. She first worked as an engineer for the General Electric Company in London – "even though the salary on offer was the lowest," she admitted, "the work they were offering was varied ... [and] I wanted to sample the bright lights of the South-East" – and her role encompassed solving production problems, organising schedules and doing research and development of cathode ray tube components. At length, GEC offered her the chance to pursue a PhD in chemistry and in 1985 Sharman enrolled at Birkbeck College in London to explore the luminescence of rare-earth ions in crystals and glasses. Two years later, she began working for Mars Confectionery as a research technologist, planning new ice-cream products. Sharman found the work fascinating, not least in the fact that it enabled her to incorporate many facets of everyday life into her work. "It's no good understanding and manipulating the chemistry of emulsifiers in your ice-cream," she wrote, "if the toffee falls off the ice-cream before it's had a chance to harden." That technical knowledge and practical experience, coupled with Sharman's uncommon ability to interact with people of different ages and backgrounds, would serve her well in the Project Juno selection process.

Astronaut Wanted, No Experience Necessary ...

As she drove home on that June evening in 1989, Sharman found herself musing on the fact that the job of *astronaut* – a career path that she had never considered possible – was fairly straightforward ... at least on paper. She was British, she was aged between 21 and 40 years old, she had a formal scientific training, she had a proven ability to learn a foreign language (Sharman spoke French and German) and she had a high standard of physical fitness. When she applied, she was asked for a few basic details about herself, completed a form and became one of more than

13,000 people who applied for the role. In the weeks which followed, the names were winnowed down to 150 and that of Sharman remained on the list. She attended a medical evaluation in London in August, which made her a member of the final 32 candidates, and was subsequently called to a meeting at Brunel University. Whilst she was there, Sharman met fellow finalist Major Tim Mace, a 34-year-old helicopter pilot and accomplished parachutist in the Royal Army Air Corps. He was, she later wrote, "a pilot, an aeronautical engineer and ... Britain's skydiving champion". (Later in his career, Mace went on to serve in South Africa as Nelson Mandela's helicopter pilot.) In *Seize the Moment*, Sharman recalled her personal conviction at the time that the smooth-talking Mace stood out as an obvious front-runner in the Juno competition.

"The stakes were much higher than any of us had realised," wrote Sharman. "The publicity would fuel the sponsorship, but for most of the applicants it would inevitably mean rejection in public." As the selection process continued into the latter part of 1989, the 32 remaining candidates were reduced in number to just 22, and later to 16. It was at this stage that her chances of selection seemed to vanish. "It seemed that the Russians had said, of the candidates being tested in Britain," Sharman explained, "that they wanted the two finalists to be of the same sex. They expressed no preference about male over female, but ... with only two women left in the final 16, the odds against both of us beating all the men were negligible." That negligibility increased when the number was reduced to ten and the only other woman in the competition was eliminated; shortly thereafter, after a visit from Soviet doctors, only six candidates remained. By the beginning of November, the four finalists were announced: Sharman and Mace, together with Royal Navy flight surgeon Gordon Brooks and Kingston Polytechnic aeronautical engineer Clive Smith. And by the *end* of November, Sharman and Mace had been picked to train for the mission and were flown into a snowy Star City, near Moscow, to begin 18 months of intensive preparations.

It was during that time that Project Juno came to a figurative and literal crossroads, with the axe of cancellation hanging ominously for some time. Right from the start, it had been a commercial enterprise, with a reported Soviet fee of around $12 million (£7.5 million). In Britain, Antequera, Ltd., was established to administer the selection process and the mission itself, but by the spring of 1990 it was evident that sponsorship was falling far short of the required level. British Aerospace, Memorex and Interflora had joined the pool of corporate sponsors, and ITV had bought the television rights, but the consortium as a whole failed to raise the entire sum. (Once aboard Mir, Sharman would participate in a televised advert for Interflora, by 'ordering' flowers for delivery to her mother.) A paltry $1.7 in commercial revenue was generated and Antequera was dissolved. In the wake of this event, the British government refused to support the mission. Thankfully, by December, word reached Sharman and Mace that Moscow Norodny Bank – a Russian-owned, but London-based, commercial venture – had agreed to underwrite the final stages of Project Juno. In her autobiography, Sharman described the bank as "our white knight". It is believed that Mikhail Gorbachev personally insisted that the mission proceed at his country's cost, in the interests of furthering international relations. Quoted by *Flight International* in June 1991, Alexei Leonov, then serving as deputy chief of the cosmonaut corps, damningly noted that Project Juno "has not added to the prestige of our space research programme", although he acquiesced that Sharman herself worked "magnificently". To be fair, the financial situation posed little difference to the daily work of the cosmonaut trainees, but it was an embarrassment for Britain that their first foray into human space exploration should have been so mired with financial difficulties and should have come so close to the brink of outright collapse.

From an international perspective, it was humiliating that one of the richest and most technologically advanced nations on Earth should have been unwilling to claw itself out of such a prostrate position of reliance upon the Soviet Union. Its scientific questionability was placed on an equal footing to the flight of Japanese journalist Toyohiro Akiyama, whose mission included televised broadcasts, the carriage of a few Japanese tree frogs and precious little else, and there was concern in the early stages of 1991 that Project Juno was an abject failure, even as a publicity stunt. Writing in mid-May 1991, aerospace journalist Tim Furniss noted that France and Germany – the latter struggling with the costly process of reunification, following the collapse of the Berlin Wall – had forked out $50 million between them to fly a pair of their cosmonauts to Mir in 1992. "Britain's hitch-hike into space," wrote Furniss, "is the subject of much mirth and, in some quarters, anger."

Aside from the politics and the finances, the choice of Helen Sharman as the first Briton in space came on 19 February 1991, when Air Vice Marshal Peter Howard – former commandant of the Royal Air Force's Institute of Aviation Medicine, aerospace physician and veteran of the world's first rocket-propelled ejection seat, who had been placed in charge of the Project Juno astronaut selection – arrived in Moscow for separate discussions with the two candidates. "They told me I was to be in the prime crew," Sharman recalled in *Seize the Moment*. "This would be subject to a medical examination in March and another immediately before the launch. They then went next door to see Tim." The Army Air Corps officer's reaction was characteristically gracious and the pair continued to work together, side by side, until, on 23 April, their formal training concluded. Less than a month remained before the launch of Soyuz TM-12.

Shoulder to shoulder with Sharman would be a pair of professional cosmonauts. In command of the mission was Anatoli Pavlovich Artsebarski, born in Prosyana, to the south-east of Kiev, within today's independent Ukraine, on 9 September 1956. He graduated from the Soviet Air Force Institute in 1977 and was selected for cosmonaut training in September 1985, alongside the man who would serve as his flight engineer on Soyuz TM-12, Sergei Krikalev. Artsebarski's first two years were spent preparing for the Buran shuttle project, but in 1987 he began working on Soyuz-TM and Mir. With Krikalev, he served on the backup crew for the joint Soviet-Japanese mission in December 1990. In *Seize the Moment*, Sharman described Artsebarski thus: "He's a good organiser, pays attention to detail and has great stamina – he's *always* one of the last to leave a party." That stamina would be amply tested during the summer of 1991, as Artsebarski and Krikalev were required to support no fewer than six EVAs, totalling almost 32 hours in the vacuum of space.

Sergei Konstantinovich Krikalev, meanwhile, was making the second of what would turn into a six-flight cosmonaut career and, at the time of writing, in early 2013, he holds the unchallenged record for having spent more cumulative time off the planet (803 days) than any other human being. Krikalev came from Leningrad – today's St Petersburg – where he was born on 27 August 1958; his father was an engineer and his mother a survivor of the Nazi siege of Leningrad. After high school, Krikalev specialised in chemistry and entered the Leningrad Mechanical Institute, graduating first in his class in 1981 with a degree in mechanical engineering. As part of his studies, he worked on the design and manufacturing of flight vehicles. Krikalev also served as an aircraft technician and learned to fly at a Leningrad aviation club. He joined Energia and was involved in the testing of space equipment and ground control operations; in fact, aged just 26, he worked on the Salyut 7 rescue team to develop contingency docking procedures with the inert station. (More than one source has commented that this work by Krikalev assured him of Soyuz T-13 commander Vladimir Dzhanibekov's support in his application to become a cosmonaut.) Krikalev also excelled in athletics, swimming and aerobatics and was a member of the Soviet national flying teams, winning the accolade of Champion of Moscow in 1983 and Champion of the Soviet Union in 1986. By this time, he had entered cosmonaut training, having been selected in September 1985, and worked for a time on the Buran programme. He was officially appointed as a 'test-cosmonaut' in February 1987 and flew his first mission aboard Soyuz TM-7.

Early on 18 May 1991, the prime crew and their backups – Alexander Volkov, Alexander Kaleri and Tim Mace – stepped smartly through their final chores, in readiness for launch at Tyuratam. Out on the launch pad stood Soyuz TM-12, atop its rocket, whose nose shroud had been painted with both the Soviet Hammer and Sickle and the British Union Jack. Aboard the bus, Sharman was approached by Alexei Leonov, who offered her a number of 'unofficial' items to carry aboard Mir: a Swiss army knife, a camera, several rolls of film ... and a handful of grass. "I pick this for you," he told her, paraphrased in *Seize the Moment*. "Take it with you into the space station. There's nothing much to smell up there and this will remind you of home." The sweet-scented grass was *polin*, a kind of wormwood, or absinthe, which grew ubiquitously over the steppe of Kazakhstan.

Liftoff occurred precisely on time at 3:50 pm Moscow Time, amidst what Sharman could only describe as a steadily increasing, though muffled, rumbling. "I could feel the vibration, but no sense of acceleration," she wrote. Intellectually, she *knew* they must have left the ground, but was "in that momentary limbo where the rocket seems to balance precariously on its thrust, surely destined to topple". Yet topple it did *not*, for the acceleration steadily began to increase as Soyuz TM-12 thundered towards orbit. Her ears and senses attuned to every nuance from the rocket, indoctrinated through months of training, Sharman *heard* and *felt* the passage of maximum aerodynamic turbulence and the separation of the rocket's tapering strap-on boosters, until, finally, the escape tower and protective fairing was jettisoned, uncovering the capsule's windows for the first time and admitting a flood of sunlight. "Dreams sometimes *do* come true," Sharman mused in her autobiography, "and I felt *so* alive."

By five minutes, the central core stage of the booster had also been expended and was jettisoned, taking with it a flurry of ice particles. And finally, at eight minutes, as objects within the cabin seemed to ponder the question of whether or not they were still governed by terrestrial gravity or instead were the subjects of orbital free-fall, came the separation of the upper stage and the abrupt suddenness of arrival in space. "It did not happen gradually," Sharman wrote of the dying moments of the third stage. "One moment, it was burning ferociously behind me, in the next it stopped completely. One moment I was being pressed hard into my seat and in the next I was not. I had been straining against the G-force without realising I had been doing so." In front of the cosmonauts' eyes, the objects finally surrendered themselves to the peculiar state of 'microgravity'. They were weightless.

Two days later, on 20 May, Soyuz TM-12 drew towards Mir for the last stages of what should have been an automated docking with the station. Then, at a distance of 200 km, a malfunction in the Kurs rendezvous system forced Artsebarski to assume manual control of the final approach, earlier than normal, and docking was successfully accomplished with the front port of Mir's multiple docking adaptor and the visitors were quickly engulfed in bear hugs and a traditional bread-and-salt welcome from the resident crew, Viktor Afanasyev and Musa Manarov. With the dismal demise of much British commercial participation in the mission, Sharman's experiment programme was almost exclusively designed by the Soviets, with a primary emphasis upon the life sciences. She wore electrodes to track her heart rate, monitored her mental co-ordination and reaction speed with the 'Prognos' series of random light patterns and took a dozen blood samples from the tips of her fingers each day. In this manner, she added 'data points' to the standard monitoring of adaptation to weightlessness. Additionally, Sharman took air samples throughout Mir, to assess the prevalence of dust in the station, and participated in a number of 'agricultural experiments'. She grew wheat seedlings and potato roots, carried 125,000 pansy seeds – which were placed in the Kvant-2 airlock, the portion of the station least shielded from space radiation, to promote genetic mutations – for British schoolchildren and supported high-temperature semiconducting experiments. Other work focused on the growth of luciferase protein crystals and the exposure of ceramic films to the harsh vacuum of space. Finally, a significant portion of Sharman's time was devoted to photography of the United Kingdom itself. Yet it was impossible to avoid the reality that much of the work was Soviet-devised, Soviet-orchestrated and Soviet-financed. "Sharman will operate 17 biotechnological, medical and technical experiments – all Soviet," wrote Tim Furniss in *Flight International*, shortly before launch. With undisguised scorn, he added that "the British element of the Project Juno mission will be Sharman herself ... "

On the day after docking, 21 May, she was in the process of one of her radio contacts with British schools and happened to mention that Mir was experiencing difficulties with its solar arrays, caused by changes in orientation. Later, the level of background noise on the station fell dramatically from the normal 75 decibels, as fans, circulating pumps and other equipment began shutting down. In *Seize the Moment*, Sharman related the eerie nature of the change, as the five cosmonauts shared dinner in the base block. "The first fans to stop were those at the extremities of the station," she wrote, "in the Kristall and two Kvant modules; shortly afterwards, the fans in the base block slowed down, then halted. From being a noisy place, the space station became silent. After the fans, the lights extinguished themselves ... all we were left with was a solitary emergency fluorescent tube ... " A computer in the orientation system had failed some time ago, preventing the arrays from properly tracking the Sun, and as a consequence Mir's storage batteries were quickly drained. When the station re-entered orbital sunlight, its orientation was adjusted in order that its batteries could be recharged. Afanasyev and Manarov, by now old hands with more than five months aboard the station, assured her that such power outages were commonplace. Fortunately, a replacement computer had been brought along by Soyuz TM-12.

In keeping with other 'standard' international visits of this era, Sharman's mission lasted a little under eight days, with undocking and the return to Earth scheduled for 26 May. She would come home with Afanasyev and Manarov, concluding their 175-day flight, aboard the old Soyuz TM-11 craft, leaving Artsebarski and Krikalev and the new Soyuz TM-12 behind. Planning at the time called for another mission – Soyuz TM-13, with an Austrian cosmonaut amongst its crew – to fly in October 1991, although the political and economic climate over the next few months was to change the Soviet Union beyond all recognition ... as well as affecting the situation of the cosmonauts themselves. Few cosmonauts would be more affected by what happened in Russia in the latter half of 1991 than Sergei Krikalev.

Early on 26 May, at 9:13 am Moscow Time, Soyuz TM-11 undocked from its berth at the Kvant-1 module and began the three-hour process of returning to Earth. As Viktor Afanasyev flew in formation with Mir and Manarov fired off photographs, Anatoli Artsebarski – a fan of the British pop/folk singer Tanita Tikaram – played out their final separation manoeuvre and departure with a tape of 'World Outside Your Window', given to him by Sharman. Shortly after noon, the irreversible retrofire burn was performed and by 12:38 pm the orbital and instrument modules had been jettisoned, leaving only the descent module to endure the searing hypersonic descent through the atmosphere. Eleven minutes later, the drogue parachutes deployed. Conditions at the landing site in north-central Kazakhstan, near the industrial city of Jezkazgan, were windy and Sharman remembered that much of the final descent was spent moving laterally with respect to the ground. Fifteen metres from touchdown, a proximity indicator illuminated and she informed Afanasyev and Manarov to brace themselves. A few seconds later, and a couple of metres above the Kazakh steppe, solid-fuelled retrorockets fired to cushion the impact. Although the capsule quickly tipped over in the high winds and bounced hard, it was nevertheless a successful landing.

It was 1:03 pm Moscow Time and Britain's first astronaut was safely home after a little under eight full days in space. At the conclusion of his second mission, Manarov became the undisputed record-holder for time in space, with a cumulative 541 days, easily eclipsing the 430 days of fellow cosmonaut Yuri Romanenko. Manarov would enter politics in the wake of his space career and served as a member of the Russian Duma. For her part, Helen Sharman embarked on a career in television and science promotion and applied unsuccessfully (as did Tim Mace) for admission into the European Space Agency's astronaut corps in 1992 and 1998. Only Viktor Afanasyev would fly again and he would command three more missions, bridging the conclusion of the Mir era and the beginning of the International Space Station. As for Sharman's Soyuz TM-12 crewmates, Artsebarski would return to Earth in October 1991, according to pre-flight plans, and would never fly again. The case of Sergei Krikalev, though, would be far different as politics in Moscow took an ugly turn in mid-August.

AN EXTENDED FLIGHT

Despite its outward appearance as a single-party, authoritarian state, the Soviet Union had undergone many changes during the middle to late 1980s, chiefly under the auspices of General Secretary Mikhail Gorbachev, who actively sought reconciliation with the West and strove to implement political and economic reform. However, this stance of greater liberalism and plurality had not prevented him from attempting to forcibly re-integrate Lithuania back into the Soviet sphere in January 1991 and threatening the independence ambitions of Latvia. By March, six former Soviet republics had moved toward separation and in an attempt to avert the collapse of the Union, Gorbachev proposed to reorganise the other nine republics into a 'confederation' of equal sovereign states, led by a common president, with a shared foreign policy and military force. With some conditions, the so-called 'New Treaty' was generally accepted by eight of the nine republics (with the exception of Ukraine) and plans were laid for the Russian Soviet Federal Socialist Republic, together with Kazakhstan and Uzbekistan, to ceremoniously sign it in Moscow on 20 August 1991.

By dawn on that morning, however, the Soviet capital was in crisis.

Communist Party hardliners had long feared that the New Treaty would induce other republics to pursue full independence and on the 18th they took control of the Kremlin, confining Gorbachev to his Crimean dacha, where he was on vacation, and preventing him from returning to Moscow to sign the document. The conspiracy had begun several months earlier, in December 1990, when KGB Chairman Vladimir Kryuchkov made a call for order over Moscow Central Television and began organising a plan of measures to be taken in the event of a state of emergency. In the days and weeks which followed, Kryuchkov brought key Soviet officials, including the defence minister, the internal affairs minister, the secretary of the Communist Party's Central Committee and others, into a wider effort to persuade Gorbachev to declare a state of emergency and enable them to "restore order".

By late July 1991, Gorbachev, together with Boris Yeltsin, newly-elected President of the Russian Soviet Federal Socialist Republic, and Kazakh President Nursultan Nazarbayev discussed the possibility of replacing the hardliners with more pliable candidates. Shortly thereafter, Gorbachev went to his dacha and a group of the hardliners demanded that he either declare a state of emergency or resign in favour of Soviet Vice-President Gennadi Yanayev. It is believed that Gorbachev refused and the dacha's KGB-controlled communications channels were cut and additional guards stationed at its gates. Ominously, the conspirators ordered a quarter of a million pairs of handcuffs and 300,000 arrest warrants to be sent to Moscow and the Lefortovo detention centre was emptied to receive an anticipated influx of new political prisoners. In the capital, they signed a new 'Declaration of Soviet Leadership', announcing the state of emergency and its mandate to "effectively maintain the regime of the state". Gennadi Yanayev declared himself as 'Acting President of the Soviet Union', during what was described as Gorbachev's period of incapacity, caused by "illness".

From the early hours of 19 August, pro-coup documents were broadcast by state radio and television and independent stations were taken off-air. Meanwhile, military units and elite paratroopers rumbled into Moscow and several key figures were arrested ... though, inexplicably, the decision was taken not to detain Boris Yeltsin – a move which ultimately proved to be the coup's undoing. Yeltsin arrived at Russia's parliament building, the White House, later that morning and declared through flyers that the coup was unconstitutional and urged the armed forces not to support it. In some of the most dramatic images from the crisis, at one stage, that evening, Yeltsin clambered atop one of the tanks outside the White House and addressed the crowd. His remarks appeared on the state media's news broadcast.

Late on the afternoon of the 20th, the pro-coup units prepared for 'Operation Grom' (or 'Thunder') to assault the White House, but fears of bloodshed led several commanding officers to call for its cancellation. An hour before the scheduled attack, trolleybuses and street-cleaning machines barricaded a tunnel against oncoming troops and at length the military units were pulled out of Moscow. From his dacha, Gorbachev refused to meet any of the conspirators and instead declared their decisions void and dismissed them all from their state offices. By the morning of the 22nd, the General Secretary was back in Moscow. In the days that followed, arrests were made and the old imperial colours of red, white and blue replaced the Soviet Hammer and Sickle, becoming the official flag of Russia in October 1991.

It was the beginning of the end of the Soviet Union – a union which stretched across much of Europe and virtually all of northern Asia – as one of the largest contiguous conglomerates of nations in human history. Gorbachev resigned as General Secretary in the days after the coup and Yeltsin decreed that all Communist Party activities in Russia must be discontinued. Moldova, Ukraine, Azerbaijan, Kyrgyzstan and Tajikistan declared independence, whilst the independence of Estonia, Latvia and Lithuania was formally recognised. Others followed. By the end of October, Armenia had announced its independence, as had Turkmenistan, leaving only Russia, Belarus, Kazakhstan and Uzbekistan. Finally, on 21 December, a new 'Commonwealth of Independent States' (CIS) came into being and on Christmas Day Gorbachev resigned from his post as General Secretary. The Russian tri-colour flag was raised above the Kremlin, replacing the Hammer and Sickle, and the Soviet Union ended. Today, the regional organisation of the CIS effectively serves as a commonwealth of nations between many of the old Soviet-bloc states, carrying powers of co-ordination in terms of trade, finance, lawmaking and security, as well as cross-border crime prevention.

In the months preceding and following the coup to "restore order" on the Soviet Union, cosmonauts Anatoli Artsebarski and Sergei Krikalev remained aware, but largely unaffected, by the unfolding events back home. That is, of course, until the tanks rolled into Moscow. Contingency plans had been established to abandon Mir in the event of an outbreak of war, but the greatest fear in the latter part of 1991 – as more Soviet republics voted with their feet and sought independence – was that Kazakhstan, within whose territory Tyuratam resided, might nationalise the launch site or refuse the Russian government permission to use it. Indeed, Kazakhstan had announced its sovereignty as a republic in October 1990 and would declare full independence as a separate nation on 16 December 1991, effective Christmas Day. As a consequence, the original crewing line-up for the end of the year was changed. Originally, Soyuz TM-13 was scheduled to launch on 2 October, with Soviet cosmonauts Alexander Volkov and Alexander Kaleri (to replace Artsebarski and Krikalev), together with Austrian astronaut Franz Viehböck. The Austrian was to return to Earth on the 10th with Artsebarski and Krikalev, leaving Volkov and Kaleri aboard Mir for five months until the spring of 1992. In the meantime, to placate Kazakhstan, a pair of ethnic Kazakhs – Toktar Aubakirov and Talgat Musabayev – had been selected earlier in 1991 to train for a short-duration 'visiting' mission on Soyuz TM-14 in mid-November. According to Mark Wade on the website www.astronautix.com, this crew would have included Soviet cosmonauts Alexander Viktorenko and Yuri Usachev, together with Aubakirov. However, as budgetary woes intensified, the November mission was cancelled and Aubakirov was shoehorned into the Soyuz TM-13 crew, taking the seat of Kaleri. The result was that Volkov would now be joined by two 'guest' cosmonauts, the Austrian Viehböck and the Kazakh Aubakirov ... and *that* would effectively leave Mir without a dedicated flight engineer. The obvious solution was to call upon Sergei Krikalev, the flight engineer already aboard Mir in the summer of 1991, to remain as Volkov's crewmate for a double-duty ten-month stay. The pair would then return to Earth in March 1992.

As the space programme writhed and contorted with these changes, other former Soviet-bloc republics flexed their own muscles. In November 1991, *Flight International* reported plans for a trio of Ukrainian cosmonauts to fly aboard Mir to monitor the area around the Chernobyl nuclear power plant and the city of Kiev for "the after-effects of the accident", more than five years earlier. "The Ukraine mission," explained Tim Furniss, "is likely to comprise a commander and flight engineer from the existing Star City cadre, and journalist Yuri Krikun, who is being sponsored by the Ukraine." It never happened. In fact, as circumstances transpired, the only cosmonaut of Ukrainian origin to fly aboard Mir in 1992 was Alexander Volkov ...

In the midst of this tumult, life went on aboard Mir. Within days of the departure of Soyuz TM-11 and Afanasyev, Manarov and Sharman, the new residents of Artsebarski and Krikalev set to work on an expansive programme of work ... and EVAs, for several tasks were scheduled, including the replacement of a kicked-loose Kurs antenna on the Kvant-1 module. Progress M-8 delivered tools at the end of May and on 14 June the cosmonauts practiced their work inside Mir, before

2-06: The Sofora girder is clearly visible, protruding from Kvant-1, in this image.

venturing into the vacuum of space of the 24th for almost five hours. The repair was exceptionally delicate, since it involved the use of small tools, including a dental mirror, and several items not intended for EVA handling. After replacing the Kurs dish, Artsebarski and Krikalev assembled a prototype thermo-mechanical 'joint' outside Kvant-2 in readiness for the installation of a 14.5 m girder, built by the Energia design bureau and known as 'Sofora'. Although experimental, and laden only with sensors, the Soviets hoped to attach a roll-control thruster system to the 20-segment Sofora if it functioned well. Four days later, they were back outside for more than three hours, installing a space exposure experiment – devised, interestingly, by the University of California at Berkeley – to examine the effects of super-heavy cosmic-ray nuclei as they passed through layers of phosphate glass. The cosmonauts recovered the thermo-mechanical joint and attached a charged particle detector.

With two EVAs accomplished, Artsebarski and Krikalev were now seasoned spacewalkers ... but their time outside had barely begun. No fewer than *four* further excursions would be performed, throughout July, to prepare the Sofora worksite and install the gigantic truss structure. The name *Sofora* derived from that of a fast-growing Asian shrub, which proved apt, for during their six-hour EVA on 15 July the cosmonauts began setting up equipment and on the 19th started the installation process in earnest. Krikalev used the controls of the Strela boom to move

Artsebarski and two boxes of Sofora parts, together with the first cubical, half-metre-wide Sofora segment, to the Kvant-1 site. "The truss was put together lying back over Soyuz TM-12 at the aft port," wrote David S.F. Portree in *Mir Hardware Heritage*, "parallel to the long axis of the Mir base block. Krikalev and Artsebarski used the four heating and assembly devices to shrink the memory-metal sleeves in the truss joints. They had difficulty seeing their work as the lighting changed, but managed to keep working during orbital night …" Neither man could use foot restraints as an anchor and they were obliged to rely upon their hands and arms to hold themselves in position. After five and a half hours, with three Sofora segments installed, the exhausted cosmonauts returned inside Mir. Four days later, on the 23rd, they installed a further 11 pieces and, finally, on the 27th, came a mammoth final EVA, lasting six hours and 49 minutes, to finish the work. Artsebarski and Krikalev fitted the final three sections of the Sofora girder, affixed it to a mounting plate on Kvant-1 and raised it so that it was almost perpendicular with the long axis of the base block. (It was actually 'sloped' by 11 degrees, like a celestial Leaning Tower of Pisa, toward the front of Mir, in order to place its apex above the station's centre of gravity.) Ominously, and perhaps propitiously, the cosmonauts attached a Soviet flag to the top of the girder, a move which Moscow TV's *Vremya* programme praised. "It is not difficult to understand," the programme reported, "Anatoli Artsebarski and Sergei Krikalev, who, on their own initiative, placed a Soviet flag atop the girder. After all, our country has *not* totally fallen apart yet …"

Many in the Soviet Union had long since seen the writing on the wall, that their once-proud nation was in mid-1991 a mere shadow of its former grandeur, but even as the Moscow coup fell apart, the effort to support Mir and her cosmonauts continued. On 20 August, at the height of the crisis, Progress M-9 was launched uneventfully from Tyuratam and, two days later, delivered supplies and equipment to sustain Artsebarski and Krikalev. The men were also shown pro-coup (from Soviet Central Television) and anti-coup (from Russian Radio) broadcasts through video feed transmitted by mission controllers. By now, the awareness that the old republics were breaking away was evident and the need to please Kazakhstan was acute. Of course, under the USSR, various individuals from what would later become post-Soviet independent nations flew into space, including two – Vladimir Shatalov and Viktor Patsayev – who had been born within the former Kazakh Soviet Socialist Republic. However, the flight of Aubakirov achieved much media interest, for this was the first flight of an ethnic Kazakh, staged under the banner of an independent republic. Toktar Ongarbayuly Aubakirov was born in the coal-mining industrial centre of Karaganda – the fourth most populous city in Kazakhstan – on 27 July 1946 and he followed an aviation career in the military, graduating from the Soviet Air Force Institute. He qualified as a test pilot and parachutist and eventually rose to become a major-general within the post-independence Kazakhstan Air Force. During his career, Aubakirov flew the first arrested landing and deck takeoff of the MiG-29K from the Soviet aircraft carrier *Tblisi*. Selected for cosmonaut training in January 1991, his flight was originally scheduled for further downstream, but was advanced in the tumultuous summer months of that year. Launching into orbit on 2 October, Toktar Aubakirov became only the fifth spacefarer of the Muslim faith.

Joining Aubakirov was Soyuz TM-13 commander Alexander Alexandrovich Volkov, whose family would later earn its place in history, when his son, Sergei, also served as a cosmonaut. The Volkovs thus became the first case in which successive generations of the same family flew into space. Yet the elder Volkov's interest in the heavens extended back much earlier. He was of Ukrainian descent, having been born in the city of Gorlovka – also known as 'Horlivka' – in the Donetsk Oblast on 27 April 1948. Two weeks before his 13th birthday, Volkov had watched, awestruck, as his fellow Soviet countryman, Yuri Gagarin, became the first man to venture into space and it was *this* which inspired the young boy to go on to train as a cosmonaut. Twice flown on the eve of the Soyuz TM-13 launch, Volkov had spent two months in orbit aboard the Salyut 7 space station in the autumn of 1985 and five more months aboard Mir in 1988-89. Upon his return from his third mission, he would have accrued more than a year of his life off the planet and would command the Russian cosmonaut corps from 1991-98.

If the decision to fly an ethnic Kazakh to Mir was something of a knee-jerk reaction on the part of the Soviet government, then the mission of an Austrian national was quite the opposite. In fact, an 'Agreement Between the Republic of Austria and the Union of the Soviet Socialist Republic on the Execution of a Common Austrian-Soviet Space Flight' was signed in February 1989. It called for an eight-day flight by an Austrian cosmonaut for a total fee of 85 million Austrian Schilling (approximately $8.2 million), but awareness and interest had been growing for more than a year before that date.

In mid-1987, the idea of a joint Soviet-Austrian mission was discussed in Vienna, during a visit by the Chairman of the Council of Ministers of the USSR, and options were considered during subsequent meetings. On 5 April 1988, the Austrian government resolved to stage the mission and within days advertisements were placed in newspapers, inviting interested parties to tender their applications and interested scientists to submit synopses for experiments. The flight was formally approved in October and more than a hundred candidates were physically and psychologically screened and their number was winnowed down to around 30. These finalists underwent parachute trials, intensive mental and physical testing under pressure and flight training through the spring of 1989. By the end of the process, only a handful remained and after further gruelling endurance and power-training tests, swimming and trampolining exercises – *and* a similarly vigorous evaluation in the Soviet Union – the final two candidates were sent to Star City in January 1990 to begin almost two years of training. The pair of individuals who survived these enormous burdens and came out fighting on the other side were physician Clemens Lothaller and the man who would go on to become Austria's first – and so far only – spacefarer: an engineer named Franz Artur Viehböck. Born in Perchtoldsdorf in Vienna on 24 August 1960, he studied electronic engineering and ultimately earned a doctorate and lectured as an assistant professor at the University of Vienna.

Throughout 1990, Lothaller and Viehböck received Russian language instruction and described their astonishment at living in Star City, with some of the most famous cosmonauts – Alexei Leonov, the world's first spacewalker, and Valentina

Tereshkova, the world's first female spacefarer, to name just two – for neighbours. The family of the late Yuri Gagarin, they realised, lived close by. Early in the following year, they began simulator training and in June-September 1991 they worked extensively with their respective crews: Viehböck, on the prime crew, was teamed with Alexander Volkov and Toktar Aubakirov, whilst Lothaller, on the backup crew, was joined by Alexander Viktorenko and Talgat Musabayev. Of course, both crews had lost their Soviet flight engineer – Kaleri on the prime crew had been bumped by Aubakirov and Yuri Usachev on the backup crew had been bumped by Musabayev – and it is quite remarkable that these close-knit groups of men managed to pull together during one of the most harrowing and uncertain times in modern Russian history.

And *history*, and particularly tradition, was of overarching importance for Austria, too. Many symbolic objects were scheduled to accompany the first of its sons into orbit: a large national flag, streamers of the Austrian Republic and the provinces and the national colours, dozens of medals, a copper cuboid to be used to galvanise more medals, 30 special-issue envelopes, a hundred special-issue stamps ... and a few unusual and somewhat quirky additions, including a cassette tape of the 'Danube Waltz', nine original (and highly valuable) Mozart musical scores and a page from the book *The Little Prince* by Antoine de Saint-Exupéry. Native foods – including Viennese rice meat, wholemeal rye bread and Styrian bacon, Salzburger Mozart balls and coffee – were carried and a battery of research equipment for the mission (dubbed 'AustroMir') was despatched to Mir aboard Progress M-9 in August 1991.

With scores of Austrian politicians and dignitaries in attendance at Tyuraram on the morning of 2 October, Soyuz TM-13 roared spectacularly into orbit at 8:59 am Moscow Time. Federal Chancellor Franz Vranitzky was there, as was Federal Minister of Science and Research Erhard Busek, Minister of Defence Werner Fasslabend and others. Two days later, in accordance with the flight plan, Alexander Volkov guided his craft to a smooth docking and a period of joint work involving three discrete nations began. Early in the mission, Viehböck spoke with Austrian Federal President Kurt Waldheim, then plunged into his research programme. This included studies of his cognitive functioning, together with life sciences and technological investigations and a protracted effort to photograph Austria itself. At length, on 10 October, Artsebarski, Aubakirov and Viehböck transferred into Soyuz TM-12, undocked from Mir and landed without incident on the desolate Kazakh steppe, 78 km east of Arkalyk, at 7:11 am Moscow Time. Conditions that morning were extremely windy and the descent module struck the ground somewhat harder than anticipated, producing an enormous cloud of dust. Although the three men were unharmed, Viehböck later described the sensation of impact as "a hammer".

Describing his voyage as "something really unique" and "wonderful", Franz Viehböck returned home to Austria ten days later, on 20 October 1991, to a rapturous welcome ... and a very special first meeting. His wife had given birth to a baby girl, Carina Marie, whilst Viehböck was aboard Mir and this made him the first spacefarer to become a parent whilst in orbit. For the Republic of Austria and its newest hero, the birth amply demonstrated that unique and wonderful things can occur both off the planet and upon its surface.

END OF AN ERA

As the fierce sound of tanks and the equally fierce sound of political rhetoric receded in the closing months of 1991, life aboard Mir continued at the same pace, although it was obvious that the situation on Earth would place the station's future fortunes very much in question over the coming years. It was already clear that new agreements with the United States – whose clauses described a series of NASA astronauts spending long durations aboard Mir from 1995, as well as a series of Shuttle docking missions – might spell the old outpost's salvation. To be fair, the Soviets (soon to be simply the 'Russians') had secured a number of deals with other nations and France and Germany had already paid a total of $50 million to fly a pair of their astronauts to Mir in 1992. And after that, there was an expectation of longer flights, under the 'EuroMir' banner. Other 'private' expeditions also remained a possibility and in early 1991 Space Travel Services, a Houston-based company, had proposed a national telephone sweepstakes competition, the first prize of which would be a trip to Mir.

It is impossible not to see that the close of 1991 truly marked the end of an era, for all of Russia's space achievements had occurred under the banner of the Hammer and Sickle and were testament to an overwhelming Communist triumph. All of these achievements – from Sputnik to Yuri Gagarin and from Valentina Tereshkova to Alexei Leonov and from the first Salyut space station to Mir itself – had been planned, orchestrated and executed by a political system which tolerated no parties other than the Communist Party, a system whose ideals of fairness and equality for all had produced a closeted, secretive society and had frequently led it to intimidate and to dominate its neighbours. *Freedom* was a word which meant little in the Soviet era; it poorly suited the ideals of a state which saw the responsibilities of its citizens to 'Mother Russia' as greater than the rights of the individual. Nor was the fragmentation of the Soviet Union to provide the answer, for many of the breakaway republics endured their own economic, political, ethnic and religious troubles in the early 1990s that produced starvation, bloodshed and further conflict.

Aboard Mir, work continued, with Alexander Volkov and Sergei Krikalev welcoming their first unpiloted visitor, Progress M-10, on 21 October. Still, the station was in need of substantial repairs and the cosmonauts spent considerable time replacing storage batteries and tending to ongoing solar array problems, whose power output had diminished to just 10 kW. Early in December, during one communications pass, Volkov was overheard speaking about two small holes in the solar arrays of the 'D-module' (Kvant-2). Elsewhere, four of Kvant-2's six gyrodynes – launched barely two years earlier – had failed, affecting the stabilisation and orientation of the entire complex. When a new Progress arrived in late January 1992, it brought replacement parts for the gyrodynes, but in this first month of the new Commonwealth of Independent States it was impossible for Earthly troubles to leave space exploration unaffected: although they did not directly impede the docking of Progress M-11 with Mir, flight controllers in Moscow had gone on strike for higher rates of pay. Tracking network cuts meant that Volkov and Krikalev spent up to nine hours, each day, out of touch with Mission Control.

2-07: Against all the odds, Mir survived the collapse of the Soviet Union.

Notably, on 27 February, Russian President Boris Yeltsin signed a decree to establish the *Rossiyskoe Kosmicheskoye Agentsvo* (RKA, the Russian Space Agency), marking a clear break with the Soviet tradition of placing power in the hands of design bureaux and councils of Chief Designers. As early as November 1991, it had been reported in the Western press that a "single space centre", along the same organisational lines as NASA or the European Space Agency (ESA), should urgently be established to avert the dissolution of the Soviet space industry. "The Energia design bureau has declared the Mir orbiting station its property," wrote Tim Furniss in *Flight International*, adding that "Kazakhstan has laid claims on the [Tyuratam] cosmodrome." The RKA would henceforth receive funding consent from the Russian Duma and the office of the president himself. Its first director (until 2004) was Yuri Koptev and its early years were filled with crises, as the old design bureaux fought to survive and preserve their own authority. (For example, the decision to keep Mir operational beyond 1999 was made by the private shareholder board of the Energia bureau, rather than by the RKA.) Decreased financial commitment in the early 1990s took its toll on Mir and by the middle of the decade, as the first long-duration American crew members began to arrive, they beheld a station which was a shadow of its former self.

Late on 20 February 1992, as their mission drew into its final month, Alexander Volkov and Sergei Krikalev ventured outside Mir for the first Russian EVA of the new era. For more than four hours, they worked to install space exposure experiments and to disassemble the apparatus used in the construction of the Sofora girder. Unfortunately, the heat exchanger in Volkov's suit clogged at an early stage, which required him to remain close to the Kvant-2 airlock and rely upon its exchanger. Russian sources later blamed the problem on the fact that the suit had been stored for several months. It was Krikalev's seventh career EVA and took his cumulative spacewalking time to 36 hours and 29 minutes, a world record which would endure for the next four years. Later, Volkov's suit was dumped overboard. During earlier operations, abrasions on the suit's gloves had allowed unusually large quantities of air to escape. Further, the liquid-cooling garment connector had come apart during the checkout of the suit, again due to advanced age, and it was decided that a spare would be used and the old one discarded. This did not go down well with the cosmonauts or the newspaper *Pravda*, who all felt that the Orlan could have been returned to Earth and sold for profit to a museum.

By now, Krikalev had also secured for himself the accomplishment of becoming one of only four men to have spent in excess of 300 days aloft on a single mission. The end of his mission drew inexorably closer, as Soyuz TM-14 roared into space at 1:54 pm Moscow Time on 17 March. Aboard was German astronaut Klaus-Dietrich Flade, together with a new long-duration crew of veteran cosmonaut Alexander Viktorenko and his rookie flight engineer, Alexander Yuryevich Kaleri, who was born in Yurmala, in today's Latvia, on 13 May 1956. Kaleri grew up with the Soviet space programme; he was five years old at the time of Yuri Gagarin's epic voyage and had decided that he wanted to be a cosmonaut as soon as he heard about Gherman Titov's day in orbit aboard Vostok 2 in August 1961. "Maybe I was *born* with that wish," he told a NASA interviewer, but his father's experience as a navigator, engineer and parachutist provided further stimuli. In 1979, Kaleri graduated from the Moscow Institute of Physics and Technology as a specialist in aircraft flight dynamics and control systems, then undertook postgraduate work on the mechanics of fluids and plasma. He was selected as a civilian cosmonaut candidate in early 1984, qualified in 1987 and was paired with Alexander Volkov on the backup crew for Soyuz TM-4, which should have led to a seat on the Soyuz TM-7 prime crew, but an injury temporarily grounded him. Again teamed with Volkov, he served on the Soyuz TM-12 backup crew with Britain's Tim Mace and eventually received assignment to the prime crew of Soyuz TM-13, only to be dropped in mid-1991 in favour of Toktar Aubakirov and moved to the next mission. Despite these disappointments, Kaleri later remarked that much of his life and career put him "at the right time in the right place". At the time of writing, he is one of only six cosmonauts to have flown as many as five times and, if one takes his ethnicity into account, he is the world's most experienced Latvian spacefarer.

Flying alongside Kaleri and Viktorenko aboard Soyuz TM-14 was Major Klaus-Dietrich Flade of the German Air Force. He was born on 23 August 1952 in Büdesheim in Rhineland-Palatinate, in the far west of Germany, and was one of the first of his countrymen to fly into space after the fall of the Berlin Wall and political reunification. Flade entered the German Air Force after school and trained as an aeronautical engineer, then studied aerospace engineering at Bundeswehr University Munich from 1976 to 1980. He subsequently received flying tuition, graduated as a test pilot in 1989 and was selected in October 1990 – alongside civilian physicist Reinhold Ewald – as a candidate for the joint German-Soviet 'EuroMir' mission, scheduled for the early part of 1992. A few days after the selection of Flade and Ewald, *Flight International* noted that the total cost of the mission would be DM38 million (around $25 million). During the eight-day flight, the German astronaut would operate 14 experiments, the primary focus of which was to prepare for Space Station Freedom and its European-provided Columbus laboratory.

Two days after launching from Tyuratam, Soyuz TM-14 docked perfectly with Mir and the second German ever to board a space station got briskly to work. Much public interest centred, of course, upon Sergei Krikalev, who had been the unfortunate subject of many erroneous 'Stranded in Space' stories, as his old nation collapsed and a new one arose on Earth in his absence. The reality, of course, was quite different and when Volkov, Krikalev and Flade touched down, near Jezkazgan, at 11:51 am Moscow Time on 25 March, they did so without incident ... with the notable exception that Russia had to pay the Kazakh authorities $15,000 in rent for airports and helicopters during the recovery operation. In spite of this, it cannot be overstated that Krikalev's landing that day truly marked the definitive end of the Soviet Union. He really *was* 'The Last Soviet Citizen', returning to a changed homeland. Three hundred and eleven days earlier, shoulder to shoulder with Anatoli Artsebarski and Helen Sharman, he had been launched from Tyuratam, in the *Kazakh Soviet Socialist Republic*, under the banner of the Hammer and Sickle, yet was returning to *terra firma* to the soil of the Republic of Kazakhstan and a very different world. It was a world of acute economic hardship and political and social strife, both on Earth and in space, and its effects would be felt with equal acuteness by Mir and her cosmonauts over the next few years.

3

The International Space Years

JOURNEYS TO THE EDGE

One summer evening, early in August 1492, three ships – the carrack *Santa María* and a pair of a smaller caravels, the *Pinta* and the *Santa Clara* – put to sea from Palos de la Frontera, in southern Spain, to begin one of the most remarkable voyages in human history. Ever since the fall of the Mongol Empire, and, later, the final collapse of Byzantine hegemony in the eastern Mediterranean, it had become increasingly difficult and dangerous for European traders to pursue overland silk and spice routes to India and China. In response, Spain and Portugal had long vied with one another to search out seafaring routes to the 'East Indies'. In 1488, an eastward possibility had opened when the Portuguese explorer Bartolemeu Dias reached the Cape of Good Hope, but it was Christopher Columbus' expedition *westwards*, in 1492, which garnered the most excitement. In the first of four epic voyages, Columbus – born in Portugal, but serving the Spanish crown – sailed via the Canaries, in the expectation of reaching Japan ... but after many weeks at sea, he dropped anchor instead in the Bahamas.

In the years that followed, Columbus explored numerous Caribbean islands, including Cuba, Hispaniola and Haiti, and even made landfall on the northern coast of Venezuela. His expeditions would arouse general European awareness of the Americas for the first time. Within a generation, Ferdinand Magellan, another Portuguese explorer in Spanish employ, would have navigated the entire eastern coast of South America and reached a strait, which today bears his name, just north of Tierra del Fuego, through which the first European passage into the Pacific Ocean was made. The discoveries in those years, more than five centuries ago, would have profound implications for the 'Old World' and although Columbus never admitted it in his lifetime, his expeditions *did* uncover a 'New World' of which Europeans were hitherto unaware. That Spanish and Portuguese colonists were amongst the first to reach these new lands is well illustrated in their names and, particularly in South America, by their official languages.

Fast forward to 1992 and a proposal by Hawaii Senator Spark Matsunaga to designate the year of the 500th anniversary of Columbus' pioneering expedition as the 'International Space Year'. In total, 29 national space agencies and ten international organisations supported the theme of the year as a Mission to Planet Earth, with a focus on space exploration and sustainable technology. It was embraced by the United Nations, whose Secretary General, Boutros Boutros-Ghali, declared on 28 August 1992 that the year would "highlight the importance of understanding the Earth as a single, complex, inter-dependent system" and "stress the unique role that space science and technology can play in promoting that understanding". That year, several key voyages of scientific exploration were despatched, including TOPEX-Poseidon, to map the topography of terrestrial oceans, Mars Observer – intended to be NASA's first mission to the Red Planet in more than a decade – and the Extreme Ultraviolet Explorer (EUVE), a major astronomical observatory. Additionally, NASA despatched no fewer than eight Space Shuttle missions, the greatest number to date in the post-Challenger era. These missions contributed enormously to humanity's understanding of the Home Planet and the cosmos. Research was conducted into the life and microgravity sciences, the atmospheric and solar physics and the geosciences and plasma physics, and representatives of nine discrete nations – Russia, the United States, Canada, Belgium, Germany, France, Switzerland, Italy and Japan – flew aboard the Shuttle and the Mir space station during the International Space Year.

At the beginning of 1992, two cosmonauts, Alexander Volkov and Sergei Krikalev, were aboard Mir; with the formal collapse of the old Soviet Union and its replacement by a loose Commonwealth of Independent States, these men would return to Earth in March to a very different world. In the meantime, NASA readied itself for its most challenging year since before the loss of Challenger. That challenge had been amplified in the case of the first Shuttle flight of the year, STS-42, whose crew composition had changed several times since the first assignment of astronauts in June 1989. The mission was particularly important, for it carried the first International Microgravity Laboratory (IML-1), inaugurating a series of – it was hoped – at least three Spacelab flights dedicated to the life and microgravity sciences, featuring co-operation from more than 200 scientists in over a dozen nations. It was also several years overdue, having been postponed from its original launch date of May 1987 by the Challenger disaster. In particular, NASA, the European Space Agency (ESA), the Deutsche Agentur für Raumfahrtangelegenheiten (DARA, the newly-unified German space agency), the Canadian Space Agency (CSA), the French Centre National d'Études Spatiales (CNES) and the National Space Development Agency of Japan (NASDA) sponsored significant research aboard IML-1.

The sensitive microgravity requirements of many of the experiments required STS-42 to operate in a so-called 'pseudo gravity gradient' orientation, whereby the Shuttle would operate in a 296 km, 57-degree-inclination orbit, with its tail directed Earthward. This would allow positioning to be effected and maintained primarily by natural forces, thus reducing the number of disruptive Reaction Control System (RCS) thrusters firings on the experiments. Many of these experiments were located inside the Spacelab long module, similar to the one carried aboard STS-40, and

3-01: Replicas of Christopher Columbus' ships – the carrack *Santa María* and a pair of smaller caravels, the *Pinta* and the *Santa Clara* – sail past Space Shuttle Endeavour on Pad 39B in the days preceding her maiden voyage, STS-49. The year 1992 represented the 500th anniversary of Columbus' first expedition to the Americas and was commemorated, in part, by International Space Year.

would be operated around the clock in two, 12-hour shifts. To facilitate planning for what was envisaged to be an early demonstration of how research would be conducted aboard Space Station Freedom, members of the IML-1 'science' crew were announced in advance of the 'orbiter' crew ... and a new astronaut designation was born. In June 1989, a pair of veteran astronauts, physician Norm Thagard and engineer Mary Cleave were named as IML-1 mission specialists and a cadre of four international scientists – physiologist Ken Money and physician Roberta Bondar, both from Canada, were joined by US physicist Roger Crouch and German physicist Ulf Merbold – were selected to train for two payload specialist positions. The remainder of the seven-member crew (known as the 'orbiter' crew, since they would be responsible for managing the Shuttle's systems throughout the mission) would be named at a later date. Scheduled for launch in December 1990, aboard Columbia, IML-1 would last at least nine days, with the possibility of a tenth day being added, making it one of the longest Shuttle missions to date.

Within a matter of months, however, the first changes to STS-42 took place. In late January 1990, NASA revealed that Cleave had resigned her position on the crew "for personal reasons" and she was replaced by another veteran astronaut, Navy flight surgeon Manley 'Sonny' Carter. In her NASA oral history, many years later, Cleave explained some of her reasoning for the change. She had flown two previous Shuttle missions, separated by almost four years, and with a PhD in environmental engineering she was shocked by the rate at which the Home Planet had changed during that interval. "Cities were grey smudges," she remembered of the often depressing views from space of human devastation. "The air looked dirtier, less trees, more roads." She opted to relocate out of the astronaut office in Houston to work at NASA's Goddard Space Flight Center in Greenbelt, Maryland, on robotic environmental projects. Many of her colleagues thought she was crazy and others advised her to remain in Houston, working in an engineering role for a year, in the hope that she would change her mind. "It was standard military practice," Cleave said. "Don't do any traumatic decisions until you think about it for a year." She did not change her mind. In May 1991, she was named as deputy project manager for a Goddard mission known as 'SeaWiFS', dedicated to studying the biological mass of the world's oceans through analysis of chlorophyll content and plankton production.

By the time Carter replaced Cleave on STS-42, two other major crew additions had been made. Firstly, on 2 January 1990, astronauts Ron Grabe, Steve Oswald and Bill Readdy were assigned as the orbiter crew, followed, three weeks later, by a surprising announcement from NASA that a new mission specialist designation had been created: the 'Payload Commander'. The agency retroactively named four astronauts, already in training for major Shuttle science missions, as payload commanders... and the first would be Norm Thagard on IML-1. In its news release, NASA announced that the role was "expected to serve as a foundation for the development of a Space Station mission commander concept" and that its purpose was "to provide long-range leadership in the development and planning of payload crew science activities". This responsibility encapsulated the development and co-ordination of training plans for the science crew, liaison with the mission's Payload Operations Control Center and principal investigators, attendance at relevant meetings and oversight of pertinent hardware and software changes. Thagard would thus be in charge of IML-1 and the organisation of its science crew and would be responsible for the conduct and accomplishment of its scientific objectives; the *mission* commander (in this case, Ron Grabe) would maintain overall authority for the safety and success of the flight.

Fellow astronaut Kathy Sullivan was also named that January as payload commander for another Shuttle science flight, ATLAS-1, and in her oral history she rationalised the thinking behind the new position. "You've got a very complex Spacelab mission with a dozen to three or four dozen experiments," she said. "The training time that the responsible mission specialist should put into that needs to be longer than the pilot and commander probably need to put into the basic orbital operations, so you're going to want to slot in mission specialists 18-24 months in advance, so they can build the relationships that are necessary with the scientific team and the payload operations team." Part of the role of the science crew was to help the simulators develop a means to model the payloads correctly for training purposes and Sullivan considered it imperative that "one of the NASA mission specialists is considered and recognised as authoritative in all those early planning decisions". With the assignment of the orbiter crew, she concluded, the balance shifted somewhat, for the mission commander was in charge, but the longer 'lead' time enjoyed by the payload commander enabled the development of the flight and the training of the astronauts to be far smoother.

For Norman Earl Thagard, the IML-1 mission would be his fourth Shuttle flight.

Born on 3 July 1943, his birthplace is given as Marianna, Florida, and in his NASA oral history he admitted that his father was a Greyhound bus driver and Marianna was the bus changeover point, "so I was born at a time when he was doing that". Thagard attended high school in Jacksonville and studied engineering science at Florida State University, gaining bachelor's and master's qualifications in 1965 and 1966. (As a high school senior, Thagard told his classmates that his aspiration was to become an electrical engineer, a jet pilot, a medical doctor and an astronaut, in *that* order ... which, incidentally, is precisely the path that his career ultimately took.) Shortly after receiving his master's degree, he entered active duty as a reservist with the Marine Corps, quickly attained the rank of captain and was designated a naval aviator in 1968. Under normal circumstances, as a designate for the Platoon Leaders Class-Aviation (PLC-A), Thagard would have been put through Marine basic school, but "the Vietnam War was on by that time and they needed pilots desperately, so they basically eliminated the requirement ... for PLC-A designates". Interestingly, Captain Thagard won his wings in the very same squadron as one Lieutenant Rick Hauck, who would also become an astronaut. After initial flight instruction at Naval Air Station Pensacola in Florida, Thagard flew the F-4 Phantom at Marine Air Corps Station Beaufort in South Carolina, and then, in January 1969, went to Vietnam.

Thagard flew 163 combat missions in total and, although he admitted to never flying over North Vietnam, his role was primarily to support the fleet in the Tonkin Gulf or provide fighter cover for B-52 Stratofortress bombers or close aerial support for ground units in the South. In later years, his opinion of the conflict would harden, but at the time, "I thought that we were very altruistic ... I saw nothing wrong with what we were doing in South-East Asia, but I saw a *lot* wrong with perhaps being killed and having people in the United States think that was a good thing ... my *just* desserts! I had a *lot* of problems with that!" Returning to civilian life in the early 1970s, he would meet with derogatory comments when he revealed that he was a former fighter pilot, which irritated him greatly, not least because, in Thagard's mind, "anybody that thinks the North Vietnamese were good people has a very poor idea about human character ... *that* was a pretty bad group we were fighting". To Thagard, the principle was the seemingly endless march of communism, and the fear that so many countries went from their previous political orientation *to* communism, but rarely *went back*. "There was a real concern," he said, "that if *that* continued, ultimately, the end result would have been all the world would be communist ... so it didn't seem to me wrong or unreasonable for us to try to keep a country from being taken over by a communist regime." Three decades later, in his conversation with NASA's oral historian, clear anger was evident in Thagard's words: that it was impossible to fight a war without adequate support – moral and otherwise – and that much of the opposition, back home in the United States, including the refusal to accept the military draft, was little more than a "self-serving" desire "to avoid risk and responsibility".

Returning to the United States, Thagard served as an aviation weapons division officer back at Beaufort and resumed his academic studies in March 1971, pursuing additional work in electrical engineering and a medical degree. At first, he attended

Florida State University, working for a PhD in electrical engineering, but it was decided soon afterwards to terminate the whole school – "the Apollo programme was winding to an end," Thagard recalled, and "engineering was kind of on a down cycle that year" – and medicine seemed the most likely alternative. Here, he found severe age discrimination: Baylor College of Medicine in Waco, Texas, warned him that applicants over 25 were rarely given serious consideration and Duke University in North Carolina told him it would take the same view. However, a trump card appeared from the most unlikely of places: Thagard's *mother-in-law*! "She knew the dean of the School of Medicine, University of Texas Health Science Center [in San Antonio, Texas]," he explained, "because she was working there at the time." She asked the dean on Thagard's behalf and was told that if he moved to Texas and established residency there, he would be admitted. More trouble was afoot after the Thagards' arrival in Texas in June 1972. "We got there on a Friday evening at about five o'clock," he continued, "and got up the next morning, Saturday, and the headlines in the *San Antonio Light* [newspaper] read 'Medical School Dean Fired'. *This* was the fellow that said if I'd move there, they'd admit me to his school!"

Thagard worked initially as an engineer, whilst awaiting responses from other medical schools. He eventually accepted a place at the University of Texas Southwestern Medical School in Dallas and received his doctorate in 1977, only to learn that NASA was taking applications for its astronaut corps. His wife actually sent off for the application pack, "before I ever even *knew* about it", and Thagard applied for *both* the pilot and mission specialist categories, although his 804 hours of jet experience fell short of the minimum 1,000 hours needed for a pilot. "Realistically," he recalled, "the pilot [way] wasn't really going to happen. I didn't meet 1,000, but ... it wasn't *just* 1,000: you also needed to be a graduate of a military test pilot school, which I also was not." The mission specialist application, though, bore fruit and a month after starting his medical internship, in the summer of 1977, paperwork for security clearances, background checks and requests to attend his local police station for fingerprint profiles began to arrive at the Thagards' home. Late in October, he received the call from NASA to come to Houston for a week of screening. He heard nothing more until January 1978. "I actually had one of those rare Saturday evenings off," he told the oral historian, "and we had some friends over to the house. The news was on and there was an announcement ... that NASA on Monday was going to announce the first new group of astronauts. I was a little embarrassed, because to me that meant that I *wasn't* taken, because it seemed to me that this is *Saturday* and they're going to announce *Monday*, they've already told the people who are selected." Two days later, on Monday morning, Thagard was at the Veterans Administration Hospital, about to start work, when he received *the* call from George Abbey which would change the direction of his life. He went on to fly two Shuttle missions before the Challenger disaster and a third in May 1989.

One of Thagard's biggest challenges came relatively late in the IML-1 training process. The hydrogen leaks endured by Columbia and Atlantis in the summer of 1990 forced several missions, further downstream, to be unavoidably delayed, and STS-42 was amongst them. By the time Columbia finally flew STS-35, NASA was keenly aware that the veteran orbiter was scheduled for a year-long overhaul, beginning in mid-1991, to add capabilities for extended duration missions. This meant two of Columbia's flights, STS-42 and STS-45, had to be shifted onto her sisters, Discovery and Atlantis. By the time the space agency issued its December 1990 manifest, STS-42 was listed to fly aboard Atlantis in December 1991, but the problems with Discovery a few weeks later threw another spanner into the works. Ultimately, by mid-1991, the mission had been reshuffled again and was now scheduled to launch aboard Discovery, no earlier than 22 January 1992. This change posed a fundamental problem. One of the reasons that Columbia had been assigned virtually all of the longer-duration Spacelab science flights, lasting around ten days or so, was because she was alone in the fleet in having the capability to house as many as five cryogenic oxygen and hydrogen reactant tanks underneath her payload bay floor to supply the electricity-generating fuel cells. (At the time, Discovery and Atlantis carried four tanks apiece.) Prior to the Extended Duration Orbiter modifications, Columbia was thus capable of supporting a maximum mission length of between nine and 14 days. As shown in NASA's manifests from December 1990 onwards, the decision to transfer STS-42 to Atlantis and, subsequently, Discovery, prompted a reduction in the length of the mission, down to just seven days. According to Roberta Bondar, who was selected in January 1990, with Ulf Merbold, as a prime payload specialist for IML-1, this change to duration, but not mission content, made the workload of the astronauts extremely frenetic.

With the presence of Bondar and Merbold, the flight truly earned its moniker of an 'International' Microgravity Laboratory. Roberta Lynn Bondar would become Canada's first female astronaut and the first qualified neurologist to enter space. In her later career, she would gain numerous honours – including the Order of Canada and the Order of Ontario – and in October 2011 became the first astronaut to receive a star on Canada's Walk of Fame in Toronto. Bondar was born on 4 December 1945 in Sault Ste. Marie, Ontario, the youngest daughter of Edward Bondar, an office manager for her home town's Public Utilities Commission, and Mildred, a business and commerce teacher. Her interest in science began at a young age and her father built a small laboratory in the basement of their home. After high school, she entered the University of Guelph to study zoology and agriculture. A master's degree in experimental pathology from the University of Western Ontario in 1971 was followed by doctorates in neurobiology from the University of Toronto in 1974 and medicine from McMaster University in 1977. A qualified and accomplished scuba diver and parachutist, Bondar was selected as one of Canada's first six astronauts – alongside Ken Money – in December 1983. At the time, she was serving as an assistant professor of medicine, with a neurology specialism.

The second IML-1 payload specialist, Ulf Dietrich Merbold, became one of only a handful of the non-career scientist-astronauts to undertake more than one space mission. In fact, when Merbold flew aboard the Spacelab-1 mission in November 1983, he was considered 'West' German by nationality and political status; by the time of STS-42, the two disparate halves of his country had reunified and he thus became the first 'German' astronaut. Merbold came from Greiz, Thuringia, less than 40 km from the birthplace of Germany's first astronaut, the 'East German' Sigmund Jähn. Both Merbold (born on 20 June 1941) and Jähn (four years his senior) had

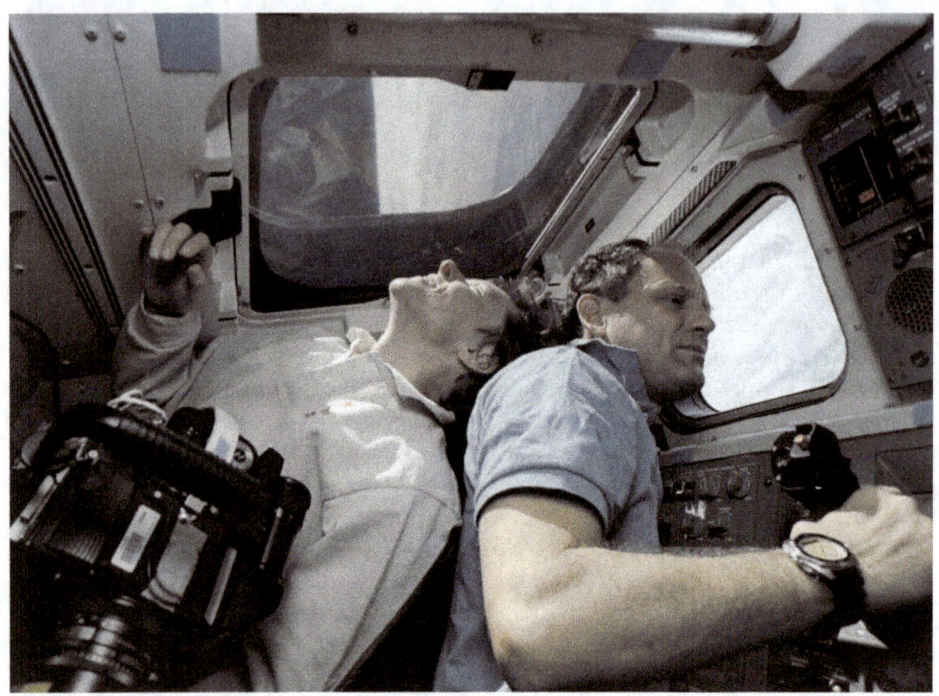

3-02: As Norm Thagard (right) gazes through Discovery's aft flight deck windows, crewmate Roberta Bondar – the first Canadian female astronaut – prepares to photograph a point on Earth through one of the overhead windows.

spent their formative years in the German Democratic Republic, the communist-led East Germany, one of the Soviet Union's satellite states. When he finished high school in 1960, not long before the construction of the Berlin Wall, the young Merbold was one of thousands who defected to the Federal Republic and the democratic West. He attended the University of Stuttgart, receiving a diploma in physics in 1968 and a doctorate in 1976, then joined the Max Planck Institute for Metals Research as a scholar of the Max Planck Society. He rose to become a faculty staff member, working on solid-state and low-temperature physics, with particular research interests in lattice defects in body-centred cubic metals. In December 1977, along with Dutch physicist Wubbo Ockels and Swiss astrophysicist Claude Nicollier, he was selected by ESA from more than 2,000 applicants as a candidate for Spacelab-1 and, in May of the following year, he commenced payload specialist training. In his subsequent career, Merbold would also fly aboard the Russian Mir space station and, in so doing, he became the first European to complete three missions.

The sheer complexity of IML-1's scientific objectives had certainly made the contributions of Norm Thagard as the first payload commander virtually indispensable. Yet there was considerable misfortune still to come. On the calm afternoon of 5 April 1991, Atlantic Southeast Airlines Flight 2311 was approaching Glynco Jetport (today's Brunswick Golden Isles Airport) in Brunswick, Georgia, after a short, hour-long flight from Atlanta. As the jet approached the runway, in clear weather conditions, eyewitnesses reported that it was flying at much lower altitude than normal – less than a hundred metres – after which it suddenly rolled sharply to the left and descended, nose-down, to crash into a patch of trees. All 20 passengers and three crew were killed. Amongst the dead were two small children, together with Texas Senator John Tower ... and IML-1 astronaut Sonny Carter. Initial investigations pointed to severe asymmetric thrust condition with the left-hand engine's propeller control unit, which led to a rapid loss of control. The pilots were unaware of the problem, until it was too late, and did not even have time to declare an emergency.

For the astronaut corps, Carter's death was devastating. "He was never without a smile," wrote Mike Mullane in his memoir, *Riding Rockets*, "and a positive word." In Mullane's mind, Carter's death was "a gross violation of the natural order" – for it was to be expected that an astronaut dying in an aircraft would do so as a *pilot*, not a *passenger*. Two weeks later, veteran astronaut Dave Hilmers was assigned to replace Carter on STS-42. It was, said Don Puddy, the head of Flight Crew Operations, a difficult decision and one which he made with "regret". When the STS-42 astronauts released their official patch, later that year, it included a single gold star, hanging over Earth's horizon ... in memory, said the crew, of "our crewmate, colleague and friend".

The choice of David Carl Hilmers was a logical one. Although he was not a scientist – he was an active-duty Marine Corps officer, with an engineering background – Hilmers had long exhibited a fascination with medicine and intended to pursue it after his astronaut career. The IML-1 mission featured life sciences as one of its two primary research aims. Hilmers came from Clinton, Iowa, where he was born on 28 January 1950. He completed high school in his home state and earned a degree in mathematics from Cornell College in 1972, then entered the Marine Corps in the summer of that year. Completion of Basic School and Naval Flight Officer School brought assignment to Marine Corps Air Station Cherry Point in North Carolina, during which time Hilmers flew the A-6 Intruder as a bombardier-navigator. Subsequent positions included air liaison officer with the United States Sixth Fleet in the Mediterranean Sea and assignments to Marine Corps air wings in Japan and California. He completed a master's credential in electrical engineering from the Naval Postgraduate School in 1977, followed by a degree in electrical engineering a year later. He was selected as an astronaut candidate in May 1980 and by the time of IML-1 had already flown three Shuttle missions. "He just never missed a beat," recalled Bill Readdy in admiration. "You couldn't throw too much information at him. The guy's just a sponge and able to absorb it all and then somehow figure out how to process it and spit it back to you."

Aside from his military background, Hilmers was very religiously conservative – in fact, his devout faith was remarked upon by Mike Mullane – and he had cultivated a long-standing interest in medicine. At one stage he intended to take an Advanced Cardiac Life Support Course. "He wanted to be a doctor," remembered Rhea Seddon. "It was like he never got a chance to be a doctor. He would come and ask me medical questions. I know he had a lot of fun with the medical training, because it was something he wanted to know more about." In his oral history, Norm Thagard recalled a conversation he had with Hilmers in the summer of 1992, a few months after IML-1. "We were discussing what we might do at NASA that would induce us to stay," Thagard said. "Dave said there was nothing that was going to induce him to stay, because he'd been accepted into medical school at Baylor College of Medicine and that's what he'd always wanted to do. He was going to do that, regardless of anything that NASA might offer him." Upon his departure from the space agency in November 1992, Hilmers entered Baylor College, in Houston, and today works as a paediatrician and nutritionist. Adjectives such as "outstanding" have been used to describe him, but the words of Don Puddy sum up Hilmers the best: "A totally unselfish person."

Despite the trauma of Sonny Carter's death, the new addition to the crew must have come as a great pleasure for STS-42 commander Ronald John Grabe ... for both he and Hilmers were members of the May 1980 astronaut class *and* had flown their first space mission together in October 1985. Grabe was born in New York on 13 June 1945 and was an accomplished Air Force fighter and test pilot. After high school, he received his degree in engineering science at the Air Force Academy and spent a year as a Fulbright Scholar at the Technische Hochschule in Darmstadt, West Germany, then returned to the United States to complete pilot training at Randolph Air Force Base in Texas. Grabe flew the F-100 Super Sabre and in 1969 was deployed to Vietnam, where he completed 200 combat missions. Later work included operational testing of the weapons systems of the F-111D Aardvark tactical strike aircraft and in 1974 he was selected for test pilot school at Edwards Air Force Base. Grabe was awarded the prestigious Liethen-Tittle Award for Outstanding Student at the school. Graduation brought an assignment as a test pilot for the A-7 Corsair and the F-111 and he worked as an exchange pilot with the Royal Air Force between 1976 and 1979, flying the Harrier and Sea Harrier at Boscombe Down in Wiltshire. His final assignment before NASA selected him in May 1980 was as an instructor at Edwards. Now, on STS-42, Grabe would be flying his third Shuttle mission overall and, accompanied by 'orbiter team' members Stephen Scot Oswald and William Francis Readdy, would be in charge of the flight deck during around-the-clock payload operations.

"Human space flight," Oswald once said, "is flying airplanes on *steroids*. It's just *really* unforgiving of neglect." The thick-set naval aviator knew from first-hand experience at NASA the intrinsic dangers of both: for within months of his selection as an astronaut in June 1985, he lost classmate Steve Thorne in an aircraft accident and seven other friends – the crew of Challenger – as they speared for orbit. In the case of the latter, the pain experienced by Oswald and fellow newcomers Carl Meade, Jay Apt and Rick Hieb was even closer, for they were in attendance in Florida, supporting the 51L families. Barely six months after selection, the eleventh class of astronauts were on a 'fast track' to space and were already being shifted into their first technical assignments. "We were going to fly in two years," Oswald remembered, "and needed to hurry up. We had about five months of generic stuff, touring around and learning to fly the T-38s, and then they put us in our jobs." On the night before Challenger's fateful launch, Oswald and Hieb had been inside the orbiter's cockpit, attending to a few final tasks.

Oswald was born in Seattle, Washington, on 30 June 1951, and remembered his first encounter with space exploration as a child, when his father took him to see Alan Shepard's Freedom 7 capsule, which was at the World Fair. After waiting in a long queue, the young Oswald finally had the chance to look inside ... and was dismayed by the *smallness* of it. His father stood 1.85 m – about six feet and one inch – and Oswald himself would grow to about the same height. The idea of someday becoming an astronaut vanished. "I just wrote that off," he said. "It's not anything I was interested in." *Flying*, though, was something different. After high school, Oswald entered the Naval Academy and earned his degree in aerospace engineering in 1973, then entered active service and was designated a naval aviator in September of the following year. His early career was spent flying the A-7 Corsair aboard the USS *Midway*, but in 1978, when Oswald was selected to go to Patuxent River and study to become a test pilot, something happened which altered his life trajectory. NASA selected its first group of astronauts for the Shuttle. Crucially for Oswald, his height did not bar him from applying.

Several aviators with whom he had previously worked – including Dan Brandenstein, Rick Hauck and Robert 'Hoot' Gibson – were selected by NASA and in early 1980 Oswald tendered his own application. He got through to the interview stage – and, interestingly, interviewed in the same group as fellow naval aviator Ken Cockrell, with whom he later flew into space – but was unsuccessful and continued his flight test career, conducting performance and propulsion work on the A-7 and the F/A-18 Hornet fighter. A couple of years later, Oswald resigned from active naval duty and joined Westinghouse Electric Corporation as a civilian test pilot. He reapplied to NASA for its May 1984 astronaut class, but again was unsuccessful. "That irritated me a bit," he told the oral historian. "I was thinking that one was going to work out okay." Another route into the space agency quickly presented itself, though, in the form of an offer to become an aerospace engineer and instructor pilot. At first, the offer did not appeal to Oswald. "I had a pretty good job flying airplanes," he recalled, "so it just didn't make sense to do." One weekend, in the middle of 1984, Oswald was at Andrews Air Force Base in Maryland, participating in a Navy Reserve 'drill', flying F-8 Crusaders. Whilst there, Oswald told his friend about the space agency's invitation. At the time, NASA planned for the Shuttle to be flying as often as twice a month by the late 1980s and the friend advised him to accept it. In November 1984, Oswald joined NASA ... and was selected as an astronaut candidate in June of the following year.

As STS-42's pilot, Oswald would work on the same team as Ron Grabe and Norm Thagard and Roberta Bondar. They dubbed themselves 'The Blue Team'. On the opposite shift, the 'Reds', were Dave Hilmers, Ulf Merbold and the team leader, former naval aviator Bill Readdy. "They were 24-hour flights," Readdy said of the IML Spacelab missions, "two-shift flights, and you needed to have a pilot on the other shift. They tended to keep the pilot and the commander on the same shift, so that meant that you had to have somebody else that was schooled in all the orbiter systems and piloting tasks on the other shift." In effect, Readdy was classed as STS-42's 'third' pilot.

In his later career with NASA, Readdy would rise meteorically through the ranks to become the agency's Associate Administrator for the Office of Space Flight – a position to which he had recently been appointed at the time of the Columbia accident in early 2003. Readdy came from Quonset Point, Rhode Island, where he was born on 24 January 1952, the son of a naval aviator who had flown in Korea.

After high school, he entered the Naval Academy, a year after Oswald, to study aerospace engineering and gained his degree in 1974. As a military aviator, Readdy flew the A-6 Intruder and served as an attack pilot aboard the USS *Forrestal* in the North Atlantic Ocean and Mediterranean Sea in 1976-80. This experience was followed by test pilot school and a spell as an instructor at Pax River, and overseas deployments to the Caribbean and Mediterranean aboard the USS *Coral Sea*. His applications to become an astronaut almost mirrored those of Steve Oswald; although Readdy was unsuccessful in his efforts to enter the space agency in May 1984 and June 1985, he resigned from active naval duty and entered NASA as a research pilot in October 1986. Based at Ellington Field in Houston – home of the Texas Air National Guard's 147th Reconnaissance Wing – Readdy served as programme manager for the Boeing 747 Shuttle Carrier Aircraft and was selected as a Shuttle pilot in the June 1987 astronaut group.

"A little footnote, there was an '85 group and then an '86 group" he recalled in his NASA oral history, "but after Challenger happened, that basically went away. There was no '86 selection. There were several of us that had interviewed for that and we got invited down to Houston to work in different support jobs until the next astronaut selection. The support jobs were all related to Return to Flight after Challenger." Readdy's support job on the Boeing 747 included the procurement and modification of the aircraft for Shuttle requirements. "Getting that all put together also gave us an opportunity to upgrade the existing Shuttle carrier airplane with better engines," he said. "It was a very involved structural modification with all the truss structure and everything else. You want to empty the airplane as best you can, so that you can support all that weight, but then you find out that because of *where* the orbiter is located on it – fairly far aft – that you need to keep the airplane in balance. We used to fly with baggage containers of pea gravel in all our ground support equipment in the forward cargo bay of the 747."

Weather concerns and a hydrogen pump anomaly with one of the fuel cells delayed the launch of STS-42 by an hour on the morning of 22 January 1992, but Discovery finally ascended majestically into space at 9:52 am EST. It was a relief for the Spacelab programme, which had its heaviest-ever booking of module flights in the international Space Year. With three missions – IML-1, the first United States Microgravity Laboratory (USML-1) in June and the joint US/Japanese Spacelab-J in September – and only three module units available, STS-42 was committed to flying in January or February in order that its hardware could be turned around in time for the Japanese mission.

By the beginning of the International Space Year, as Steve Oswald prepared to become one of the newest members of the unique fraternity of spacefarers, it seemed that few 'records' remained to be broken. He was not the oldest or youngest astronaut, nor the first from a particular engineering school or military arena. Instead, by his own admission, he was the first astronaut to "use the bathroom" on the crew access level of the launch pad structure. Ordinarily, crew members wore modified diapers during launch, to alleviate the pressure of fluid movement whilst lying on their backs for several hours. During the standard pre-launch countdown dress rehearsal – despite having consumed only coffee and orange juice at breakfast – the incessant blood flow from his legs into his torso took its toll and Oswald *knew* he needed to relieve himself. "Of course, you're out for two and a half to three *hours*," he told a Smithsonian interviewer. "About the time I was going to start to use the diaper, somebody would always talk on the intercom and it would break the spell. Like most of us, after 40 years of being told to *not* pee while you're lying on your back, it's a *hard* thing to do." When the dress rehearsal ended, Oswald had been the first to rush out of the orbiter's side hatch ... and straight into the little boys' room. "I was in that bathroom in a *heartbeat*," he said, "and *that's* how I made history!"

Due to the heavier return weight of the orbiter, with the IML-1 Spacelab module in the payload bay, STS-42 was always scheduled to return to Edwards Air Force Base in California, despite NASA's 1991 decision to place the Kennedy Space Center's Shuttle Landing Facility on an equal footing in terms of priority. The IML-1 experiments were devoted entirely to materials and life science in the microgravity environment and featured significant collaboration from US, European, Canadian and Japanese researchers. Of the life sciences complement, the European-built Biorack sought to understand the fundamental functions of organisms in the microgravity environment, including cell proliferation and differentiation, genetics, gravity sensing and membrane behaviour. To further this research, Biorack carried three incubators, a glovebox and a cooler/freezer to grow, handle and store hundreds of biological samples. Embryonic mouse limb cells were studied in an effort to characterise the similarities observed between skeletal malformations in rodents and human children, with a focus on helping to clarify the processes by which bones heal in microgravity. Previous Soviet long-duration experience had already suggested that bone damage during extended space flights – particularly as far afield as Mars – would be difficult to heal and a contributory factor in the breaking of weakened bones is the loss of calcium through prolonged microgravity exposure. Other experiments utilised the eggs of African clawed frogs and fruit flies, together with yeast, bacteria, lentil roots and plant shoots and the *Physarum polycephalum* slime mould, to understand the role of gravity in embryonic and cell development. The effect of radiation on soil samples and the eggs of stick insects was also closely studied and Germany's Biostack facility analysed the influence of cosmic rays on bacteria and fungus spores, together with thale cress seeds and shrimp eggs.

Elsewhere, housed inside a Spacelab Double Rack, was the Gravitational Plant Physiology Facility, which provided nothing less than a small space-based botanical laboratory, equipped with centrifuge chambers, floodlights, videotape recorders and plant compartments. It supported investigations into the gravity-sensing mechanisms of oat seeds and the reactions of wheat specimens to the effect of light stimulation.

With Canadian astronauts Roberta Bondar and Ken Money having trained extensively for IML-1 payload science activities, the contribution of their nation to the mission was correspondingly important. Canada's Space Physiology Experiments focused on the adaptation of the human organism to the weightless

3-03: Bill Readdy studies a checklist as he measures the veins in his lower right leg on Discovery's middeck. He uses an electronic monitor and a pair of large blood pressure cuffs to measure changes in blood volume. This view highlights the relative smallness of the middeck, with lockers behind Readdy's back and the astronauts' phonebox-sized sleep stations at far right.

environment, including the vestibular apparatus, the body's sense of position, energy expenditure, cardiovascular adaptation, eye-motion oscillations and back pain. The latter was devised in response to a typical incidence of spinal lengthening by 5-7 cm and back pain in over two-thirds of all astronauts.

In support of these experiments, members of IML-1's payload crew utilised 'The Sled', in the centre aisle of the Spacelab module, and wore a helmet instrumented with accelerometers to measure head motions and visors to provide visual stimuli. The sled could be moved backwards and forwards predictably, at a constant speed over the same distance, or at varying distances or at varying speeds with sudden stops and starts. As crew members underwent the test, their response was monitored to gauge their ability to visually track moving objects. Studies of the adaptation of the 'otolith' – the gravity-sensing part of the inner ear – and its effect upon the nervous system, together with head and eye movements, were also performed in the sled and in a swivelling chair, part of NASA's Microgravity Vestibular Investigations. Despite a tripped circuit breaker, which caused the chair to stop working a few seconds into its first run, the experiment produced pleasing results which "quantified the human vestibular function in the microgravity environment". Additional experiments required the crew to drink water, especially enriched with stable, non-radioactive isotopes of oxygen and hydrogen, to enable researchers to determine energy expenditure through post-flight urinalysis.

In the microgravity science arena, NASA experiments in vapour-driven protein crystal growth were undertaken, together with German Cryostat investigations which employed 'liquid diffusion' and offered researchers the flexibility of a temperature-controlled facility. Specifically involved in the experiments was Beta-galactosidase, a key enzyme found in the intestines of human and animal infants, which assists in the digestion of milk products. (This enzyme was the first protein ever crystallised in space in November 1983.) Also under study was the satellite tobacco mosaic virus.

NASA's Fluids Experiment System and Vapour Crystal Growth System carried a range of investigations which grew crystals of triglycine sulphate and mercury iodide, as well as performing laser diagnostic recording and creating more than 300 three-dimensional structural holograms for post-mission analysis. Several of the mercury iodide experiments were supported by CNES, the French national space organisation, and record-sized crystals measuring approximately $16 \times 16 \times 8$ mm were yielded. The Japanese Organic Crystal Growth Facility sought to produce semiconducting crystals of tetrathiafulvalene (TTF, a compound with applications in molecular electronics) and nickel, whilst demonstrating the effectiveness of an epoxy cushioning material to damp accelerations otherwise known to disrupt the growth process, and the European Critical Point Facility explored the behaviour of fluids when they reached the precise temperature-pressure stage at which the difference between vapour and liquid became indistinguishable. Samples of sulphur hexafluoride – primarily used on Earth in the electronics industry – as part of a number of experiments were processed for as long as 60 hours during IML-1. Radiation and acceleration measurements were performed to determine the influence of Shuttle motions upon sensitive experiments.

Beyond the Spacelab module, at the rear end of Discovery's payload bay, was a 'bridge' of Getaway Special canisters. On STS-42, the bridge was originally manifested to carry a maximum load of 12 canisters, but two were dropped due to technical difficulties and only ten eventually flew. (To satisfy the centre-of-gravity constraints, a pair of 'ballast' canisters made up the full dozen.) The experiments came from various nations, including China. The latter was a debris motion study and materials experiment and marked the first time that a Chinese payload had ever flown on the Shuttle. Elsewhere were US investigations into the growth of brine shrimp, a German convection study, a Swedish alloy experiment, a Japanese study of gas bubbles in liquids and Australia's Endeavour ultraviolet telescope.

Despite the decision, late in 1990, to reduce the length of STS-42 from ten to seven days, an option was preserved to extend the mission by up to 24 hours if on-board consumables remained acceptable. As circumstances transpired, the crew's consumables usage remained below planned levels and an additional day was flown in order to complete the collection of scientific data in the IML-1 module. Touchdown eventually took place on Edwards' concrete Runway 22 at 8:07 am PST (11:07 am in Florida) on 30 January, completing a voyage of a little over eight full days.

MEETING OF MINDS

Two remarkable men met on the morning of 2 April 1992. One of them was on his second day as NASA Administrator, whilst the other was an astronaut, newly returned from space. Today, Daniel Saul Goldin – NASA's longest-tenured Administrator, who served until December 2001 – is remembered both positively and negatively; positively for having transformed the troubled Space Station Freedom from a project on the brink of cancellation to a fruitful, revitalised endeavour which saw the Russians courted as full partners, and negatively in that the United States' aspirations to explore beyond Earth orbit with humans were indefinitely set back in favour of faster, better and cheaper robotic craft. The other man, Charles Frank Bolden Jr, had already made history as one of the United States' first black spacefarers and at the time of writing is NASA's first African-American Administrator. Yet the first exchange between the two men was not quite the meeting of minds that Goldin had anticipated.

"I'm Dan Goldin," he began. "I'm the new NASA Administrator and I want you to come and work for me."

The pair shook hands, but Bolden had no desire to leave his job at the Johnson Space Center in Houston for one at NASA Headquarters in Washington.

Goldin thought for a moment. "Well, when you get finished with your debriefs, come and talk." Shortly thereafter, Bolden went to Washington and was impressed by the space agency's new chief – "The guy was a visionary," he reflected – to such an extent that he accepted Goldin's offer to become Assistant Deputy Administrator. Within six months, however, Bolden would be informed of his selection to command another Shuttle mission, scheduled for November 1993. When he learned of the mission's content, Bolden did not want it ... for STS-60 would be the first flight in the new co-operative enterprise with Russia and its crew would include a cosmonaut. As an active-duty colonel in the Marine Corps, Bolden had spent his entire adult life *hating* Russia and all it stood for. He was partially appeased by the offer to at least *meet* the two cosmonaut candidates – Vladimir Titov and Sergei Krikalev – and immediately liked them both. Titov spoke no English and, recalled Bolden, "couldn't even ask for water", although his resilience in the face of this struggle and throughout a year of arduous training was admirable. Years later, Bolden would remember this period of his life as the most memorable, for he made unlikely friends and forged the early bonds of an international partnership between two former foes which endures to this day.

Twenty years ago, at the beginning of 1992, several of the crewing woes which affected STS-42 also impaired Bolden's STS-45 mission. Originally, the flight had a science crew of NASA mission specialists Kathy Sullivan and British-born Mike Foale, together with two career scientists as payload specialists: biomedical engineer Byron Lichtenberg and physicist Mike Lampton. Both Lichtenberg and Lampton had a long association with NASA, having been selected in 1978 as payload specialist candidates for Spacelab-1.

Byron Kurt Lichtenberg came from Stroudsburg, Pennsylvania, where he was born on 19 February 1948. His yearning to become an astronaut started from an early age; reading science fiction books and seeing the first pioneers set off on their missions of exploration in the early 1960s. Lichtenberg knew that they were military test pilots with advanced engineering or science credentials and he dedicated himself to following a similar career path. He received a bachelor's degree in aerospace engineering from Brown University in 1969 and entered the Air Force, flying the F-4 Phantom and F-100 Super Sabre supersonic fighters and the A-10 Thunderbolt ground attack aircraft. During the course of the Vietnam conflict, Lichtenberg flew 238 combat missions, for which he received two Distinguished Flying Crosses and ten Air Medals. The pull of the space programme remained strong, however, and he left active duty in 1973 to return to academia. However, he remained a military reservist, serving with the Massachusetts Air National Guard and providing close aerial support in the A-10. A master's degree in mechanical engineering from Massachusetts Institute of Technology came in 1975 and a doctorate in biomedical engineering – from the same institution – in 1979. "I realised if I was a fighter pilot with a doctorate," he explained years later, "I would have a better chance of getting into space." Whilst undertaking his doctoral studies, Lichtenberg applied, unsuccessfully, for both the 1978 *and* 1980 NASA astronaut selections. At around the same time, he was selected by the IWG as a candidate for the US payload specialist slot on the Spacelab-1 flight.

Lichtenberg went on to fly on Spacelab-1, with Mike Lampton as his backup, and in May 1984 the two men were assigned to support a future Shuttle flight, carrying the first in a series of Earth Observation Missions (EOM-1), then scheduled for launch in the summer of the following year. As early as February 1985, however, *Flight International* noted that NASA had cancelled EOM-1, reassigned a number of its NASA 'core' crew members to other missions and had merged it with the *second* mission in the series, EOM-2. In December 1985, Belgian physicist Dirk Frimout and

US physicist Charles 'Rick' Chappell were named as Lichtenberg and Lampton's backups. By this time, the renamed 'EOM-1/2' mission had slipped until the middle of August 1986. A summertime launch would have benefitted one of the EOM-1/2 experiments – the European Space Agency's Metric Camera, mounted in the roof of a Spacelab short module – by offering a better chance of good weather over primary land masses in the northern hemisphere. All hope for EOM-1/2 was lost on the cold morning of 28 January 1986 when Challenger exploded and the Shuttle fleet was grounded.

In the year after Challenger, former astronaut Sally Ride chaired a task force to formulate a new strategy for NASA. One of its four key recommendations was for the implementation of a 'Mission to Planet Earth', designed to foster new technologies for the study of our home world, its atmosphere and climate. It was at around the same time that EOM-1/2 underwent a figurative and literal metamorphosis, changing both its configuration and its name. Instead of the short pressurised module, single Spacelab pallet and Mission-Peculiar Equipment Support Structure (MPESS), it would fly on a pair of pallets and its control and command systems would be housed in a pressurised 'igloo', not dissimilar to that flown aboard STS-35. Of course, EOM-1/2 had been planned as long ago as 1983, and was not originally conceived to be part of Mission to Planet Earth, "but the nature of its scientific work," said Kathy Sullivan, "really genuinely did align with the purposes of that new programme, which was gaining momentum and clarity".

The mission received an impressive new name: the Atmospheric Laboratory for Applications and Science (ATLAS). Early plans called for a series of up to ten ATLAS flights, launched every 12-18 months, to undertake research in atmospheric chemistry, solar and space plasma physics and ultraviolet astronomy. As part of NASA's Mission to Planet Earth, ATLAS sought to characterise solar energy inputs across an entire 11-year activity 'cycle' of the Sun, recording seasonal elements of atmospheric change over the long term. Of additional importance was assessing the extent of human impact upon the atmosphere, through agriculture, forestry and heavy industrial processes. It had already become clear by the late 1980s that chlorofluorocarbons, together with halons and naturally occurring chemicals, such as methane and nitrous oxide, and increasing concentrations of carbon dioxide, were playing a significant role in the depletion of Earth's stratospheric ozone layer and impacting global atmospheric temperatures. The ATLAS-1 experiments were designed to explore the features, gaseous constituents and solar effects upon the troposphere (which extends from the surface to an altitude of about 20 km), the stratosphere (20-50 km), the mesosphere (50-85 km) and the thermosphere (85-690 km). Mounted on the two pallets were 12 instruments to support 14 experiments from the United States, France, Germany, Belgium, Switzerland, the Netherlands and Japan. Several were scheduled to be reflown on subsequent ATLAS missions. "Reuse of these facilities," noted NASA's ATLAS-1 press kit, issued in March 1992, "also will allow scientists to expand their base of knowledge to provide a more accurate, long-term picture of Planet Earth and its environment."

Of these experiments, NASA's Far Ultraviolet Space Telescope (FAUST) sought to acquire spectra of high-temperature celestial objects at far ultraviolet wavelengths (between 1,300-1,800 angstroms), as part of an effort to gain insight into the life cycles of distant hot stars, faint diffuse galactic features, large nearby galaxies, quasars and stellar nebulae. Unfortunately, on its previous flight, Spacelab-1 in November 1983, a fogged film ruined many images and an investigation recommended that, when it flew next, the telescope should record photons electronically, rather than on film as time exposures, to better pinpoint the cause. It did, however, achieve 95 percent of its scientific objectives and took the first far ultraviolet image of the Cygnus Loop, a relatively close supernova remnant. Its activities on ATLAS-1 were unfortunately hampered by a blown fuse which left it without power and its aperture door stuck open, preventing it from being reactivated in flight.

Elsewhere on the ATLAS-1 pallets, the Active Cavity Radiometer (ACR), the Measurement of Solar Constant (SOLCON) and Measurement of Solar Spectrum (SOLSPEC) were also veterans of Spacelab-1 and were designed to precisely track the total amount of solar energy received by Earth's atmosphere and its impact upon our planet's environment to further the study of the solar-terrestrial relationship. Of particular note, ACR had a direct counterpart aboard NASA's Upper Atmosphere Research Satellite (UARS), launched in September 1991. Through comparison with SOLCON data, its role on the ATLAS missions was to support extended solar irradiance experiments, in order to calibrate UARS' instruments over the long term, and establish radiation scales at the solar total flux level. Meanwhile, SOLCON itself was a self-calibrating radiometer, tasked with measuring the absolute value of the solar 'constant' and monitoring its long-term variation. Working in close conjunction was SOLSPEC, whose trio of spectrometers observed the variability of solar irradiance with an accuracy of better than 0.1 percent.

Other former Spacelab-1 experiments carried over onto ATLAS-1 included NASA's Imaging Spectrometric Observatory (ISO), which comprised five spectrometers, housed in a single unit, and examined the presence and relative abundances of oxygen, nitrogen and sodium in the mesosphere. ISO formed part of an extensive project to build a comprehensive database of the atmosphere's vertical structure and its energy-transfer processes. Meanwhile, the European Grille Spectrometer studied the dynamic behaviour of the gaseous constituents of the stratosphere, mesosphere and thermosphere. Unfortunately, this device – so-called because it employed a special 'grille' as a window for part of its optical system and as a mirror for the other – had been only partially successful on Spacelab-1, due to lengthy launch delays and unfavourable observing conditions, but would more than prove its worth on ATLAS-1. Among other achievements, it would detect far higher quantities of chlorine in the upper atmosphere than scientists had previously anticipated. One of its key sets of measurements was the Energetic Neutral Atom Precipitation (ENAP), which observed very faint emissions arising from fluxes of energetic neutral atoms in the thermosphere.

Particularly visible on the ATLAS-1 pallets were the black spheres of the Space Experiments with Particle Accelerators (SEPAC), designed by Japan's Institute of Space and Astronautical Science, based in Tokyo. It comprised an 'electron gun' to investigate the charged-particle dynamics of the ionosphere and during the mission was used to emit a stream of xenon plasma to 'clamp' the electrical potential of the Shuttle to the plasma potential of the upper atmosphere, as part of efforts to gain a clearer understanding of aurorae, the nature of the planet's magnetic and electric fields and the effects on the orbiter itself. During Spacelab-1, SEPAC's electron beam assembly – capable of operating at voltages of between 500 volts and 7.5 kilovolts at 1.6 amps – failed to function in a high-power mode, although it returned pleasing results. "We were all pretty jazzed up about this," said Kathy Sullivan of SEPAC. "The idea was to have the orbiter oriented so that the aperture of the instrument would inject these electrons roughly along the magnetic field line down towards the atmosphere, near the polar regions." It was, in her words, a 'dose-response' experiment; just like in medicine, it injected a known dose of energy into the atmosphere and measured the brightness of the resultant aurora-type glow. "If I

3-04: Appropriately backdropped by Africa's Atlas Mountains, whilst Atlantis flew high above Mali, the ATLAS-1 hardware was the first dedicated Shuttle research mission to explore the Sun, the atmosphere and their respective interactions. The black spheres of the SEPAC experiment are particularly prominent, whilst on Earth the Iguidi 'sea' of sand dunes, together with the edge of a large sandstorm, can be clearly seen.

know I put in this many kilowatts of energy and I measured that luminosity," she reasoned, "maybe I can start to get a clearer understanding of how the energy of the incoming solar particles couples into the atmosphere and creates auroral luminescence." To Sullivan, SEPAC was "the biggie".

Working in conjunction with the electron-gun investigation were experiments which sought to measure atmospheric constituents, 'trace' molecules – including carbon dioxide and ozone – and other gaseous emissions, together with the impact of solar inputs on the chemistry of the upper atmosphere. The Atmospheric Trace Molecule Spectroscopy (ATMOS) was designed to determine compositional structures between the altitudes of 20-120 km – from the base of the stratosphere to the low thermosphere – and examine the 'partitioning' of solar energy within those various gaseous levels. During ATLAS-1, data from both ATMOS and Grille revealed clear evidence of aerosol bands resulting from the Mount Pinatubo eruption. The Atmospheric Emissions Photometric Imager (AEPI) investigated transport processes within the ionosphere (a region embedded in the mesosphere) and worked with ISO and SEPAC in observing the optical properties of artificially-induced electron beams, naturally-occurring aurorae and atmospheric 'airglow' and Shuttle-generated emissions. The Millimetre Wave Atmospheric Sounder (MAS) focused on temperature and ozone distribution in the stratosphere, mesosphere and lower thermosphere and provided critical correlative measurements with UARS' Microwave Limb Sounder experiment. Also closely linked with UARS was the Solar Ultraviolet Spectral Irradiance Monitor (SUSIM), which operated concurrently with a similar experiment aboard the satellite to improve the absolute accuracy of solar irradiance measurements. Finally, the Atmospheric Lyman-Alpha Emissions (ALAE) measured the relative abundances of hydrogen and deuterium in the atmosphere.

Flying again as part of efforts to calibrate ozone-monitoring sensors aboard the National Oceanic and Atmospheric Administration's NOAA-9 and NOAA-11 satellites, together with NASA's Nimbus-7, the Shuttle Solar Backscatter Ultraviolet (SSBUV) instrument was also aboard ATLAS-1. Although it had flown on three previous occasions, STS-45 represented the first SSBUV mission in the wake of the September 1991 eruption of Mount Pinatubo in the Philippines. Its results from ATLAS-1 onwards enabled scientists to confirm an approximately ten-percent depletion of ozone in the northern hemisphere and at mid-latitudes, probably triggered by the presence of residual aerosols in the upper atmosphere after the eruption. Moreover, the effects of cold stratospheric temperatures in the winter of 1992-93 would also be seen in subsequent SSBUV data.

Sullivan, Foale, Lichtenberg and Lampton were assigned as the ATLAS-1 science crew in September 1989, with launch anticipated in March 1991. Backing up the two payload specialists were Dirk Frimout and Rick Chappell. A few months later, in May 1990, the other members of the crew were named, with Bolden in command, joined by pilot Brian Duffy and a third mission specialist, Dave Leestma. Like IML-1 before them, the ATLAS-1 crew would operate two 12-hour shifts, around the clock, with Bolden, Duffy, Sullivan and Lampton on the 'Blue Team' and Leestma, Foale and Lichtenberg on the 'Red Team'. The shift planning was logical, reasoned Sullivan, because it enabled a proper spreading of expertise across both teams. (She,

Bolden, Leestma and Lichtenberg had all flown before, whereas Duffy, Foale and Lampton were 'rookies'.) Although not restrained to a single shift, Bolden aligned his schedule with that of the Blues, since it worked well into the mission's planned re-entry and landing timelines.

Charles Frank Bolden Jr was born in Columbia, South Carolina, on 19 August 1946. "I never wanted to be an aviator," he told the NASA oral historian. "I saw a programme on television, called *Men of Annapolis*, when I was in eighth grade; fell in love with the uniform, fell in love with the fact that they seemed to get all the good-looking girls!" After high school, where his father served as the head football coach, Bolden entered the Naval Academy and graduated with a degree in electrical engineering in 1968. His original intent was to become a Navy frogman or an underwater demolition expert, but after graduation he entered the Marine Corps, inspired by an infantryman, Major John Riley Love, who had been his first company officer at the academy. "I knew that infantry officers *died* real quick when they went to Vietnam," Bolden said. The devastating Tet Offensive had recently occurred and life expectancies for infantrymen were expressed in *months*, rather than years, but he pressed on through Basic School, then decided to change course and enter Marine aviation instead. He underwent flight instruction, received his wings in May 1970 and later flew 100 sorties – many of them in the hours of darkness – over Vietnam in the A-6 Intruder.

The lure of test pilot school was strong, although Bolden met with many rejections. Upon his return to the United States from Vietnam, he served as a selection and recruitment officer in Los Angeles and took various assignments at Marine Corps Air Station El Toro, California. After half a dozen attempts, he was finally accepted by the test pilot school at Patuxent River in Maryland and graduated in June 1979. Years later, he was convinced that a master's in systems management, earned from the University of Southern California in 1977, was instrumental in this success. At around this time, Bolden picked up a NASA application form, but did not fill it in, preferring not to waste his own time and that of the Marine Corps. However, when he *met* some of the group who *had* been selected, Bolden decided to try his hand and was interviewed in Houston. At the end of May in 1980 – on his wife's *birthday*, of all times – he received the call from George Abbey that would change his life. Twelve years later, with two flights as a Shuttle pilot behind him, he was ready to lead his first astronaut crew into orbit.

As a first-time Shuttle commander – and only the second black astronaut to accomplish the feat – Bolden took up an option from the NASA operational psychologists to have his crew participate in a personality evaluation, based along the lines of the Myers-Briggs psychometric questionnaire, to ensure that the team would function at their peak in terms of performance and cohesion. "We were going to fly two out of four guys who'd been sitting around for a decade, waiting," said Sullivan. "Quite a different mix of folks. I think Charlie knew he wanted to look at everybody to have a sense of how best to move them and drive them, support, encourage and propel them." The tests revolved around six personality types, two of which described focused, goal-oriented individuals, which Sullivan noted was a trait representative of about 15 percent of the *general* population ... but around 98 percent of the *astronaut* population. Such personality characteristics for astronauts are hardly surprising, but Sullivan and Bolden considered the psychometric tests enlightening in that they uncovered each crew member's strategies for handling periods of calm and extreme stress.

Kathryn Dwyer Sullivan was born in Paterson, New Jersey, on 3 October 1951, although her family moved to California when she was six years old and she attended Californian schools, finally entering the University of California at Santa Cruz to study Earth sciences. Whilst there, she spent a year as an exchange student at the University of Bergen in Norway. Sullivan received her undergraduate degree in 1973 and a doctorate in geology from Dalhousie University in Halifax, Nova Scotia, in 1978. Her PhD research focused on remote sensing of the seafloor in the Mid-Atlantic Ridge, the Newfoundland Basin and fault zones off the coast of southern California using acoustic and geophysical instruments.

When NASA issued its call for astronaut candidates, Sullivan applied, but she also had an offer of a post-doctoral position in deep-sea marine geology, using the Alvin submersible. "Two fabulous things were in front of me," she recalled, "either of which just seemed tremendous things to get to be involved in. It made my mother a little crazy that I was either going to 10,000 feet *down* in the ocean or 200 miles *up* off of the planet!" She was summoned to Houston for her interview and then accepted into the astronaut corps. In October 1984 she became America's first female spacewalker and in April 1990 she was amongst the crew of STS-31, which deployed the Hubble Space Telescope.

The Shuttle's hydrogen leaks in the summer of 1990, and the inevitable impact on future missions, meant that STS-45 was moved off Columbia and reassigned to Atlantis, with launch shifted from March 1991 until March 1992. Unperturbed, the seven astronauts, together with Frimout and Chappell, continued their training. Then, with just six months to go, in early September 1991, Mike Lampton was abruptly removed from the flight, due to a medical issue, and was replaced by Frimout. In her NASA oral history, Sullivan recalled that Lampton's ailment began around Christmas 1990, but as he "got life-threateningly ill" over the following months, Lennard Fisk, NASA's Associate Administrator for Space Science and Applications, decided that he should be replaced. It hit the rest of the STS-45 crew particularly hard. As the mission's payload commander, Sullivan remembered the cross-evaluation process which led to the formal announcement of Frimout as Lampton's replacement. "Both very competent," she said of the two men. "There wasn't really any high-level distinguishing factor there."

Ennobled as a viscount in the wake of his Shuttle mission, Dirk Dries David Damiaan Frimout came from Poperinge, Belgium, where he was born on 21 March 1941. He received elementary schooling in his home town and studied electrical engineering at the University of Ghent, gaining his degree there in 1963 and his doctorate in applied physics in 1970. He subsequently undertook post-doctoral research in atmospheric science at the University of Colorado. Frimout's early career was spent at the Belgian Institute for Space Aeronomy, working on stratospheric balloon and sounding rocket projects, before joining ESA in 1978 as the Experiment Co-ordinator for Spacelab-1.

As Belgium's first astronaut, Frimout earned overnight fame on STS-45. Launch was originally scheduled for 23 March 1992, but was halted when higher than acceptable concentrations of liquid hydrogen and oxygen – 860 parts per million, far higher than the maximum allowable 500 ppm – were detected in the orbiter's aft compartment during operations to slow-fill the External Tank. Efforts to troubleshoot the problem failed to reproduce the cause and this led engineers to the conclusion that it was the result of plumbing in the main propulsion system being improperly conditioned to the propellants. Launch was postponed by 24 hours and although liquid oxygen concentrations peaked above the maximum allowable limit, they rapidly recovered in what NASA later described as "anticipated and acceptable". Atlantis set off at 8:13 am EST on the 24th.

For the first-time fliers, it was an awesome experience. "I was really impressed with just how much raw power there is in that vehicle," Brian Duffy told the NASA oral historian. "You can think about what it might be like, but you don't actually physically *feel* it until you go do it. The simulator is great in that it can give you vibration and some little sense of motion, but it doesn't give you that acceleration. I had flown Mach 2 in an F-15 in my career, many times, and had thought *that* had been pretty fast ... and we blew through Mach 2 in nothing flat and we *still* had a long way to go to accelerate!" After the separation of the boosters, Atlantis continued to climb, under the thrust of her main engines, and although there were few 'good' visual references it was obvious to Duffy that their velocity was immense. Every few seconds, he would glance at the acceleration tapes on his instrument panel and would register momentary astonishment as it zinged its way higher and higher up the scale. This cacophony of controlled violence seemed to end when the main engines shut down, as planned, some eight minutes into the mission, and the entire cockpit fell deathly quiet. "You go from the most violent place you've ever been in your life," Duffy said, "to the most peaceful place you've ever been in your life ... in a couple of *seconds*."

Born in Rockland, just south of Boston, Massachusetts, on 20 June 1953, Duffy grew up watching the contrails of jet aircraft from nearby South Weymouth Naval Air Station and aspired for a career in aviation. After completing high school in Massachusetts, he entered the Air Force Academy and earned a degree in mathematics in 1975. "To be a 19-year-old ... in the back seat of an F-4 [Phantom]," he said, "going 600 miles an hour made me realise *that* was something I really wanted to do." Undergraduate pilot training followed at Columbus Air Force Base, Mississippi, and Duffy spent the next several years flying the F-15 Eagle fighter out of Langley Air Force Base, Virginia. He was deployed overseas to Japan, earned a master's credential in systems management from the University of Southern California in 1981, completed test pilot school as a Distinguished Graduate in 1982 and directed F-15 testing at Eglin Air Force Base in Florida. By his own admission, Duffy loved it.

Then, one Friday afternoon in the late spring of 1985, at the end of a weekly pilots' meeting the squadron commander concluded with exciting news: NASA had issued a call for Shuttle astronaut candidates. As Duffy looked over the rigorous academic and professional requirements, he was astonished to realise that he satisfied them all. "And I thought, I *couldn't* let that eight-year-old kid from so many years ago down to find out that I was qualified to apply and not do it," he recalled. "Would I be able to live with myself?" Selected in June 1985, he brought his family to Houston later in the summer and by the end of the following January the Duffys had built a new home in El Lago and had just retrieved their boxes from storage. At around the same time, Challenger exploded, stalling the Shuttle for almost three years. As circumstances transpired, Duffy was the last member of his class to fly into space and it suited him just fine.

"There used to be a policy that when you flew the first time, you got what we called a 'flight promotion'," he told the NASA oral historian, "where, say, if you launched as a lieutenant-colonel, you would land as a colonel; you'd get a one-rank promotion." In the wake of Challenger, the Department of Defense cancelled this policy and Duffy's class was the last group of astronauts to whom this flight promotion would apply. Duffy knew how his promotions were due to fall in time and asked chief astronaut Dan Brandenstein to consider him for a later flight assignment. As circumstances transpired, he was the last member of his class to enter space. On 24 March 1992 he launched as a lieutenant-colonel in the Air Force and landed nine days later as a colonel.

Within minutes of reaching a 296 km orbit, the seven astronauts divided themselves into their two teams to begin the activation of the ATLAS-1 experiments. Inclined at 57 degrees to the equator, Atlantis' orbit carried her approximately as far north at Juneau in Alaska and a little further south than Tierra del Fuego in Argentina, enabling atmospheric scientists to gather data from the tropics to the auroral regions and over diverse geographical areas, from rainforests to deserts and oceans to landmasses. ATLAS-1 activation was completed within three hours, although a 13-minute delay to the launch shifted 'shadow times' by just over 1.5 degrees, which required the adjustment of experiment observation timings.

Neither of the two 12-hour teams saw themselves to be in 'competition' with the other. "Charlie didn't ever really use such a device like that to drive performance," remembered Sullivan. "Commitment to each other, commitment to the mission [were] the intrinsic factors that he exemplified and reinforced. He wouldn't have needed to set up some fake game for me to make me do anything better." Dave Leestma, who was in charge of the Red Team, broadly agreed, although he admitted to some light-hearted competition during training. "Maybe one day," he told the NASA oral historian, "the Red Team has their sim and the Blue Team has it the next day. When you're done, you ask your training team: 'Well, how did we do compared to those guys?' There's always that natural competitiveness." As a mission specialist, Leestma was effectively in charge of the flight deck during each of his team's 12-hour shifts and therefore was responsible for executing a number of manoeuvres, flying the vehicle in order to direct the ATLAS-1 instruments. Primary responsibility for the instruments themselves, and the science, fell to the members of the payload crew: Foale and Lichtenberg on Leestma's shift, Sullivan and Frimout on Bolden and Duffy's Blue Team.

David Cornell Leestma came from Muskegon, Michigan, where he was born on 6 May 1949, although he attended high school in California and entered the Naval

Academy to study aeronautical engineering. He graduated first in his class in 1971 and was assigned to the frigate USS *Hepburn*, before receiving orders to report to the Naval Postgraduate School for a master's degree, also in aeronautical engineering. Leestma subsequently moved into flight instruction, received his aviator's wings in October 1973 and underwent training on the F-14A Tomcat fighter. During the next few years of his naval career, he participated in three overseas deployments to the Mediterranean and North Atlantic, flying off the USS *John F. Kennedy*, and in 1977 was sent to Naval Air Station Point Mugu in California as part of the air testing and evaluation squadron. Leestma served as an operational test director on the Tomcat, with his particular focus on the development of new tactical software and programmable signal processors. Whilst aboard the *Kennedy*, his operations officer advised him to apply for NASA's 1978 astronaut intake – and even sent off for the application form on his behalf – but Leestma was too busy with his other duties and put such thoughts to one side. By the end of 1978, he was back on shore duty and when the space agency announced its intent to select more astronauts, he dusted off the application form and submitted it. He was summoned to Houston, part of the very first group to be interviewed, in February 1980, and heard no more for several months. "And then, in late May of 1980, I got this call, early in the morning," he told the NASA oral historian. "I was in California, so it was two hours earlier than here [in Houston]. As I found out later, they start calling around eight o'clock in the morning and it was around six-thirty or something ... and it was George Abbey on the phone. Now, I didn't know that if George calls you, then *you're in*, and if somebody else calls you on the board, then you're *not* in!" Leestma was told to report to JSC in early July, but remembered that selling his house in California and getting his military orders from the Navy took longer than expected. He ended up reporting a week late, but wound up as the first member of his class to fly into space.

The rear of the flight deck was a hive of activity during both shifts, with the relatively small space packed with hand-held cameras, photomultipliers and filters. Twenty-four hours into the mission, the first firing from the coffee-can-like SEPAC electron beam generator was scheduled to occur in the early stages of a Red shift. "Us Blue guys were all down below" in the middeck, recalled Sullivan, "mucking around with dinner and starting to get changed for sleep. We knew it was coming and we were eager to see it, but we thought we should get out of their way, let them get set up for this and get into this."

All at once came a cry from upstairs: *"Oh my God, look at THAT!"*

"There's a cardinal rule on spaceflights," said Sullivan, "or at least on all *my* crews, which is 'There shall be no sentences from the flight deck ending in *that*', as in 'What the hell was *that*?', because you'll terrify the guys down below, who can't see anything." As a consequence, the Blues floated up to Atlantis' flight deck to witness the first firing. The spectacle reminded Sullivan of something from a sci-fi movie, as an oscillating blue blob of energy – looking like "some luminescent blue creature ... about to ooze out" – lingered atop the SEPAC canister and abruptly shot into the upper atmosphere.

"It goes by quickly," she recalled. "It was starting to curve away. You could see

Meeting of minds 227

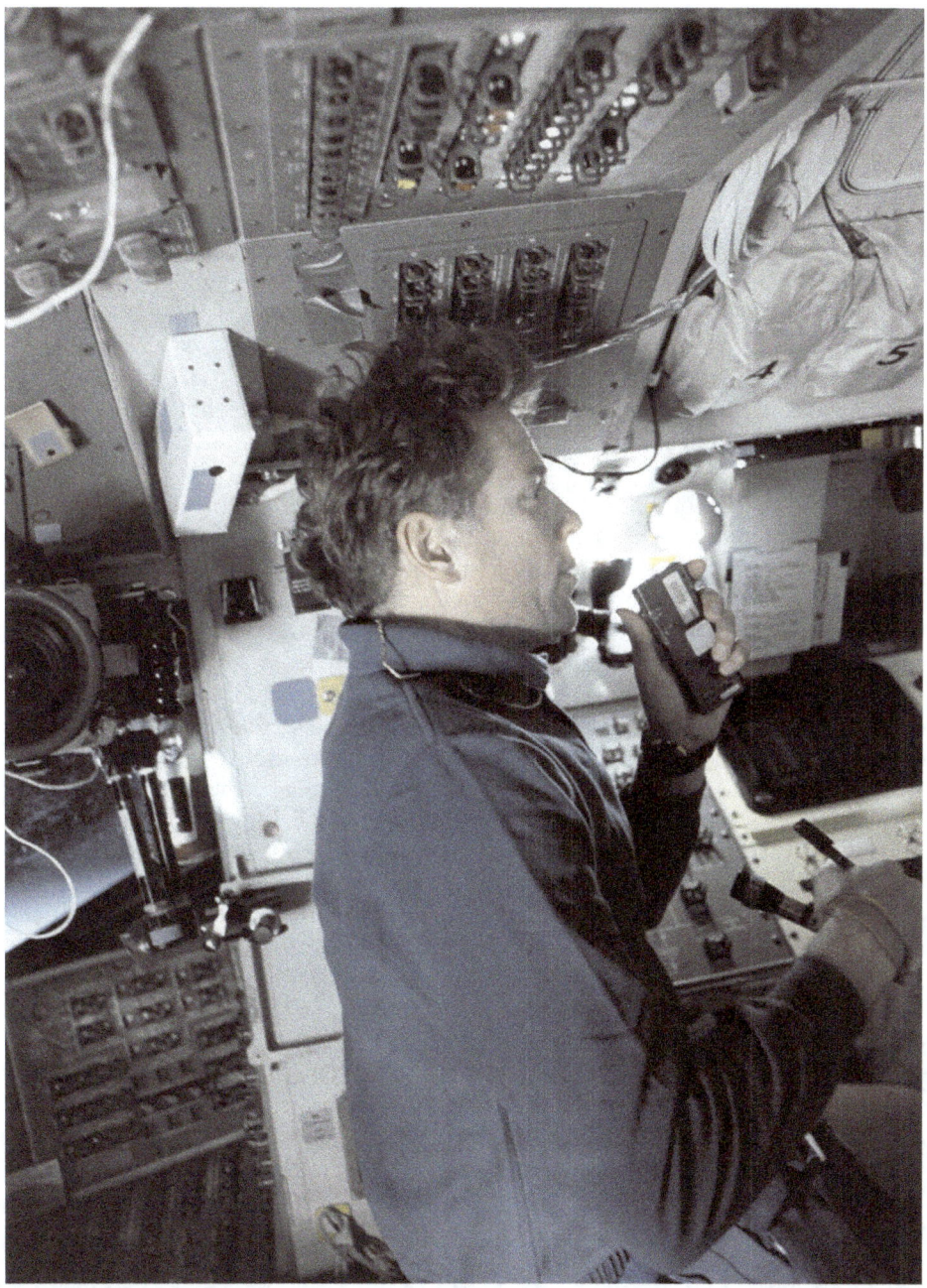

3-05: Byron Lichtenberg became one of relatively few non-career payload specialists to make a second space mission on STS-45. He is pictured at the commander's station on Atlantis' forward flight deck.

the curvature of the magnetic field line. You could just see it begin to spiral along. All this material you drilled into your head in college physics ... and *now* you're seeing, in front of your eyes, the curvature of the magnetic field lines, the electron gyro-radius as this thing spirals around it." The Red Team's exclamation was surely understandable when faced with such a captivating sight. SEPAC reminded Leestma of a phaser or some futuristic space gun from *Star Wars*. Interestingly, the STS-45 crew participated in the presentation of the Irving G. Thalberg Memorial Award to *Star Wars* creator George Lucas whilst in orbit. They also carried a real Oscar statuette aboard Atlantis.

For Sullivan, although she admitted to being a fan of Lucas' films, "*our* photon torpedoes", from SEPAC, "were *much* better!" Yet they were not quite as long-lived. Without warning, on only its second or third firing, SEPAC's electron beam assembly abruptly arced and shorted out. "The fuse died," Sullivan concluded, grimly. "They got two or three doses off. We were distraught." However, the SEPAC team announced that the data already gained was more than sufficient for their research needs.

Despite the disappointing loss of SEPAC, the remainder of the ATLAS-1 mission proceeded without incident and produced spectacular results. Atmospheric science stations as far afield as India, Indonesia, Japan and New Zealand made joint observations with the Shuttle. Kathy Sullivan, for whom STS-45 was her final mission, found time during her last days in orbit to marvel at the view of the aurora, a brilliant red and purple mass, extending all the way from Africa to Australia. It reminded Sullivan of "a huge, richly brocaded theatre curtain", albeit that on this occasion the 'curtain' hung *upwards* from Earth into space. The scientific part of her brain was momentarily overtaken by one of childlike wonder.

Today, two decades after STS-45, Sullivan's involvement with atmospheric research and environmental protection continues; since May 2011 she has been assistant secretary of commerce for environmental observation and prediction and deputy administrator of the National Oceanic and Atmospheric Administration. Her career with NOAA stretches back to the end of her astronaut career. Her friend Sylvia Earle had been the administration's chief scientist from 1990-92 and proposed Sullivan as her potential successor. Sullivan had considered leaving NASA for some time and in August 1992 she succeeded Earle and received formal Senate confirmation in March of the following year.

Colin Michael Foale – who earned a CBE in the Queen's New Year's Honours List in December 2004 – came from Louth, in Lincolnshire, England. He was born on 6 January 1957, the son of a Royal Air Force pilot father and an American mother. "I grew up with the sound of jets," Foale later told the NASA oral historian, "and I lived in exotic places and I developed a taste for not so much adventure, but new vistas, new places, new things. I quickly decided that I wanted to fly." Frequent visits to his mother's family in Minnesota introduced him to the space programme as a child, when he saw John Glenn's Friendship 7 capsule in the state fair. "My father didn't discourage in any means my interest in being a pilot or an astronaut," he reflected. "I'm not sure he credited very realistic my dreams and aspirations there, especially since Britain did not have – and *still* does not have – a human spaceflight programme. However, there was quiet support."

A test piloting and military aviation career beckoned, but in his mid-teens Foale was misdiagnosed with a vision issue and it was this turning point which altered his focus toward the sciences. He entered Queen's College, Cambridge, and would earn a first-class degree in physics in 1978. By the time the misdiagnosis was uncovered and he realised that his vision was perfectly fine for military aviation, NASA's astronaut requirements had changed. Test piloting credentials were no longer mandatory and he could apply as a scientist. After his degree, he completed a doctorate in laboratory astrophysics in 1982 and made his first move towards NASA the following year.

In June 1983, Foale entered the Johnson Space Center in Houston, Texas, as a payload officer in Mission Control. He unsuccessfully applied for admission into the astronaut corps in 1984 and 1985, but in the aftermath of the Challenger disaster he altered the focus of his application essay from describing his own dreams to considering the managerial realities faced by NASA. On his third attempt, he was accepted as an astronaut candidate in June 1987. Two years later, he was one of the first members of his class to be selected for a space mission and in March 1992 – only days before the launch of STS-45 – Foale was named as payload commander of ATLAS-2, scheduled for the spring of 1993. It would be the start of an illustrious, six-flight astronaut career which would see him become the first person of British nationality to perform an EVA, complete a long-duration flight, service the Hubble Space Telescope and command the International Space Station.

Aside from the scientific rigours of their mission, the crew of STS-45 also carried the Shuttle Amateur Radio Experiment (SAREX), enabling ground-based 'hams' to speak directly to the astronauts. Leestma, Sullivan, Duffy and Frimout were all licenced amateur radio operators and the 57-degree inclination of the mission permitted worldwide contact possibilities, including high-latitude areas not normally accessible to the Shuttle. At various stages when the SAREX equipment was running – whilst flying over China or New Zealand, for example, or central Africa or Tierra del Fuego – the astronauts would be inundated with amateur calls. It was impossible to speak for very long, admittedly, because the Shuttle might lose signal or another ham might break in, but Sullivan established contact with an Australian man and his seven-year-old daughter on one occasion. On another, Dave Leestma woke her up to speak to someone at Palmer Station, on Antarctica's Anvers Island. She and Leestma succeeded in speaking to hams on all seven terrestrial continents. On yet another occasion, Brian Duffy was unexpectedly patched through to an amateur radio shack at JSC and from thence to his wife, Jan, sitting at the kitchen table of their home in El Lago.

At one stage, Duffy had contacted at least one ham from almost every continent, save Asia. "We had a night pass coming up right along, just offshore, parallel to the islands of Japan," recalled Sullivan. "As Duffy began his communications session, it seemed that a hundred *thousand* voices came back!"

Duffy grabbed one call sign. "Roger, got you," he said ... and promptly turned off the radio and pulled out the antenna. He had achieved his final continental contact. Job done.

"Duffy, you are *cruel*," Sullivan scolded. "There are 99,999 *really* disappointed people on the ground there!"

To be fair, Duffy's responsibilities on STS-45 were heavy; as the pilot of the Blue Team, he helped to run the flight deck for his shift and, with Bolden, was charged with bringing Atlantis safely back to Earth.

Elsewhere, on the middeck, the Investigations into Polymer Membrane Processing experiment focused upon the physical and chemical processes underway during the formation of polymer membranes in the microgravity environment. Supporting the precipitation process, the STS-45 flight of the experiment offered new insights into the potential use of polymer membranes in industrial separation processes on Earth, including the enrichment of the oxygen content of air, the desalination of water and human kidney dialysis. The Space Tissue Loss investigation, meanwhile, examined the growth of muscle, bone and endothelial cells, whilst the Air Force's Visual Function Tester used a piece of equipment akin to a pair of binoculars to measure the visual acuity of Lichtenberg and Frimout. The two payload specialists had been tested prior to launch, tested themselves each day during STS-45 and would be tested again after Atlantis' return to Earth. Other military-sponsored experiments included the Radiation Monitoring Equipment, which observed gamma rays, electrons, neutrons and protons in the crew cabin, and CLOUDS – the Cloud Logic to Optimise the Use of Defense Systems – whose objective was to quantify variations in apparent cloud cover as part of efforts to develop meteorological observation models. Outside the orbiter, in Atlantis' payload bay, a single Getaway Special canister carried a crystal growth investigation, which produced samples of the important semiconductor gallium arsenide.

Exceptionally economical use of cryogenic consumables meant that on 29 March the Mission Management Team opted to add an extra day in orbit. The mission was therefore extended from eight to nine days, with return to Earth rescheduled for 2 April. The hour-long re-entry back to the Kennedy Space Center in Florida on that day came as a surprise to Duffy. "[We came] back in, pretty much all at night," he told the NASA oral historian, "because we were going to land just after dawn in Florida, so the whole thing was in the dark." This fact alone generated a spectacular return to Earth.

Periodically, Duffy's gaze would flicker from his instruments and over to the dramatic fire show outside his window: the nose glowed carnation-pink and it seemed that Atlantis was surrounded with a yellow and orange blaze. Every so often, he would look over at Charlie Bolden, in the commander's seat, watching his gauges, and would realise how different the real thing was from the simulator. "The biggest surprise, however, was physically," Duffy said, "and part of it was because ... it was the first time in which I'd gone to extended weightlessness to now back into the G-field, into gravity, and it's not just to 1 G – it's to 1.4 or 1.5 G and it's sustained for a long period of time, and you're coming in at a 40-degree [nose-up] angle. I actually had to put my hand on the glare shield and hold my torso up to keep it from slumping down and forward during the entry."

It has often been remarked that the pilot's key role at the end of a mission was to deploy the landing gear. "I'm calling altitudes and airspeeds to [Bolden] as we're coming in for the final approach," Duffy said, "because a lot of things are happening very quickly." At 6:23 am EST, a little under nine full days since leaving the Cape, Atlantis' tyres kissed Runway 33 of the Shuttle Landing Facility to conclude STS-45.

The weeks which followed were consumed by traditional post-flight tours and, having transported Frimout into space, one notable focus was a journey to the Kingdom of Belgium. Kathy Sullivan was unable to attend, because she was already involved in the confirmation hearings for her appointment to NOAA, but for the rest of the crew it was both a pleasure and a privilege. They had spoken to Prince Philippe whilst aloft and Duffy remembered the post-flight trip very well. "We were just treated like royalty," he reflected. "It was nice for the spouses, too, because they don't get the rewards that we get when we fly."

SUCCESS AND FAILURE

"Ready. Ready. *Grab!* "

The words of Rick Hieb echoed through the silent Mission Control Center.

The view through Space Shuttle Endeavour's aft flight deck windows on the evening of 13 May 1992 was quite different from anything ever seen before. Not only was this the maiden voyage of NASA's newest orbiter – a vehicle which, but for the loss of Challenger, might have remained a set of structural spares – but it also involved the first EVA with as many as three people. This mission, STS-49, commanded by chief astronaut Dan Brandenstein, had long been anticipated to be the most visible Shuttle flight of 1992, but it demonstrated that human space flight retains the ability to deliver unexpected surprise. When the crew was announced, their mandate was to retrieve the Intelsat 603 communications satellite, delivered into an improper orbit by the Commercial Titan booster in March 1990. Spacewalkers Hieb and Pierre Thuot would venture into Endeavour's payload bay to attach a new rocket motor, after which the Intelsat would be boosted into its 35,600 km geostationary orbit, ahead of its pivotal role in covering that year's Summer Olympics in Barcelona.

After the Intelsat activities, a further two spacewalks – the first with Kathy Thornton and Tom Akers, the second with Thuot and Hieb – would rehearse Space Station Freedom construction techniques. Thornton's inclusion made her only the third woman, after Svetlana Savitskaya and Kathy Sullivan, to perform an EVA. It was a role for which she had previously trained in preparation for her first Shuttle mission, STS-33, in November 1989. "I absolutely insisted that she be the EVA person," her STS-33 commander, Fred Gregory, later recalled in his NASA oral history, "over great protest ... If we had not insisted, probably a person of her size would never have done something like this. Kathy [Sullivan] was a larger woman who could fit into the suits ... but Kathy Thornton was not, so we really had to force the issue." Doubtless, Dan Brandenstein was in full agreement that Thornton – nicknamed 'K.T.' in the astronaut office – was the most appropriate choice. She would go on to fly as part of the EVA team which first serviced Hubble in December 1993.

Kathryn Ryan Cordell Thornton secured a minor record of her own on STS-33, becoming the first and only woman to fly aboard a classified Department of Defense Shuttle mission. Many have speculated that her assignment came about because she had served as a research physicist at the Army's Foreign Science and Technology Center, before becoming an astronaut, and had worked on a NATO-awarded post-doctoral fellowship at the Max Planck Institute for Nuclear Physics in Heidelberg, West Germany. She was born on 17 August 1952 in Montgomery, Alabama, and after high school entered Auburn University to study physics. She received her degree in 1974 and a PhD from the University of Virginia, five years later. After completing her post-doctoral research in West Germany, she was employed by the Army in Charlottesville, Virginia, in 1980, and entered NASA's astronaut corps in May 1984.

The other three astronauts involved in the STS-49 EVAs were male. Rick Hieb was already in training to fly STS-39 at the time the Intelsat crew was assembled in December 1990 and Tom Akers had returned only weeks earlier from the Ulysses deployment mission, STS-41. The man in charge of the team – designated 'EV1' and wearing red stripes around the legs of his pure-white space suit for identification – was Pierre Joseph Thuot, nicknamed 'Pepe'. He came from Groton, Connecticut, where he was born on 19 May 1955. After high school in Virginia, he entered the Naval Academy to study physics. He graduated in 1977 and entered naval flight training in July of the same year. After receiving his wings in August 1978, Thuot worked as a radar intercept officer in the rear seat of the F-14 Tomcat fighter and was later deployed to the Mediterranean aboard the USS *John F. Kennedy* and the Caribbean aboard the USS *Independence*. He later attended the Navy Fighter Weapons School ('Top Gun') at Naval Air Station Miramar in California and the Naval Test Pilot School at Patuxent River, Maryland, graduating in 1983 as a test engineer. By the time of his selection into NASA's astronaut corps in June 1985, Thuot had gained a master's degree in systems management from the University of Southern California and accrued considerable experience as a flight instructor at Pax River.

When he flew STS-36 in the spring of 1990, Thuot became the first of his class to be assigned a mission and the first to actually fly. Mike Mullane remembered Thuot as a fast mover and a fast thinker. "Pepe was a 24-volt guy in a 12-volt world," he wrote. "He reminded me of a hummingbird in the way he darted at whatever he was doing, whether he was turning the page of a checklist, punching in a phone number or flipping cockpit switches." From a personal perspective, as a teenager, I contacted Thuot to ask him about his career. One comment in particular stood out and proved illustrative of his work ethic. "Whatever you do in life," he told me, "always make sure that you enjoy what you're doing and *aim high*. Pick challenging goals and work hard to achieve them."

If everything ran as timelined, STS-49 would thus be the first Shuttle flight to feature as many as three spacewalks and include two teams of spacewalkers; both of which were critical prerequisites if NASA was to execute as many as five EVAs on each mission to service the Hubble Space Telescope and build Space Station Freedom. On the face of it, retrieving and repairing Intelsat, for all its drama, offered a backward glance at the Shuttle's pre-Challenger heyday and was unusual, for in the

3-06: STS-49 commander Dan Brandenstein speaks to the gathered crowds during the rollout ceremony for Space Shuttle Endeavour in April 1991. Seated behind Brandenstein is NASA Administrator (and former astronaut) Dick Truly.

wake of the disaster it had been mandated that the reusable orbiters would henceforth not be used for commercial missions. STS-49 was thus the last of its kind. At the same time, as Space Shuttle Program Director Bob Crippen explained in June 1990, it offered "an opportunity for expanding our experience base in the planning, training and performance of EVA" by "helping preparations for Freedom".

Others agreed that such a mission was useful for other purposes. It was "a throwback to the good old days," said Endeavour's first processing manager, John Talone, "when we used to go out and do these kinds of things." Added NASA's Associate Administrator for Space Flight, former astronaut Bill Lenoir: "It's a mission we wanted to do. It gave me the opportunity to have *real* work that *really* mattered; that was going to get measured, where we either succeeded or failed."

The Intelsat 6 series represented the eighth generation of communications satellites, designed by Hughes for the International Telecommunications Satellite Organisation – originally an inter-governmental consortium, but since July 2001 a private company, known as Intelsat, Ltd. – which were capable of providing 33,000 telephone circuits and four television channels. Five were built between 1983 and 1991 and a half-scale model of the satellite today resides in the lobby of Intelsat's Washington, DC, headquarters. The 4,200 kg cylindrical satellites were spin-stabilised at 30 revolutions per minute, with a 'de-spun' segment to house the communications payload and direct it towards a desired location on Earth. Originally scheduled to be flown aboard the European Ariane 4 booster and, in the pre-Challenger era, also the Shuttle, Intelsat 6 was a wide-body satellite, measuring 3.6 m in diameter and 5.2 m tall, expanding to a height of 11.7 m when its concentric solar arrays and communications payload were fully operational in geostationary orbit.

These huge satellites were fed by a twin-propellant system of nitrogen tetroxide and monomethyl hydrazine, which fed radial and axial thrusters for station-keeping and attitude control. The outer surfaces of the satellites were coated with photovoltaic solar cells, which provided around 2,600 watts of electrical power, whilst nickel-hydrogen pressure vessel batteries supported operations in Earth eclipse. The communications payload carried 38 C-band and ten Ku-band transponders. The third Intelsat 6 – codenumbered '603' – was launched atop a Commercial Titan III from Cape Canaveral Air Force Station in Florida on 14 March 1990, but the rocket's second stage failed to separate properly and the satellite could only be released by means of jettisoning its attached Orbus-21 perigee kick motor. This left it effectively unable to achieve geostationary orbit and the $157 million Intelsat – which had not been insured, but was instead 'self-insured', using the company's own funds – was left stranded in an inoperable low-Earth orbit.

In the weeks and months following the malfunction, Hughes entered into a contract with NASA, worth in excess of $90 million, for a Shuttle flight to reboost Intelsat 603. Two possible options quickly gained prominence: either to carry a new perigee kick motor into orbit and attach it to the satellite to reboost it into geosynchronous transfer orbit or retrieve Intelsat and bring it back to Earth for refurbishment. Concerns about the extent to which the satellite's surfaces might degrade over two years were allayed by the test flight of several solar array sample 'coupons', attached to Discovery's RMS during the STS-41 mission in October 1990. These were exposed to the harsh atomic oxygen environment for a minimum of 23 hours, with few ill-effects. Two months later, in December, the seven-member STS-49 crew was named to conduct the audacious retrieval mission. For Daniel Charles Brandenstein, it was not only to be his fourth space voyage, but his third involving rendezvous and proximity operations: as commander of STS-51G in June 1985 his crew deployed and retrieved the Spartan astronomy satellite and on STS-32 in January 1990 the Long Duration Exposure Facility was plucked out of orbit. Brandenstein was born in Watertown, Wisconson, on 17 January 1943. After attending high school in his home town, he enrolled at the University of Wisconsin in River Falls and received a degree in mathematics and physics in 1965.

Aviation had always been at the back of his mind and he considered America's fledgling manned space programme as "the *ultimate* form of aviation" and thus a goal for the future. He read the biographies of the Original Seven Mercury astronauts and identified the main requirements: active-duty military officers and test pilots with degrees in science or engineering. To Brandenstein, mathematics and physics "were always my favourite courses" and "as close to engineering as you could get". The decision over which branch of the armed services to join came in his final year of college: Air Force pilots landed on several kilometres' worth of runway, whereas naval 'aviators' brought their jets screaming onto a couple of hundred metres of *pitching steel* in the middle of the ocean! "Looking at what looked to be the most interesting and challenging," Brandenstein told the NASA oral historian, "the naval aspect of aviation caught my fancy."

He entered active Navy duty in September 1965, was attached to the Naval Air Training Command for flight instruction and received the designation of a naval aviator a little under two years later. Flying A-6 Intruders, Brandenstein participated in two cruises to Vietnam between 1968 and 1970, aboard the USS *Constellation* and the USS *Ranger*, logging 192 combat missions. Subsequent work focused on operational tests of weapons systems and tactics for the A-6 and he was selected (alongside Rick Hauck) for the Navy's Test Pilot School at Pax River in 1971. After graduation, he conducted electronic warfare systems tests in a variety of fighter aircraft and deployed to the Western Pacific aboard the USS *Ranger*, again flying A-6 jets, from March 1975 until September 1977. Each step in his education and experience had guided him closer to the astronaut corps, but by this point the Shuttle was on the horizon and in the summer of 1977 Brandenstein was invited to Houston – part of the first group in a total of ten groups which would be interviewed between August and November of that year – for an extensive series of tests. Less than six months later, his wife pulled him, dripping wet, out of the shower one Monday morning to take a call from NASA which changed his life.

More than a dozen years later, and since April 1987 the head of the astronaut office, Brandenstein found himself in command of the first flight of a new Space Shuttle *and* a rendezvous and retrieval mission with EVAs which promised to be filled with drama. "One of my first concerns when we first got assigned and started working with Hughes on the mission," he told the NASA oral historian, "was if we try and grab it, if we *bump* it, is it going to go out of whack and float away? Part of the requirements from the customer were that we didn't touch any sensitive area, which left you a very small ring that ... had a limited accessibility and that was supposed to be the way we grabbed it."

The mechanism by which Thuot and Hieb were intended to seize Intelsat was a so-called 'capture bar', designed and built by engineers in the Crew and Thermal Systems Division at JSC. Weighing 73 kg, it measured 4.6 m long by about a metre wide and a metre tall and included detachable beam extensions and a steering wheel. As Thuot rode on the end of Endeavour's RMS mechanical arm, he would be positioned close to the base of Intelsat 603 and after grappling the satellite would lower it delicately into a Hughes-built cradle assembly. "There was a lot of analysis done," continued Brandenstein, "and we were assured that because it was spinning slightly and it had a lot of mass, we could bump it and it would stay pretty much in place and wasn't going to be a problem." Throughout 1991 Thuot and Hieb trained underwater and on the air-bearing table, to such an extent that they could follow the procedure with their eyes closed.

Endeavour arose from a series of already extant Shuttle spares, assembled before the loss of Challenger to facilitate repairs or possibly the creation of an entirely new vehicle in the event of an accident. The $389 million contract to build the spares – which consisted of an aft-fuselage, a mid-fuselage, two halves of the forward fuselage, vertical stabiliser and rudder, wings, elevons and an aft body flap – was awarded to the Shuttle's prime contractor, Rockwell International, in April 1983. Three years later, the destruction of Challenger added a new level of urgency to these plans and led directly to a decision to assemble the spares into a new vehicle, initially designated Orbiter Vehicle-105 (OV-105). However, there were powerful political voices which opposed the idea. White House Chief of Staff Donald Regan and several members of Congress argued that the cost of developing the spares into a new Shuttle could be better spent on an entirely new spacecraft. Nevertheless, in September, the construction of OV-105 was approved and in July 1987 NASA awarded a $1.3 billion contract to Rockwell. Construction was completed within three years and the orbiter was formally powered-up to begin systems testing on 6 July 1990. Bruce Melnick and Bill Readdy were assigned as astronaut office representatives for Endeavour's construction. "We were there for all the programme reviews and the integration tests," Readdy recalled in his NASA oral history, "saw it from being built up as a forward fuselage, then an empennage and a set of wings, to being built and then tested and delivered from Palmdale to the Cape."

By this time, OV-105 had received the name 'Endeavour'. It was spelt in the English fashion, since it paid homage to Captain James Cook, whose own vessel, HMS *Endeavour*, sailed to the South Pacific in 1768-71 to observe the transit of Venus, part of ongoing scientific investigations to measure the distance between Earth and the Sun. During the course of the voyage, Cook reached Tahiti and Hawaii, charted New Zealand for the first time and surveyed the eastern coast of Australia.

In response to the tremendous outpouring of student grief in the wake of

Challenger, Republican Congressman Tom Lewis of Florida initiated a resolution to enable students to name the new orbiter. Lewis' bill, passed in Congress in October 1987, inaugurated the 'NASA Orbiter Naming Program'. More than 71,000 students, representing 6,154 schools across the United States, submitted their entries during the course of 1988. The guidelines dictated that the name must have previously belonged to an exploratory or research seagoing vessel, that it must be appropriate for the new Shuttle, that it must capture the spirit of America's mission in space and that it should be easy to pronounce for radio transmission. Three finalists were eventually reached by the NASA judges and the agency's Educational Programs Officer Muriel Thorne: Endeavour, Horizon and North Star. Of these, Endeavour was by far the most popular entry, accounting for almost a third of all state-level winners in the competition and in May 1989 the new orbiter was formally named by President George H.W. Bush. When the STS-49 crew came to design their crew patch, they not only included Captain Cook's *Endeavour* ... but also exhibited the colours of the two winning schools – Senatobia, Mississippi (Division I, elementary) and Tallulah Falls, Georgia (Division II, secondary) – atop the ship's masts.

On 25 April 1991, less than two years after Bush named Endeavour, the sparkling new Shuttle was rolled out of Rockwell's Palmdale facility in California. She was delivered to the Kennedy Space Center on 7 May. Her targeted maiden launch in May of the following year quickly became mired with difficulty, as "hundreds of problems" were identified by NASA: faulty cables and connectors, contaminated propellant lines, incorrectly fitted insulation blankets and even a biscuit, mistakenly dropped in the fuselage. At their worst, in the late summer of 1991, up to 70 electronic, hydraulic or mechanical problems were being reported each week, prompting NASA to announce that it expected to delay STS-49 from May until at least July 1992. The cannibalisation of parts during Endeavour's construction to address problems with her sister ships Columbia and Atlantis compounded the delay. Then, in March 1991, cracks in Discovery's 43-cm disconnect doors prompted inspections of Endeavour and uncovered a similar flaw. However, noted *Flight International*, these cracks were "considered to be an inherent design error, rather than a result of poor manufacturing".

With these problems in mind, it is quite remarkable that the anticipated delay to STS-49 did not transpire and she was rolled out to Pad 39B on 13 March 1992 to begin final preparations for launch. In physical appearance, Endeavour differed very little, outwardly, from her sister ships. Internally, though, she carried Advanced General Purpose Computers, with twice as much memory and three times as much processing speed as the older versions, as well as being smaller, lighter (at just 29 kg) and requiring less power (around 550 watts). The High-Accuracy Inertial Navigation System was intended to eventually replace earlier inertial measurement units, with one HAINS flying alongside two of the older devices on STS-49. Endeavour was also fitted with three improved Tactical Air Navigation systems, a pair of enhanced Master Events Controllers and solid-state trackers. Her Auxiliary Power Units were enhanced over previous models, as were her gas generators, fuel pumps, redundant seals and new materials. Many of these upgrades were designed to be more reliable than earlier systems, utilising lower power and requiring far less maintenance. She also featured a drag chute to improve landing safety and was fitted with the plumbing and electrical connections to enable Extended Duration Orbiter (EDO) missions of up to 28 days. A fifth cryogenic oxygen tank and a fifth hydrogen tank beneath her payload bay floor supported this provision. "On the rest of the orbiter fleet," noted a NASA release, "Columbia also has five tank pairs and Atlantis and Discovery each have four tank sets." Although the official plans at the time called for Columbia alone to fly EDO missions, Endeavour would perform one such flight, later in her career.

Three weeks after rollout to the pad, on 6 April, her three main engines underwent the standard Flight Readiness Firing (FRF). This test firing had historically been performed before each orbiter's maiden voyage to demonstrate the engines' capability to throttle and gimbal as they would during flight. Preparations for the FRF proceeded in a manner not unlike a real countdown:

"T-minus 12, 11, ten, nine ... we have a Go for engine start ..."

At four seconds, Endeavour's engines roared to life at 120-millisecond intervals, reaching 90 percent of their rated thrust and hitting the 100-percent mark precisely at T-zero.

"... two, one, zero ..."

The Shuttle visibly flexed, as if she ached to break free of her shackles and climb, crewless, toward the heavens.

"... engines are now at 100 percent of rated power ..."

And so they were. Vast clouds of steam billowed from the pad. Three seconds later, engineers simulated the retraction of the External Tank umbilical and the Solid Rocket Boosters' hold-down posts and after a further 15 seconds of stable thrust, shutdown commands were issued to all three engines. In total, the FRF lasted 22 seconds and was a great success, but for a couple of technical issues. High vibration levels were detected in one of the engines' high-pressure liquid oxygen turbopumps, whilst another exhibited a loud 'popping' noise shortly after shutdown, indicative of hydrogen ingestion into the fuel injector. Prudently, on 8 April NASA decided to replace all three engines with a set previously earmarked for Atlantis' STS-46 mission in July, although a second FRF for Endeavour was not considered necessary.

Originally scheduled for the evening of 4 May 1992, launch was postponed until the 7th in order to permit greater conditions of daylight for photographic coverage of the ascent. According to *Flight International*, the change represented an overruling of the Flight Readiness Review by new NASA Administrator Dan Goldin. Had Endeavour flown on the 4th, she would have been committed to an hour-long 'window' lasting from 8:34 pm until 9:27 pm, whereas the move to the 7th placed T-0 at a point earlier in the evening.

After a 34-minute delay, caused by marginal weather conditions at one of the Transoceanic Abort Landing (TAL) sites, STS-49 duly thundered into space at 7:40 pm EDT on 7 May and during the next couple of days Intelsat controllers manoeuvred their satellite into a 'control box', within reach of the Shuttle's orbit. These manoeuvres also served to reduce Intelsat's rotation from 10.5 to around 0.65

3-07: Astronauts Rick Hieb, Tom Akers and Pierre Thuot stabilise and secure Intelsat-603 during the world's first three-man EVA. Their efforts ultimately succeeded in retrieving the errant communications satellite, preparatory to the installation of a new rocket motor and a successful boost into the geosynchronous orbit.

rpm. By the 10th, as they approached to within 13 km of the satellite, Thuot and Hieb completed their procedures of suiting-up and were assisted into the airlock by crewmate Akers. Shortly thereafter, at 4:25 pm EDT, they opened Endeavour's outer hatch into the payload bay – which was then in the pitch black of orbital darkness – and Thuot fastened himself into a foot restraint on the end of the RMS, being deftly operated by STS-41 veteran Bruce Melnick. After he had been manoeuvred to face the base of the satellite, Thuot extended the capture bar into position ... and the latches failed to latch.

He tried again, without success.

A third attempt was similarly fruitless.

From his station on Endeavour's aft flight deck, Brandenstein noticed that Intelsat 603 was beginning to oscillate and drift somewhat, "so I got in my chase-it mode, because I had to keep him aligned". When Thuot's third attempt failed, Brandenstein had used a "tremendous" amount of propellant and instinctively knew that the chances of success were slim at best. The RMS exacerbated the difficulty, because its joints were being driven into positions which they could not support. "We decided, though consultations in the ground, to get out of there and try another day," Brandenstein recollected. "That was a pretty low point, because when we left, it had a pretty good rate [of oscillation]. We thought we'd lost this $150 million satellite ... and Pierre was particularly depressed because, obviously, he thought it was his fault."

Thuot and Hieb returned inside Endeavour after three hours and 43 minutes and later that evening Hughes engineers confirmed that they had managed to stabilise Intelsat. Next day, at 4:30 pm EDT on 11 May, the spacewalkers were back outside for a second attempt. "Instead of doing it at night, we were going to wait and do it in daylight," Brandenstein said. "We decided we weren't going to even make an attempt until everything was just perfect. Pierre went in and the rotation slowed down." From Hieb's position, it looked as if Thuot had completed the capture ... but, alas, the satellite again began to oscillate. His alignment was unquestionably correct, but the $1 million capture bar refused to seat itself properly and Intelsat wobbled. A few weeks after the mission, Thuot explained to this author that the satellite "was much more dynamic than our training had led us to believe".

As the disappointed spacewalkers returned inside the cabin for the second time – this time after five and a half hours – they at least knew that Hughes engineers could regain control of Intelsat for another attempt. However, although the propellant reserves allowed for it, *three* separate rendezvous on a single Shuttle mission had never been attempted and Brandenstein recommended a day off to plan for the third attempt.

In an interview for the Smithsonian, Rick Hieb remembered that the evening of the 11th was a sombre time. All at once, Kevin Chilton, the STS-49 pilot and the only rookie member of the crew, joined Hieb on the flight deck and the pair entered an impromptu brainstorming session ... a session which would mark a significant turnaround in the fortunes of a mission which seemed snake-bitten with ill-fortune.

Kevin Patrick Chilton today holds the record for having achieved the highest rank of any military astronaut; a career Air Force officer until his retirement in February

2011, he eventually rose to become a four-star general, holding several directorial posts at the Pentagon and later commanding the Air Force Space Command and the US Strategic Command. Chilton came from Los Angeles, California, where he was born on 3 November 1954. After high school, he entered the Air Force Academy and earned a degree in engineering sciences in 1976, followed by a master's in mechanical engineering, on a Guggenheim Fellowship, from Columbia University in 1977. Chilton earned his wings at Williams Air Force Base in Arizona, qualified in the RF-4C – the reconnaissance version of the Phantom jet – and was deployed to Korea, Japan and the Philippines.

He converted to the F-15 Eagle in 1981 and served as a squadron pilot, attending squadron officer school the following year at Maxwell Air Force Base in Alabama and emerging with the Secretary of the Air Force Leadership Award as the top graduate. Over the next several years, Chilton was a weapons officer, instructor pilot and flight commander, before entering test pilot school at Edwards Air Force Base in California in 1984. He graduated first in his class and in August 1987 was selected by NASA as an astronaut candidate. By this point in his career – aged only 32 – Chilton already held the rank of major and by the time of STS-49, five years later, he had reached lieutenant-colonel and was only months away from promotion to full colonel.

As Hieb and Chilton talked, other members of the crew floated upstairs to join the discussion. The main concern was *where* to manually grab Intelsat. The top of the satellite, where the delicate antennas were located, was not ideal, and it was Bruce Melnick who suggested an EVA with not two spacewalkers... but *three*. No excursion in history had ever involved more than two members, partly due to safety concerns and partly because of the sheer practicality of getting three people into the tiny airlock. On the other hand, Endeavour carried four suits for Thuot, Hieb, Thornton and Akers, so in theory it was a possibility. "When Bruce said that," recalled Hieb, "a big mental switch flipped over, at least for me. In my mind, having a third set of hands out there meant that we would be successful, although we weren't yet sure how."

Mission Control knew that the astronauts were still awake, because Endeavour's monitors had not been turned off. At length, the crew turned them off and continued talking in the dark, but eventually called the ground to report Melnick's idea. Years later, Brandenstein remembered that it was Chilton who sketched out a three-person EVA scenario and held it in front of the television camera to allow mission controllers to see it. "The big choke point," Brandenstein said, "was can you put three people in the airlock to get them outside?" In the Houston water tank, fellow astronauts Story Musgrave, Jim Voss and Michael 'Rich' Clifford donned suits and demonstrated the techniques and geometries involved in accomplishing the feat. It was doable. On the evening of 12 May, Capcom Sam Gemar radioed Mission Control's approval to the crew.

Late on 13 May, the third attempt got underway. Truss members belonging to the Assembly of Station by EVA Methods (ASEM) – a demonstration payload scheduled to be used during EVA tests later in the mission – were removed and arranged into a triangular structure for Thuot, Hieb and Akers to anchor their feet.

Brandenstein positioned the orbiter directly beneath Intelsat 603 and controllers verified that its surface temperatures would not exceed the 160 degrees Celsius touch limit of the astronauts' gloves. With Hieb close to the starboard payload bay wall, Akers in the centre, attached to an ASEM strut, and Thuot on the end of the RMS on the port side, the astronauts could do little but watch as Endeavour drew closer. They studied its slow rotation for about 15 minutes, until, on Hieb's call, they moved in for the capture.

All at once, Thuot spotted a slight wobble. He called the attempt off.

Shortly thereafter, they tried again. This time, at last, the three men grabbed the satellite and held it firmly. The time was 7:55 pm EDT. "I actually thought the other two guys had stopped it from rotating," Thuot said later, "so little force had I applied. Very gently, the thing came to a stop." From the flight deck, Dan Brandenstein asked them if they had a good grasp. On Thuot's response in the affirmative, the commander was able to advise ground controllers, with more than a hint of relief: "Houston, I think we've got a satellite!"

With Intelsat snared, the astronauts removed the steering wheel and installed an extension to the capture bar, which Bruce Melnick then grappled using the RMS. The satellite was then positioned above its 10,400 kg Orbus-21 solid-fuelled perigee kick motor, which sat vertically in its cradle. After closing four docking clamps to secure the pair, and attaching two electrical umbilicals between Intelsat and the motor itself, the spacewalkers set a pair of deployment timers and retreated to Endeavour's airlock. Meanwhile, Kathy Thornton prepared to activate the springs to deploy the payload. At first, it did not move. "They had made a change in the wiring of the deploy system," recalled Brandenstein, "and the change never made it through the process [and] never got into the checklist. Fortunately, somebody in Mission Control apparently knew about it. They just quickly called up a different switch sequence and she did that sequence and it went." Deployment occurred at 12:53 am EDT on 14 May and the satellite vacated the payload bay at a little under 0.2 m/sec. Less than an hour later, the three spacewalkers repressurised the airlock and returned inside the cabin.

Speaking a decade or more after the flight, Dan Brandenstein regarded those few days of STS-49 as "one of those missions from hell" and for newly-appointed NASA Administrator Dan Goldin it was "a baptism of fire". Nevertheless, at 1:25 pm EDT on the following day, 15 May, Intelsat 603's new motor ignited perfectly and it was on its station in geostationary orbit by the 21st. As well as becoming the first Shuttle crew to accomplish as many as three EVAs in a single mission – a record which they would break with a *fourth* excursion – the triumphant three-man spacewalk established itself as the longest in history. Their eight hours and 29 minutes outside would remain unbroken until March 2001.

By now, the difficulties had prompted the Mission Management Team to extend STS-49 by 48 hours from its planned seven-day duration. On 14 May, the fourth EVA got underway when Akers and Kathy Thornton ventured outside for the ASEM station tests. Originally scheduled to involve two EVAs – one by Thornton and Akers and the second by Thuot and Hieb – the Intelsat 603 retrieval forced the cancellation of one walk.

Activities included the construction of a pyramid-shaped truss, the unberthing of an MPESS – manoeuvred by the RMS – and efforts to evaluate the ability of spacewalkers to work at positions 'above' and 'forward' of the payload bay, including 'over the nose' of the Shuttle. The MPESS contained two node boxes for the pyramid, a releasable grapple fixture and interface plate and a truss leg and strut dispenser. Five crew rescue techniques were to be trialled, including a lasso-like 'astro-rope', a seven-section telescoping pole and a hand-held propulsive device. The latter, according to NASA's STS-49 press kit, was "a redesigned hand-held manoeuvring unit from the Skylab programme", in which pressurised nitrogen jets were employed as thrusters.

During their seven hours and 43 minutes in the payload bay, Thornton and Akers completed the construction and disassembly of the ASEM attachment fixture, tested the propulsive device, affixed six of eight legs onto the MPESS ... and, unexpectedly, were called upon to manually stow Endeavour's Ku-band antenna, which had experienced a positioning motor failure. According to NASA's post-mission report, this EVA was planned to be RMS-intensive, although the mechanical arm was used to accomplish only a single ASEM task and the spacewalkers' timeline was further impacted by the Ku-band activity.

Returning inside Endeavour's airlock after the excursion, the astronauts of STS-49 could now boast four EVAs – lasting a grand total of 25 hours and 27 minutes – which had snatched success from the fangs of defeat. The physical appearance of the four spacewalkers in their snow-white suits was also quite distinct from previous missions, all of which had featured no more than two members. In order to distinguish them, Thuot (designated 'EV1') wore red stripes around his suit legs, whilst Hieb (EV2) wore a pure-white suit ... and, for the first time, Thornton (EV3) wore dashed stripes around her suit legs and Akers (EV4) wore red diagonal hatches around his suit legs. In spite of their remarkable achievement, only relatively minor glitches plagued the spacewalkers – a failed joint on one of the portable foot restraints, a loud noise over the headsets when power tools were being used and a battery problem, amongst others – and their suits held up exceptionally well.

The Intelsat rendezvous commitment and ASEM work left little opportunity for additional experiments during STS-49, although the Commercial Protein Crystal Growth hardware was on the middeck. This included a new refrigerator/incubator module with a pre-programmed temperature profile, which was automatically reduced from 40 degrees Celsius to 22 degrees Celsius over a four-day period. One of the expected results of this decrease was to improve the 'ordering' of the protein crystals. Endeavour was also used passively for the Air Force's investigation, in which electro-optical sensors on the Hawaiian island of Maui tracked the orbiter's progress and recorded signatures from thruster firings, water dumps or other induced phenomena.

Endeavour's return to Earth on 16 May 1992, concluding her first flight in a 25-voyage spacegoing career, brought with it a number of test objectives, the most visible of which was the deployment of the Shuttle's long-awaited drag chute on the runway. Measuring 11.8 m in diameter, when fully unfurled, the chute was designed to reduce steering problems and relieve stress on the Shuttle's brakes and tyres, which had suffered particularly significant damage on STS-51D in April 1985. In the post-Challenger era, options to improve landing safety were extensively evaluated. These included the installation of a special 'skid' on the landing gear. In the event of a blown tyre, the latter was meant to preclude the chance of a second tyre failing; effectively providing a 'roll-on-rim' capability for a predictable rollout pattern. There was also an arresting barrier emplaced at the end of the runway.

Design requirements for the drag chute – which was tested extensively at NASA's Dryden Flight Research Center at Edwards Air Force Base in California – included an ability to bring a 112,500 kg orbiter to a halt in less than 2,500 m with an 18.5 km/h tail wind and maximum braking at a ground speed of 260 km/h. Housed in a cylindrical container, just beneath the Shuttle's vertical stabiliser fin, the chute was to be manually deployed by the pilot after main gear touchdown and ahead of nose gear touchdown. It would then be jettisoned when the rollout speed had dropped to around 110 km/h. During re-entry, the orbiter's main engines were repositioned some ten degrees 'lower' than normal, in order to eliminate the chance of damage to the chute.

Air trials at Dryden in the summer of 1990 saw it fitted to a modified NB-52 carrier aircraft – piloted by former astronaut Gordon Fullerton – which tested it at landing speeds of between 260 km/h and 370 km/h, with no negative effects. These trials enabled engineers to predict that it would reduce the Shuttle's landing rollout distance by some 300-600 m. Early in January of the following year, NASA modified its production contract for Endeavour with Rockwell by $33.3 million to include the design, fabrication and installation of the chute. Since STS-49 was to be its first space mission, plans called for it to be deployed after all six wheels – main gear and nose – were firmly on the runway. After an uneventful re-entry, Endeavour's main gear touched concrete Runway 22 at Edwards at 1:57 pm PDT (4:57 pm EDT), followed by the nose a few seconds later. Then came the chute.

At the pilots' command, pyrotechnics blew the door away from the chute compartment and a mortar fired, driving out firstly a 3 m pilot chute, then the main canopy, which 'reefed' to 40 percent of its total diameter for a few seconds to lessen the initial structural loads on Endeavour herself. It trailed the orbiter by 27.2 m on a 12.6 m riser. Following the successful operational of the reefing line cutter, the chute blossomed to its fully inflated condition. Photographic analysis of the landing illustrated that the reefed chute rode at a higher angle than anticipated and the door trajectory differed slightly from the NB-52 tests; additionally, its behaviour and closeness to the orbiter's centreline were later attributed to the effect of the aerodynamic flow for the fully-open speed brake.

Summing up his fourth and final Shuttle mission – for he would retire from NASA in October 1992 – Dan Brandenstein was happy with Endeavour's performance on her maiden voyage. This was particularly important in view of the Intelsat 603 difficulties and the relative paucity of additional time to tend to systems glitches. "They built a beautiful vehicle," he said, "because it's based on all the other things that diverted our attention on that flight. It was really nice that Endeavour performed like an old pro." Indeed, NASA's official post-mission report highlighted 36 anomalies, none of which were of sufficient concern to impact the successful conduct of the mission.

In the aftermath of STS-49, the crew themselves would highlight the fact that their mission raised awareness of the *need* for more EVA experience in the years before the start of construction of Space Station Freedom. At one stage, in the late 1980s, as many as four EVAs *per week* were envisaged; an astonishing estimate which NASA Administrator Dick Truly deemed totally unacceptable. Yet as the plans for the station matured, it was obvious that the construction process would be EVA-intensive ... and *that* required different ways of working and training. "We have to take a good look at the time it takes to do a job," Brandenstein said. "We need better ways to train so that the learning curve isn't quite so steep." Pierre Thuot added that the Intelsat 603 retrieval task was something they "couldn't train for fully".

After their return, Kathy Thornton and Tom Akers – who would go on to service Hubble together in December 1993 – took an active role in developing new EVA methods in the WET-F in Houston. "Even in the tank, you still have the resistance of water," Brandenstein recalled, "so you can kick your feet and swim. In zero gravity, we've got movies of Tom going to that instinct. You can see him kicking his legs and *nothing's* happening. Also, if you move something in the water, as soon as you *stop* moving it, the *water* stops it. But in zero gravity, you start moving something and it just keeps moving until you come back on it. They made some significant changes in the tank training procedures." The first flight of Endeavour's career had gone spectacularly well and played a significant role in shaping the missions – and the assembly of the space station – which would follow.

"TRAIN LIKE YOU FLY"

Although Endeavour was equipped with the electrical and plumbing connections to support the EDO capability, NASA's plan in the early 1990s was that Columbia – the only other orbiter with as many as five liquid oxygen and liquid hydrogen cryogenic reactant tanks underneath her payload bay floor – would serve as the primary vehicle for such missions. Extended flights of up to 28 days had been in the back of the space agency's mind since the pre-Challenger era. They were considered a useful means of evaluating astronaut group dynamics in a closed environment, as well as providing the scope to perform scientific research for longer than previously achievable. By the spring of 1990 it was decided that Columbia would be decommissioned for about six months in the summer of the following year, with an expectation that the initial EDO missions would fly for 13-16 days, extendable to 28 days by the mid-1990s.

Early in January 1991, NASA modified its Shuttle production contract with Rockwell International by an additional $93.5 million to accommodate the EDO work. "Under the terms of the modification," noted a space agency news release, "Rockwell is required to modify the Columbia orbiter to extend the mission duration of flights from ten days to 16 days, plus a two-day contingency." The improvements included new environmental control and life-support systems, better waste collection facilities and additional gaseous nitrogen and crew stowage provisions. The response from within NASA and the scientific community was audibly favourable: by the eve

3-08: The Extended Duration Orbiter (EDO) pallet, laden with four tanks for liquid oxygen, two for liquid hydrogen and two for helium, carried the potential to increase the length of Columbia's missions to a maximum of 16 days, plus two contingency days. Between June 1992 and February 2003, Columbia would stage no fewer than 13 EDO missions, including STS-80, which accomplished the longest Shuttle flight of all time, at 18 days.

of STS-50, four more Spacelab flights had been placed on the manifest – SLS-2, IML-2, USML-2 and SLS-3 – with an option to fly a 28-day ASTRO-2 mission in late 1994.

The EDO modifications got underway within weeks of Columbia's return to Earth from STS-40; after several weeks of standard 'de-servicing' activities in Florida, she was moved from Bay 2 of the Orbiter Processing Facility to the new Bay 3 – formerly known as the Orbiter Maintenance and Refurbishment Facility – for fit checks in readiness for cross-country transportation to Rockwell's Palmdale plant. On 10 August 1991, secured atop the Boeing 747 Shuttle Carrier Aircraft, she left Florida and after two refuelling stops and two days on the ground due to bad weather, arrived in California on the afternoon of the 13th. For almost six months, Columbia underwent more than 150 modifications, including the 1,630 kg EDO pallet, built by Rockwell, and a new Regenerative Carbon Dioxide Removal System (RCRS). The EDO pallet measured 4.6 m wide and was designed to occupy an upright position at the rear end of the payload bay. It carried four liquid oxygen tanks, two liquid hydrogen tanks and two helium tanks, with the option to add more for yet longer flights. When fully loaded, the pallet could store 1,420 kg of liquid oxygen and 167 kg of liquid hydrogen, which pushed its total weight to almost 3,180 kg. As already mentioned, Endeavour had also been equipped with EDO provisions during her assembly, together with the option to mount *two* pallets in her payload bay to support 28-day missions, although this capability was removed for weight-saving reasons in 1996.

Meanwhile, the RCRS provided a method for removing the crew's exhaled carbon dioxide from the cabin in a more efficient manner than had been previously achievable with lithium hydroxide canisters. It worked by means of adsorption, instead of absorption. "It uses amine, coated on very small beads, almost like a powder," said Frank Samonski, then-chief of NASA's Environmental Control and Life Support Systems Branch, "but it pours like sand into a multi-layered bed, like a heat exchanger. When you pass gas containing carbon dioxide and moisture over it, the moisture activates this coating and the carbon dioxide molecules stick to the coating. It's not a chemical bond and can be broken by exposure to vacuum. One bed is 'online', collecting CO2, and the other is 'desorbing' to space vacuum. Through a series of valves, you switch those beds and dump the CO2 overboard in a cyclic manner. It's not good for long-term missions like the space station, but it's just right for the Shuttle."

As well as the pallet, two extra nitrogen tanks were added to Columbia during those six months to support the crew cabin and several more stowage lockers were installed in the orbiter's middeck. Other changes included the drag chute and beefed-up synthetic rubber tyres to replace the earlier, natural rubber ones. The vehicle was returned to the Kennedy Space Center on 9 February 1992 to begin preparations for the first EDO flight, a 13-day voyage, scheduled for June. "I think we've all been aware that living together and working together for 13 days is certainly a challenge," said Dick Richards, the commander of STS-50, whose two previous missions had both lasted around four days. "We'll be sharing the same sleeping quarters. We have only one restroom on-board. We all have to co-operate to make life as good as possible."

In the wake of his second flight, STS-41, Richards had complained to chief astronaut Dan Brandenstein about the short nature of Shuttle flights. "I just raved to NASA about the fact that I can't believe we're going up there and spending four days here," he told the space agency's oral historian. "So they said, okay, we've got this flight coming off called United States Microgravity Laboratory ..." The rest was history.

Richards' words highlighted a key objective of EDO: to examine the astronauts' physical and psychological performance in close quarters over the span of a long mission. When the STS-50 stack was rolled out to Pad 39A on 3 June, his crew had been in training for around 18 months. His pilot, Kenneth Dwane Bowersox, nicknamed 'Sox' in the astronaut corps, would secure a record as the youngest person ever to serve as a Shuttle pilot. (In October 1995, he would also secure the record as the youngest-ever Shuttle commander.)

Bowersox came from Portsmouth, Virginia, where he was born on 14 November

1956. For several years, his family lived in Oxnard, California, since his father was stationed at Port Hueneme naval base. Growing up in the early days of America's manned space programme, Bowersox was quickly hooked. Childhood friend John Jarvis later recalled that the boys would beg their parents to be allowed to skip school in order to watch the Gemini missions on television. For Bowersox, it stretched back even before Gemini. "When I was about seven years old," he told a NASA interviewer in 2002, "I was driving around in a car with my father, listening to the radio, and on the radio was a broadcast, describing John Glenn orbiting the Earth. *That* sounded like a pretty neat thing to do. From that point, I've just wanted to be an astronaut."

Achieving such an exalted goal was far easier said than done, of course, but Bowersox had a helping hand in the form of a school reading assignment sheet, which listed the education and experience requirements. At the time, these revolved around military aviation, test pilot training and advanced engineering credentials. "And so I said, well, that's what I'm going to do, and I took that path," Bowersox said. In 1978, he graduated from the Naval Academy with a degree in aerospace engineering and entered active naval service. A master's degree in mechanical engineering from Columbia University followed in 1979. Designated a naval aviator two years later, he flew the A-7E Corsair off the USS *Enterprise* and logged over 300 carrier-arrested landings. He completed test pilot school in 1985, flew the A-7E and F/A-18 aircraft at the Naval Weapons Center at China Lake, California, and was selected by NASA as a pilot astronaut candidate in June 1987.

Bowersox's involvement in STS-50 was a curious one. When his name was announced for STS-50 in December 1990, it was as one of the mission specialists, serving as flight engineer to Richards and another pilot, John Casper. This practice of initially flying pilot astronauts as flight engineers was not new and had been seen on a number of dual-shift missions in which a 'third' pilot was needed to supervise the orbiter's flight deck. Circumstances changed in the summer of 1991, when veteran Shuttle commanders Mike Coats and Bryan O'Connor retired from NASA. The result was a shortfall in the number of available veteran pilots to rotate to the commander's seat. In August, the space agency announced that two astronauts previously assigned as pilots – John Casper on STS-50 and Jim Wetherbee on STS-46 – would be removed from their posts and promoted to command a pair of subsequent missions. Both Casper and Wetherbee had previously flown once as a pilot, thus satisfying a requirement to command a Shuttle flight. This enabled flight engineers Bowersox and Andy Allen to fill their shoes by rotating into the pilot's seat on STS-50 and STS-46. In turn, *their* flight engineer seats were taken by two other astronauts: Ellen Baker for STS-50 and Marsha Ivins for STS-46.

Ellen Louise Baker was born in Fayetteville, North Carolina, on 27 April 1953, but was raised in New York City, the daughter of Dr Mel Shulman and politician Claire Shulman. She completed high school in Queens in 1970 and earned a degree in geology from the University of Buffalo at the State University of New York in 1974 and a doctorate in medicine, four years later. She trained in internal medicine at the University of Texas Health Science Center in San Antonio and was board-certified in 1981. Her career with NASA commenced that same year, as a medical officer at the

Johnson Space Center, and she later served as a physician in the Flight Medicine Clinic, before selection into the astronaut corps in May 1984. She first flew into space in October 1989. With the removal of Casper and the arrival of Baker, the rest of the STS-50 crew remained together. In addition to Richards, Bowersox and Baker, the 'science crew' comprised mission specialists Bonnie Dunbar and Carl Meade and payload specialists Larry DeLucas and Vietnamese-born Gene Trinh.

They would supervise 31 experiments in crystal growth, combustion science, fluid dynamics and biotechnology aboard the Spacelab module, housed in Columbia's payload bay. Designated as the First United States Microgravity Laboratory (USML-1), more than one scientist considered its purpose to be a precursor to the work scheduled for Space Station Freedom. Around-the-clock science activities – carried out by a 'red' team of Dunbar and DeLucas, led by Bowersox, and a 'blue' team of Meade and Trinh, led by Baker – would utilise the microgravity environment to produce new materials and protein crystals in support of new pharmaceuticals. Moreover, the USML-1 experiments were expected to help the United States to maintain its 'lead' in microgravity research and applications. Terrestrial spinoffs were expected to be the production of newer, faster semiconductors for the next generation of high-speed computers and the construction of more efficient chemical catalysts for converting petroleum into gasoline. Some of this work would be undertaken within a laboratory-style 'glovebox' and other research facilities lined the walls of the Spacelab module in refrigerator-sized racks.

As with STS-42 and STS-45, the scientific importance of USML-1 demanded that one of its mission specialists – in this case, Bonnie Jeanne Dunbar – would serve as the payload commander, responsible for primary oversight of the research goals and accomplishments. She was assigned to the mission in September 1990, about three months ahead of the other crew members. "We think we're ready," she told journalists as the STS-50 crew arrived in Florida on 22 June 1992, three days ahead of launch. "We're really looking forward to an aggressive 13-day flight." On this mission, Dunbar would actually break her own record for the longest Shuttle mission, jointly set as a member of the STS-32 crew, two years earlier, when she spent almost 11 days aloft. Upon her return from STS-50, she would be able to claim *another* record, by having the most number of hours for a female spacefarer.

In his NASA oral history, Dick Richards recalled that Dunbar's role in USML-1 ran deep: through her efforts, she had enabled NASA to gain funding for the mission. As chair of the agency's Microgravity Materials Science Assessment Task Force in 1987, during the post-Challenger hiatus, Dunbar realised that much of the planning for Space Station Freedom centred on Spacelab missions which had a life sciences bias, with little emphasis on the development of furnaces for directional solidification or materials research. In fact, the only Spacelab missions under consideration to focus on the physical sciences were those which involved European and Japanese participation. Dunbar was keenly aware that such work had been pioneered aboard Skylab in the early 1970s and she felt that it was "a real loss of our investment and scientific discoveries of the future if we didn't build facilities for [the new] station". She worked closely with fellow astronaut Sally Ride, who was leading a team to prepare a strategic roadmap for NASA's future, and one of the products of their efforts was the

United States Microgravity Laboratory. "We flew USML-1 ... with all new facilities that were destined for the International Space Station," Dunbar told the NASA oral historian. Those facilities included an entirely new variety of furnaces, developed by Teledyne Brown Engineering, Inc., along with a fluid physics module and a range of life sciences experiments and lower-body negative pressure apparatus. "What was thrilling for me was to be asked to be payload commander of that flight," she said. "In five years, I had an opportunity to see it go from a concept in a report to an actual flight of brand-new equipment and hardware."

As the payload commander, Dunbar worked extensively on the mission timelines and would later derive satisfaction from the fact that 'hand-overs' between the two USML-1 shifts typically lasted no more than 15 minutes. It illustrated the excellence of their training. Much of that training was initially centred on the various institutions which developed the experiments: at the University of Alabama at Birmingham for protein crystal growth, at the University of Wisconsin at Madison for plant growth and at the Jet Propulsion Laboratory in Pasadena for fluid physics. "You understand what the researcher wants to find, first," Dunbar explained, "because you're their hands, eyes and ears and if you don't understand what they're looking for, then when the unexpected comes up you also don't recognise that, either. You're part of the research team." By December 1991, six months before launch, the Spacelab crew were running fully integrated simulations; picking the busiest day of the flight, choreographing it to the minute and running through their tasks. "You train like you fly," she said, "and you fly like you train. You don't have a chance to start over on orbit."

After rollout to Pad 39A, Columbia underwent routine preparations, together with several new ones, in which technicians rehearsed loading cryogenic reactants into her storage tanks and EDO pallet to establish 'realistic' timelines for how long these procedures would take. The orbiter looked identical to her sister ships, outwardly, but internally she had changed markedly. "Columbia may be the oldest," said Shuttle Test Director Eric Redding, "but with all the modifications, it's the best orbiter we have. I don't really expect an increase in frequency of problems." The only anticipated problems over the weekend of 20-21 June seemed to be related to the weather. Lightning, hail and heavy rain materialised, although Air Force meteorologists predicted a 70 percent chance of favourable conditions at launch time on the 25th. However, overcast skies ultimately led the Air Force's Mike Adams to declare that he could not be sure. "We might be able to get the Sun to burn off those clouds and give us a break," he said. "To be honest, it's too close a call. I'd have to put my two dollars on Mother Nature."

Then, on the 23rd, engineers' attention was drawn to an erratic sensor, whose spurious readings threatened a week-long delay. The sensor was located close to the combustion chamber on one of Columbia's three main engines and was responsible for monitoring the temperature of liquid oxygen as it entered the External Tank. It also provided an early warning to hedge against possible oxidiser leaks. If the sensor was at fault, then it could be replaced in time for launch, but if the *main engine* was the cause of the problem, a longer delay to remove it would be unavoidable. Fortunately, blame fell on the sensor, which was replaced and satisfactorily tested.

By the morning of the 25th, Richards and his crew were strapped into their seats aboard Columbia. "We'll see you in a couple of weeks," he radioed the launch control team. For a time, it seemed as if the Florida weather would not co-operate, as bands of cloud and rain marched across the pad, alternately raising and dashing hopes for an on-time liftoff. Five minutes later than planned, STS-50 thundered off the pad at 12:12 pm EDT and quickly disappeared behind a vast bank of cloud. The ascent to orbit was nominal ... with one caveat: in an event which would portend Columbia's loss on STS-107, a little more than a decade later, a chunk of foam, some 60 × 25 cm, fell from the left bipod ramp of the External Tank and left a 20 × 10 cm dent in one of her thermal protection tiles. "Instead of hitting the leading edge of the wing," as would happen on STS-107, "it went about a foot below the leading edge of the wing," recalled Richards in his NASA oral history. "Of course, we launched through an overcast, so we never knew that it even happened. We took photographs of the tank, but the photographs weren't downlinked to anybody. The first thing we found about it was our post-flight walk-around after landing. Ken Bowersox and I went out and we said 'Oh, look at *that*!'" Only years later would they fully realise that their dented tile might have actually turned into something far worse.

Safely in orbit, Dunbar's red team was responsible for powering up USML-1 as their blue counterparts bedded down for the night. Making her third Shuttle flight, Dunbar came from Sunnyside, Washington, where she was born on 3 March 1949. Her grandparents had arrived from Scotland and homesteaded in Oregon and the young girl grew up on a ranch, raising Hereford cattle in eastern Washington State. She often listened as her elderly grandfather told her why he came to the United States: to own his own soil. His perspective was that life was an adventure and the key to unlocking that adventure was education. Her father had similar, yet also different, ideals. "He was a Marine in World War II in the South Pacific," Dunbar recalled, and had "fought for [his] sons *and* daughters to be able to become what they want to become, if they wanted to work hard enough to do it."

Her parents instilled this ethic into her and she remembered that the first books she received as a child were a set of encyclopaedias. If anything appeared on the television which the young girl did not understand, her parents would direct her to the shelf and ask her to read the encyclopedia entry to the rest of the family. "It was fun," Dunbar recalled. "Reading was mandatory to be able to participate in this game." The Soviet launch of Sputnik in October 1957 switched Dunbar on to the idea of space travel for the first time and she became engrossed in the work of H.G. Wells and Jules Verne. In eighth grade at school, her principal asked her what she wanted to do in life and the girl, embarrassed to admit that she wanted to be an astronaut, told him instead that she wanted to design and build spaceships. He did not laugh. Nor did he belittle her dream. "You'll have to know algebra," he told her. She had no idea what *algebra* actually was, but in later years would come to appreciate that algebra, geometry, trigonometry and calculus were effectively the 'keys' for an engineering career. Dunbar had little idea how challenging it would be to *enter* such a career.

After completing high school in Sunnyside, she entered the University of

3-09: With the white room positioned against her middeck side hatch, Columbia stands on Pad 39A, ready for STS-50, her 12th voyage into orbit.

Washington to study engineering and was advised by Dr James Mueller, the dean of ceramic engineering, to change her major from aeronautics to either materials or ceramic engineering. It would prove a pivotal moment in her future career and it was Mueller who introduced her to many key engineering minds who were working on research for the Shuttle's thermal protection system. "That then put me on a course of becoming more involved in the real parts of space," Dunbar recalled. "They had summer positions open and so one of the summer positions at the University of Washington was doing X-ray diffraction research on some of the fibres that NASA was considering." Very few women studied engineering and Dunbar was received with difficulty by some students and professors. "My response to any negative attitudes," she concluded, "was to *ace* the class!" Much of her work was not subjective; it involved work whose results were either *right* or *wrong* and, as one supportive professor told her: If you've got it *right*, you've got it *right*! Gender made no difference.

Graduation in 1971 was followed by two years as a systems analyst with Boeing and a call from Mueller to invite her back for a master's degree in ceramic engineering. From 1973 until 1975, Dunbar worked on the mechanisms and kinetics of ionic diffusion in sodium beta-alumina, part of a project to develop a high-energy-density battery, primarily for the automotive industry. In fact, Dunbar would later give a lecture to the Ford Motor Company, who were performing similar research. By the summer of 1975, she had completed the requirements of her master's degree and was made a quite astonishing offer: the rigorous quality and extent of her research meant that she had effectively completed most of the requirements for a PhD, awardable in just *six months*. However, Dunbar was broke and opted instead for a six-month visiting scientist position in England at the Atomic Energy Research Establishment in Oxford, studying the wetting behaviour of liquids on solid substrates. Years later, she would profoundly regret not going ahead and completing her doctorate. Back in the United States, she joined Rockwell International to work on the Shuttle's thermal protection system and later received the Engineer of the Year Award for it. She applied, unsuccessfully, for admission into the astronaut corps, but, in 1978, entered NASA's Johnson Space Center as a payload officer and flight controller, providing guidance and navigation for the Skylab re-entry. She was finally selected as an astronaut in May 1980 and three years later completed her doctorate in mechanical and biomedical engineering at the University of Houston. Her PhD focus was an evaluation of the effects of simulated orbital flight on bone strength and fracture toughness. In October 1985 she flew her first Shuttle mission, a joint Spacelab flight with West Germany, and a little more than four years later, aboard STS-32, played a pivotal role in the retrieval of NASA's Long Duration Exposure Facility from orbit.

Dunbar's workmate on the USML-1 red shift was a biochemist whom the London *Sunday Times* labelled in January 1999 as one of 18 world-class scientists with the potential to shape the 21st century. Lawrence James DeLucas was born on 11 July 1950 in Syracuse, New York. Whilst at the University of Alabama at Birmingham, he earned bachelor's and master's credentials in chemistry in 1972 and 1974, a second bachelor's degree in physiological optics in 1979 and doctorates in optometry in 1981 and biochemistry in 1982. By that time, DeLucas had also spent several years as a research associate and began a career at the university's Vision Science Research Center, the Comprehensive Cancer Center and the Center for Macromolecular Crystallography, the latter of which he is currently an associate director. He has carried out research for NASA in the area of protein crystal growth and holds a number of adjunct professorships, both at his own and other institutions. Selected as one of four USML-1 payload specialist candidates in August 1990, DeLucas would later serve as NASA Chief Scientist for the International Space Station.

Although four payload specialist candidates had been chosen – DeLucas and Trinh, together with Joe Prahl, a mechanical engineer from Case Western Reserve University in Cleveland, Ohio, and Al Sacco, a chemical engineer from Worcester Polytechnic Institute in Worcester, Massachusetts – only two would be able to fly aboard USML-1. DeLucas and Trinh's names were formally revealed by NASA as prime crewmen in May 1991. The relationships between NASA crew members and payload specialists has always been an interesting one. "You're a family when you're up there," said Dunbar, "so if there are any [interpersonal] problems to resolve, you need to resolve them on the ground." After one training session, she invited the four payload specialists out for dinner ... and was surprised by the response.

"What? *Together?*"

It was a comical moment, perhaps, but it underlined in Dunbar's mind the importance of starting out as a team and bonding into something as close as possible to a family. "You have to get through to that so when you get on orbit, you know *exactly* what the other person is going to do and how they're going to react." For the second payload specialist, Eugene Huu-Chau Trinh, the NASA members of the crew had more knowledge, for he had served as a backup crew member on the Spacelab-3 mission in April 1985. Trinh was born in Saigon, South Vietnam, on 14 September 1950, but moved with his parents to France as a young child. He received a baccalaureate degree from Lycee Michelet in Paris in 1968 and travelled to the United States. In 1972, Trinh earned a degree in mechanical engineering and applied physics from Columbia University, followed by master's credentials in 1974 and 1975 and a PhD in applied physics in 1977. As Project Scientist for the Drop Physics Module, he was selected as a payload specialist candidate for Spacelab-3 in 1983 and on USML-1 became the first person of Vietnamese-French-US citizenship to enter space.

With the arrival of the STS-50 crew in orbit on the afternoon of 25 June 1992, former astronaut Brewster Shaw, then serving as chair of NASA's Mission Management Team, quipped that "we're going to find out whether these kids are *really* friends or not". He had previously flown with Dick Richards and need not have worried. Echoing the sentiment of the remainder of the crew, Ken Bowersox later described his colleagues as "a joy to be around". The red team, plus Richards, who, although not attached to a specific shift, tended to align his work schedule with the reds, progressed smartly through the USML-1 activation process. Ten minutes ahead of schedule, at 3:42 pm EDT, Bonnie Dunbar opened the hatch into the Spacelab, floated inside and turned on the lights.

Activation of the module was complete by 6:00 pm and as she busied herself with preparing it for 13 days of research, Bowersox and DeLucas set to work unstowing and photographing a pair of Protein Crystal Growth (PCG) specimens on Columbia's middeck. It was hoped that data from these experiments would yield new knowledge about the proteins' molecular arrangement and assist in the production of more nutritious foods, highly resistant crops and better medicines with fewer adverse side effects. Moreover, the extended nature of USML-1 meant that slower-growing crystals could be produced. After their return to Earth, the three-dimensional structures of the crystals were to be carefully mapped. Crystals of proteins grown in terrestrial laboratories are large enough to study, but usually contain numerous flaws, caused by gravity-induced convection, buoyancy and sedimentation. Space-grown crystals, on the other hand, are of greater purity and have more highly-ordered structures which significantly enhances their X-ray analysis. On USML-1, protein crystal growth experiments were performed both within the middeck incubator and inside the glovebox of the Spacelab module.

The importance of this research to the pharmaceutical industry was illustrated by the presence of Larry DeLucas, who was already recognised as one of the United States' leading experts in crystallography. Consequently, USML-1 would be the first time that experiments of this nature had been performed with a human expert present to improve the production processes with a special controllable furnace. Previous experiments had been automated and the only astronaut interaction had been the mixing of the chemical solutions. "Someday," DeLucas said before the flight, "I feel confident we will find a drug that might help to prevent the complications of diabetes, maybe prevent hypotension, maybe find a cure for some types of cancer." It was considered possible that such research might eventually lead to treatments for emphysema and HIV-AIDS. Yet DeLucas was under no illusions; more than a decade of further work and a permanent space station would be essential in carrying such plans through to fruition. "My investigators don't need one crystal so we can make a breakthrough," he said. "We need a *constant* supply of crystals."

In addition to the protein crystal growth investigations, a Crystal Growth Furnace (CGF) was employed to solidify a wide range of materials – mainly semiconductors – which form the basis of electronic devices, including computers, timepieces and communications equipment. The furnace could process samples at temperatures of up to 1,260 degrees Celsius and enabled scientists to investigate the factors affecting crystal growth and the best methods for producing them. Post-mission analysis of the crystals from CGF revealed distinguishable differences from Earth-grown specimens and showed them to be of far higher quality. During USML-1, the furnace used 'directional solidification', whereby the solidification front proceeded in a specific direction along the sample, and 'vapour crystal growth', in which part of the sample was heated to make it sublime and the vapour was then allowed to flow towards and condense upon a substrate base in a cooler part of the sample ampoule. Specimens of cadmium telluride, mercury-zinc telluride, gallium arsenide and mercury-cadmium telluride, all of which have found applications in medical infrared sensors, night-vision goggles and telescopes, were produced.

The furnace itself could handle up to six samples at a time, each housed in its own quartz ampoule within a rotary changer. These samples were processed under computer control, although the experiments' principal investigators on the ground could uplink changes or adjustments whenever necessary. A flexible glovebox provided access to the interior of the furnace, but physical handling of the samples was only done if absolutely necessary, so as to minimise the risk of contamination to the Spacelab environment. Bonnie Dunbar started the first CGF experiment, late on 25 June, when she initiated an 18-hour solidification run with six samples of mercury-cadmium telluride. Unfortunately, a circuit breaker tripped a few days later. Carl Meade followed a procedure and successfully revived it, but a subsequent mercury-zinc telluride crystal had to be stopped halfway through its growth cycle. Meade then inserted a cadmium-zinc telluride sample for 92 hours. Later in the mission, the CGF grew the longest gallium arsenide crystal (almost 16.5 cm) ever produced. Moreover, the crystal contained a deliberate trace impurity, known as a 'dopant', to enable scientists to precisely engineer its electronic properties. On Earth, gravity-driven convection makes it very difficult to control the placement of dopants and if placed imprecisely the resultant crystal can yield inconsistent material properties. Another mercury-cadmium telluride sample was also removed from the CGF on 5 July, after six hours of growth. Its wafer-thinness enabled crystallographers to examine its structure with relatively little processing. By the end of the mission, the CGF had operated for 286 hours and produced seven crystals – three more than planned – and achieved a peak temperature of 870 degrees Celsius.

Elsewhere, crystals of 'zeolites' – complex arrangements of silica and alumina, which occur both naturally and synthetically – were grown both in Columbia's middeck lockers and in the Spacelab glovebox. Zeolites have open crystalline structures which are selectively porous, enabling them to function as molecular sieves, capable of 'adsorbing' elements or compounds and forming an integral part of catalysts and ion-exchangers. On several occasions, Dunbar spoke directly to backup payload specialist Al Sacco, the principal investigator of the experiment, to discuss mixing procedures for test runs. "A person up there optimising the mixing of the crystal solution," said USML-1 Mission Scientist Don Frazier, "is a milestone in research."

All of this research almost distracted from the long-duration nature of the Shuttle's first EDO mission. The Rockwell-built 'wafer' pallet of tanks began supplying cryogenic reactants about 24 hours after launch and performed admirably, with the exception of a slight leak on the evening of 26 June. The rate of oxygen seepage was too small to pose a problem to the mission, but it was decided to use the leaking tank first, in order to reduce wastage. This procedure was completed the following day. The RCRS carbon dioxide removal system also suffered problems, failing six times in total. Lithium hydroxide canisters were installed in the Shuttle's cabin and aboard the Spacelab module. It was subsequently determined that faulty sensors which showed the positions of internal valves were causing the device to shut itself down prematurely. On 30 June, Mission Control transmitted repair instructions to the middeck teleprinter. Richards and Bowersox executed a 32-step, five-hour repair procedure – pulling out lockers, unplugging connectors, splicing wires – until the RCRS steadily belched and wheezed its way to fully functional status.

As the orbiter crew members, Richards and Bowersox were responsible for maintaining Columbia in a gravity gradient attitude, with her tail directed Earthwards and her nose some 12 degrees off the direction of travel. This provided a stable platform and reduced the need for disruptive thrusters firings. They also worked with SAREX and they later recalled chatting to a 'ham' in New Zealand about the All-Blacks rugby team there, to someone in Southern California whose transmitter drowned out everyone else, and to a guy in Argentina who said he could patch a transmission through the public telephone system and route a call to Richards' wife, Lois. It sounded like a great idea. "By the time I figured this thing out on Day Eight," Richards told the NASA oral historian, "I was about one rev away from having him patch me through to my wife, which I thought would be kinda cool, because she'd be on her mobile phone, driving around town." Unfortunately, Lois Richards had her phone switched off and the impromptu contact never took place. When he got back to Earth, he told her that on his *next* flight, if he got chance to call, she was to keep her phone switched *on* ...

Another interesting SAREX contact came from a Polynesian sailing vessel, heading from Hawaii to French Polynesia, trying to recreate ancient navigation techniques. As the astronaut and the sailor talked, Richards was able to estimate his position and give him an update on cloud cover to the north of his route. A day or so later, the pair made contact again and the sailor invited the astronaut and his wife to the Pacific Asian Arts Festival in the Cook Islands. A month after landing, in what Richards described as "one of the cooler trips I ever had", he and Lois were aboard a replica Polynesian vessel in the Cook Islands.

In general, the mission was proceeding smoothly, both in terms of the USML-1 research and the performance of Columbia herself. The text and graphics machine on the middeck suffered a paper jam and so all printed messages – including the RCRS repair procedure – had to be sent instead to the Shuttle's teleprinter. As time wore on, the crew tried to pace themselves, with Bonnie Dunbar hopeful that they would return to Earth as refreshed as when they left. Saddled with the burden of command, Richards took it upon himself to insist that crew members got to sleep at the end of their shifts; he was well aware of previous Spacelab missions, in which astronauts had stayed awake for 14 hours or more, working a troublesome experiment. Richards' rule was simple: "I don't care if the lab is coming *unglued* You'd better train the guy that's going to relieve you, because even if that's your experiment, you're going to bed. The other guy has got to be able to do it. You've got an hour to hand over and then you're going to bed."

Now, as STS-50 passed its halfway point, Richards was more than happy with its progress. Larry DeLucas also described himself as "relaxed" after a scheduled period of off-duty time. Independence Day was observed and two days later, on 6 July, they quietly exceeded the 11-day Shuttle record set by STS-32. To commemorate their achievement, Mission Control played the Zodiacs' 1960s hit, 'Stay'.

Remarkably, in spite of their relaxed state, the astronauts enthusiastically pursued their hectic around-the-clock schedule of research. In addition to the crystal growth experiments, a wide range of investigations examined the behaviour of fluids under differing influences, including the application of heat, in the hope that they might be used to produce high-technology glasses, ceramics, semiconductors, metals and alloys from ingredients mixed in a liquid state and cooled into solids. It was already known that fluid motions on Earth often introduces defects which restrict certain materials from achieving their full potential as lenses, computer chips, turbine blades and other products. In microgravity, the influence of this buoyancy-driven convection is drastically reduced and other, more subtle forces begin to prevail in fluid mechanics.

On 28 June, Larry DeLucas began the Interface Configuration Experiment (ICE) inside the glovebox. This apparatus closely mirrored the kind of gloveboxes used in terrestrial research laboratories: it contained a pair of openings, through which a crew member could insert his or her hands, and a third port for the installation of experiment samples. Rugged gloves and finer surgical gloves were employed, dependent upon the degree of precision required. Most of the sample ampoules scheduled to be used with the glovebox had magnetic bases, enabling them to 'stick', and a large plastic window on the top of the unit allowed the astronaut to see the interior. Four still and video cameras provided coverage, which could be transmitted to Earth in real time. After mounting cameras for the ICE task, DeLucas filled an experiment vessel with an immersion fluid – hydrogenated terphenyl and an aliphatic hydrocarbon – to allow researchers to better determine the behaviour of fluids in

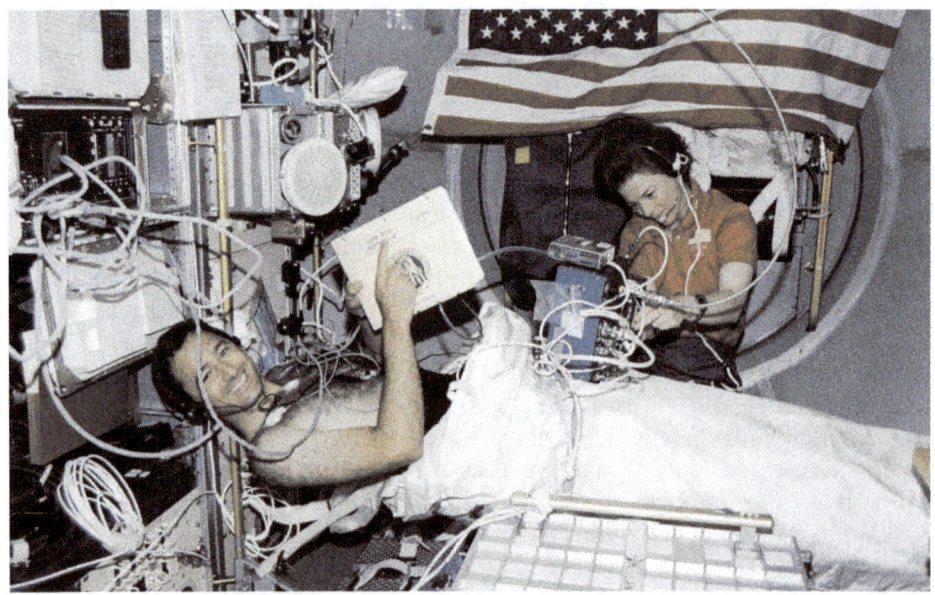

3-10: Larry DeLucas participates in a Lower Body Negative Pressure investigation, with Bonnie Dunbar on hand to provide assistance, during STS-50. This view was taken inside the USML-1 Spacelab module.

differently-sized containers. On Earth, fluids behave in particular ways, but their flows in microgravity must be properly understood in order for better fuel tanks and containers for biological or human waste to be manufactured.

The Surface Tension Driven Convection Experiment (STDCE) and the Drop Physics Module (DPM) were fundamental facilities in the execution of this work. The former explored how fluids reacted in microgravity conditions when temperature differences existed along their interfaces, whilst the latter enabled 'containerless' studies of droplet behaviour in a weightless environment. During USML-1, the STDCE used a lightweight silicone oil, which was not susceptible to surface contamination, and more than 13 hours of video footage were recorded. Three runs of the experiment, each lasting around eight hours, were planned, but four were eventually conducted. For most of the mission, Gene Trinh supervised the experiment. The only problem was an inability of the carbon dioxide laser to reach the required power level and the accidental introduction of bubbles into the sample chamber. Elsewhere, the DPM research saw water, glycerine and silicone oil droplets subjected to various external forces. Speakers mounted inside the experiment chamber allowed the droplets to be rotated, oscillated, merged, 'split' and suspended under acoustic pressure. It was hoped that such research could lead to new methods of 'encapsulating' living cells within membranes to protect them from harmful antibodies, potentially leading to revolutionary medicines. Drops were suspended, singly, for up to three hours at a time and at one stage, late in the mission, Dunbar was even able to 'bounce' droplets with the DPM speakers.

Flame behaviour was investigated, with the ignition of polyurethane foam cylinders, sealed in clear Lexan, and electrical wires were heated and burned within the glovebox. Paraffin and stearic acid candles were ignited to examine the physical appearance of flames in microgravity; they flared into spherical balls with glowing yellow cores. Typically, within ten seconds or so, presumably owing to the presence of soot, they turned blue and assumed hemispherical shapes, measuring around 1.5 cm across, then extinguished themselves within a minute or so. This data was consistent with short-duration studies aboard parabolic aircraft and drop-tower testing in the United States and Japan. Together with the USML-1 research, it led to new theories about the change in flame shape from spherical to hemispherical.

The final major group of USML-1 experiments focused on biotechnology, with the Generic Bioprocessing Apparatus supporting more than a hundred studies of molecules and small organisms and 'Astroculture' demonstrating a prototype system for providing plant nutrition. Housed in a middeck locker, Astroculture was tended by Ken Bowersox, who darted downstairs from the flight deck during his shifts to ensure that its water and nutrients were satisfactory and check on its lighting and humidity controls.

As STS-50 neared its end, it was already being hailed as hugely successful. Early on 8 July, Carl Meade's blue team oversaw the deactivation of the Spacelab, but the progress of Tropical Storm Darby over the Baja peninsula threatened the option to land at Edwards Air Force Base. It had been classified as a Category 2 hurricane on the 6th, but was downgraded to a tropical storm on the 7th and by the 9th had been downgraded yet further to a tropical depression. Still, it left Edwards rain-sodden.

Undeterred, the crew moved smartly through their de-orbit checklist, but to no avail. Mission Control waved off their landing attempt for 24 hours. Columbia's OMS engines were pulsed later that day for 30 seconds to better align their orbital path for subsequent landing opportunities. Flight Director Jeff Bantle had expressed preference to land in California, rather than the Kennedy Space Center, since Columbia's heavier-than-normal weight of 102,500 kg made an Edwards touchdown safer. When the rain in California did not clear by 9 July, Bantle opted to divert the astronauts to Florida.

With Spacelab closed up, the extra few hours aloft gave the astronauts some time to catch up on their notebooks, complete their flight recordings and gaze through the windows at the grandeur of Earth. For Larry DeLucas, this latter pastime brought great pleasure. At one stage, he disappeared into the airlock to fetch some film for a camera, closed his eyes for a moment ... and fell asleep. Bonnie Dunbar, who went looking for him, could see nothing but a pair of feet sticking out from the midst of a floating mass of stowage bags. "If you were tired and you were floating," she later recalled, "you didn't have to lay down or sit down. You just sort of nod off once in a while."

Thanks to Tropical Storm Darby, STS-50 turned into Columbia's first return to KSC, within sight of the launch pad that she had vacated two weeks before. Weather conditions were ideal, with light winds and a slight chance of patchy ground fog. The Shuttle streaked like a meteor over Houston, heading for Florida, and touched down on Runway 33 at 7:42 am EDT. For the first time, Bowersox deployed the drag chute between main gear and nose gear touchdown, and, combined with her new tyres, Columbia was brought smoothly to a halt in less than a minute. The flight had lasted 13 days, 19 hours, 30 minutes and four seconds, establishing itself as the fourth longest mission in US history and the longest Shuttle voyage to date. Speaking to journalists on the runway, Dick Richards apologised on behalf of blue team crewmates Baker, Meade and Trinh, none of whom had slept for 24 hours. "They just need to get home for a well-deserved rest, because it's a very tiring process, not just staying up over 24 hours, but getting used to gravity again," he explained.

In general, the crew had returned in good physical and psychological condition from their two weeks in orbit. Throughout the mission, they utilised equipment from the EDO Medical Project, which included the Lower Body Negative Pressure suit: an inflatable cylinder, 1.2 m tall, which sealed around the waist and drew fluids into the legs to offset the upward fluid shift which tends to occur upon arrival in the microgravity environment. They also drank a litre of heavily-salted water before re-entry to help to keep body fluids in place until they could adapt to terrestrial gravity at a more leisurely pace. Richards, who did no exercise on his previous flights, took his physical wellbeing much more seriously this time. "During re-entry," he said, "you *have* to have a strong cardiovascular response, particularly during the final phase."

"WEIRD SCIENCE!"

One day in 1987, ESA astronaut Claude Nicollier came to Jeff Hoffman with an invitation to an interesting meeting. It was about an Italian project called the Tethered Satellite System (TSS).

"Jeff," Nicollier said, "you're a physicist. This is something that might intrigue you."

A year had passed since the loss of Challenger and several scientist-astronauts were taking sabbaticals or pursuing advanced degrees. Hoffman was enrolled at Rice University on a master's course in materials science, focusing on crystal growth, when Nicollier invited him to the TSS meeting. Nicollier was about to be sent to the Empire Test Pilots' School in Boscombe Down, Hampshire, in the United Kingdom, to work on the development of ESA's proposed Hermes spaceplane, and when the two men attended the meeting Hoffman was quickly hooked. So began a nine-year involvement with one of the strangest Shuttle experiments of all time – an experiment which astronaut Marsha Ivins once described to this author as "weird science".

Weird is certainly an apt choice of descriptor for the TSS, which encompassed nothing less than a satellite trawled 'on a string' through the electrically-charged ionosphere, as part of efforts to demonstrate the electrodynamics of a conducting tether to convert kinetic energy into electrical energy. Originally proposed to NASA in the early 1970s by Guiseppe Colombo of Padua University and Mario Grossi of the Smithsonian Astrophysical Observatory, it was hoped that the concept might ultimately lead to systems featuring electricity-generating tethers, using Earth's magnetic field as a power source. Moreover, by 'reversing' the direction of current in the tether, the force created by its interaction with Earth's magnetic field could potentially place objects into motion, boosting the velocity of a spacecraft without propellant, thereby counteracting the effects of atmospheric drag.

Following Colombo and Grossi's initial proposal, the Facilities Requirements Definition Team met in 1979 to consider the applications of tethered satellites and its report, published the following year, strongly endorsed a Shuttle-based research mission. A memorandum of understanding was signed in 1984, which called for NASA to develop the deployment mechanism and Italy to build the satellite. A science advisory team provided guidance in preparation for a formal Announcement of Opportunity, in April of that same year, for experiments. A total of 12 experiments were selected, featuring participation from the Italian Space Agency, NASA and the Air Force's Phillips Laboratory.

The TSS Deployer Core Equipment and Satellite Core Equipment (DCORE/SCORE), provided by Carlo Bonifazi of the Italian Space Agency, was responsible for controlling electrical current between the satellite and the orbiter. It featured an electron accelerator with two electron beam emitters, each capable of ejecting up to half an amp of current from the system. The Research on Electrodynamic Tether Effects (RETE), supplied by the Italian National Research Council in Rome, employed a pair of instrumented booms to measure the electrical potential in the plasma 'sheath' formed around the satellite during deployment. Franco Mariana of the Second University of Rome was principal investigator for the Magnetic Field

Experiment for TSS Missions (TEMAG), which used two magnetometers to map fluctuations in magnetic fields around the satellite. Elsewhere, the University of Genoa sought to investigate the extent to which the TSS could 'broadcast' from space, with magnetometer emplacements around the world and extreme-low-frequency receivers at the Arecibo Radio Telescope in Puerto Rico primed to track emissions and plasma wave directions. Finally, the University of Padua supplied a pair of experiments to analyse the satellite's oscillations over a range of frequencies in real time.

The United States' side of the experiment payload included the Research on Orbital Plasma Electrodynamics (ROPE), provided by NASA's Marshall Space Flight Center of Huntsville, Alabama, which studied the behaviour of ambient charged particles in the ionosphere and ionised neutral particles in the vicinity of the satellite itself. Peter Banks of the University of Michigan at Ann Arbor provided the Shuttle Electrodynamic Tether System (SETS) to evaluate the capacity to collect electrons by determining current and voltage and measuring the resistance to current flow in the tether. The Smithsonian Astrophysical Observatory conducted research into electromagnetic emissions and dynamic noise from the TSS, whilst the Tether Optical Phenomena (TOP) experiment featured hand-held, low-light-level television cameras for visual data. Other experiments provided theoretical electrodynamic assistance and measured the levels of the Shuttle's own electrical potential in comparison to the ambient space plasma.

In the late 1980s and early 1990s, the TSS was a particularly appealing concept for the designers of Space Station Freedom and its successor, today's International Space Station, as a means of compensating for atmospheric influences on the gigantic outpost. Additionally, it was hoped that the electrodynamic tethering concept might lead to the development of new devices to trail scientific platforms far below orbital altitudes in difficult-to-study regions of the atmosphere, including the fragile ozone layer above the South Pole. Other possibilities included providing a foundation for extremely low-frequency antennas that were capable of penetrating land and seawater, and perhaps generating artificial gravity or accelerating payloads into higher orbits.

At the time of the Challenger disaster, future Shuttle manifest projections ended in August 1988 and did not include the TSS, suggesting that the mission would have been conducted at some point after that date. The satellite itself was a 1.6 m sphere, weighing 517 kg, with an outermost skin of aluminium alloy, coated with an electrically-conducting layer of white paint. It was, however, far more than just an oversized metallic football. Piercing its shell were windows for Sun, Earth and charged-particle sensors, a connector for the umbilical tether and access doors for its on-board batteries. Extending from one side of the TSS was a long, fixed instrument boom, whilst a shorter antenna sprouted from the other side. To assist with thermal control, the interior of the spherical shell was painted black. If one were to open the shell, like an egg, two compartments would be revealed: a Payload Module for the scientific experiments and a Service Module for the subsystems. Additionally, in the centre of the shell was a tank of pressurised nitrogen, which provided propellant for the satellite's 12 cold-gas manoeuvring thrusters. According to Italian prime contractor Aeritalia – today part of Alenia Aeronautica – the TSS required more than a million man-hours of work to prepare it for final integration.

If the satellite itself represented a technological marvel, then the 2-mm-thick tether which connected it to a supporting mast on a Spacelab pallet in the Shuttle's payload bay was an equally impressive creation. Surrounding its Nomex core was electrically-conducting copper wire, insulated with Teflon and coated with ultra-strong braided Kevlar-29 and an outer 'jacket' of braided Nomex to protect it from abrasion and the corrosive effects of atomic oxygen in low-Earth orbit. During deployment operations, the tether was unreeled from a 2,027 kg mechanism affixed to the Spacelab pallet and an MPESS. This regulated the length, tension and rate of deployment of the tether and was capable of unreeling at a maximum speed of about 7 km/h.

Essentially, the structure took the form of a four-sided erectable 'tower', not dissimilar in appearance to a small broadcasting pylon, which unfolded slowly from a storage canister using a series of rollers. As the canister rotated, fibreglass batons popped out of their stowed, bent-in-half positions to form cross-members – 'longerons' – which supported the vertical segments. The tower was deployed to a height of 11.8 m above the payload bay to minimise the risk of the satellite impacting the Shuttle as it was being deployed and recovered. "The complexity of the experiment," said astronaut Andy Allen, who served as pilot on the first TSS mission and later commanded the second, "is extreme." Aside from the risks, the sheer audacity of the mission is amply illustrated by the numbers: when the tether was unreeled to its full length of 20.6 km, the TSS/orbiter combination would be a *hundred times longer* than any other spacecraft in human history, its electrical potential was anticipated in the region of 5,000 volts and its maximum current output was expected to be one ampere.

That a European astronaut was one of the first to draw attention to the TSS concept within the astronaut corps offers a nod toward its international flavour, for Claude Nicollier was the first non-US citizen to fly aboard the Shuttle as a fully-fledged mission specialist. "My first dream as a child was to become a pilot," he told a NASA interviewer years later. "My second dream was to become an astronomer." In his professional life, he did both. Born in Vevey, Switzerland, on 2 September 1944, the son of a civil engineer, Nicollier graduated from high school at the Gynmase de Lausanne in 1962. He studied physics at the University of Lausanne and earned his degree in 1970. Several years' worth of postgraduate research at the university's Institute of Astronomy and the Geneva Observatory followed and Nicollier earned a master's credential in astrophysics in 1975. By this time, he had also been a Swiss Air Force pilot for almost a decade and he joined the Swiss Air Transport School in Zurich in 1974 to fly the DC-9 commercial airliner. A research fellowship in airborne infrared astronomy at ESA's Space Science Department at Noordwijk in The Netherlands began at the end of 1976 and in July of the following year Nicollier was selected – alongside West German physicist Ulf Merbold and Dutch physicist Wubbo Ockels – as one of the first European astronauts. Initially assigned to prepare for a payload specialist position on the first Spacelab mission, in May 1980 Nicollier began training as a mission specialist candidate. At the time of

3-11: The Tethered Satellite – described by STS-46 astronaut Marsha Ivins as "weird science" – inches out from its extendible mast at the beginning of deployment operations. The satellite and its tether reached a distance of 256 m from Atlantis, before a stuck bolt prevented further deployment.

the Challenger disaster, he was scheduled to fly on STS-61K in August 1986, the EOM-1/2 mission, which later morphed into ATLAS-1.

As circumstances transpired, Nicollier and Hoffman ended up flying together on the first TSS flight. They were named as mission specialists for STS-46 in September 1989, working towards a scheduled launch date in May 1991. Also named was Franklin Chang-Díaz, the first Hispanic-American astronaut. With its increased emphasis on the recruitment of minorities in the late 1970s, NASA recognised that whilst blacks and women were applying for the astronaut programme, Hispanic applicants were fewer in number. In September 1979, the agency acted and specifically called for interested volunteers. "Many qualified Hispanics are hesitant to apply," admitted Jose Perez, deputy chief of NASA's Equal Opportunity Programs Office. "I would like to encourage those persons, and others, to call or write NASA for an application." One young man who answered this call, Franklin Ramón Chang-Díaz, originated from Costa Rica in Central America. He was born on 5 April 1950 in San José, the son of Ramón Chang-Morales and María Eugenia Díaz De Chang. Though his parents were both Costa Rican citizens, his father was of Chinese descent, whilst his mother was wholly Hispanic, and the family had also lived in Venezuela for a time. After studying at La Salle School he moved to the United States as a teenager in August 1968, with $50 in his pocket, hoping to become an astronaut. "I was captivated by Sputnik as a child," he told an interviewer. "I felt that, someday, humans would travel to distant planets and I decided that I wanted to be one of those travellers. I would be a space explorer."

It seemed an impossible dream for a youth who spoke no English. "My family never prevented me from doing that," he continued, "but they couldn't really help me. We were not a well-to-do family. Even though my parents put us in the best, most expensive schools and we got a first-rate education, my parents were not rich. Neither one of them finished college, so I was expected to make my own way as soon as I finished high school. I couldn't expect to receive a college education on my father's dime, but I was expected to *have* a college education." Encouraged by his parents, and his grandfather, to go to the United States, Chang-Díaz enrolled at the public high school in Hartford, Connecticut, and learned English through total immersion in the language. He failed his classes in the first two quarters, but his third and fourth quarters were outstanding. One of his teachers, Alan Winter, took notice of his efforts and began coaching the teenager and preparing him for university. Chang-Díaz succeeded in securing a scholarship for the University of Connecticut as an engineering student, but his lack of US citizenship presented an obstacle. "Well, *that* was a bucket of cold water," he remembered. "I went back to the high school and related this story to the teachers, who apparently wrote a *petition*. The Connecticut legislature met and decided to offer me one year of the scholarship and let me pay the lower, in-state tuition, because they had already offered the scholarship." Chang-Díaz was obliged to take loans and work in the university's physics department to support himself through the remaining three years. To him, the story of those years was all about America – "the ability to get ahead by hard work" – and he got his bachelor's degree in mechanical engineering in 1973.

By this stage in his life, he was gravitating towards energy and nuclear fusion, figuring that *this* power source would be critical for getting future astronauts to Mars. It offered a small insurance policy in the event that he did not succeed in his aspiration to become an astronaut. Chang-Díaz entered graduate school at MIT and in 1977 earned a PhD in applied plasma physics, with a research focus on the problems of controlled thermonuclear fusion. After the completion of his doctorate, he joined the technical staff of the famed Charles Stark Draper Laboratory, working on the design and integration of control systems for fusion reactor concepts. Two years later, NASA announced its intention to select Hispanic candidates for the astronaut programme and he tendered his application. When he received the call, in May 1980, that he had been selected, Chang-Díaz "went running out the door and across the street. I almost got run over by a *cab!* I was in a *totally* different world. My life changed completely from that day on." At first, being a scientist, and not a test pilot, posed another hurdle. It seemed to Chang-Díaz that his lack of flying credentials made him a less attractive candidate to draw a mission assignment. "That didn't seem right to me," he said, "and I kept working to remain both a scientist *and* an astronaut. In the end, I won out." By the end of his career with NASA in 2005, he had flown *seven missions*, creating a joint record with fellow astronaut Jerry Ross. Their record stands to this day.

Little could they have foreseen at the time, but Hoffman, Nicollier and Chang-Díaz would fly together on *both* missions of the TSS. A few months after being announced as members of the STS-46 crew, in January 1990, Hoffman was named as the payload commander, but was unable to commence full-time training until he returned from his ASTRO-1 flight in December. Years later, he remembered the excitement of the TSS effort. "It was something that nobody had ever done before," he told the NASA oral historian. "It was like learning how to go to the Moon. How do you do it? Nobody knew how to control a tethered satellite. In fact, the way that it had originally been designed, they had thought that this was going to be an easy thing to do. It was going to be completely automatic and all you would do is push a button. It would go 'up' and then you'd push a button and it would come back." The satellite *did* have an attitude-control mechanism, but there was no control over pitch or roll controllability ... and this raised concerns in Hoffman's mind. What if the tether went 'slack', he asked, or if a malfunction required them to halt it at mid-deployment. Would there be a chance of loss of control?

"The problem is when the thing really gets going, it's coming out at several metres per second," he recalled. "That's pretty fast. If you just slam the brakes on, it's going to go wildly unstable. They basically had never designed for all these contingencies. As we did more and more simulations, and we learned more and more about the system, we came up with more and more scenarios where you *need* these manual capabilities." Hoffman worked closely with TSS project management, which proved very accommodating in organising the necessary financial resources to install extra capabilities on the satellite.

In addition to the three mission specialists, NASA also assigned veteran astronaut Robert 'Hoot' Gibson as commander of STS-46, indicative, perhaps, of the need for a pilot's perspective from the outset. It would mark Gibson's fourth Shuttle mission ... but fate would dictate otherwise. In the summer of 1990, he was abruptly removed from the mission and dropped from T-38 flight status for a year, apparently as a punitive measure following "violation of a policy which restricts high-risk recreational activities for astronauts named to flight crews". On Saturday 7 July, he participated in an air race at the civilian airshow in central Texas, when his aircraft collided with another. Gibson nursed his machine to the ground, but the other pilot was killed. Two days later, Don Puddy, the director of Flight Crew Operations, formally grounded him from flight status. "Our high-risk activity policy defines plain and simple guidelines for astronauts assigned to flight crews," Puddy explained in a NASA press release. "They are intended to preserve our crews as assigned and apply regardless of the time prior to launch." At the time of the incident, the launch of STS-46 had slipped slightly in the manifest and was tentatively scheduled for the late summer of the following year.

Five months after Gibson's grounding, in December 1990, fellow astronaut Loren James Shriver – recently returned from commanding the Hubble deployment mission, STS-31 – was assigned to fill his shoes. Shriver was born in Jefferson, Iowa, on 23 September 1944. Unlike many of his peers, who either grew up with military parents or dreamed of aviation and the prospect of travelling into space, Shriver had no such inspiration. "When I was a young boy," he told the oral historian, "we lived on a farm in Iowa ... so I often wonder how I got started in wanting to be a pilot." Not until he was in his mid-teens did the Space Age commence and, even then, it did not drive his ambition. Nevertheless, an interest in engineering and aviation developed out of nowhere at high school. Despite having been accepted to study at Iowa State University, the problem of paying for a college education led him to the Air Force Academy. He initially went through 'prep school' and underwent training at Lackland Air Force Base in San Antonio, Texas, before entering the academy in 1963. "The Air Force Academy can't grant a master's degree," he recalled. "It's only authorised to give bachelor's-level degrees, but I was among a group of several guys that took advance courses and they were credited by Purdue University." Consequently, after receiving an undergraduate credential in aeronautical engineering from the Air Force Academy in 1967, Shriver worked feverishly to complete the remainder of his master's in astronautical engineering at the civilian university. By January of the following year, master's degree in hand, he left Purdue and entered pilot training at Vance Air Force Base in Enid, Oklahoma. After four years there as a T-38 flight instructor, in 1973 he qualified in the F-4 Phantom and was sent to Thailand.

The prospect of becoming an astronaut had still not figured on Shriver's radar; by his own admission, he was simply following areas of aviation which fascinated him, although test pilot school beckoned and he spent much of 1975 at Edwards Air Force Base. He later served on the joint test force for the new F-15 Eagle fighter. When the NASA call for astronauts materialised, around 60 percent of the pilots at Edwards applied. His wife was none too keen, but eventually agreed and he submitted his application. Shriver himself was fascinated by the notion that this winged spacecraft was billed as being capable of launching into orbit on a *weekly* basis. He was called to Houston in late September 1977 and remembered being somewhat bemused the following January when he was informed, matter-of-factly, of his selection into the most elite flying fraternity in the world: NASA's astronaut corps.

Shriver shared many of his crewmates' concerns about the controllability of the TSS during deployment. Like Hoffman, he knew that the danger of the tether slackening was a serious worry; although gravity gradient forces would keep it tight during part of its deployment, the risk was very real in the early stages. "The dynamics," said Shriver, "could do most anything and it could go frontwards and then backwards and side to side." The obvious implication was that the Shuttle itself could be endangered. "The trouble was finding any computer system that NASA had that could model that," he continued. "Within the last couple of months of going to fly, [we] had some stand-alone trainers that began to show some of that in a reasonable manner, but we never did have really good training set-ups in the Shuttle mission simulator. There were some approximations, but they were never that complete or that good."

As for his crewmates themselves – and this was his first Shuttle mission to feature an international team of astronauts – Shriver had nothing but the most extreme praise. With an Italian-built primary payload, NASA always intended to fly an Italian payload specialist and in September 1991 physicist Franco Malerba was selected for the position. Together with Claude Nicollier and five Americans, this made STS-46 only the fourth Shuttle mission to feature representatives of three discrete nations. (They even informally added Jeff Hoffman's English-born wife, Barbara, to the crew roster.) "It never ceases to amaze me," said Shriver, "how quickly crews coalesce into a highly-functional unit. Once you get worked out who is going to be doing what ... then the plans start to fall in place and you go off and start training and everybody knows what they need to do."

Franco Egidio Malerba was born in Busalla, Genoa, on 10 October 1946, and would earn renown on the mission as Italy's first citizen to enter space. He completed high school (*Maturita classica*) in 1965 and entered the University of Genoa, from which he graduated with distinction (*laurea 110/110 cum Laude*) as an electronics engineer, with a telecommunications specialisation, in 1970. Malerba earned a doctorate in biophysics from the same institution, four years later, having undertaken research work for the Italian National Research Council and at the National Institutes of Health in Bethesda, Maryland. Whilst in the United States, he obtained his private pilot's licence and later served as a reserve officer in the Italian Navy (the *Marina Militare*), lecturing in science and working on transmission systems. His early professional career encompassed experimental work on membrane biophysics and biological membrane modelling and in the summer of 1977 he was selected by ESA as one of four payload specialist candidates for Spacelab-1. Malerba helped to develop the French-led Phenomena Induced by Charged Particle Beams (PICPAB) experiment for Spacelab-1 as a staff member at the European Space Research and Technology Centre (ESTEC) at Noordwijk in The Netherlands. He also worked on microprocessor systems, computer network engineering and telecommunications technologies. When the Italian Space Agency (ASI) was founded in 1988, Malerba took a leading role as a staff member and he was selected the following year, alongside fellow physicists Umberto Guidoni and Cristiano Batalli-Cosmovici, to train for a payload specialist position on the mission.

The TSS deployment and science-gathering operation demanded that the STS-46 crew worked in two teams – nicknamed 'red' and 'blue' – for around-the-clock activities. On the red team were Hoffman, Chang-Díaz and Marsha Sue Ivins, who had already worked for NASA as an engineer for a full decade before she was selected as an astronaut candidate in May 1984. Ivins grew up with dreams of one day flying into space. She was born in Baltimore, Maryland, on 15 April 1951. "When I was ten years old," she told an interviewer, shortly before her final Shuttle mission, "Alan Shepard made the first flight in the American space programme." Ivins remembered that the event captured her imagination and she refused to allow the fact that all of the astronauts were *male* and *military pilots* to get in her way. She realised that they were also engineers and, "for no other reason", went to the

University of Colorado at Boulder to study aerospace engineering. Ivins received her degree in 1973 and began working at the Johnson Space Center in July of the following year. Her initial role was as an engineer for the displays and controls of the Shuttle, which was then in its earliest stages, and in 1980 she became a flight engineer in the Shuttle Training Aircraft. She applied for the astronaut programme on *three* occasions, was unsuccessful in 1978 and 1980, and made the cut four years later.

The blue team for STS-46 consisted of Claude Nicollier and Franco Malerba, together with pilot Andrew Michael Allen, born on 4 August 1955 in Philadelphia, Pennsylvania. He graduated from Archbishop Wood Catholic High School and entered Villanova University to study mechanical engineering. Upon receipt of his degree in 1977, Allen entered the Marine Corps, trained as a pilot and spent several years flying F-4 Phantom fighters in Beaufort, South Carolina. Subsequently selected for fleet introduction of the F/A-18 Hornet, Allen was reassigned to Marine Corps Air Station El Toro in California from 1983-86, serving as a squadron operations officer and completing both the Marine Weapons & Tactics Instructor Course and the Navy Fighter Weapons School ('Top Gun') at Miramar. In 1987, shortly before his selection by NASA as an astronaut candidate, Allen was completing his studies at the Naval Test Pilot School at Patuxent River, Maryland. Among his earliest technical assignments at NASA were improvements to the Shuttle's nosewheel steering, brakes and tyres and the design and implementation of the drag chute. Ironically, on STS-46, which was Atlantis' last voyage before a major series of modifications and refurbishments, Allen would fly aboard the final Shuttle mission to fly *without* the new chute.

STS-46's seven days in space were to be tightly packed with activities; the activation and checkout of the TSS was scheduled to begin only hours after orbital insertion, followed by experiments with the TOP investigation and the actual deployment on the fourth day. Within hours of deployment, the satellite was expected to reach its maximum distance of 20.6 km from the Shuttle, whereupon it would be reeled back in to about 2.5 km for further experiments, then finally retrieved and docked back onto the top of the mast. As if these operations were not complex enough, the TSS formed one of *two* primary payloads on STS-46. The other was ESA's European Retrievable Carrier (EURECA), a unique, box-like satellite, loaded with 15 experiments in protein crystallisation, space biology, fluid mechanics, solar physics, aeronomy and climatology and electric propulsion. At the time of the Challenger disaster, it was scheduled for its first launch in March 1988, but was extensively delayed. After deployment from Atlantis' RMS mechanical arm at an altitude of 425 km on the second day of the mission, EURECA was to employ its own thrusters to ascend to around 515 km, spending the next nine to ten months untended, ahead of retrieval by a subsequent Shuttle crew in the early summer of 1993. Europe paid NASA $14.1 million to launch EURECA and an additional $3.9 million to recover it from orbit.

The project received approval in May 1982, with a minimum 80 percent commitment from ESA, and an originally planned maiden voyage four years later. West Germany agreed to provide up to 42 percent of the financial support for EURECA – the highest stake of an individual ESA member nation – and in July

1985 its MBB-Erno organisation accordingly gained the prime contract for its development.

With international participants also including Belgium, Denmark, France, Italy, the United Kingdom and The Netherlands, EURECA measured 2.5 m long and 4.5 m in diameter and weighed almost 4,500 kg. It was constructed from high-strength carbon-fibre struts, connected by titanium nodes, to form a framework of cubic elements which provided relatively few thermal distortions, enabled higher alignment accuracy and was easy to assemble and maintain. It had a pair of deployable, five-segment solar arrays – capable of producing a total of five kilowatts of electricity

3-12: The European Retrievable Carrier (EURECA), its solar arrays unfurled in wing-like fashion, is raised into its pre-deployment position by Claude Nicollier, ahead of release from Atlantis. Visible in the foreground, still stowed aboard its pallet, is the spherical Tethered Satellite, which would itself be deployed, later in the mission.

– and was designed (as its name implies) to be 'reusable', with at least one other nine-month mission scheduled for the mid-1990s. This was subsequently cancelled, in part due to its hefty $250 million predicted price tag. As a result, EURECA flew only once and is today housed at the Swiss Transport Museum in Lucerne.

During its mission, the satellite was managed by ESA's Space Operations Centre in Darmstadt, Germany, which exercised control through a pair of ground stations in Kourou, French Guiana, and Maspalomas in southern Spain. It would actually be in direct contact for a relatively short period during each day and was thus designed to operate highly autonomously, detecting and responding to systems problems and isolating and recovering from faults. Its scientific payload was complex and varied: protein and semiconductor crystallisation; analysis of the impact of space radiation on spores, seeds and eggs; Italian and German materials science furnaces, capable of reaching temperatures of up to 1,400 degrees Celsius; a high-precision thermostat; a pair of solar physics investigations; a study of aerosol and trace gas densities in the mesosphere and stratosphere; a wide-angle gamma ray and X-ray telescope; an inter-orbit communication device to relay data through ESA's Olympus satellite; observations of meteoroids and cometary dust; an advanced gallium arsenide solar panel; and an experimental electric propulsion thruster. Magnetic torquers and EURECA's own attitude and orbit-control system were designed to ensure that accelerations would be kept to a minimum, thereby avoiding disturbance to the experiments.

With two ambitious satellites thus tucked inside her cavernous payload bay, Atlantis rocketed off the pad at 9:56:48 am EDT on 31 July 1992. The countdown clock had been held at T-5 minutes for 48 additional seconds, when the ground launch sequencer verified that the fuel isolation valve position for one of the Shuttle's Auxiliary Power Units (APUs) was closed. Upon achieving orbit, the astronauts began preparations to deploy EURECA. Six hours into the mission, Nicollier completed the checkout of the RMS, with no anomalies, and positioned it above the payload bay in an 'overnight park' position, preparatory to its deployment on 1 August.

However, mission events began to slip when a series of intermittent data problems were encountered with EURECA whilst attached to the RMS. At the 'lower hover' position above the payload bay, the orbiter's payload data interleaver lost its communications link with the satellite, although the solar arrays and antenna were successfully unfurled. It was decided to postpone the deployment by 24 hours, whilst ground controllers commenced troubleshooting, and it was not until the late evening of 1 August that Nicollier once again returned EURECA to its pre-deployment position. At length, at 3:07 am EDT on the 2nd, Nicollier released the satellite and Shriver and Allen began the procedure to withdraw Atlantis to a safe distance. The pilots maintained a station-keeping position at a distance of 300 m until the checkout of EURECA had been satisfactorily concluded.

For a time, during this station-keeping, Allen was alone on the flight deck, and when Atlantis slipped into orbital darkness it was almost impossible to see the satellite. As a result, he flew using the radar, making sure that he kept an appropriate distance from EURECA. "On a night pass, we have about ten minutes where we lose radio contact with the ground," Allen told an interviewer for the Smithsonian, years later. This occurred whilst crossing the gap in the coverage of the 'East' and 'West' TDRS geostationary relay satellites. "We had just entered one of those periods, so there was no contact with Mission Control and no view of the satellite." All at once, the radar started picking up 'range-rate' data, indicative of a changing position of EURECA relative to Atlantis. "At first I thought it was a problem with the radar, but the rate kept increasing," he continued. "It finally got to the point where I figured if it was really true, it was either going to hit us in a few seconds or I had to do something about it."

Allen started firing off a number of "pretty aggressive" bursts from the Shuttle's thrusters, whose cannon-like reports quickly drew a concerned Loren Shriver up to the flight deck. Neither man could see EURECA in the darkness and, even with flashlights aimed through the windows, their view was poor. Eventually, contact was re-established with Mission Control. "What had happened," said Allen, "was that the control centre in Germany had tried to rotate the satellite, but had sent the wrong command. Unbeknownst to Mission Control and to the crew, they fired *all* the jets in such a fashion as to send EURECA towards us at four to five feet per second. They couldn't have planned it better if they had wanted to *hit* the orbiter!" Moreover, flight rules forbade them from using the Shuttle's thrusters within a distance of about 220 m of the satellite, for fear of damaging its delicate solar arrays. Allen had gotten well within that range, but thankfully no damage to either craft was sustained.

In the meantime, the RMS was kept in a poised-for-recapture position, to hedge against the possibility of a contingency retrieval of the satellite. Eventually, it was decided to leave the mechanical arm uncradled for a further day, in order to use its wrist camera to acquire imagery of one of the TSS scientific experiments. The RMS was finally cradled early on 3 August. However, the gremlins were not yet finished with EURECA. Seven and a half hours after deployment, its thrusters were ignited for what should have been a 24-minute burn to position the satellite into its operational orbit. Unfortunately, this firing was cut to just seven minutes, due to the occurrence of unexpected attitude data. Ground controllers worked the issue over the following couple of days, uploading corrected values into the satellite's Low-Attitude Conical Earth Sensor (LACES), until, on 6 August, it was successfully boosted into its correct orbit.

The delayed departure of EURECA had already prompted the Mission Management Team to approve an additional (eighth) day to the flight and pre-deployment activities associated with the TSS had been correspondingly pushed back. Original planning called for the deployment to occur with Atlantis' payload bay facing away from Earth – tail slanted 'upward' and nose pitched slightly 'down' – such that the satellite would initially unreel at an angle of about 40 degrees behind the Shuttle's flight path. Departing at a relatively slowpoke pace, initially around 8 m/sec, but gradually increasing to 2.6 km/h, it would be halted temporarily at a distance of 1.6 km to reduce the deployment angle from 40 degrees to five degrees, putting the satellite in the same plane, directly overhead. Deployment would then resume and upon reaching a distance of 5.5 km, a quarter-revolution-per-minute spin would be imparted via the TSS's attitude thrusters, thus kicking off a lengthy series of scientific investigations.

The next stage would see the satellite extended to a peak velocity of around 8 km/h, reaching a distance of 15 km from the Shuttle, after which the rate of tether unreeling would be slowed. Five and a half hours after first motion from the top of the TSS mast, at a maximum distance of 20.6 km, the spin would be briefly stopped in order for measurements of tether dynamics to be taken. Of specific interest to the physicists were exploring the validity of theoretical concepts that tethers became increasingly more stable with increasingly length, together with analysing the effects of induced disturbances, such as 'bobbing', pendulum-like vibrations (known as 'librations') or backwards-and-forwards skipping-rope-type motions. One particular test, to be exercised at a distance of some 4.1 km, involved adjusting the Shuttle's autopilot to cause a slight 'wobble' of up to ten degrees in any direction – five times more than normal – before executing a thruster firing to damp them out. This 'Ten-Degree Deadband' test would be employed to judge disturbances caused through looser attitude control on the part of Atlantis herself.

Working on the theoretical prediction that payload instability reached its most acute whilst the tether was at its shortest deployed extent, the retrieval speed of the TSS was planned to decrease at a distance of about 7.2 km. At length, it would be halted again at 2.4 km out, preparatory to several hours of final science operations, followed by the final retrieval and reberthing back onto the deployment mast in the payload bay. Since many of the dynamic features of the tether/satellite combination were unknown, a guillotine feature provided for the cutting of the payload at any stage.

Many of these plans were thrown into some disarray, when problems were encountered with the TSS from the outset. Four days into the mission, at 12:12 pm EDT on 4 August, the mast was deployed without incident and – despite a troublesome umbilical which initially refused to separate – the first attempt to release the satellite started a little over five hours later. It was quickly aborted by the crew, due to excessive side-to-side motion in the tether. Following a lengthy check of the reel mechanism and vernier motors, a second attempt at 6:51 pm was successful and the tether deployed smoothly to a distance of 179 m. It was stopped at 7:47 pm, because of possible buried winding on the reel, and was reeled in a few metres, then reeled back out at a slightly higher rate to a total extent of 256 m.

From the back of Atlantis' cabin, the view was spectacular. "When the Sun set," recalled Jeff Hoffman to the NASA oral historian, "and everything turned red, it was just glorious." Ninety minutes later, deployment resumed, but stalled again after two minutes at a length of 257 m and was powered down to survival levels to maintain the satellite's battery lifetime during a crew sleep period. Next day, at precisely 9:00 am EDT on 5 August, the tether was retracted to 224 m, whereupon it refused to move in either direction.

"I remember clearly looking through the camera" at the tether, Hoffman said. "All of a sudden, it started to get all these wiggles in it." *Wiggles* were worrisome, implying a sudden slackness in the tether and suggestive of either a break or a jam. "The tether hadn't broken," he continued. "We could see that. The tether had jammed. The satellite had a jet of nitrogen gas to pull it away, and that was still on, but now it had bounced back." The nitrogen burst was pushing the satellite in a sideways direction and Hoffman and Nicollier struggled to regain control. Loren Shriver knew that its current extent "was right in the middle of the so-called unstable zone"; moreover, mission rules prohibited the tether from exceeding a 45-degree 'red-line' angle, beyond which it would have to be cut loose for safety reasons. As the mission commander, Shriver was determined to prevent that from happening. "Every once in a while, I listen to the audio from that," Hoffman reminisced. "It was certainly the wildest time that I've ever been in space. We were really up against the wall. We got very close to the red line, where we would've had to cut the tether."

A contingency EVA was another option and the cabin pressure was duly reduced to enable Hoffman and Chang-Díaz to begin the necessary 'pre-breathing' period. The plan called for Hoffman to climb the TSS mast "and basically pull it in, hand over hand, and Franklin was going to wrap up the tether". One hundred and fifty metres above the payload bay, at length, the satellite's motions were calmed. The location of the jam was believed to reside in the upper or lower tether control mechanism and, mindful of the problems associated with the actual deployment, a consensus was reached to clear the glitch and bring the satellite back into the payload bay. "The motor that extends the boom was actually more powerful than the motor which reels in the tether," said Hoffman, "so maybe if there was a kink in there, by extending the boom, that would be able to pull the kink free."

At 3:52 pm EDT on 5 August, the procedure got underway. Firstly, the mast was retracted by one panel, in order to allow the crew to visually check for any tether slackness, but nothing was detected. This indicated that the problem existed in the upper tether control mechanism. The astronauts then re-extended the mast with its reel brakes engaged, which successfully cleared the jam and enabled them to retrieve and reberth the satellite at 6:54 pm. A little over an hour later, the mast was finally retracted and the satellite secured.

The maximum deployed distance had been 257 m and although the predicted 45-volt electromagnetic flux was developed, the levels of induced voltage proved insufficient to excite the physical process for many of the mission's primary (Category I and II) scientific objectives. Some Category III tasks – including studies of electron beam propagation, beam-gas cloud interactions and Shuttle surface glow – were achieved and the basic concept of the tethered satellite was successfully demonstrated.

As the mission entered its homestretch, STS-46's secondary payloads took precedence ... and several of these required Atlantis to reduce her orbit to one of the lowest ever achieved by the Shuttle: a circular path, just 229.3 × 228.5 km in altitude. The twin burns of the OMS engines needed to achieve this orbit provided optimum conditions for a 40-hour-long experiment known as the Evaluation of Oxygen Interaction with Materials (EOIM). This was intended to obtain accurate reaction-rate measurements of the interactions between proposed Space Station Freedom materials with the harsh atomic oxygen environment of low-Earth orbit, as well as the effects of solar ultraviolet radiation. Despite a momentary power outage, the experiment was highly successful. So too was the Two-Phase Mounting Plate

(TEMP), the first demonstration of a mechanically pumped two-phase ammonia thermal control system for future station applications.

The low altitude proved awe-inspiring, particularly for Jeff Hoffman. "You're down where the atomic oxygen is rather thick," he explained. "The entire Shuttle was just *glowing* bright orange, just spectacularly beautiful … kind of an orange-white ethereal glow. It was so bright, you could see it with your naked eyes."

Elsewhere, the Consortium for Materials Development in Space's Complex Autonomous Payloads (CONCAP) – sponsored by NASA's Office of Commercial Programs – encompassed a pair of experiments, housed inside GAS canisters. The first exposed different types of high-temperature superconducting thin films to the atomic oxygen environment, whilst the second focused upon the electroplating of metals in microgravity conditions. More than 350 candidate space structure materials formed part of the Limited Duration Space Environment Candidate Materials Exposure (LDCE) project, whose GAS canister lids were periodically opened and closed throughout the mission to expose them to various conditions. Atlantis herself was the subject of observation by the Air Force's electro-optical site on the island of Maui and by the Ultraviolet Plume Experiment on the Strategic Defense Initiative's Low-Power Atmospheric Compensation Experiment (LACE),

3-13: Loren Shriver gobbles a handful of sweets as he enjoys the view through one of Atlantis' windows.

whilst a large IMAX payload bay camera acquired imagery of the TSS and EURECA deployment operations.

Seven astronauts in a relatively closed area of the flight deck and middeck did make for somewhat crowded conditions, although Loren Shriver admitted that the three-dimensionality of operating in microgravity did not cause undue difficulty. "It was a little more crowded on the flight deck," he told the NASA oral historian, "during the busy periods, but everybody had a specific function for being up there: whether it was doing Shuttle-related stuff, or flying the Shuttle, as I was doing, or operating the systems of the satellite itself. Everybody had a specific purpose and it worked out fine. In zero-G, you tend not to notice the congestion as much, just because you can occupy more of the space than you can on the ground."

After eight intensive days in orbit, Atlantis swept onto KSC's Runway 33 at 9:11 am EDT on 8 August, completing the final Shuttle mission to land without the benefit of the drag chute. In the months after STS-46, she would be extensively upgraded and her next landing, in November 1994, would do so with a chute. The limited data return from the TSS proved sufficient to warrant calls for a follow-on mission, but in the days after landing a Board of Investigation – chaired by Darrell Branscome, the Chief Engineer at NASA's Langley Research Center in Hampton, Virginia – convened to assess the problems. Later in August, the board reported back to Jeremiah Pearson, the Associate Administrator for Space Flight, concluding that the tether had snagged on a bolt in the deployment mechanism and corrective action should not impair the prospect of a repeat flight. When the Tethered Satellite flew next, in February 1996, it would be accompanied by most of the original STS-46 crew ... although its outcome would be markedly different.

If 1992 had been dedicated as 'International Space Year', it was certainly shaping up to be nothing less than a true representation of that title. The Russians had hosted cosmonauts from France and Germany aboard the Mir orbital station, whilst Germans, Canadians, Belgians, Italians, Swiss and astronauts born in the United Kingdom and Vietnam had also flown aboard the Shuttle. That international flavour was amply illustrated through the post-mission adventures of the STS-46 crew. "We went on a very nice European tour," Jeff Hoffman told the NASA oral historian, "because Claude was ESA's first mission specialist astronaut. After the ESA part of the tour was over, Claude's Swiss friends invited us to Switzerland for a week, so it was great. We were being driven around in Mercedeses and Rolls-Royces and staying at five-star hotels. I remember a couple of the wives ... had never even been out of the country and they were just totally blown away." By the time Hoffman returned from those tours, he had been assigned to a new mission: one which would be no less 'international' in its flavour and one which would play a pivotal role in securing the Shuttle's ascendancy from the post-Challenger doldrums. In a very real sense, the sheer complexity, enormous risk and spectacular accomplishment of that mission would mark it out as one of the finest human space flights of the decade.

INTERNATIONAL MISSION

Traditions frequently arise from the most unlikely of places. One tradition which had become entrenched in the astronaut office since the early Shuttle era was the practice of a crew's spouses being responsible for organising and hosting farewell parties and other events. Of course, astronauts became as close as family with their crewmates, to such an extent that invitations to dinner or excursions with each other's kids were commonplace. In the early days, when missions were male-dominated, the parties were organised by wives, but as women began flying aboard the Shuttle, the party-makers tended to be both male and female. It made little difference, of course, and much celebration was planned and executed both before and after missions, with few obvious problems.

Then there was STS-47.

In spite of its numerical designation, Endeavour's second space voyage was actually the 50th Shuttle mission and proved historic for other reasons, too. Among the seven-strong crew was the first married couple ever to fly into orbit together. Mark Lee and Jan Davis had been assigned as mission specialists for STS-47 in September 1989 and married secretly during training, but only revealed this fact to NASA management shortly before the flight. Although they would eventually divorce in 1998 (the ramifications of which, some have speculated, possibly contributed to eliminating Lee's assignment to an International Space Station assembly mission) the marriage was notable for its novelty ... and for the fact that STS-47 included only three crew spouses. With Lee and Davis married and pilot Curt Brown and mission specialist Mae Jemison both single, it was left to Rhea Seddon – an astronaut herself and wife of STS-47 commander Robert 'Hoot' Gibson – together with Eleanor 'E.B.' Apt (wife of mission specialist Jay Apt) and Akiko Mohri (wife of Japanese payload specialist Mamoru Mohri) to handle many of the partying arrangements.

Of course, in the eyes of the media, Lee and Davis were the main attractions. Mark Charles Lee came from Viroqua, Wisconsin, where he was born on 14 August 1952. After high school, he entered the Air Force Academy and graduated in civil engineering in 1974, then trained as a pilot at Laughlin Air Force Base in Texas. He flew the F-4 Phantom for several years at Kadena Air Base in Okinawa, Japan, as part of a tactical fighter squadron, and entered Massachusetts Institute of Technology in 1979 to earn a master's degree in mechanical engineering. His specialism was in graphite-epoxy advanced composites. Lee was then assigned to Hanscom Air Force Base in Massachusetts, resolving mechanical and material deficiencies affecting the combat readiness of Airborne Warning and Control System (AWACS) aircraft. In 1982, he upgraded to the new F-16 Fighting Falcon jet and served as an executive officer and flight commander. Two years later, in May 1984, he was selected as a member of NASA's tenth group of astronaut candidates. He first flew in May 1989 and within months was named as a mission specialist – and, from January 1992, as payload commander – of STS-47.

His crewmate and future wife, Nancy Jan Davis, was born in Cocoa Beach, Florida, on 1 November 1953. Surnamed 'Smotherman' at birth, she later assumed her stepfather's patronymic and received much of her schooling in the Huntsville, Alabama, area. She earned two bachelor's degrees: one in applied biology from Georgia Institute of Technology in 1975 and a second in mechanical engineering from Auburn University in 1977. She then joined Texaco in Bellaire, Texas, to work as a petroleum engineer in tertiary oil recovery, before moving to NASA's Marshall Space Flight Center in Huntsville in 1979 as an aerospace engineer. Master's and doctoral credentials in mechanical engineering, both from the University Alabama in Huntsville, followed in 1983 and 1985, respectively. Her career with NASA continued to blossom and in 1986 she led a structural analysis team with a focus on two of the agency's Great Observatories: the Hubble Space Telescope and the Advanced X-ray Astrophysics Facility (later to be renamed the Chandra X-ray Observatory). Shortly before her admission into NASA's astronaut corps in June 1987, Davis worked as the lead engineer for the redesign of the attachment ring that linked the Shuttle's Solid Rocket Boosters (SRBs) with the External Tank.

STS-47 was repeatedly delayed, as the Shuttle manifest writhed and contorted in response to numerous technical issues, not the least of which were the crippling hydrogen leaks in the summer of 1990. The primary payload for the flight was Spacelab-J, a co-operative life and microgravity science mission organised by NASA and the Japanese National Space Development Agency (NASDA, forerunner of today's Japan Aerospace Exploration Agency, JAXA). Originally conceived by NASDA in 1979 as the 'First Materials Processing Test', it received 103 experiment proposals from the Japanese science community and in March 1980 a total of 62 finalists were chosen for consideration. Thirty-four experiments were ultimately chosen by NASDA and principal investigators were named in July 1984. At around the same time, the mission was formally proposed to NASA as a potential Shuttle payload and a Launch Services Agreement between the two space agencies was signed in March 1984. However, the 34 Japanese experiments did not fill the Spacelab module and NASA developed seven of its own investigations and two joint ones with NASDA to complement the themes of the mission. "On Spacelab missions, astronauts do the science," said NASA's Spacelab-J Program Manager Gary McCollum, before the flight. "This mission is typical of how we will routinely work in space for much longer periods when Space Station Freedom begins operations later this decade."

Twenty-four experiments focused upon materials and processes in the microgravity environment, with emphasis upon the production and analysis of protein crystals, electronic components, fluid dynamics, glasses and ceramics, metals and alloys. The frequent-flying Protein Crystal Growth investigation sought to yield crystals by the vapour-diffusion and liquid-to-liquid diffusion processes, whilst four high-temperature furnaces – the Gradient Heating Furnace, the Image Furnace, the Crystal Growth Furnace and the Continuous Heating Furnace – were employed to melt and solidify a variety of materials.

The Gradient Heating Furnace facilitated the exploration of crystal formation processes in semiconductors, ceramics and alloys and featured three temperature zones, which allowed gradients to be 'moved' up to a maximum 1,100 degrees Celsius along the length of a sample. During the Spacelab-J mission, the furnace was employed to process specimens of lead-tin-telluride for potential electronic applications, including fire security and imaging systems. The Image Furnace supported investigations which used the 'floating-zone' growth procedure, whereby a liquid 'zone' was moved through a material during the crystal-formation process, and its samples included indium-antimonide, a compound whose Earth-bound applications span a broad area of the infrared technology arena, ranging from thermal imaging cameras to missile guidance systems and telescopes. Other experiments sought to understand flow processes in a viscous, gold-laced glass sphere. Finally, the Continuous Heating Furnace provided high temperatures of up to 1,300 degrees Celsius, together with a rapid-cooling capability, to two samples in tandem. Specific materials heated and cooled in the furnace included compounds of aluminium-lead-bismuth, silver-copper and silver-yttrium-barium-copper. The choice of the Continuous Heating Furnace for these samples meant that two could be heated as two others were being cooled, thereby achieving increased processing efficiency. A Large Isothermal Furnace uniformly heated large samples, including tungsten, to maximum temperatures of 1,600 degrees Celsius, then rapidly cooled them through helium purging. Protein crystal growth formed an additional important thrust of the research conducted in the Spacelab module and middeck.

The remaining 19 experiments focused upon life sciences and employed a variety of organisms, including the seven astronauts, together with female frogs, the osteoblastic (bone-forming) cells of rats, kidney cells, fungi, chicken embryos, a pair of Japanese carp – one of which had its gravity-sensing organ, the otolith, purposely removed before flight to enable comparisons of space adaptation processes with its twin – and the mutative effects of cosmic radiation upon the larvae of fruit flies. Notably, the frogs' eggs were fertilised in orbit and examined at various developmental stages, from the embryonic level to tadpoles and to adulthood.

As the United States and Japan moved together as partners in the Space Station Freedom development effort – soon to become the International Space Station – a key investigatory component of Spacelab-J was to identify the underlying cause of space sickness, which is known to affect around a quarter of all astronauts and cosmonauts. Aboard Endeavour was the Autogenic Feedback Training (AFT) experiment, previously flown on Spacelab-3 in April 1985, which included electronic instrumentation to record physiological data such as sweat, pulse, heart and respiration rates. The experiment offered clear insights into the effects of crew workload and behavioural responses to environmental stress; 'baseline' information which proved important when planning future long-duration missions.

For NASDA, the presence of a Japanese payload specialist aboard Spacelab-J was of fundamental importance in these studies. Mamoru Mark Mohri, a chemist by profession, received scans of his spine and legs before and after the mission, from which comparisons would be drawn about changes in the muscle volume of his calves and thighs, changes in fat and water content in his spinal bone marrow and changes in the volume, shape and water content of his spinal vertebrae. Mohri subjected himself to blood and urine specimens to assess his stress levels, the extent of bone-muscle atrophy in the peculiar microgravity environment of low-Earth orbit and the impact of the adaptation process on his normal body biochemistry. He also participated in advanced studies of space sickness, including tracking flickering light targets, whilst anchored in different orientations within Spacelab.

Mohri was born in the town of Yoichi, famed from its fruits, its wines and its whiskies, on Japan's second-largest island, Hokkaido, on 29 January 1948. He attended school in the local area and later studied chemistry at Hokkaido University, from where he received his bachelor's degree in 1970 and his master's credential in 1972. Mohri earned his doctorate in chemistry from Flinders University in Adelaide, South Australia, in 1976, and spent the following decade as a faculty member at Hokkaido University's Department of Nuclear Engineering. By the early 1980s he had risen to the position of associate professor, with extensive interests in surface physics and chemistry, high-energy physics, ceramic and semiconducting thin films, environmental pollution and biomaterials spectroscopy.

At the time of the naming of the principal investigators for Spacelab-J in July 1984, a call was issued for astronaut candidates. Mohri applied and in July of the following year he and aerospace engineer Takao Doi and physician and physiologist Chiaki Naito (later Mukai) – were selected from 533 qualified candidates reviewed by NASDA. Four months later, NASA's final pre-Challenger manifest listed Spacelab-J on Mission 81G in February 1988. The loss of the Shuttle and her crew led many flights to be suspended indefinitely and in 1987 Mohri was appointed as an adjunct professor of physics in the Center for Microgravity and Materials Research at the University of Alabama at Huntsville. During his two years in Huntsville, he was involved in microgravity experiments in alloy solidification and liquid behaviour at the Marshall Space Flight Center's drop tower. In April 1990 he was named as the prime Japanese payload specialist for Spacelab-J.

By the time that Mohri officially joined the crew, the NASA members of the science team had already been announced. In September 1989, veteran astronaut Mark Lee was named as the Spacelab-J payload commander – with primary oversight and responsibility for the mission's scientific objectives – together with fellow mission specialists Jan Davis and Mae Jemison. Selected alongside Davis in June 1987, Mae Carol Jemison was the first African-American woman to undergo NASA astronaut training and on STS-47 became the first black female to venture into orbit. She was also designated as the first 'science mission specialist'. "Under new NASA guidelines for missions requiring a payload specialist not provided by the customer," noted a Spacelab-J summary brochure, "NASA selects a mission specialist to fill those duties ... Since the chosen NASA astronaut will be performing payload specialist duties and is a trained mission specialist, the term 'science mission specialist' has been developed." Serving as Jemison's backup in this unique role – a role never seen again in the Shuttle era – was a University of California at Riverside biochemist and protein crystal growth specialist named Stanley Koszelak.

Born in Decatur, Alabama, on 17 October 1956, Mae Jemison was the daughter of a maintenance supervisor father and an elementary school teacher mother. Unlike many youngsters of her generation (and ethnicity), Jemison always assumed that she would someday reach space. Her interest in the natural sciences was inspired from an early age and in kindergarten, when asked by her teacher what she wanted to do, the young Jemison responded that she wanted to be a scientist.

The teacher looked puzzled. "Don't you mean a *nurse*?" she asked.

"No," replied Jemison, "I mean a *scientist*." There was nothing *wrong* with being a nurse, of course, but Jemison simply wished to follow a different career ... a career which was not readily accessible to women at that point in time. The arts were also a fascination and she pursued dancing – ballet, jazz, modern, traditional African – and even auditioned for the role of Maria in 'West Side Story'. She did not get the part, but her dancing skills were sufficiently impressive for her to be given a place as a background dancer. As she approached the end of her secondary education, Jemison was faced with a choice: she could become a professional dancer or enter medical school. Her mother helped resolve the issue with wise words of advice: "You can always dance if you're a doctor," she said, "but you can't doctor if you're a dancer."

Jemison left Chicago's Morgan Park High School in 1973 and entered Stanford University to study chemical engineering. As an undergraduate, she met resistance both as a woman in an engineering discipline and as a *black* woman in an engineering discipline. "Some professors would just pretend I wasn't there," she reflected, years later. "I would ask a question and a professor would act as if it was just so dumb; the dumbest question he had ever heard. Then when a white guy would ask the same question, the professor would say 'That's a very astute observation'." Upon receipt of her undergraduate degree in 1977, Jemison proceeded to Cornell Medical College and gained her medical degree in 1981. Following her internship, she worked as a general practitioner and in 1983-85 joined the Peace Corps, based in war-torn Liberia and Sierra Leone. Whilst there, she provided medical care, wrote self-care manuals and helped to research vaccines.

On one occasion, a volunteer was diagnosed with malaria. As the disease progressed and worsened, Jemison became convinced that it was meningitis and called for an Air Force hospital aircraft, based in Germany, to extract the casualty. A medical evacuation cost $80,000 and embassy personnel asked Jemison if she had the authority for such a call. "I don't need anyone's permission," retorted the 26-year-old physician, "for a *medical* decision." She accompanied her patient to Germany, remained with them for 56 hours, and was later able to report a full recovery.

Jemison had been preceded into space by four African-American men – Guy Bluford, Ron McNair, Fred Gregory and Charlie Bolden – and like them had been inspired by one high-profile black actress: Nichelle Nichols, who played Lieutenant Uhura in *Star Trek*. In June 1987 Jemison was selected into NASA's 12th class of astronaut candidates, alongside future Spacelab-J crewmates Jan Davis and Curtis Lee Brown Jr. The latter, who served as STS-47 pilot, came from Elizabethtown, North Carolina, where he was born on 11 March 1956. "My dream as a small kid was to fly," he once said. "I fell in love with aircraft and flying movies and things about flight and built all the little airplanes that kids always did when I was growing up." After completing high school in his hometown, Brown entered the Air Force Academy to study electrical engineering and earned his degree in 1978. He was commissioned as a second lieutenant and underwent flight instruction at

Laughlin Air Force Base in Texas, later flying A-10 Thunderbolt aircraft in South Carolina.

Brown's work subsequently earned him a position as an instructor pilot on the A-10 at Davis-Monthan Air Force Base in Arizona. Graduation from the Air Force's Fighter Weapons School at Nellis Air Force Base in Nevada in 1983 was followed by duties as an A-10 weapons and tactics instructor. He was selected for Test Pilot School at Edwards Air Force Base in June 1985 and upon completion of this rigorous and demanding course in mid-1986 he worked as a test pilot for the A-10 and the F-16 Falcon until his selection as an astronaut candidate. When he was named as the pilot of STS-47 in August 1991 – alongside mission commander Robert 'Hoot' Gibson and flight engineer Jay Apt – Brown became the last of his class to receive a Shuttle flight assignment. Yet his NASA career would become a stellar one in terms of the number of times that Brown ultimately flew: in December 1999 he became the first person to have served six times as a Shuttle pilot or commander. It is a distinction that, even today, after the retirement of the orbiters, he shares with only one other astronaut, Jim Wetherbee.

Seated to Brown's left side on the flight deck for STS-47 was Robert Lee Gibson, who had grown up with the moniker 'Hoot', as had his father. "I always tell people that it comes from 'not worth a hoot'," he once told an interviewer, but in reality it

3-14: STS-47 payload commander Mark Lee translates through the connecting tunnel between Endeavour's middeck and the Spacelab-J module. On this flight, Lee and Jan Davis became the first married couple to fly together on the same mission.

originated from Edmund Richard Gibson (not a relation), a famous rodeo champion who turned into a cowboy film star in the 1920s and 1930s; *his* nickname of 'Hoot-Owl' came from co-workers and, later, evolved simply into 'Hoot'. "So after that," his astronaut namesake continued, *"everybody* whose name is Gibson usually picked up the name 'Hoot'." In fact, when he progressed into the Navy Fighter Weapons School – the famous 'Top Gun' – Gibson chose the nickname for his radio callsign. Born in Cooperstown, New York, on 30 October 1946, he completed high school in Huntington and entered Suffolk County Community College on Long Island to study for an associate degree in engineering science. He later earned a bachelor's degree in aeronautical engineering from California Polytechnic State University in 1969.

Today, he is renowned in space circles as a five-flight veteran, commander of the first Shuttle-Mir docking mission and former chief of the astronaut office, but underpinning each of those accomplishments has been Gibson's lifelong passion for aviation. His father was a test pilot and inspector for the Civil Aeronautics Administration and built his own private aircraft in the garage, whilst his mother was one of the few women to fly general aviation aircraft in her day; in her youth, she and two friends bought a J-2 Taylor Cub. With such an impressive pedigree, it is hardly surprising that their son should have charted his own course for the skies and beyond. As a boy, Gibson travelled frequently with his father on CAA business and, on one occasion, the pair were at an airport in Phoenix, sitting in a Beechcraft Bonanza with just a single yoke, and Paul Gibson handed the controls to his son to perform the takeoff. The boy was just ten years old. "I was so proud that he trusted me," Gibson recalled years later. "He was my inspiration." That was just the start. Gibson soloed in a Piper Colt on the "windy, rainy, solid overcast" day of his 16th birthday and gained his private pilot's licence at 17.

After completing his bachelor's degree, Gibson entered the Navy and received basic and primary flight instruction in Florida and Mississippi, then advanced training in Kingsville, Texas, and eventually moved to Naval Air Station Miramar in California for assignment to the F-4 Phantom fighter. "I was in awe of the F-4," he told Robin White in an interview for *Air & Space* magazine. "It looked so big and heavy and the wings seemed so small. I was reluctant to slow it down. I was sure it would fall out of the sky, but it was just totally rock-solid on approach to the carrier." From April 1972 to September 1975 he served aboard the USS *Coral Sea* and the USS *Enterprise*, flying the F-4 over Vietnam during two tours of duty. When his commanding officer asked him if he wanted a *third* tour, Gibson was not enthusiastic – "I was extremely ready to hit the beach," he told White – until he learned that the tour would involve operational deployment of the new F-14 Tomcat. If Gibson was in awe of the old F-4, then this new fighter functioned on a totally different level. On one occasion, with just 30 hours' experience in the Tomcat, he faced a thousand-hour F-4 veteran for a training dogfight. "We called *Fight's On*," Gibson recalled, "and 30 seconds later I was sitting in his six [behind him]. We ran the engagement three times. The results were *always* the same. An F-14 with a *nugget* at the stick could out-manoeuvre, out-turn and out-fight a Phantom flown by an old hand!"

Completion of the Top Gun course was followed by assignment as an F-14 instructor pilot and graduation as a test pilot in June 1977. Barely two months later, he was summoned to Houston, part of the second group of astronaut applicants to undergo a week of intensive physical and psychological testing, and was selected in January of the following year. In addition to his military aviation, Gibson also participated in air races – one of which would lose him the command of a Shuttle crew in the summer of 1990 – and secured a number of world speed and altitude records. In May 1981, Gibson married fellow astronaut Rhea Seddon and in July of the following year, their first child, Paul, was born. *Time* magazine promptly labelled the first child born to two American astronauts as 'Astrotot'.

One of his reasons for joining NASA, obviously, was to fly the Shuttle, which he considered to be the highest and fastest form of aviation possible. He also hoped that after several tours of Vietnam, landing on aircraft carriers and making it through Top Gun and test pilot school would give him plenty of stories for his grandchildren. "But, man, when I went into space, *that* wiped out everything," he told an interviewer in July 2005. "*That* was the biggest thrill that you could ever have. I wasn't scared, but about a week before I was set to launch, I asked myself, 'Are you *sure* you wanna do this?'" Naturally, after training for more than five years, the answer was a resounding *Yes*, regardless of the risk. And Gibson was certainly attuned to taking calculated risks. In July 1990 he lost his place in command of STS-46 when he participated in an air racing accident, in which one pilot was killed. On STS-47 Gibson would be embarking on his fourth Shuttle mission.

Like several previous Spacelab flights, the STS-47 crew was divided into two teams, each working 12-hour shifts to operate the multitude of experiments around-the-clock. The 'red' team consisted of Brown, Lee and Mohri, whilst the 'blue' team comprised Apt, Davis and Jemison, with Gibson, as the mission commander, free to anchor his schedule across both. The biological nature of many of the experiments meant that from around 30 hours before launch many of the 'time-sensitive' items – the frogs and the carp, for example – had to be carefully loaded into the Spacelab module, by now in a *vertical* orientation on Pad 39B. Consequently, a technician was lowered in a special sling-chair down the connecting tunnel between Endeavour's middeck and the Spacelab module. Other samples, including the seed cultures for the protein crystal growth investigations, were loaded even later, barely 14 hours ahead of the 12 September 1992 liftoff.

"You are aware that you are sitting on a controlled explosion," Mae Jemison remembered of her first and only launch into space, "but you also realise that you've taken all the precautions. You trust the people you have been working with and you know they have worked to try to keep things safe. After that, you have to leave it alone. If you keep worrying about it, then you're not going to be able to do your job." Endeavour rose perfectly from Earth at 10:23 am EDT that morning and entered a circular orbit of 300 km, inclined 57 degrees to the equator, whereupon Gibson and Brown set to work readying their vehicle for a scheduled seven days of scientific research. Led by payload commander Mark Lee, the activation of the Spacelab-J module was begun a little over two hours after launch and – according to NASA's post-flight summary – experimental work began "almost immediately to ensure maximum exposure to the microgravity environment". Jay Apt's blue team retired to bed for their first sleep period at around this time and one of the earliest experiments to begin running was the Space Acceleration Measurement System (SAMS), an instrument designed to record low-level vehicle motions during orbital operations, to help to characterise the environment for microgravity experiments. Its main unit was situated near the rear of the Spacelab module, with sensor heads located close to major experiment facilities.

For the next seven days, the crew of STS-47 stepped smartly through their scientific work, with hardware located not just in the pressurised module, but also inside Endeavour's middeck and outside on a specialised 'bridge' in the payload bay. The Israeli Space Agency Investigation About Hornets (ISAIAH) sought to observe the effects of microgravity on combs constructed by oriental hornets, which are known to have a unique ability to build combs in the direction of gravity. Elsewhere, the Solid Surface Combustion Experiment tested flame spreading along an instrumented sample of filter paper in a test chamber of 35 percent oxygen and 65 percent nitrogen at a pressure of 1.5 terrestrial atmospheres. Outside, behind the Spacelab, in Endeavour's payload bay, was a bridge of nine Getaway Special (GAS) canisters, supporting a range of student and government investigations. The Boy Scouts of America supplied a variety of experiments into capillary pumping, cosmic radiation, crystal growth, fibre optics and fluid physics, whilst others explored enzyme crystallisation, the thermal conductivity of fluids in microgravity conditions, materials processing, the behaviour of bread yeasts and crystal growth. The experiments involved students and investigators from Sweden, Canada and the United Kingdom.

Of particular note was the Boy Scouts' experiment, which formed part of 'Project POSTAR', a portmanteau of 'post' and 'star', which involved the participation of Explorer Posts and Sea Explorer Ships. It got underway in late 1978, when TRW purchased a GAS canister for $10,000, but was extensively delayed in the wake of the Challenger accident. The United Kingdom experiment – which originated from the girls-only Ashford School in Kent – was the result of a competition organised by Independent Television News. After its return, the experiment was displayed in the London Science Museum.

As STS-47's flight engineer, Jay Apt was in charge of Endeavour's flight deck whilst the blue team was on duty, and he found that his geographical training enabled him to look up from his work and instantly recognise where he was. Apt had seen the grand peaks of the Himalaya during his first mission, STS-37, but now in a higher-inclination orbit, was able to see a much broader swath of Asia: the vast Taklimakan Desert, the giant depression of Lake Baikal and the forbidding boundary of the Ural Mountains. But there was something else that Apt wanted to see ... something which eluded him, but which remains a popular misconception to this very day. "We spent several passes looking for the Great Wall of China with no luck," he wrote in his 1996 book *Orbit*, co-authored with Michael Helfert and Justin Wilkinson. "Although we can see things as small as airport runways, the Great Wall seems to be made largely of materials that have the same colour as the surrounding soil. Despite persistent stories that it can be seen from the Moon, the Great Wall is almost invisible from only 180 miles up!"

For Mamoru Mohri, each day of the STS-47 mission felt like he was an extra for 'Alice in Wonderland'. "Everything I saw no longer fit within its known parameters," he explained in a later NASA interview, "so as a scientist I could explain what was occurring around me and yet I was very much entranced."

Economical expenditure of consumables enabled the Mission Management Team to authorise an additional (eighth) flight day and it was late on 19 September that the final deactivation of the Spacelab-J module got underway. At 7:52 am EDT on the 20th, Endeavour's OMS engines roared silently into the vacuum for 153 seconds, committing the vehicle to a hypersonic descent back through the atmosphere, towards a touchdown at the Kennedy Space Center. Concluding her second mission, the orbiter landed safely at 8:53 am on concrete Runway 33. Gibson and Brown deployed the drag chute satisfactorily, this time *ahead* of nose gear touchdown, bringing STS-47 to a close after almost eight full days.

Notwithstanding Mae Jemison's desire *not* to be remembered as the first black female spacefarer, her presence on the STS-47 crew led to great public interest. Her first view from orbit, she later noted, was a glimpse of her hometown, Chicago, and she is said to have carried with her a photograph of another famous black female resident of the Windy City: Bessie Coleman, the first African-American woman aviator.

When Jemison returned to her *alma mater* a few weeks after the mission, she was keen to share that human imagination and human ability were only constrained by themselves; and *her* imagination and ability had certainly expanded well beyond the sphere of what others expected of her as a youngster growing up in the early years of civil rights in America. Her accomplishments before STS-47 had already been recognised by a plethora of awards: Woman of the Year in 1990 for the Gamma Sigma Gamma fraternity, one of McCall's Ten Outstanding Women of the 'Nineties in 1991, a Black Achievement Trailblazer in 1992 and an inductee of the National Women's Hall of Fame in 1993. It was more than *ethnicity* which marked Jemison out as something special, for she was also one of relatively few *women* to reach space ... and an example of a person for whom hard work and an unswerving devotion to the importance of education and following dreams could pay enormous dividends. "When I grew up in the 1960s," she said, "the only American astronauts were *men*. Looking out the window of that Space Shuttle, I thought if that little girl growing up in Chicago could see her older self now, she would have a *huge* grin on her face!"

Never were truer words spoken.

A GLASS HALF-EMPTY?

Much to the chagrin of her crew, thousands of satisfied scientists and newly-appointed Kennedy Space Center Director Bob Crippen, the 13th voyage of Columbia was saddled with the age-old question: *Is the glass half-full or half-empty?* It is a question often posed in situations which are sometimes perceived to be unworthy of the expenditure of effort and risk involved. To be fair, the payloads carried into orbit by STS-52 on the afternoon of 22 October 1992 were a mixed bag, with a small Italian satellite, three bridge-mounted experiments and a myriad of other investigations – including a *quilt* affixed to the Shuttle's Canadian-built RMS mechanical arm.

For astronaut Lacy Veach, one of three mission specialists on STS-52, the glass was always full. His Hawaiian birth and keen awareness of its historic, Pacific-voyaging canoes prompted him to carry into orbit an adze stone tool from his grandfather. The stone originated from the Keanakako'i quarry, located high on the slopes of Mauna Kea, and Veach would photograph it floating in Columbia's windows, with its Hawaiian place of origin visible far below. Several years later, shortly before his untimely 1995 death from cancer, Veach told an interviewer that "you can never believe the beauty of Island Earth until you see it in its entirety from space".

The major payload – if it could be so called – for Jim Wetherbee's STS-52 crew was the second Laser Geodynamics Satellite (LAGEOS-II), part of a collaborative venture between NASA and the Italian Space Agency (ASI). It was essentially a large aluminium sphere, 60 cm in diameter, with a dense brass core which gave it a heavier-than-it-looked mass of 410 kg. It was built in Italy at a cost of $160 million, using blueprints from NASA's Goddard Space Flight Center in Greenbelt, Maryland, which had launched a near-identical satellite almost two decades earlier. Like LAGEOS-I, it was a 'passive' spacecraft, dedicated to 'geodesy'. Laser beams transmitted from the ground would impinge upon several hundred nearly-equally-spaced, corner-cube retroreflectors, embedded within the satellite's shell. By measuring the round-trip travel time of the beams, it was expected to determine the distances between ground stations with an accuracy of a few centimetres. Data from LAGEOS-I had already shown that the Pacific tectonic plate, upon which lie the Hawaiian Islands, is slowly drifting in a northwesterly direction – at a rate of just a few centimetres per year – towards Japan. Movement, or *lack* of movement, on this scale is particularly important, for it enables geophysicists to understand the processes ongoing underneath our feet and allowing a greater level of predictability of earthquakes of volcanic activity. "The satellite may be small," said LAGEOS-II Project Scientist Miriam Bartuck, "but the data returned is *big-time science*."

The solid, outermost layer of our planet – measuring up to 100 km thick in places – is known as the 'lithosphere' and is composed of enormous rigid 'plates', which float upon the semi-liquid region of the 'mantle', known as the asthenosphere. These plates move with respect to each other, but generally at speeds no greater than a handful of centimetres every year. This motion, dubbed 'plate tectonics', is the means by which North and South America are presently drifting away from Europe and Africa and, although these motions are incredibly slow, they can prove cataclysmic when collisions occur. For example, the Himalayas were formed when India struck southern Asia. Before launch, it was hoped that LAGEOS-II would address the tectonic model which theorised that the coastal strip of southern California, lying to the west of the San Andreas fault, is steadily migrating northwards and will collide with Alaska in 150 million years' time.

Present scholarly discussion centres on a giant 'super-continent' ('Pangaea', Greek for 'all lands'), which existed in the middle of a huge ocean some 200 million years ago. Within the next 20-30 million years, this gigantic land mass broke apart into smaller continents and by 150 million years ago cracks started to emerge in the most northerly of these new regions. Molten rocks from deep within the mantle poured up through the cracks and served to push the plates further apart. Carrying the fragments of the super-continent with them, they left newly-created patches of sea floor and, over many millions of years, the continent that we know as North and South America ended up in its present position, more than 4,800 km from Eurasia. Modern maps offer tantalising glimpses into Pangaea's appearance, suggesting that parts of West Africa might 'slot' neatly into the eastern coastline of Brazil, although it is the lines of the continental 'shelves' which are considered a more reliable marker of where the land masses pulled apart.

In order to track these incredibly slow tectonic motions, the entire aluminium skin of LAGEOS-II was coated with 426 retroreflectors, giving it an appearance not

3-15: After several months of limited rainfall, the Kalahari Desert entered the summer monsoon season in this view of gathering thunderheads, peeking above storm clouds, taken from STS-52. The summer monsoon was known to typically last from November until March, contributing almost all of the annual rainfall to the environmentally-sensitive Kalahari region.

unlike a gigantic golf ball. Each retroreflector had a flat face and a prism-shaped back. Most of them were made from suprasil – a fused silica glass – and four from germanium. The satellite's aluminium and brass composition evolved from a series of manufacturing trade-offs: it needed to be heavy enough to minimise the effects of non-gravitational forces, yet light enough to be inserted into a high, stable Earth orbit. Preparations to launch LAGEOS-II moved into high gear on 29 September 1992, when it was inserted into Columbia's payload bay at Pad 39B. Attached to two boosters – the spin-stabilised Italian Research Interim Stage (IRIS) and the LAGEOS Apogee Stage (LAS), one 'above' the satellite and the other 'below' it – the combination was an unusual sight to behold. Within the payload bay, it was covered by a large protective sunshade.

The history of the IRIS is an interesting one and has its roots in the creation of Italy's National Space Plan, unveiled in the early summer of 1983. The booster was originally conceived as a means of supporting the Italian Italsat communications satellite system, but was always intended for a secondary role as an upper stage in Shuttle operations, targeted for customers who did not require as much power as was provided by McDonnell Douglas' Payload Assist Modules. Before Challenger, the booster was baselined for two Shuttle flights: the deployment of LAGEOS-II – then planned for 1988 – and the deployment of Italy's Satellite for X-ray Astronomy (SAX). Its solid-fuelled rocket motor was successfully test-fired in July 1986, six months after the loss of Challenger, but changes to the SAX mission and its eventual metamorphosis into a joint Italian-Dutch venture, BeppoSAX, meant that LAGEOS-II was the IRIS' only Shuttle payload.

Right up until the eve of the STS-52 launch, NASA defended its decisions (made before the loss of Challenger) to launch such a small satellite as a 'primary' payload. The questioning did not simply relate to the presence of LAGEOS-II, but also the first United States Microgravity Payload (USMP-1), another small cargo which critics argued should have flown on another mission, alongside a 'major' payload. When one peruses the immediate post-Challenger Shuttle manifest, however, it is clear that LAGEOS-II was always considered to be a 'primary' or 'major' payload. In NASA's January 1990 manifest it was baselined alongside a demonstration of the Flight Telerobotic Servicer and a SPARTAN free-flying astronomy satellite and by the February 1991 manifest it was joined by USMP-1 and the retrieval of EURECA. In response to the critics, NASA was quick to explain that it had contracted with ASI to launch LAGEOS-II before the Challenger accident and intended to fulfil its commitment. As for USMP-1, the agency's Deputy Associate Administrator for Space Science, Al Diaz, summed up its importance best with a short, clipped rhetorical question: "How do you determine how much Nobel Prize science is worth?"

Diaz was referring in particular to one of USMP-1's three research investigations: the Lambda Point Experiment (LPE), which sought to test a complex theory of the thermal conductivity of liquid helium, which had won mathematician Kenneth Wilson the Nobel Prize in 1982. Wilson predicted that, under the right pressure and other conditions, liquid helium would conduct heat a thousand times more efficiently than copper, but his theory could not be realistically tested on Earth, due to limitations imposed by gravity. Nevertheless, Wilson's work was expected to have wide-ranging applications, from the study of hurricane dynamics to the development of new superconductors. When activated, the LPE experiment examined liquid helium as it changed – through a transitional phase known as its 'Lambda Point' – from a 'normal' fluid into a substance known as a 'superfluid'. In this latter phase, the helium moved freely through small pores which blocked other liquids and conducted heat more effectively than copper. The LPE experiment sought to cool a liquid helium sample far below its Lambda Point at minus 268 degrees Celsius, after which its temperature would be slowly raised and measured with an accuracy of less than a billionth of a degree throughout the entire transitional period.

Joining LPE on the USMP-1 payload were two other experiments: the Materials for the Study of Interesting Phenomena of Solidification on Earth and in Orbit (a French-language acronym which spelt 'MEPHISTO') and the Space Acceleration Measurement System (SAMS), a variant of which flew previously on STS-47. The former was a joint US-French effort to study the behaviour of metals and semiconductors as they solidified to help determine the influence of gravity on the point at which the liquid met the solid (its so-called 'solid-liquid interface'). It was hoped that such research would lead to the development of more resilient metallic alloys and composites for future aircraft engines and turbine blades. On STS-52, MEPHISTO was loaded with three identical samples of a tin-bismuth alloy, which were processed by one fixed and one moving furnace. On Earth, buoyancy-driven convection and differences in hydrostatic pressure affect how materials solidify, as well as masking several key stages in the solidification process. During USMP-1, sensors affixed to the furnace accurately measured the temperature and shape of the solid-liquid interface and determined the rate at which it moved through the tin-bismuth sample. The sophisticated SAMS accelerometer monitored and recorded the impact of orbiter thrusters firings upon LPE and MEPHISTO to characterise the quality of the microgravity environment in which these experiments were performed.

The USMP-1 experiments were mounted on a pair of MPESS, which straddled the payload bay, like a bridge. The 'front' MPESS provided electrical power, data, communications and thermal control services to the payload, whilst the experiments themselves were attached to the 'rear' one. Managed by NASA's Marshall Space Flight Center, the USMP payload was effectively autonomous and required little crew interaction, other than switching it on and then off. Moreover, it also offered exciting opportunities for principal investigators on the ground to control their experiments remotely via 'telescience'.

Notwithstanding the importance of the science offered by LAGEOS-II and USMP-1, the question of a half-empty glass lingered as Jim Wetherbee led his crew out to Pad 39B on the morning of 22 October. "I decided when I was about ten years old that *this* is what I wanted to do for a job," Wetherbee told an interviewer before one of his Shuttle missions about his decision to become an astronaut. "I don't really know why." Perhaps it was one of the dramatic Gemini missions, with their rendezvous and spacewalking activities, and he remembered as a child sneaking into a classroom to listen to a small, nine-volt transistor radio. Instead of scolding Wetherbee, the teacher allowed him to sit at the back of the class and plot Gemini's orbital progress on a map of the world to show the other children. To James Donald Wetherbee, it was *exploration* which most excited him.

Born in Flushing, New York, on 27 November 1952, he grew up *tall* in more ways than one ... in fact, even today, at 1.93 m – six feet and four inches – he remains one of the tallest spacefarers ever launched. He attended high schools in New York and earned a degree in aerospace engineering from the University of Notre Dame in 1974, then entered the Navy, hoping to become a pilot and land on aircraft carriers. Certainly, aviation was important to him, since his father had served as an Army Air Corps aviator in the Second World War and later became an American Airlines captain and chief pilot. In his youth, becoming an astronaut was a dream, but Wetherbee recognised that the chances of bringing that dream to fruition were very low. He received his aviator's wings in December 1976 and trained initially in the A-7E Corsair II, flying for three years aboard the USS *John F. Kennedy* and accumulating more than a hundred carrier landings. Wetherbee's next step was test pilot school, from which he graduated in 1981, and he was serving as a project officer for the weapons delivery systems and avionics integration of the new F/A-18 Hornet fighter. He later flew the Hornet as an operational pilot. It was his wife, Robin, who spotted NASA's call for astronauts and encouraged Wetherbee to apply. Selected in May 1984, with the rank of a lieutenant, Wetherbee was the youngest and most junior of the pilots ... yet he would fly more often than any of them. He had drummed for the marching band at Notre Dame as an undergraduate and had put his kit away when he entered the Navy – "I *couldn't* take the drums on an aircraft carrier!" – but little did he know that in the summer of 1987 he would be recruited by Hoot Gibson and Brewster Shaw to join the 'Max Q' rock band. A little more than a year later, he would be recruited again ... only this time to his first Shuttle crew. In January 1990, he flew as pilot on STS-32.

After his return from that mission, Wetherbee was assigned – along with fellow astronauts Loren Shriver and Dave Hilmers – as a backup crew for the time-critical STS-41 Ulysses deployment in October 1990. NASA had stopped naming backup crews several years earlier, but in February 1990 a respiratory ailment in one of the STS-36 crew members forced a delay of several days. Ulysses' narrow launch window made such a delay unacceptable and thus two crews trained in parallel. Fortunately, the backups were not needed and in December 1990 Wetherbee was named as pilot of STS-46, the Tethered Satellite mission. This plan changed dramatically in August 1991, when he was moved to command STS-52. His place as STS-46 pilot was taken by Andy Allen, whose own place as a mission specialist on the same flight was taken by Marsha Ivins. The reason for the shuffle appears to relate to the retirement of two flight-experienced Shuttle commanders (Mike Coats and Bryan O'Connor) in the summer of 1991. This left the astronaut corps decidedly 'short' of pilots sufficiently qualified to rotate into the commander's seat.

Joining Wetherbee on STS-52 were four NASA astronauts who had all flown within the last two years: Mike Baker served as the pilot, with Lacy Veach, Bill Shepherd and Tammy Jernigan as mission specialists. The sole member of the crew who would be experiencing orbital flight for the first time was Canadian physicist Steven Glenwood MacLean, who had been named as a payload specialist in

February 1992. The majority of his time during the ten-day mission was devoted to operating a battery of scientific and technical experiments, including a space vision system being developed by his country as a contribution to the Space Station Freedom programme.

MacLean – later to serve as President of the Canadian Space Agency (CSA) – came from Ottawa, Ontario, where he was born on 14 December 1954. His parents hailed from Nova Scotia and he spent many summers as a child "in the centre of timberland", exploring the cliffs or the forests, and nurturing a fascination with the natural world. MacLean attended primary and secondary schools in his hometown and earned a bachelor's degree in 1977 and a doctorate in 1983, both in physics, from York University in Toronto. Whilst an undergraduate, MacLean was a member of the Canadian national gymnastics team and throughout his years of doctoral research he taught part-time at York University. He was a visiting scholar at Stanford University in 1983 and pursued extensive research in laser physics, with particular focus upon electro-optics, laser-induced fluorescence of particles and crystals and multi-photon laser spectroscopy. He was selected as one of Canada's first six astronaut candidates in December 1983 and commenced formal training as a potential Shuttle payload specialist in February of the following year.

"When I wake up in the morning, I still feel like I'm floating six inches off the ground," MacLean recalled in an interview, years later. "I feel so lucky, so privileged that I was selected back then." Even when Canada put out the announcement for astronaut applicants in late 1983, he did not feel that he would meet the requirements and did not immediately complete an application. A telephone call from a colleague ultimately changed his mind. "You know, you should really do this," said the colleague. "This is something that maybe you could do." So MacLean did. And he was selected.

By the time of the Challenger accident, MacLean had been named as the prime payload specialist for Mission 71F, scheduled for launch in March 1987. He would have been the second Canadian citizen to venture into orbit, following Marc Garneau, and would have operated the second set of Canadian Experiments (CANEX-2). With the reshuffling of Shuttle mission priorities in the post-Challenger era, Roberta Bondar flew IML-1 in January 1992 and MacLean became the third Canadian spacefarer. It mattered little, of course. During the post-Challenger down time, he served as Program Manager for Canada's Advanced Space Vision System and Astronaut Advisor to the Strategic Technologies in Automation and Robotics (STEAR) effort.

Launch of STS-52 was originally scheduled for 15 October 1992, but slipped by a week because of suspected cracks in the liquid hydrogen coolant manifold in one of Columbia's three main engines. NASA managers decided that it would be quicker to simply replace the engine, rather than conduct painstaking X-ray analysis of the damaged area. Even though their launch was thus delayed, when Wetherbee and his crew finally set off at 1:09 pm EDT on 22 October they set a new post-Challenger record for the shortest interval between two flights by the same Shuttle. Columbia had returned to Earth from her previous mission, STS-50, only 105 days earlier. Nevertheless, launch on the 22nd was delayed by almost two hours due to unacceptable crosswinds at the Shuttle Landing Facility, which would have been used if problems forced Wetherbee and Baker to perform a Return to Launch Site abort during the ascent.

However, after discussions between members of the Mission Management Team, chaired by former astronaut Brewster Shaw, it was decided that although the 37 km/h crosswinds exceeded flight rules – which stipulated speeds no higher than 27 km/h – they were safe enough for STS-52 to go. Ascent Flight Director Jeff Bantle's reservations were overruled and Shaw decided to proceed with the launch attempt, based on simulations which confirmed that Wetherbee and Baker would be able to brake Columbia safely to a halt on the runway if necessary. "We accepted Jeff's recommendation," Shaw said at the post-launch press conference, "based on his interpretation of the guidelines and made a management decision that went in a different direction." Columbia made a perfect ascent and entered an orbit of 302 × 296 km, ready for ten days of intensive activity.

Although the mission was not quite as long as the two weeks that her STS-50 crew had spent aloft, several Extended Duration Orbiter components remained aboard: the most notable being the Regenerative Carbon Dioxide Removal System, which behaved erratically on its first flight, but which performed perfectly on STS-52. The astronauts activated it on the evening of the 22nd and it flawlessly scrubbed the cabin atmosphere of exhaled carbon dioxide for the remainder of the flight. For Steve MacLean, the nature of Columbia's orbit enabled him to see only a small portion of his homeland. "We were on an equatorial orbit [at 28.45 degrees]," he explained later. "I could see kind of the bottom of the Great Lakes, just on the edge of the horizon. Even then, because I knew they were there, I told myself I could see them, because they're right on the edge of what you can see." On his second Shuttle mission, in 2006, MacLean flew to a higher inclination of 51.6 degrees and was able to travel directly over virtually every major Canadian city.

The first major objective was the deployment of LAGEOS-II, under the direction of mission specialist Tammy Jernigan. The satellite and IRIS and LAS upper stages were housed within a protective sunshade cradle, composed of a series of machined aluminium frames and chrome-plated steel longeron and keel trunnion fittings, coated with Mylar insulation. The sunshade measured 2.4 m long by 4.6 m wide across the width of the payload bay. At the base of the cradle was a turntable with two electric motors to impart the required spin rate, which could be set in the range 45-100 revolutions per minute, dependent upon the stability needs of LAGEOS-II, and a spring-ejection mechanism to release the satellite and its attached motors.

"Houston, we see a good deploy," Jernigan told Mission Control as LAGEOS-II and Columbia parted company at 10:57 am EDT on 23 October, almost a full day into the mission. Forty-five minutes later, the IRIS booster – built by Italy's Alenia Aerospazio company – ignited to deliver the satellite into a 297 × 5,923 km orbit, inclined 41 degrees to the equator. At Jernigan's shoulder, Lacy Veach was at the controls of the RMS mechanical arm and used its cameras to view the IRIS burn. Then, at around 1:30 pm, the satellite's own LAS motor fired to insert itself into a near-circular orbit of 5,617 × 5,950 km orbit, inclined at 52.6 degrees. With deployment completed, Wetherbee and Baker manoeuvred Columbia into a lower

3-16: Astronaut Lacy Veach commemorated his Hawaiian roots by carrying a Stone Age adze among his personal kit on STS-52. The adze originated from the Keanakako'i quarry, located high on the slopes of Mauna Kea, and is pictured floating in Columbia's overhead windows, with Hawaii's Big Island visible far below.

orbit of 287 km to better support USMP-1 operations for the remainder of the mission.

Unlike LAGEOS-I, which had been inserted into a 109.9-degree orbit, the inclination of its successor was chosen to enable it to provide more complete coverage of the seismically active Mediterranean basin, as well as California. It was also hoped to clear up several irregularities that had cropped up with LAGEOS-I's position; these seemed to be linked to the erratic spinning of the satellite. LAGEOS-II underwent a month-long checkout to precisely calculate its orbit by laser ranging, after which scientific operations commenced in earnest. In fact, its orbit turned out to be so stable that it will not re-enter Earth's atmosphere for another eight million years! It should remain 'operational' for around half a century and the only limiting factor is the slow degradation of its retroreflector prisms. In view of the longevity of the mission, NASA asked planetary scientist Carl Sagan – who chaired the team which produced plaques for the Voyager spacecraft – to create a stainless steel disk for LAGEOS-II. This contained three 'maps' of Earth: the first as our planet appeared 268 million years ago, the second as it appears in the current era and the third as it may be around eight million years into our future. Interestingly, the 'future' map included the eventual drift of the strip of California to the west of the San Andreas fault way out into the Pacific Ocean.

TELESCIENCE ON TRIAL

When probed by journalists about the risks associated with launching such a small cargo on the Shuttle, Jim Wetherbee was philosophical. "If we want to get this great science," he explained, "we must take a certain amount of risk. I think this science is worth going after." With the deployment of LAGEOS-II behind them, the crew focused their energies on more than a week of medical, scientific and technological experiments in the payload bay and cabin. Although USMP-1 was the most visible payload, the crew actually had little interaction with its operation. They activated it around two and a half hours after launch and although SAMS operated autonomously, the LPE and MEPHISTO experiments benefitted from regular adjustments sent up by ground controllers, using 'telescience'.

The Lambda Point Experiment underwent a day of calibration of its high-resolution thermometers, which verified their capability to measure heat capacity. By the midpoint of the ten-day mission it had received no fewer than 600 commands remotely transmitted from ground controllers ... and by the end of the flight Principal Investigator John Lipa of Stanford University reported that the experiment had acquired more than 300 percent of its expected temperature data in order to test Kenneth Wilson's Nobel-winning theory. Meanwhile, the MEPHISTO researchers also calibrated their experiment within the first 24 hours of the mission, heating it up to 343 degrees Celsius and checking out its tin-bismuth specimens. Close to both experiments, SAMS performed its own measurements, with Wetherbee and Baker executing a series of thruster firings to induce vibrational disturbances and thus calibrate the detector heads. Such data was expected to aid the development of future microgravity experiments, particularly for Space Station Freedom.

As USMP-1 operated in the payload bay, Columbia's astronauts were busy with a multitude of other experiments in the middeck. Steve MacLean had his hands full with CANEX-2 – the first set of Canadian Experiments having been performed by his fellow countryman Marc Garneau on Mission 41G in October 1984 – to such an extent that Jim Wetherbee asked Mission Control to give him some free time. "Steve's doing a great job," he explained. "He wants to do it all. It's my call to offload him a little bit."

CANEX-2 comprised ten space science, technology, materials processing and medical experiments, all provided by the Canadian Space Agency. Of these, one of the most exciting was the Space Vision System (SVS), which sought to improve astronauts' perception of large structures under unfavourable viewing conditions. It was already known that, in space, visual acuity is compromised by frequent periods of extreme darkness and lighting and few reliable points of spatial reference. This had led many astronauts to express concerns about the difficulty of accurately judging distances and range rates. During STS-52, MacLean evaluated a prototype vision system for Space Station Freedom proximity operations, which provided him with data on position, orientation and motion of several small 'targets'. He employed a television camera, which monitored a series of dots on the Canadian Target Assembly (CTA) – a 2-m-long, 82 kg 'domino' – as the Space Vision System provided a real-time display of its position and changing orientation. Said

MacLean's crewmate, Lacy Veach: "This could be as revolutionary as instrument flying was to aviation."

Veach was at the controls of the RMS arm for the CTA tests. On 29 October he lifted the domino and rolled it from side to side, then 'up' and 'down', whilst MacLean closely monitored its movements with the vision system. The following day, Veach and Jernigan conducted another series of tests and then on the 31st the domino was finally jettisoned and Wetherbee flew Columbia in formation at a distance of 42.6 km. Steve MacLean described the performance of the vision system and the target as "excellent" and, as planned, Wetherbee pulsed the orbiter's thrusters to monitor its dynamic response.

Other CANEX-2 experiments included a high-temperature furnace, the Queens University Experiment in Liquid Metal Diffusion (QUELD), which had the capacity to heat up to 900 degrees Celsius and which examined the diffusion of bismuth and tin. The Material Exposure in Low-Earth Orbit (MELEO) comprised a 'quilt' of around 350 materials samples affixed the mechanical arm, which were analysed after the mission to determine the effects of harsh atomic oxygen exposure. These materials included candidates for Space Station Freedom's Canadian-built mechanical arm and were proposed for inclusion in Canada's Radarsat mission. In other experiments, a Canadian spectrophotometer measured the light-absorption characteristics of the upper atmosphere and MacLean photographed the Shuttle's tail and flight surfaces during thrusters firings to understand the underlying mechanisms of atomic oxygen 'glow'. Experimental heat pipes were evaluated for future thermal-control systems and Bill Shepherd supervised the Crystals by Vapour Transport Experiment, provided by Boeing, which yielded large, high-quality, electro-optical crystals of cadmium-telluride. Medical investigations focused upon the effects of hormone therapy on changes in the bones of a dozen adolescent male albino rats (*Rattus norvegicus*).

The rats were matched by body mass into pairs within a pair of Animal Enclosure Modules in the Shuttle's middeck. One member of each pair was given subcutaneous injections of an anti-osteoporotic protein, two days before launch, followed by a second injection of a bone marker called 'calcein'. The other rats were untreated. Other medical investigations included the ongoing Canadian study of back pain in astronauts. Already, research conducted by Roberta Bondar on the IML-1 mission in January 1992 had revealed that back pain tended to be most acute in the first few days in orbit and was possibly attributable to spinal lengthening in the range of 7.5 cm. This, investigators hypothesised, may have been due to increased water content and thus the height of the discs which separate the spinal vertebrae. The resulting tension on soft tissues, such as muscles, nerves and ligaments, might then lead to the occurrence of back pain. During STS-52, MacLean carefully measured his height and used a diagram to record instances, locations and intensities of back pain.

After a mission which Bob Crippen described as "chocked full of work", Columbia re-entered the atmosphere early on 1 November. A few moments of drama were caused by the failure of a cockpit indicator, which was needed to display data on the positions of the orbiter's elevons and other aerosurfaces. Wetherbee and Baker switched to small digital displays as a backup measure and temporarily regained control of the indicator, before it failed a second time. Post-mission inspections would blame a faulty fuse. Columbia touched down safely on Runway 33 at the Cape at 10:05 am EDT, wrapping up a mission of nine days, 20 hours, 56 minutes and 13 seconds. Interestingly, STS-52 had lasted almost exactly the length of time predicted in NASA's pre-flight press kit.

The performance of the new drag chute was still a cause for concern. Six weeks earlier, during the rollout of Endeavour at the end of the STS-47 mission, it had 'dragged' the orbiter slightly to one side on the runway. During the STS-52 touchdown, Mike Baker deployed the chute a couple of seconds before the nose gear rotated down to the runway surface and the astronauts noticed a perceptible 'tug' into the wind as the canopy billowed into its fully-reefed configuration. In fact, the chute pulled Columbia about 4.6 m to the 'left' of the centreline on the 90-m-wide runway. Nonetheless, at the post-landing press conference, Jim Wetherbee expressed his confidence in the design and gave it his stamp of approval. "It didn't cause me any concern," he said. "If we were landing on a very narrow runway like [an emergency] runway over in Africa, and it pulled even more, then it would be cause for a little bit more concern."

UNDER THE 'VEIL'

Guy Bluford did not expect to fly into space again after STS-39.

The United States' first black astronaut had suffered a herniated disk in his back in November 1990, whilst training for his third space mission, and despite physical therapy he continued to endure problems. Numbness in his right shin meant that he could not stand for more than 30 minutes, before his leg started aching. Discussions within senior management about removing Bluford from the STS-39 crew were hardly ideal and after an operation he was able to rejoin his fellow astronauts within two weeks. The mission – arguably the most complex Department of Defense flight ever attempted, with multiple rendezvous operations and a full plate of scientific research – was a huge success, but it gave Bluford food for thought about his future. "I really enjoyed the work in the office and the camaraderie with my fellow astronauts," he told the NASA oral historian, "but my problem with my back ... reminded me that it may be time to leave."

By his own admission, though, Bluford had made no definite plans, when all of a sudden in August 1991 he received a call from Don Puddy, the head of Flight Crew Operations, to inform him of his selection as a mission specialist on STS-53, the final classified Shuttle flight in support of the Department of Defense. Although STS-39 had carried military payloads, it operated in an 'unclassified' capacity, whereas STS-53 would ferry a top-secret satellite into orbit; a satellite whose precise nature and purpose remained the subject of conjecture for several years. Joining Bluford were Dave Walker, the commander of the crew, together with pilot Bob Cabana and mission specialists Jim Voss and Rich Clifford. The crew represented all four primary military services of the United States: Walker was a naval aviator, Bluford an Air Force engineer, Cabana a Marine Corps aviator and both Voss and Clifford were from the Army.

Michael Richard Uram Clifford was the first member of NASA's 1990 astronaut class to receive a flight assignment and to reach orbit. (Interestingly, he was also among the first of five groups to be interviewed in Houston in September 1989.) Born in San Bernadino, California, on 13 October 1952, he graduated from high school in Ogden, Utah, and entered the Military Academy at West Point. Clutching his science degree in 1974, Clifford was commissioned into the Army as a second lieutenant and served initially with the 10th Cavalry in Fort Carson, Colorado, before entering the Army Aviation School in 1976, from which he graduated as the top of his class. Designated as an Army Aviator, he was based in Nuremberg, West Germany, as a service platoon commander, and completed a master's degree in aerospace engineering from Georgia Institute of Technology in 1982. Clifford spent some time as a faculty member in West Point's Department of Mechanics and in December 1986 graduated from Naval Test Pilot School. Although he applied unsuccessfully for admission into the astronaut corps on two occasions, Clifford entered NASA's Johnson Space Center in July 1987 as a Shuttle Vehicle Integration engineer and worked extensively on design certification and integration of crew escape systems. He was also an executive board member of the Solid Rocket Booster Post-flight Assessment Team. Clifford was selected as part of NASA's Group 13 astronaut class – 'The Hairballs' – in January 1990.

Although he was the sole first-time flier on STS-53, Clifford was assigned a number of critical responsibilities, with relation to several payloads. One of these was the Battlefield Laser Acquisition Sensor Test (BLAST), provided jointly by the Army's Space Command at Colorado Springs, Colorado, the Space Technology Research Office in Adelphi, Maryland, and the Night Vision Electro-Optics Directorate at Fort Belvoir, Virginia. Its purpose was to demonstrate the technologies associated with using a spaceborne laser receiver to detect energy from ground-based test locations at the Air Force's Maui Optical Site in Hawaii and Malabar Test Facility in Florida. At these and portable instruments at various Department of Defense field locations, low-power visible lasers tracked the Shuttle and their optical signals were captured by the BLAST hardware, which was mounted in one aft flight deck window. Unfortunately, of the 20 targets, only two were successfully tracked, due to problems ranging from inclement weather, ground hardware difficulties and restrictions imposed by the presence of an unidentified aircraft in the observation area. Clifford also oversaw the Fluid Acquisition and Resupply Experiment (FARE), developed by NASA's Marshall Space Flight Center, which investigated the processes of fluid transfer in the microgravity environment. Iodine-treated water, blue food colouring and a wetting solution (known as 'Triton X-100') to mimic the consistency of a propellant were to be transferred between containers. Eight successful experimental runs were performed.

In command of STS-53 was David Mathieson Walker, who retained his aviator's nickname of 'Red Flash', in honour of his sandy hair ... and it was *this* nickname which contributed to an informal name for the entire crew: the 'Dogs of War'. Walker came from Columbus, Georgia, where he was born on 20 May 1944. After attending high school in Florida, he entered the Naval Academy, received his degree in 1966, and immediately underwent flight instruction at the Naval Aviation

Training Command in Florida, Mississippi and Texas. Walker was designated as a naval aviator in December 1967 and served two tours in Vietnam, flying the F-4 Phantom from the USS *Enterprise* and USS *America*, during which he gained the Distinguished Flying Cross. Following his return to the United States in 1970, he entered test pilot school at Edwards Air Force Base and later served as an experimental and engineering test pilot at the Naval Air Test Center in Maryland, working on the evaluation of the new F-14 Tomcat fighter. Walker also performed flight tests of a leading-edge slat modification of the Phantom jet and, in 1975, after replacement pilot training on the Tomcat, he served as a fighter pilot on two overseas deployments to the Mediterranean aboard the *America*. As an aviator, Walker was top-notch, but in the words of Mike Mullane, who flew with him as a member of the T-38 chase team in support of STS-1, "he was *too* cocky, the type of pilot who thinks he's bulletproof even when he's sober". Crewmate Bob Cabana later paid tribute to Walker as "kind of a throwback to the 1960s".

STS-53 was the ninth and final dedicated Department of Defense mission, a point reinforced by NASA Administrator Dan Goldin in the weeks before launch. "Nine DoD primary payloads have been carried into space by the Shuttle since 1985," he noted. "The fact that complex mutual objectives have been achieved by two federal organisations, chartered with often divergent goals, is a wonderful and remarkable demonstration of inter-agency co-operation at its best." Added the Air Force's Assistant Secretary for Space, Martin Faga: "STS-53 marks a milestone in our long and productive partnership with NASA. We have enjoyed outstanding support from the Shuttle programme. Although this is the last dedicated Shuttle payload, we look forward to continued involvement with the programme with DoD secondary payloads."

Therefore, as the last crew of a military Shuttle flight, the 'Dogs of War' received their own 'dog names' and 'dog tags'. In recognition of his hair, Dave Walker became 'Red Dog', crew-cut Marine Bob Cabana was 'Mighty Dog', Army man Jim Voss was 'Dog Face' and new boy Rich Clifford became 'Puppy Dog'. As for Guy Bluford, who was in Europe on a NASA public relations tour when Walker handed out the names, he quickly received 'Dog Gone'. (This nickname possibly also lent a lighthearted nod to Bluford's status as the most flight-experienced, and oldest, crew member.) At one point, Walker even bought an old decrepit car and painted their dog names on the side. "The trainers and support people in Mission Control all had dog names," remembered Bluford. "We would drive around JSC in the Dogmobile." Paper dog mascots were stored in the Shuttle's middeck lockers, but the mission itself was conducted in all seriousness.

Central to that mission was the deployment of a classified payload, known only as 'DoD-1'. The reader will recall from Chapter 1 that the Department of Defense had long since begun to distance itself from the Shuttle, for two fundamental reasons: firstly, the reusable vehicle had proven far more complex and difficult to process for flight than originally envisaged and was therefore unable to fly as 'routinely' as billed in the mid-1970s, and secondly, its nature as a piloted craft exposed it to far more public interest than a faceless expendable booster. The first classified Department of Defense mission, in January 1985, had several details about its top-secret payload

3-17: In spite of the relatively 'open' nature of STS-53, the presence of classified cargoes meant that – like many of its predecessors – the mission produced few images of the payload bay or its contents. In this carefully crafted view, only a tiny portion of Discovery can be seen, with the Moon visible at left.

revealed by a national newspaper, several months before launch. With this in mind, it is understandable why the military and intelligence communities sought an exit strategy from the Shuttle at the earliest opportunity.

Discovery, which had by now established a reputation as the 'fleet leader', was undertaking her 15th mission. At the end of her previous flight, STS-42, in January 1992, she was removed from service for six months of refurbishment. During this time, she remained at the Kennedy Space Center and a total of 78 modifications were implemented, most notably the addition of the drag chute, a capability for redundant nosewheel steering, improved Auxiliary Power Units and upgraded avionics. At the end of July, she commenced STS-53 pre-flight processing in earnest, with the installation of her three main engines. Rollout of the Shuttle stack to Pad 39A occurred on 8 November ... with the five STS-53 astronauts riding along on the crawler transporter. "We spent most of the day riding the crawler," remembered Guy Bluford, "carrying the Space Shuttle Discovery with our payload out to the launch pad. It was a wonderful ride!"

Launch on 2 December was originally scheduled for 6:59 am EST, but was delayed by more than an hour, due to the presence of ice on the vehicle and a wing-load indicator violation. When these issues had been cleared, the helium outlet regulator pressures on two main engines exceeded the uppermost allowable limit, but a waiver was approved for these excesses and the countdown proceeded. Discovery sprang from the pad at 8:24 am and her ascent to orbit was nominal in all respects.

The astronauts' first order of business was the classified DoD-1 payload, which Guy Bluford recalled was deployed towards the end of their first day in orbit. According to NASA's post-flight summary, DoD-1 left the bay at 2:18 pm, a little under six hours after launch. Two decades later, its precise nature remains unclear, although most observers have reached a consensus that it was probably the third launch of a second-generation Satellite Data System (SDS)-B military communications satellite, not dissimilar to those launched by the crew of STS-38 in November 1990 – described in Chapter 1 of the present volume – and STS-28 a year earlier, in August 1989.

With DoD-1 gone, the crew turned their attention to the multitude of military and other experiments in Discovery's middeck and payload bay. Outside were the Orbital Debris Radar Calibration Spheres (ODERACS), an investigation which sought to release six spherical objects into space for the purpose of evaluating the tracking capabilities of the Haystack Radar in Tyngsboro, Massachusetts, a facility operated by MIT's Lincoln Laboratory on behalf of the Air Force. Its data was to be used by NASA as a tool to measure the amount of debris in low-Earth orbit and the radar had the ability to measure objects as small as a single centimetre in diameter from a distance of more than 1,000 km. The stainless-steel ODERACS spheres, organised in pairs measuring 5 cm, 10 cm and 15 cm in diameter and weighing between 0.5 kg and 5 kg, were to be ejected from a Getaway Special (GAS) canister to orbit the Earth at an altitude of 377 km.

Original plans called for the spheres to be released on the 31st orbit, two days into the mission, after which tracking would be attempted by the Haystack Radar and several other worldwide installations, including the Kwajalein Radar in the Marshall Islands, the Eglin Radar in Florida and the FGAN radar, near Bonn in Germany. Additionally, a number of telescopes were scheduled to follow their progress. The spheres' expected lifetime in the harsh environment of low-Earth orbit, before being pulled back into the atmosphere, was anticipated to be two or three months. Deployment of ODERACS was the responsibility of Rich Clifford, who was to have commanded the GAS canister door to open on the 31st orbit, two days into the mission, and visually verified the spheres' departure. Unfortunately, a dead gas control decoder battery prevented the GAS canister's door from opening and there was no means of recharging it from the crew cabin. An EVA by Voss and Clifford to recharge the battery was considered, but time constraints did not allow for the development of an engineering solution to the problem. As a result, the spheres were not deployed and were reassigned to Discovery's next flight, STS-56.

It must have been a disappointment for Clifford, whose first voyage into orbit had otherwise far exceeded all of his pre-launch expectations. As the only rookie member of the crew, he slept each night in the orbiter's airlock – "a good place," he said in an interview for the Smithsonian, "because there's not a lot of noise in there" – and this made for a couple of humorous incidents during the seven-day mission. The airlock was a relatively small space and Clifford's head poked out into the middeck. On his first night, he forgot to attach his Velcro 'pillow' behind his head before sleep. "So when people got up in the middle of the night," he reflected, "and there's a constant stream to the bathroom the first day, as everybody gets rid of body fluid in weightlessness, they'd cross the cabin, create an air current and my floating head would *bounce* against the sides of the airlock hatch."

Clifford mentioned it to his crewmates the following morning.

"Why didn't you put on your pillow?" they asked. "Then you won't hit the sides."

Clifford tried, but the first couple of nights remained an annoyance. At length, he put Velcro on the back of his shirt and fastened himself to the forward wall of the middeck. It looked peculiar, with the effect of microgravity drawing his body into a foetus-like orientation, but it suited him well for the remainder of the mission. Luckily, Clifford's next flight, in April 1994, was an around-the-clock Spacelab mission and benefitted from telephone-booth-sized personal sleep stations.

His experience of weightlessness itself was wonderful, though, and midway through the flight he and Jim Voss – the two Army members of the crew – had a wrestling match on the middeck. "We were like a flying furball in the middle of the cabin," he said, "trying to get a foothold of each other. Every time you pushed one way, you'd go off in the opposite direction. It was the exhilaration of doing something you'd probably never get a chance to do again. We were like kids." After a while, Marine Bob Cabana joined in. Watching from the sidelines, 'Old Man' Guy Bluford – who admitted that he had never had a chance to wrestle on any of his three earlier missions – also got involved.

Elsewhere in the payload bay was the combined Shuttle Glow Experiment and Cryogenic Heat Pipe Experiment (GCP). The former originated from the Geophysics Directorate of the Air Force's Phillips Laboratory in Albuquerque, New Mexico, and employed the Arizona Imaging Spectrograph (AIS) to explore the interaction of the Shuttle with the ambient atomic oxygen environment in infrared, visible and ultraviolet wavelengths, with particular focus upon the mysterious patterns of 'glow' on its flight surfaces. During STS-53, the AIS observed Discovery's aft-mounted Orbital Manoeuvring System pods and thruster firings, waste water dumps and the operation of the flash evaporator system. Meanwhile, the Cryogenic Heat Pipe represented a joint NASA-DoD investigation to evaluate advanced technologies to reject excess heat from infrared sensors, instruments and spacecraft. The experiment activated a cryogenic heat pipe and operated it at a temperature range of up to 213 degrees below zero Celsius, but unlike previous heat pipe concepts – which ordinarily used water, ammonia or Freon – this one employed liquid oxygen. It was activated about nine hours into the mission and featured a series of successful tests of two different pipes. The Glow Experiment was hampered by an absence of suitable conditions of orbital darkness until 7 December. Nevertheless, 20 of its planned 23 mission requirements were met.

Inside the crew cabin, the five astronauts tended to several experiments, including several flown on previous missions: the Space Tissue Loss, Visual Function Test, Cosmic Radiation Effects and Activation Monitor, Radiation Monitoring Equipment, CLOUDS and the HERCULES tracking system. The latter consisted of a modified Nikon F-4 camera with an attached geolocation device to record images of terrestrial features and determine their latitude and longitude. During orbital 'daytime' observations, the astronauts could use any Nikon-compatible lens, but during nighttime passes they attached an image intensifier developed by the Army's Night Vision Laboratory. Developed by the Department of Defense Space Test Program, HERCULES demonstrated geolocation accuracies of about 1.5 km on its first mission. Another study, the Microcapsules in Space, produced time-releases antibiotic microcapsules as part of studies into the practicability of producing improved pharmaceutical drugs in the microgravity environment.

Discovery's return to Earth on 9 December was entirely normal, although the planned landing at the Kennedy Space Center was called off due to deteriorating weather conditions in the vicinity of the Shuttle Landing Facility. Consequently, the crew waited an additional orbit and were directed to Edwards Air Force Base in California instead. Dave Walker and Bob Cabana performed the de-orbit burn at 2:43 pm EST to commence an hour-long hypersonic dive into the atmosphere, heading towards a landing strip on the opposite side of the planet. Twelve minutes later, passing 'entry interface', one of Discovery's forward Reaction Control System thrusters had an oxidiser leak. Touchdown on the concrete of Runway 22 occurred perfectly at 3:43 pm EST (12:43 pm in California) and the drag chute deployed without incident, although there remained evidence of slight tugging of the orbiter to the left of the centreline. After landing, ground crews confirmed oxidiser concentrations of a couple of parts per million in the area of the forward RCS pod and it was decided to monitor the situation for two hours before authorising the removal of the astronauts. When the concentrations remained unchanged, a fan was used to clear the area of fumes and Walker's crew exited Discovery after a quiet mission which marked the end of an era for dedicated Department of Defense operations.

A CHANGED MISSION

When the STS-54 crew released their official crew patch in the summer of 1992, they paid tribute to two important payloads aboard their mission. The first was NASA's fifth Tracking and Data Relay Satellite (TDRS-F), a frequent Shuttle flier which would form the latest component in a critical geostationary-orbiting constellation to maintain near-continuous voice and data communications between the Shuttle and other important scientific spacecraft with ground stations. Attached to a Boeing-built IUS booster, TDRS-F would be deployed from Endeavour's payload bay late on the first day of the mission. Within the bright-red circular frame of the STS-54 patch, a fearsome bald eagle held an enormous, eight-pointed star in its talons and was about to add it to a collection of four stars already in place. According to the crew, this represented the placement of the fifth TDRS into orbit, alongside its four cousins. Behind the eagle, the glorious blue and white Earth was juxtaposed with the unfathomable blackness of space; a blackness which, on this patch, was conspicuously devoid of stars. "The blackness of space," noted the patch description, "represents our other primary mission of carrying the Diffuse X-ray Spectrometer to orbit to conduct astronomical observations of invisible X-ray sources within the Milky Way Galaxy."

In command of the mission was John Howard Casper, an Air Force colonel who had unsuccessfully applied for NASA's first class of Shuttle astronauts in January 1978, before being selected six years later. He was born in Greenville, South Carolina, on 9 July 1943 and after high school entered the Air Force Academy to study engineering science. He earned his degree in 1966 and completed a master's qualification in astronautics from Purdue University early the following year. Following initial flight instruction, Casper received his wings at Reese Air Force Base in Texas, trained on the F-100 Super Sabre and was despatched to Vietnam, where he flew more than 200 combat missions. Upon his return to the United States, he continued to fly the F-100, as well as the F-4 Phantom, and was assigned as an exchange pilot to a tactical fighter wing at RAF Lakenheath in England. Casper then attended test pilot school at Edwards Air Force Base, graduated in 1974 and headed the F-4 Test Team, performing weapons separation and avionics testing. He later worked at the Pentagon as deputy chief of the Special Projects Office, developing Air Force positions on requirements, operational concepts, policy and force structure for tactical and strategic programmes. In February 1990, Casper flew his first mission as pilot of STS-36 and in December of that year was named to his second, the 13-day STS-50 Spacelab flight. However, due to the retirement of two experienced Shuttle commanders in the summer of 1991, it was decided to rotate Casper and another experienced pilot, Jim Wetherbee, out of their previous assignments and into command positions. As a result, Casper lost the pilot's seat on STS-50, but gained the plum of a command on only his second mission.

The credentials of his other four crewmates were equally impressive. Pilot Don McMonagle and mission specialist Greg Harbaugh had served together aboard STS-39, whilst Mario Runco had returned recently from STS-44. The sole 'rookie' member of the crew was Susan Jane Helms, the second member of the 1990 class of astronauts – 'the Hairballs' – to receive a flight assignment and reach orbit. Today a lieutenant general in the Air Force, commander of the 14th Air Force (Air Forces Strategic) and of the Joint Functional Component Command for Space at Vandenberg Air Force Base in California, she holds the record for achieving the highest rank of any female military astronaut. On STS-54, she also became America's first active-duty military female spacefarer.

Born on 26 February 1958 in Charlotte, North Carolina, Helms was the daughter of a career Air Force colonel and after high school in Portland, Oregon, she entered the Air Force Academy to study aeronautical engineering. Upon receipt of her degree in 1980, she received her military commission and trained as an F-16 weapons separation engineer with the Air Force Armament Laboratory at Eglin Air Force Base in Florida. Helms rose through the ranks and within two years was the lead engineer. "I'm not pilot-qualified," Helms told the NASA oral historian, "so my entrance into the Air Force didn't have anything to do with the flying bug. I have very bad eyesight, so I never had the opportunity to even become an Air Force pilot. I was in the Air Force about seven years before I realised that there were jobs available where engineers could fly."

In 1984, by now holding the rank of captain, she was selected to pursue a master's degree in aeronautics and astronautics at Stanford University and in 1985-87 accepted an assistant professorship at the Air Force Academy. Helms' next assignment was to Air Force Test Pilot School at Edwards Air Force Base in California, from which she qualified as a flight test engineer and later worked at Canadian Forces Base Cold Lake in Alberta, Canada, as project officer on the CF-18 Hornet. She was serving in Canada when she learned of her selection as an astronaut in January 1990. "I'd like to be able to say that I was born with this burning desire to become an astronaut," Helms reflected, "but it didn't really happen that way. What ended up happening was I sort of grew into the idea over several years."

Within months of entering NASA, she was persuaded to join the ranks of the all-astronaut rock band, 'Max Q', playing keyboards. Helms had taken lessons for 11 years, as well as having played concert drums and xylophone in marching bands and choirs and as part of a jazz combo. In an interview with Michael Cassutt, she described her tastes as "pop, Top 40, everything *but* country and western" and fellow Max Q member Kevin Chilton described her as "hugely talented" and capable of listening to a song on the radio and playing it. "She was able to teach us harmonies," said Chilton. On STS-54, Helms carried a mini-keyboard as part of her personal kit and managed to tap out a one-finger version of the Air Force anthem 'Wild Blue Yonder' whilst in orbit.

In addition to TDRS-F, Endeavour was carrying the twin detectors of the Diffuse X-ray Spectrometer, mounted on Hitchhiker plates on opposite walls of the forward payload bay. This instrument originally formed part of a much larger Spacelab payload, known as the Shuttle High Energy Astrophysics Laboratory (SHEAL), and would have flown alongside the Goddard Space Flight Center's Broad Band X-Ray Telescope. However, the completion of the latter, ahead of schedule, and the deletion of another instrument from the ASTRO-1 mission, caused it to be moved forward on the Shuttle manifest and it flew in December 1990. As for the DXS, it was moved around as a secondary payload on a couple of flights, before coming to rest on STS-54.

Designed to acquire the first-ever spectra of the diffuse low-energy 'soft' X-ray background in the energy band between 0.15-0.284 keV, the DXS comprised a pair of large-area lead-stearite Bragg crystal spectrometers. Each contained a curved

3-18: Greg Harbaugh waves to his crewmates inside Endeavour, prior to performing a task during the EVA with Mario Runco.

panel of Bragg crystals mounted above a position-sensitive proportional counter, across which a spectrum would be dispersed to enable all portions to be measured simultaneously. However, whilst all wavelengths were observed at the same time, the various wavelengths came from different directions in the sky, and therefore the spectrometers 'rocked' backwards and forwards to obtain complete spectral coverage along an entire arc of the sky. The two spectrometers and their associated instrumentation were built at the Space Science and Engineering Center at the University of Wisconsin at Madison, under the direction of Principal Investigator Wilt Sanders. Unique in its ability to 'sort' detected X-rays by wavelength, the DXS identified large quantities of hot gas in the 'interstellar medium' close to our Solar System.

Although classed as a secondary payload, the importance of the instrument was such that provision was included in the STS-54 manifest for a one-day extension to the flight, "if DXS requires additional time to achieve mission success". Although scheduled to fly for less than six days, Endeavour was equipped with enough consumables to support a seven-day 'basic' mission, plus two additional days to cater for unforeseen contingencies, such as weather or other difficulties.

As 1992 drew to a close, STS-54 appeared to be a relatively 'vanilla' mission, scheduled to run for just under six days. Then, on 25 November, NASA announced its decision to add EVAs onto three future Shuttle missions to "fine-tune the methods of training astronauts for assembly tasks in space" and "increase the spacewalk experience levels of astronauts, ground controllers and instructors". The agency noted that such excursions would only be added on the proviso that they did not impact primary mission objectives ... and the first flight to benefit from the change was STS-54. It had long been recognised that an immense number of EVAs, later to be nicknamed 'The Wall of EVA', would be needed to assemble and maintain Space Station Freedom and the difficulties experienced by Dan Brandenstein's STS-49 crew during their effort to capture the Intelsat 603 satellite in May 1992 had led them to lobby for more expertise within the astronaut corps.

In fact, on the first day of 1993, of the 90 or so astronauts on active flight status, only eight had EVA experience. For STS-54, Greg Harbaugh and Mario Runco had undergone 'generic' spacewalk training, in case they had to go outside and manually close the payload bay doors, but their work moved swiftly into high gear in the final days before and after Christmas as plans were finalised for a five-hour EVA. Their tasks included moving around Endeavour's payload bay with and without 'large objects' – including each other – as well as completing close alignment tasks and installing equipment. It was mandatory for their EVA to conclude at the scheduled time, because it was assigned a lower priority than the DXS observations, which had to be suspended whilst Harbaugh and Runco were outside. After the mission, it was intended that the spacewalkers would repeat their activity in the Houston water tank to help improve future training practices.

Liftoff of STS-54 on 13 January was delayed by a little more than seven minutes to await the resolution of a Launch Systems Evaluation and Advisory Team violation. Under the combined thrust of her three main engines and twin Solid Rocket Boosters, Endeavour speared for the heavens at 8:59 am EST and successfully entered a 28.45-degree-inclined orbit shortly afterwards. Six hours and 13 minutes into the mission, high above the Pacific Ocean, to the north of Hawaii, the TDRS-F payload and its attached IUS were released and Casper and McMonagle manoeuvred the orbiter to a safe separation distance.

An hour after deployment, the first stage of the IUS ignited to achieve geosynchronous transfer orbit and the second stage fired some five hours later to circularise the orbit. Thirteen hours after launch, the TDRS separated from the IUS and underwent a complex process of unfurling its solar arrays, space-to-ground communications boom and its C-band and single-access antennas. The numerically renamed 'TDRS-6' thereby became the fifth operational satellite in a constellation which supported the Shuttle and important scientific missions, including Hubble and the Compton Gamma Ray Observatory. TDRS-B had been lost in the Challenger accident, whilst TDRS-1 – launched in April 1983 – had far exceeded its seven-year design life and was continuing to provide limited service. Another satellite, TDRS-3, was only partially functional, which left just TDRS-4 and 5 in fully-operational status ... and *this* meant that there was no spare available in the event of problems. The successful arrival of TDRS-6 in orbit 35,600 km above the equator thus provided this backup capability. Over the course of several weeks, TDRS-3 was manoeuvred out of its slot at 62 degrees West longitude to a new location at 171 degrees West and was replaced in its original position by TDRS-6.

On Endeavour's seventh orbit, the DXS instrument began scanning. Despite problems due to high particle counts, which triggered a high-voltage shutdown, an additional 15 orbits of data collection were authorised to complete its objectives. The spectrometers acquired a total of more than 80,000 seconds of good data.

Elsewhere, inside Endeavour's crew cabin, depressurisation was executed on the third day of the flight in anticipation of Harbaugh and Runco's EVA. The two men entered the floodlit payload bay at 5:48 am EST on 17 January, closely monitored by Susan Helms, who acted as the 'intravehicular' crew member. During the excursion – only the 20th EVA in Shuttle programme history – the spacewalkers translated themselves around the payload bay, both with and without large items and climbed into foot restraints without the benefit of handholds. "To simulate carrying a large object," noted NASA's pre-mission press kit, "the astronauts will carry one another." They also worked with the IUS tilt table at the rear of the bay and returned to the airlock after four hours and 27 minutes. Yet the spacewalk brought the Shuttle programme's total to barely 110 hours, far short of the 400 hours anticipated for construction of Space Station Freedom, and in response to this problem NASA added two further EVAs to STS-51 and STS-57.

Despite their numerical order, STS-51 was scheduled to occur *after* STS-57, with launch originally scheduled for July 1993. In the first week of February, NASA announced its intention to "continue extravehicular activity tests" with spacewalkers Carl Walz and Jim Newman. "The addition of the spacewalk to STS-51 will allow us to continue refining our knowledge of human performance capabilities and limitations during spacewalks," said Ron Farris, the chief of the EVA Section at the Johnson Space Center. "This EVA constitutes a continuing commitment by NASA to advance our preparations for future EVA missions, such as the Hubble Space Telescope servicing and Space Station Freedom assembly flights." Two weeks later, on 17 February, an EVA by astronauts David Low and Jeff Wisoff was added to STS-57, then scheduled for late April. The scope of this mission was described in a manner which hinted at its importance for the forthcoming Hubble servicing flight: with "procedures using the Shuttle's mechanical arm" planned to "involve work by astronauts on a platform at the end of the Shuttle's arm". Moreover, when combined with STS-54 and the spacewalks already planned for the first Hubble servicing mission, STS-61 in December, this meant that NASA was aiming to fly *four* EVA flights in 1993 ... tying a record previously set in 1984. "In a sense," said Ron Farris, "it will be a banner year for EVA and will be somewhat representative of the EVA efforts required to build and maintain Space Station Freedom."

Spacewalking induced a peculiar sensation in Mario Runco. Years later, describing the experience to a Smithsonian interviewer, he related a free moment in the EVA, waiting for Harbaugh to finish up a task. "I was standing, facing outboard on a work platform," Runco said. "The platform locks your feet down and frees your hands for work." During underwater training before launch, he enjoyed bending over backwards at the knees – "sort of like doing the limbo" – and expected it to be a comfortable stretch, relieving all of the pressure points induced by his suit. "But in space, the viscosity of the water *wasn't there* to slow me down," he continued, "so when I relaxed to stand up straight again, the suit 'twanged' forward at what seemed like an incredible velocity. It really felt like I would come right out of the foot restraint and go tumbling off into space, even though I knew I couldn't."

Gazing directly into the ethereal blackness of space, Runco was brought face to face with what he could only describe as a first-hand look at God's handiwork.

Aside from the drama of the EVA, the five astronauts oversaw a range of medical and biological experiments in the middeck, monitoring the effect of microgravity exposure upon the skeletal muscles of rodents, growing seeds of *Arabidopsis thaliana* – a small, cress-like plant, with white flowers – and supporting 28 commercial investigations into biomedical testing and drug development, control of ecological life-support systems and the agricultural manufacture of biological-based materials. The Solid Surface Combustion Experiment, itself a frequent Shuttle flier, rode in a middeck locker and studied flame propagation over solid fuel substances. On a somewhat lighter note, a collection of children's toys were flown as an educational resource. The 'Physics of Toys' investigation involved schools in the hometowns of four of the astronauts (Harbaugh was from Cleveland, Ohio, Runco from The Bronx, New York City, Helms from Portland, Oregon, and McMonagle from Flint, Michigan) – and was specifically focused on elementary children. 'Live' demonstrations on 15 January were led by Georgia-born John Casper and the toys involved a car and track, klacker balls, a basketball, magnetic marbles, swimmers, a mouse, gravitrons and a balloon helicopter. The experiment was sponsored by Carolyn Summers of the Houston Museum of Natural Sciences, who was also responsible for the 'Toys in Space' project, flown on Shuttle mission 51D in early 1985.

Endeavour's departure from orbit on the morning of 19 January proceeded without incident and the orbiter touched down in Florida at 8:37 am EST, concluding a mission of just a few minutes shy of six full days.

"RED LIGHTS IN THE COCKPIT"

The first flight of NASA's 1993 manifest had gone well. Next up should have been Columbia on STS-55 in February, but as unfolding circumstances would demonstrate, the gremlins were about to strike the mission and would ultimately eliminate the space agency's hopes of flying eight Shuttle missions that calendar year. The planned nine-day STS-55 flight was to be an international voyage, featuring the co-operation of hundreds of scientists, engineers, technicians and managers from the United States and Germany. However, although it might be expected that this co-operation would extend to space centres, university campuses and laboratories around the world, one would probably be forgiven for not adding to this list of prestigious establishments ... the *parking lot* of a Safeway supermarket.

Yet that is exactly where the progress and scientific accomplishments of STS-55 were discussed, one day in May 1993, between a shopper and one of his friends. Not too much unusual about that, perhaps, with the possible exception that the shopper's 'friend' happened to be astronaut Steve Nagel, commander of Space Shuttle Columbia, who had arranged before launch to give him a call, if time permitted, on his hand-held amateur radio. "I gave him the number for Mission Control," Nagel recalled, years later, "to call a contact back there to update that time." Houston enabled the communications session and the two friends spoke for four or five minutes, until Columbia passed out of range.

A few minutes later, Nagel's friend was approached by another shopper.

"Who were you talking to?"

The friend grinned. "If I told you, you wouldn't believe me!"

In fact, they had been talking about a mission which, for the newly unified Federal Republic of Germany, was an exceptionally expensive scientific venture, costing more than half a billion dollars to devise and bring to fruition. In order to accomplish the mission, Columbia was outfitted with the Spacelab module in her cavernous payload bay and was loaded with no fewer than 88 experiments in medicine, radiation physics, materials science and technology. "I'm not fluent in German," the mission's payload commander Jerry Ross once said. "Fortunately, most of the international science people work in English anyhow, because you've got all the other languages in Europe that have to find some common language and, fortunately, *that's it*. I tried to learn some German, but I'm not good in foreign languages to start with and trying to do all the other things I was doing there wasn't time to learn much German."

Designated 'Spacelab-D2' – for 'Deutschland' – the mission was the second to be sponsored by the former West Germany. An earlier voyage, Spacelab-D1, had been undertaken with great fanfare and success in late 1985, shortly before the loss of Challenger, and both it and the D2 mission were expected to pave the way for collaborative research on Space Station Freedom. At one stage, two additional German Spacelabs – D3 and D4, the first comprising a pressurised module and the second a 'train' of two pairs of linked pallets, possibly including the German Infrared Radiation Laboratory (GIRL) – were also manifested, but later cancelled for budgetary reasons. Of these, Spacelab-D3 came closest to fruition, with a formal reservation in July 1989 for a mission with "at least two West German payload specialists" to "conduct several experiments to prepare for operations on-board the Columbus module attached to the Freedom Space Station".

However, unlike the months-long station missions, Spacelab flights were expected to last barely a week or more. Nagel described them as miniature marathons. "You want to load them up as much as you reasonably can," he said, "because you want to get as much for your money, but they're like sprints. You go hard at it for a week or ten days and you can't do that on the space station. You've got to have some time to back off a little bit, because you can't keep a sprint up for a long time." Nagel was the only astronaut to fly both German Spacelabs and had experienced, first-hand, the feverish, dual-shift, around-the-clock nature of such missions. His crewmate, Jerry Ross, who had been named as payload commander for D2 in April 1991, agreed that training was excessively hectic and "not a very viable way to do business". By the spring of 1993 Ross and the other members of the D2 science crew – mission specialist Bernard Harris, a medical doctor, named in August 1991, and German payload specialists Ulrich Walter and Hans Schlegel and their backups Gerhard Thiele and Renate Brümmer – had been training for the best part of two years and spent the majority of that time travelling backwards and forwards between Europe and the United States. "We spent a long period of time away from family and friends," Ross told a press conference in February. "That is not the nicest way to have to do business."

Ross described his duties on Spacelab-D2 as amongst the most demanding of his entire astronaut career. "The payload commander is basically the guy that's responsible for interfacing with the payload sponsors and the crew to make sure that what the payload sponsors want to happen on-orbit is things that the crew can physically do, both from the interfaces to the payloads, the checklists and the timeline. I had to do all the co-ordination, all the dealings – *everything* – from the safety community, the medical community, the science community. I had to work all that in addition to trying to get three rookies ready to go fly on 90 very different, very complex experiments."

Since Nagel had previously served aboard Spacelab-D1, the Germans were hopeful that he would receive the D2 command. So was Ross, who had flown with Nagel on the STS-37 mission. "I was hoping he may come back and we could fly together again, so I could harass him some more," he joked. "The system thought that would be a good thing to do to have that kind of continuity and he came back as commander." However, the D1 and D2 missions were not entirely similar; according to Nagel the complement of experiments on the second flight was more heavily

3-19: Had a 'Mode One Egress' been commanded on 22 March 1993, the STS-55 astronauts would have been required to evacuate Columbia and board a series of slide-wire baskets to travel to the relative safety of a fortified bunker, near the launch pad.

biased in favour of the life sciences, as well as being correspondingly longer: ten days as opposed to seven. This life sciences bias was further accentuated by the inclusion of a physician, Bernard Harris, on the D2 crew. "The Germans had requested a medical doctor mission specialist," recalled Ross, "and since I wasn't one, they were certainly hoping that the next one would be. While I didn't personally think that a medical doctor was mandatory, I did think that it was not a bad idea, because probably over 50 percent of the work we were going to do was life sciences."

Bernard Anthony Harris Jr was born in Temple, Texas, on 26 June 1956, and became only the sixth African-American to be selected as an astronaut candidate when he entered NASA's ranks in January 1990. In his later career, he also became the first black spacewalker. Harris completed high school in San Antonio, Texas, from which he emerged with a strong interest in the sciences. He entered the University of Houston and earned a degree in biology in 1978, followed by a medical doctorate from the School of Medicine of the Texas Tech University Health Sciences Center in 1982. After an internal medicine residency at the Mayo Clinic, he completed a National Research Council Fellowship at NASA's Ames Research Center in 1987 – with an emphasis upon musculature physiology and disuse – and trained as a flight surgeon at Brooks Air Force Base's Aerospace School of Medicine in San Antonio the following year. His fellowship at Ames led to a position at the Johnson Space Center as a clinical scientist and flight surgeon, in which post Harris worked on space adaptation syndrome countermeasures. When he completed his astronaut candidate training in the summer of 1991, Harris was one of the first three members of his class, alongside Rich Clifford and Susan Helms, to draw a Shuttle flight assignment.

The two German payload specialists for Spacelab-D2 had been selected alongside three other candidates by the Deutsche Zentrum für Luft- und Raumfahrt (DLR, the German Aerospace Centre) in August 1987. After basic training, Ulrich Hans Walter and Hans Wilhelm Schlegel, together with physicist Gerhard Thiele and meteorologist Renate Brümmer, commenced preparations for Spacelab-D2 in October 1990. Walter came from Iserlohn, Germany, where he was born on 9 February 1954. After completing secondary education and two years of service in the German Federal Armed Forces (the *Bundeswehr*), he entered the University of Cologne to study physics. Walter received his diploma in experimental physics in 1980 and his doctorate in 1985, with a focus on solid state physics. He subsequently moved to the United States to accept post-doctoral positions at the Argonne National Laboratory in Chicago and, later, at the University of California at Berkeley. Schlegel, born in Überlingen on 3 August 1951, grew up close to the East German border and remembered that his parents fled to West Germany. He was chosen in 1968 as an American Field Service exchange student and graduated from Lewis Central High School in Iowa. Whilst in the United States, he stayed with the Duchman family and described the experience as offering him the opportunity to be "part of a bigger team".

Upon his return to West Germany, Schlegel studied physics at the University of Aachen and earned his diploma degree in 1979. By this time, he had also served a spell as a *Bundeswehr* paratrooper – reaching the rank of a second lieutenant – and maintained his status as a reservist. For much of the 1980s, he was a member of the academic staff at the University of Aachen, conducting research in the fields of electronic transport properties and the optical properties of semiconductors. Schlegel later married Heike Walpot, the fifth DLR astronaut candidate from the August 1987 selection, but unlike Ulrich Walter his experience as an astronaut continued beyond Spacelab-D2. In August 1998, he was selected by ESA as a member of its astronaut corps and went on to fly another Shuttle mission which delivered the Columbus module to the International Space Station. For Spacelab-D2, Schlegel described the roles of himself and Walter as "the lengthened arm" of scientists on the ground.

The Extended Duration Orbiter pallet, which might have extended Spacelab-D2's time in orbit to as long as two weeks, was not carried on STS-55, although according to Steve Nagel the Germans *were* offered the option. They turned it down, in favour of flying an external experiment 'rack' at the rear end of the payload bay. That rack was the Unique Support Structure (USS), which carried a battery of astronomy, atmospheric physics and materials science investigations. One of these was a reflight of the Modular Optoelectronic Multispectral Scanner (MOMS-2), which had been extensively modified since its first mission on Spacelab-D1. It was capable of acquiring high-resolution imagery of Earth's surface for remote-sensing purposes. Although the relatively short duration of STS-55 limited its full potential, a subsequent variant flew to the Russian Mir space station in 1996 and provided better data coverage over long periods of time.

MOMS' multispectral capabilities enabled geologists to better discriminate between different classes of vegetation and rock or soil surface cover. Columbia's 28.45-degree orbital inclination, coupled with less-than-ideal cloud cover and solar illumination, initially restricted observations to the Americas and parts of Africa. However, the 'daytime' ground track drifted westward later in the flight and images were also obtained of the Near East, India, Southeast Asia, Indonesia, Australia and the islands of the Pacific. NASA's Earth Observations Laboratory at JSC also pushed up the instrument's scientific yield by providing real-time information on individual targets, using data from geostationary and polar-orbiting satellites. This allowed MOMS-2 to take more high-resolution pictures over cloud-free regions. Among the notable successes on Spacelab-D2 were images of irrigation ditches, rice crops and roads in Vietnam, as well as underwater reef structures off the Egyptian coastline and high population centres. 'Outgassing' from Mexico's Colima volcano also revealed reddish and pinkish hues, distinguishable from the vegetation of nearby fields.

Although Colima had not erupted since 1913, the MOMS-2 imagery suggested that it had experienced a number of smaller, 'quieter' upheavals since then. Columbia's crew supplemented the MOMS-2 data with their own photographs, taken with hand-held Hasselblad and Linhof large-format cameras, which also featured the redistribution of ash from Lascar Volcano in the Chilean Altiplano. This volcano had erupted on 20 April 1993, only days before STS-55 launched, and the photographs clearly highlighted several plumes of wind-blown material. The chance to photograph a recently-erupted volcano was serendipitously presented to the crew, who were supposed to have been in orbit a full two months earlier, but were kept on the ground by a string of mechanical problems and a dramatic on-the-pad shutdown of Columbia's three main engines on 22 March.

When the flagship orbiter returned from her previous mission, STS-52, in November 1992, engineers dove into preparing for Spacelab-D2, then scheduled for late February of the following year. All went well for most of the processing flow. The experiments had already arrived in Florida and after a slight delay in latching the Shuttle's payload bay doors, the Spacelab-D2 module and USS pallet – together weighing a hefty 11,340 kg – was installed and the vehicle transported to Pad 39A on 7 February.

It was at this stage that Columbia's problems really began. Concerns had arisen in late January that the three main engines contained obsolete versions of tip-seal retainers on the blades of their high-pressure liquid oxygen turbopumps. These seals served to minimise the flow of gas around the tips of the turbine blades and, in so doing, enhanced their overall performance. Each seal was held in place by its own 'retainer'. The uncertainty was that NASA could not conclusively determine from the STS-55 pre-flight test data and paperwork whether the retainers were of the new or old variety. If they were old ones, they needed to be checked before each flight, but if they were new ones they could fly several times without inspection. To err on the side of caution, Columbia's main engines were removed and their retainers were examined. Fortunately, the retainers *were* of the newer type. By the time that this work had concluded, and the engines were back in place, it was the end of February and NASA was obliged to postpone the launch until "no earlier than" 14 March.

Despite ridicule heaped upon the episode by the media, both the space agency and the engine builder, Rocketdyne, stood by the decision on safety grounds. To accommodate the three-week delay, one of the USS-mounted experiments – the Galactic Ultra-Wide-Angle Schmidt System (GAUSS) – needed to have its camera film changed and two other payloads needed to have their batteries recharged. Then, on 2 March, during a flight readiness test, a 3 cm flex hose in Columbia's aft compartment burst and spilt several litres of hydraulic fluid. Although the hose was capped within seconds, it was decided to remove and check all 12 hoses and three were replaced. By the 9th, the task was complete.

In the meantime, the Navy, Air Force and the Hughes Space and Communications Company had agreed to postpone the launch of an Atlas rocket carrying a military satellite in order to accommodate another STS-55 launch attempt in late March. General Dynamics, which built the Atlas, surrendered their 'slot' on the Eastern Test Range on the 21st to Columbia and *this* offered a confidence boost to the Germans, who were reportedly paying upwards of a million dollars *per day* to keep their hardware in a state of flight readiness. The mission's $560 million price tag had already raised eyebrows in the Federal Republic and the enormous cost of reunification since the fall of the Berlin Wall in 1989 had imposed restrictions upon the country's space ambitions. By this time, the Spacelab-D3 mission had already been cancelled, in favour of limited participated in a pan-European 'Spacelab-E1', which would ultimately also be lost to budget cuts.

Despite a month of delay, even STS-55's new target date fell victim to problems when another Eastern Test Range reservation – that of a Delta rocket from Cape Canaveral Air Force Station, carrying a military navigation satellite – was scrubbed on 18 March 1993, due to high winds at the launch site. In accordance with range policies, the Air Force was given a second chance on the 21st, which pushed Columbia back to the 22nd. The attempt to launch the mission that day proceeded smoothly, aiming for liftoff at 9:51 am. All seven astronauts were suited and transported to the pad, took their seats aboard the orbiter and marched smartly through their checks. With 31 seconds remaining on the countdown clock, as planned, Columbia's computers assumed primary control.

Six and a half seconds before liftoff, the ignition of the three main engines got underway with a low-pitched rumble. Within three seconds, dramatically, and accompanied by gasps from the assembled spectators and dignitaries, all three abruptly shut down. Subsequent analysis would point to the incomplete start of the No. 3 engine; a liquid oxygen preburner check valve had suffered a leak in the final seconds, causing the purge system to be pressurised above its maximum allowable limit of 50 psi. The culprit: a tiny fragment of rubber, trapped in an engine propellant valve ...

With unburnt hydrogen still lingering underneath the hot engine bells, the risk of a fire or explosion was very real. Seconds after the shutdown, the 'white room' was automatically moved into position, alongside Columbia's crew access hatch, preparatory to an evacuation. Although the astronauts had trained to escape from a pad abort and slide to a fortified bunker, the danger of having them evacuate Columbia and run through an invisible hydrogen holocaust prompted controllers to instruct them to remain aboard the orbiter. For 40 tense minutes, Steve Nagel and his men remained strapped into their seats as ground personnel deactivated electronic components. Paramedics were rushed to the pad as a precautionary measure, but for the astronauts all was well – albeit eerily *still* – within the swaying cabin. "I'd convinced myself," Nagel said later, "and all my crew that we were going to fly." When the brief noise of the engines died, he called over the intercom to announce the abort to his crewmates. "I wouldn't call it *fear*," he said later about those adrenaline-charged moments. "There's a couple of moments wondering what's happened, because all you see on-board are red lights, indicating an engine shutdown. You know the computers shut down the engines, but you don't know why or exactly what went wrong with them."

The situation was equally tense in the Launch Control Center, adjacent to the Vehicle Assembly Building, from where ashen-faced Launch Director Bob Sieck watched the proceedings. "Your initial reaction," he said in the aftermath of the abort, "is to make sure there are no fuel leaks or that there's nothing that's broken that's causing a hazardous situation. Really, it was one of those nice, boring countdowns ... until the last few seconds. What *did* work, and worked very well, were the safety systems on-board. As a result, the crew is safe and the vehicle is on the pad and safe as well."

Nonetheless, the STS-55 crew emerged from the orbiter looking visibly shaken by the incident. It was becoming increasingly likely that a delay of at least several weeks would be needed to remove, replace and test the engines and on 30 March NASA announced that the next Shuttle mission – STS-56, a time-critical flight, scheduled for early April with the ATLAS-2 payload of atmospheric science and solar physics experiments – would fly next. At the time of the abort, Columbia's sister Discovery had been sitting on adjacent Pad 39B for a week and preparations for STS-56 were proceeding smoothly. The need to get ATLAS-2 into orbit in early April, almost exactly a year after the voyage of ATLAS-1, provided scientists with an ideal opportunity to "observe changes in Earth's ozone during the seasonal transition between spring and summer in the northern hemisphere", according to Space Shuttle Program Director Tom Utsman. In the meantime, Columbia's engines were replaced with a set earmarked for the next mission of Endeavour.

Provisionally scheduled to fly on the night of 6-7 April, STS-56 was itself delayed after an otherwise uneventful countdown. During the pre-planned 'hold' at T-9 minutes, discussions about a higher-than-expected temperature on the anti-flood valve of Discovery's No. 1 main engine had not been completed and the resumption of the count was postponed by an hour. Eventually, at 1:23 am EDT, controllers restarted the clock. At T-31 seconds, a 'Go' for autosequence start was issued and the orbiter's computers assumed primary control of the vehicle's critical functions. The safe and arm devices of the Solid Rocket Boosters were armed and the five STS-56 astronauts – Ken Cameron, Steve Oswald, Mike Foale, Ken Cockrell and Ellen Ochoa – braced themselves for the roar of Main Engine Start. All at once, at T-11 seconds, the clock stopped and the attempt was called off.

Subsequent investigation revealed that the hydrogen high-point bleed valve's 'closed' indication was apparently not present at T-21 seconds. Although its absence was later verified to be an instrumentation glitch, and not an actual failure, it created a breach of Launch Commit Criteria and mandated a scrub. In the wake of the event, a leak check of the valve turned up nothing amiss and NASA managers felt sufficiently confident to reschedule the launch for two days later on 8 April. The second attempt was successful and Discovery roared into the dark Florida sky at 11:29 pm EDT.

Several hours later, the payload bay doors were opened to reveal STS-56's twin primary cargoes: the ATLAS-2 laboratory and a unique retrievable satellite which the finest acronym-makers within NASA had carved into the name 'Spartan'. The 'Shuttle Pointed Autonomous Research Tool for Astronomy' was a 1,000 kg cube-shaped facility, designed to accommodate instrumentation for astrophysics or solar physics research. Built by the Goddard Space Flight Center of Greenbelt, Maryland, at the relatively inexpensive price of $3.5 million, it measured 3.2 × 1.07 × 1.22 m and was designed to be reused and flown at intervals of approximately six to nine months. Data from its instruments would be stored on tape recorders and pointing and stabilisation achieved through a three-axis attitude control mechanism. Deployed and retrieved by means of the Shuttle's RMS arm, Spartan first flew on Mission 51G in June 1985 with a battery of high-energy astrophysics detectors and was headed for orbit a second time aboard Challenger's ill-fated final flight in January 1986 to perform ultraviolet observations of Halley's Comet.

A replacement for the lost Spartan was constructed in the wake of Challenger and the carrier flew several times in the 1990s. On STS-56, it would mark the first voyage of the so-called 'Spartan-201' experiment package, with a focus upon studies of the Sun. This comprised two instruments: an Ultraviolet Coronal Spectrometer, built by the Smithsonian Astrophysical Observatory of Harvard University, to determine the velocities of coronal plasma within the 'solar wind' and a White Light Coronagraph, assembled by the High Altitude Observatory of the National Center for Atmospheric Research, to measure visible light in order to determine the temperatures and densities of coronal electrons. Once deployed from the Shuttle, the Spartan undertook a pre-programmed mission; it lacked command and control capability and was effectively left to undertake its tasks, retrieved by the Shuttle after two days and returned to Earth for the recovery of data and refurbishment in readiness for its next flight. In total, Spartan-201 flew five times between April 1993 and November 1998 and its spacecraft 'bus' is today displayed in the National Air & Space Museum in Washington, DC.

Deployment of Spartan-201 occurred on the third day of the STS-56 mission. In preparation for this event, three burns of Discovery's RCS thrusters were executed and the payload was released by the RMS arm at 12:11 pm EDT on 11 April. Shortly thereafter, the astronauts manoeuvred the Shuttle to a separation distance of about 275 km 'behind' Spartan, allowing the satellite to commence its mission. Two days later, and about four hours ahead of the scheduled retrieval time, Discovery drew closer to Spartan, passing directly 'beneath' it and taking up a position about a hundred metres away, ready for capture. The Spartan was grappled by the RMS at 1:20 pm EDT on 13 April. A little more than half an hour later, the satellite was secured onto its berth in the payload bay.

Due to the Shuttle manoeuvres and their impact on the proper pointing of the pallet-mounted instruments, ATLAS-2 operations were temporarily suspended during the Spartan deployment and recovery efforts. A pair of laser-ranging and range-rate devices were employed during deployment and retrieval. One of these was used for short-range tracking, whilst the second – known as the Mini Eyesafe Laser Infrared Observation Set (MELIOS) – supported longer-range tracking. Overall, their combined performance was described as "good" and "satisfactory", in spite of low battery voltage in MELIOS which possibly triggered inaccurate distance readings during the deployment. The STS-56 crew later admitted that their workload during this process was exceptionally heavy and as a result they had given a low priority to making distance and range-rate calls.

Although the flight of Spartan-201 was perhaps the most visually impressive spectacle of the mission, Discovery's primary payload was a reflight of the ATLAS package, first carried a year earlier on STS-45. On this occasion, however, there were a number of differences, for 'ATLAS-2' comprised just a single Spacelab pallet, rather than two, carrying only a handful of the original instruments. Moreover, no payload specialists were required aboard ATLAS-2 – or, in fact, its successor, ATLAS-3, in November 1994 – and science operations were handled principally by Mike Foale on the red team and by Ellen Ochoa on the blue team. Foale, of course, had an extensive history with the ATLAS programme, whilst Ochoa was making her first flight.

3-20: Spartan flies freely, high above Crete, in the minutes after deployment.

Ellen Lauri Ochoa went on to great heights in NASA leadership later in her career, rising from the astronaut corps to become deputy chief of the astronaut office, head of Flight Crew Operations and deputy director of the Johnson Space Center. In November 2012, she became only the second woman to assume the directorship of the storied Houston facility. Ochoa came from Los Angeles, California, where she was born on 10 May 1958, and would later attribute great credit to her mother, Rosanne, "partly because of the way she raised me and my four siblings," she said in a NASA oral history, "and partly because of her love of learning". Rosanne attended college on a part-time basis for two decades and finally graduated a couple of years after her future-astronaut daughter. As a student, Ochoa excelled in science, mathematics and engineering, although all three were typically excluded to women at the time, and graduated at the top of her class with a degree in physics from San Diego State University in 1980. A year later, she gained a master's credential in electrical engineering from Stanford and was working towards her doctorate when she learned that several friends had applied for NASA's astronaut corps.

It was a turning point in Ochoa's professional life. Following the completion of her doctorate in 1985, she worked at Sandia National Laboratories and served as head of the Intelligent Systems Technology Branch at NASA's Ames Research Center, investigating optical hardware for information processing. During this time Ochoa was a co-inventor on three patents for an optical inspection system, an optical object-recognition method and a technique for noise removal in imaging. An unsuccessful application for NASA's 12th astronaut class in 1987 was followed by success in January 1990. As the nation's first Hispanic female astronaut, Ochoa was also an accomplished flutist – in fact, this was one possible career option that she pursued in her youth – and she would carry her flute into orbit on STS-56 and perform for her crewmates.

Three of those crewmates had flown before, whilst Kenneth Dale Cockrell was, like Ochoa, a rookie astronaut. A native of Austin, Texas, he was born on 9 April 1950 and went on to fly five Shuttle missions and served for a year as chief of the astronaut office. His eyes were focused upon the skies from a very young age – "I remember at age five, seeing an airplane fly over our house ... in Pittsburgh," he told a NASA interviewer, "and this airplane, for some reason, struck a chord in me" – and created a dream of aviation. Much of Cockrell's subsequent schooling was self-directed toward the goal of joining the military as a pilot: he earned a degree in mechanical engineering from the University of Texas at Austin in 1972 and was commissioned into the Navy later that year. Initial flight instruction at Naval Air Station Pensacola in Florida was followed by an assignment to the carrier USS *Midway*, during which time he flew the A-7 Corsair in the Western Pacific and Indian Oceans. He also gained a master's degree in aeronautical systems from the University of West Florida in 1974. Cockrell was selected for test pilot school in 1978 and after graduation he was based at the Naval Air Test Center for the next four years, working on a variety of aircraft. Subsequent assignments included officer positions on the staff of the commander of the USS *Ranger* and the USS *Kitty Hawk* Battle Group. He then returned to operational duty in an F/A-18 Hornet squadron and completed two cruises on the USS *Constellation* in 1985-87.

Dovetailed into this naval career, Cockrell had a keen eye on NASA's astronaut programme and the arrival of the Shuttle – a unique *winged* spacecraft – really inspired him. Years later, he joked that "I think I'm tied for the record for the number of times applying and to get the job". He interviewed alongside Steve Oswald and Franklin Chang-Díaz. "It's kind of interesting," said Cockrell, who flew into space with both Oswald and Chang-Díaz, "that Franklin and I both got off the same airplane to attend our first interview together. We arrived in Houston and were standing in the terminal building for the car from the hotel to pick us up and struck up a conversation, realised what we were both here to do, and *that* was in 1979!" Although Chang-Díaz made the cut the following year, and Oswald in 1985, Cockrell was unsuccessful. He failed to secure an interview in 1984 and was interviewed in 1985 *and* 1987, both without success, but eventually made the cut in 1990. Perseverance definitely paid off and in his subsequent roles at NASA, including headship of the astronaut corps in 1997-98, Cockrell advised potential applicants to "pick a discipline ... that interests you a lot" and regarded the astronaut business as "a secondary profession" for individuals who already excelled in their primary professions.

Cockrell resigned from active naval duty in 1987 and joined the Aircraft Operations Division at NASA's Johnson Space Center as an aerospace engineer and research pilot, performing air-sampling and other high-altitude research. In later life, he would credit his close friend and future STS-56 crewmate Steve Oswald with having helped him secure the Aircraft Operations post and guiding him through the transition from post-Navy to civilian and NASA life. Although selected as a pilot candidate, Cockrell served as a mission specialist on his first Shuttle flight and was in charge of the orbiter's systems during the 12-hour red shift with Mike Foale. Meanwhile, Ken Cameron and Steve Oswald joined Ellen Ochoa on the 12-hour blue shift.

Activation of the ATLAS-2 payload was complete within four hours of reaching orbit. The Spacelab pallet carried six instruments previously flown aboard ATLAS-1: the Atmospheric Trace Molecule Spectroscopy (ATMOS) and Millimetre Wave Atmospheric Sounder (MAS) for atmospheric research and the Measurement of Solar Spectrum (SOLSPEC), Solar Ultraviolet Spectral Irradiance Monitor (SUSIM), Active Cavity Radiometer (ACR) and Measurement of Solar Constant (SOLCON) for solar physics. In conjunction with ATLAS-2, the Shuttle Solar Backscatter Ultraviolet (SSBUV) instrument was mounted on Discovery's payload bay wall. A new instrument, the Shuttle Ultraviolet Experiment (SUVE), was housed in a Getaway Special canister and sought to observe extreme ultraviolet solar radiation and its effect upon Earth's ionosphere. Developed by students at the Colorado Space Grant Consortium, SUVE involved 14 colleges and universities, funded by NASA. In the following months, the data from these instruments bore fruit. Of particular note was ATMOS, which suffered transmitting problems, yet still managed to capture 103 orbital sunrises and sunsets and measured up to 40 gases influencing global ozone levels. More than six hours of data were recorded on the instrument's new recorder, which had a 44-gigabyte storage capacity.

Discovery's night launch enabled ATLAS-2 to exploit a unique opportunity to make detailed measurements of the stratosphere over the Arctic. In total, during the nine-day flight, atmospheric data was gathered over 94 percent of the planet. Specifically, the ATMOS investigation, which was focused predominantly on the southern hemisphere during the ATLAS-1 mission, was now directed to observe the northern hemisphere. Although a problem was experienced with the instrument's high-data-rate transmission, a new downlink format was employed to prevent any science loss. Meanwhile, the MAS investigators faced their own trials when their experiment suffered a glitch with its pointing system; this was eventually rectified through real-time software modifications. In the solar sciences, SOLSPEC – controlled remotely from the Institut Royal Meteorologique de Belgique in Brussels – collected data to correlate with a similar experiment aboard the EURECA freeflyer, whilst SUSIM worked in conjunction with a near-identical instrument aboard the Upper Atmosphere Research Satellite to comprehensively analyse ultraviolet radiation from the Sun. Lastly, ACR met both of its pre-flight goals to make correlative measurements with UARS and to perform the maximum possible number of solar constant measurements and SOLCON (also controlled from the Brussels command centre) showed close agreement between its data and that from the EURECA instruments. Although not strictly a member of the ATLAS-2 payload, SSBUV formed an integral part in the mission's scientific objectives, with its data suggesting a marked decrease in atmospheric ozone levels at northern mid-latitudes and performing the first observations of tropospheric sulphur dioxide concentrations over urban and industrial areas of the eastern United States, Europe and eastern Asia.

Several other investigations within the crew cabin were reflights of previous experiments: the military HERCULES observation system, the RME and CREAM radiation-monitoring devices, the use of Discovery as an electro-optical target for the Air Force's tracking site on the Hawaiian island of Maui and the Shuttle Amateur Radio Experiment. (The latter was employed to make several contacts in the United States, the United Kingdom, Portugal, South Africa, Australia – and, uniquely, on 10 April, the first Shuttle amateur radio contact with cosmonauts aboard the Russian Mir space station.) One new payload was the Commercial Materials Dispersion Apparatus Instrumentation Technology Associates – known collectively by the acronym of 'CMIX' – which explored the usefulness of the microgravity environment for research in drug development, biotechnology, basic cell biology, protein and inorganic crystal growth, bone and invertebrate development, immune-system deficiencies, manufacturing processes and fluid physics. Four small (brick-sized) laboratories were housed in a middeck refrigerator/incubator and their contents included mouse bone marrow cells, brine shrimp, fish eggs, mustard seeds and mushroom spores. A set of rats were also flown as part of a rodent investigation.

Originally scheduled to return to Earth on 16 April, after eight days, the STS-56 crew were waved off by 24 hours, due to unacceptable weather conditions at the Shuttle Landing Facility. Discovery's payload bay doors were reopened and some additional ATLAS-2 research was undertaken in the unanticipated additional day. Finally, at 4:34 pm EDT on the 17th, Cameron and Oswald fired the OMS engines for three and a half minutes to drop their craft out of orbit and set it on course for a touchdown, an hour later, on the opposite side of the planet, in Florida. With pinpoint grace and precision, Cameron touched down on Runway 33 at 5:37 pm, completing a voyage of slightly more than nine days.

As her sister ship returned home, Columbia had returned to Pad 39A, for another launch attempt of the long-delayed STS-55 mission with Spacelab-D2. In the month since the pad abort, several time-critical experiments in the laboratory were removed and replaced and a new liftoff date of 24 April – just seven days after Discovery's return – was expected to set a new landing-to-launch record for the Shuttle programme. (The previous record at that time was ten days, set in the pre-Challenger era.) "From a programme standpoint, we have no concerns" about the short interval, said former astronaut Brewster Shaw, chair of NASA's Mission Management Team. "If somebody is not ready, we will wait until they are, but if everyone is comfortable with flying, then we will go fly."

Preparing for his mission, Steve Nagel received some advice from fellow astronaut Jim Wetherbee, who commanded Columbia's previous mission, STS-52. He jokingly told Nagel that when he turned the ignition key, "you're supposed to jiggle it a bit to get it started", but humour aside, fate still had one more card to play before STS-55 could fly. Nine hours before liftoff on the 24th, an intermittent power problem was detected by technicians in one of the orbiter's inertial measurement units. The device was replaced and launch was rescheduled for the 26th. It was fortuitous that the discovery was made, for the failure of a similar device on STS-44 had mandated cutting that mission short. "We're disappointed," admitted Rudolph Teuwsen, a spokesman for the German space agency, "but it would have been *far* worse to have launched and then come back without completing the science."

Two days later, at 10:50 am EDT, Columbia thundered perfectly in orbit, setting a new record for only nine days between a Shuttle landing and a subsequent launch. Shortly after reaching orbit, the astronauts also found that there was not only *one* Jerry Ross aboard the mission ... but *two*. Steve Nagel had gotten hold of the pictures he took of Ross grinning through Atlantis' aft flight deck windows during his EVA on STS-37 and had them blown up in size. Nagel arranged for the picture to be fitted into one of Columbia's aft flight deck windows, before that was covered by a protective panel for ascent. Nagel knew that Ross – seated at the rear of the flight deck for the climb to orbit – would be responsible for removing the protective panels and would hopefully get a kick out of it.

When Ross removed the panel and saw his grinning face staring back at him, he burst out laughing. Nagel, it seems, had momentarily forgotten his prank, as he and crewmates Tom Henricks – a veteran of STS-44 – and Charles Joseph Precourt busied themselves with readying Columbia for orbital operations. Like Ken Cockrell, recently returned from STS-56, Precourt had been selected by NASA in January 1990 as a pilot candidate, but owing to the dual-shift nature of Spacelab-D2 had been assigned his first Shuttle flight as a mission specialist. Interestingly, both Cockrell and Precourt went on to serve as head of the astronaut corps later in their respective careers. Born in Waltham, Massachusetts, on 29 June 1955, Precourt attended high school in his hometown of Hudson and graduated with distinction in aeronautical engineering from the Air Force Academy. Whilst there, in 1976, he participated as an exchange student at the French Air Force Academy.

Undergraduate pilot training at Reese Air Force Base in Texas was followed by roles as a T-37 and T-38 instructor and maintenance test pilot. Precourt flew operationally in the F-15 Eagle at Bitburg Air Base in Germany in 1982-84 and completed test pilot school at Edwards Air Force Base in California a year later. He received awards as both the most outstanding undergraduate pilot and the most outstanding test pilot instructor. Precourt flew test missions in the F-15E, F-4 Phantom, A-7 Corsair and A-37 Dragonfly. In 1988 he gained a master's degree in engineering management from Golden Gate University and had just completed another master's credential in national security affairs and strategic studies at the Naval War College when he was selected as an astronaut candidate in January 1990.

As STS-55's flight engineer, Precourt was in charge of orbiter systems on the all-rookie blue team, which comprised himself, Bernard Harris and Hans Schlegel. Meanwhile, the red team of Nagel, Henricks, Ross and Ulrich Walter took the lead on activating Spacelab-D2 and its experiment payloads within the first few hours of achieving orbit. Although the seven men worked in alternate shifts, Ross admitted that "there *was* a Crew C ... the Shuttle crew as a whole" and as payload commander he saw to it that time existed at changeover intervals for them to gather and discuss the experiments, any problems, summaries of flight notes. "I think a couple of times during the flight I gave little pep talks," he said. "I also made sure they'd go to sleep on time, so they'd get the right amount of rest, even though I *wasn't*. I probably didn't get more than five hours sleep a night for the whole time that we were up there and when I got back on the ground I was just flat wiped out." The phone-box-sized sleep stations were like coffins, according to Ross, and any chance to sleep fitfully was lost in the incessant noise from the middeck: other people eating and working, tapping and banging around. Ross – by his own admission a "kinda hyper" personality – found that his brain moved constantly at what seemed to be a million miles per hour.

Had Challenger not been lost in January 1986, it is likely that Spacelab-D2 would have flown sometime in 1988-89, but the resultant down time in the wake of the disaster unavoidably pushed it into the early part of the following decade. When chief astronaut Dan Brandenstein approached Ross in early 1991 to serve as payload commander for the mission, there was a moment's hesitation.

"Dan," said Ross, "you want a *scientist* for that. You don't want me, an old engineer."

"No, we want you," replied Brandenstein. "We want somebody that will work well with the Germans and get the flight pulled together." *Pulling together* the flight was easier said than done, for Spacelab-D2's multitude of experiments imposed severe timing constraints and Ross admitted before launch that simply keeping up with the timeline would pose a challenge. The official announcement of Ross as payload commander came only two weeks after his return from STS-37. "I didn't do a lot of the normal post-flight activities," he remembered, "because I went to Germany probably within three weeks of landing to start working on STS-55."

In many ways, Spacelab-D2 mirrored the research undertaken by Spacelab-D1, with investigations into the life sciences, biological sciences, materials and fluid sciences, technology and – mounted onto the Unique Support Structure at the rear of Columbia's payload bay – astronomy and Earth observations. In addition to the German and US involvement, several experiments came from the European Space Agency, together with French and Japanese investigators. Many of the materials science experiments were housed in three custom-built facilities: the Materials Science Experiment Double Rack for Experiment Modules and Apparatus (MEDEA), Werkstofflabor and the Holographic Optical Laboratory (HOLOP). Others, which required direct exposure to the vacuum of space, were attached to the USS pallet in the Materials Science Autonomous Payload (MAUS) and the Atomic Oxygen Exposure Tray (AOET). The two MAUS experiments investigated complex boiling processes and the diffusion phenomenon of gas bubbles in salt melts, whilst AOET exposed more than a hundred material specimens to the harsh atomic oxygen environment to obtain data on reaction rates, as part of efforts to develop materials for Europe's Columbus space station laboratory.

Within the Spacelab module, three separate furnaces in MEDEA supported long-duration crystallisation studies, high-temperature directional solidification runs and carefully controlled the temperatures of metallic specimens with a high-precision thermostat. At one stage, the largest gallium arsenide crystal ever grown in space was produced inside MEDEA, measuring 20 mm in diameter. On Earth, this material has seen frequent applications for electronic components, including light-emitting diodes, semiconducting lasers, photo-detectors and high-speed switching circuits. The second materials facility, Werkstofflabor, housed a quarter of Spacelab-D2's 88 experiments and successfully produced a 'monotectic' alloy of bismuth-aluminium. Elsewhere, crystals of materials which could be used to fashion stronger turbine blades were yielded. HOLOP employed the technique of 'holography', using laser light to explore and make more easily visible the physical processes of heat and mass transfer and cooling in transparent materials. More than 600 telescience commands were transmitted to HOLOP from the Microgravity Life Support Centre at Cologne-Porz during the ten-day mission.

To emphasise Spacelab-D2's German flavour, the Payload Operations Control Centre had shifted – as it had on Spacelab-D1 – from Houston to Oberpfaffenhofen, near Munich. During the first mission, several functions still had to be controlled from the United States, because Oberpfaffenhofen's data-transmission facilities were insufficient to handle all of the communications traffic from the Shuttle. Eight years later, however, the situation had changed and satellite-transmitted data was received by ground stations on German soil and forwarded to the control centre.

Also inside the Spacelab module were three painful-sounding facilities: Anthrorack, Biolabor and Baroreflex. Televised images from various stages of the mission offered the distinct impression that these facilities formed a modern-day torture chamber, with the four members of the science crew – Ross, Harris, Schlegel and Walter – being subjected to saline injections and blood draws. Due to the need for them to be in peak condition throughout the flight to support a return to Earth, the orbiter crew of Nagel, Henricks and Precourt were immune to these punishing tests.

'Anthrorack' supported nearly two dozen investigations and would perform the first-ever comprehensive, integrated screening of the human body in orbit. Simultaneous measurements were taken of their respiratory, cardiovascular and endocrine systems and an ultrasound system was demonstrated. Two particular studies were NASA's Baroreflex Experiment and Cardiovascular Regulation Experiment, which focused upon the relationship between post-flight cardiovascular 'deconditioning' and the baroreflex, which is responsible for maintaining appropriate blood pressure in the body. Such deconditioning was typically accompanied by a 'drop' in blood pressure when astronauts stood up after landing and was closely associated with the decreased workload of the heart in space. The experiments evaluated the theory that light-headedness and a reduction in blood pressure after a mission may be due to the fact that the baroreflex's regulatory ability is reduced after microgravity exposure. To perform the experiments, the astronauts wore a silicon-rubber neck cuff, through which pulses of pressure and suction were transmitted to baroreceptors, whilst heart-rate changes were monitored.

All seven men participated in the relatively new technique of 'saline loading', by drinking heavily salted water in order to rapidly expand their intravascular volumes and enabling medical scientists to better understand the mechanisms responsible for post-flight deconditioning and orthostatic intolerance. "I think we will find that space adaptation syndrome comes in phases," said Bernard Harris, the only physician member of the STS-55 crew. He would later conclude that in his judgement acclimatisation to the new environment occurred in at least three stages, spanning undetermined lengths of time, and typically began with dizziness and nausea and gradually moved to a loss of blood volume and a decrease in bone and muscle mass.

With Space Station Freedom – later to become the 'International Space Station' – steadily rising from the drawing boards and into a full international partnership, several Spacelab-D2 experiments focused upon new robotics and telescience systems. Of these, Tom Henricks operated the Crew Telesupport Experiment, which successfully achieved a two-way communications link with Mission Control ... using a device which looked uncannily similar to a child's Etch-a-Sketch toy. The other experiment positioned a series of accelerometers around the Spacelab module to examine the influence of crew motions on sensitive payloads. However, the 'toy' which really attracted media interest was the Robotic Technology Experiment (ROTEX), a six-jointed robot arm, fitted with tactile and torque sensors, laser range-finders and stereo and fixed cameras. During the flight, it was used to build a small tower of cubes and retrieve, connect and disconnect an electrical plug as part of demonstrations of its autonomous capability. Operating within its own sealed work area, ROTEX was used by the astronauts and – for the first time – was controlled via telescience from the ground.

Aside from a few minor problems, including a leak from the waste water tank, the flight proceeded exceptionally smoothly. An overheating experiment freezer on the middeck also posed difficulties, prompting the crew to switch their biological samples to a backup freezer, but this gave Ross cause for concern. "The freezers we were carrying ... were failing at a fairly rapid clip," he told the NASA oral historian. "It became apparent to me that probably 40-50 percent of the science was counting on that freezer working and bringing back those samples that we collected over the ten days on-orbit." On Ross' recommendation, a backup freezer was carried – forfeiting two middeck lockers in doing so – and this proved highly fortunate when the *prime* freezer failed a couple of days after launch.

Years later, Steve Nagel believed that the freezer glitch convinced NASA to bring Columbia home on 6 May, lest the backup also fail. Since the early stages of the mission, US and German mission managers had sought to conserve as much electrical power as possible to create an 'extra' day and easily accumulated 25 hours' worth of additional margin in the Shuttle's cryogenic fuel cells. On the morning of 3 May, the crew was given a go-ahead to extend their mission from nine to ten days. Early on the morning of the 6th, the red team took charge of the Spacelab deactivation and the astronauts bade the German control teams "aufwiedersehen", before shutting the hatch and returning to the middeck to prepare for re-entry. The de-orbit burn was scheduled for 8:00 am EDT, but was called off due to unacceptable weather at the Kennedy Space Center. The astronauts were waved-off for one orbit and, when conditions failed to improve, were diverted to Edwards Air Force Base in California.

Nagel was convinced that Columbia could have remained aloft for at least another 24 hours, but the importance of Spacelab-D2 as a science mission and the presence of only one freezer for all of the biological samples was troublesome. "The Germans were hanging it all on one freezer to save these samples," he said, "and I'm sure that's why, when it came time to land, we were weathered out of Florida and went to Edwards." As a consequence, the de-orbit burn was successfully completed at 9:29 am EDT (6:29 am in California) and Columbia swept perfectly to a landing, precisely an hour later. The mission had lasted a little under ten days, but had scored a remarkable achievement for the reusable fleet: across its first 55 voyages, the Shuttle had spent a cumulative 365 days aloft. By the end of the Shuttle era in July 2011, this figure had more than quadrupled to 1,300 days over 135 missions.

"WE DIDN'T HAVE A CLUE"

Early in 1993, an unusual cargo was loaded aboard Space Shuttle Endeavour. Known as 'Spacehab', it was not dissimilar in appearance to the European-built Spacelab modules, but with several key differences: it was smaller, consuming no more than a quarter of the payload bay, and it was designed and built not by governments, but by private enterprise. It also differed from Spacelab in overall physical shape; it was cylindrical, but with a flat roof, and this shape was exploited by the crew of the first Spacehab mission – STS-57, commanded by veteran astronaut Ron Grabe – and commemorated in their official patch. When the module rose from Earth for the first time on 21 June 1993, it marked the culmination and realisation of a decade-long dream for aerospace engineer Bob Citron, who founded the Spacehab company in 1983 and incorporated it the following year.

"People often ask me why I started Spacehab," Citron recalled on the website www.astrotech.com, "and my response usually goes something like this: It took a small group of wide-eyed dreamers and determined space enthusiasts who believed we could pull it off. We didn't have a clue about the enormous problems we would encounter and the nearly insurmountable technical, financial and institutional roadblocks that would stand in our way. Nobody had done anything like this before." The primary goal of Citron – who died from prostate cancer in January 2012 – was to create the world's first privately funded company to support human space missions, using the cavernous payload bay of the Shuttle as a carrier of commercial pressurised research modules.

The need for such provision was self-evident. In the Shuttle's pre-Challenger heyday, more than a dozen missions were planned each year and a primary thrust of the Reagan Administration's 1983 space policy was for the commercial exploitation of the microgravity environment. Already, middeck lockers were being used to carry out experiments in crystal growth and pharmaceutical development, but the limited volume meant that their commercial viability was restricted. The Spacehab module – accessed, like Spacelab, by a tunnel adaptor connected to the middeck airlock hatch – measured 2.8 m long, 3.4 m high and 4.1 m across the width of the payload bay and could increase the Shuttle's pressurised research volume by 31 m^3, effectively quadrupling the available working and storage space. The module, which weighed approximately 4,300 kg and could house a total payload of more than 1,300 kg, provided environmental provisions, electrical power, temperature control and experiment command and data functions. It could carry a maximum of 61 middeck-sized lockers and up to two large racks, in either 'single' or 'double' configurations.

After incorporating Spacehab in 1984, Citron admitted that the company was "on the verge of failure on a number of occasions during its first years", until he brought in critical professional management personnel and "things started to happen" through negotiations with NASA, the Italian Alenia Spazio and German MBB-Erno organisations and Martin Marietta and McDonnell Douglas. By the end of 1985, only weeks before the Challenger tragedy, Spacehab was tentatively scheduled for its first flight in 1987 and a lease of $5 million per mission was quoted by *Flight International*. Early plans called for the assembly of three modules, but centre-of-gravity issues gave NASA cause for concern and threatened to affect the placement of other cargoes in the payload bay. Ultimately, McDonnell Douglas was selected as the lead contractor and a decision was made to build two flight modules and an engineering test model. Other worries lingered, but in late 1986 Spacehab signed a Space System Development Agreement, in which NASA agreed to fly five inaugural missions. By October 1987, 42 firm requests had been received, with lockers priced at $300,000 for non-government users and $100,000 for government agencies and contractors. The concept was now growing from initial designs into reality. In September 1989, Patent No. 4,867,395 for a 'Flat End-Cap Module for Space Transportation System' was awarded to Spacehab, Inc., by the US Patent Office.

Hopes to fly the modules on very early post-Challenger missions received a rude awakening, however, and it was expected to be at least three years after the resumption of Shuttle operations before a first flight could be realistically expected. According to NASA's April 1988 schedule, Spacehab-1 was listed as a primary payload on STS-51 in June 1991, co-manifested with the retrieval of EURECA and the deployment of Italy's LAGEOS-II satellite. This schedule slipped and, at length, the first module was not unveiled until early 1992 at the custom-built Spacehab Payload Processing Facility (SPPF) at the Kennedy Space Center. By this time, under the Commercial Middeck Augmentation Module procurement, initiated in February 1990 and formally signed the following December, NASA had agreed to lease a total of 200 Spacehab lockers at a cost of $184.2 million. Over the coming years, the SPPF would host more than a hundred astronauts and cosmonauts, enabling them to train on real experiment hardware.

The formal unveiling of Spacehab coincided closely with the assignment of the first crew members for STS-57. The names of payload commander David Low and mission specialist Janice Voss were announced in February 1992. (It is a tragic coincidence that, like Bob Citron, both astronauts later succumbed to cancer; Low of colon cancer in March 2008 and Voss of breast cancer in February 2012.) Within weeks of their announcement to STS-57, four more astronauts were named to the crew: Ron Grabe in command, Brian Duffy as pilot and two further mission specialists, Nancy Sherlock and Jeff Wisoff. For Duffy, who was in quarantine at

KSC, a few days ahead of his STS-45 mission at the time, it came as a special surprise. "Jeez, I sure hope I like this," he remembered asking himself, "because I just signed up for *two* of them!" In all seriousness, Duffy appreciated the fact that he would be entering STS-57 training pumped up, fresh and ready to go. At the time, launch was provisionally scheduled for late in April 1993.

Janice Elaine Voss came from South Bend, Indiana, where she was born on 8 October 1956. Space exploration as a real possibility first entered her head whilst in the sixth grade, when she read Madeleine l'Engle's novel *A Wrinkle in Time*; this guided her deeper into science fiction and from thence into science generally. "I just found the whole thing so fascinating," she told the NASA oral historian before her last Shuttle mission, "that I've never been interested in doing anything else." Her love of learning was such that, in 1998, with four Shuttle missions and a handful of degrees to her credit, she once told a group of pre-teen girls that her next personal goal was to learn the piano. In high school, she pursued mathematics and science. Later, she earned a degree in engineering from Purdue, whilst working on a NASA co-operative programme in computer simulations with the Engineering and Development Directorate at the Johnson Space Center. Reflecting on Purdue, Voss was well aware of the fact that the exalted college had produced more astronauts than any other US educational institution, outside of the military academies. After gaining her degree in 1975, she completed a master's qualification in electrical engineering and returned to JSC in 1977 as a crew trainer, teaching re-entry guidance and navigation. Ten years later, she gained her doctorate in aeronautics and astronautics from Massachusetts Institute of Technology and worked for a time for Orbital Sciences. Finally, on her fifth application to NASA, she was accepted as an astronaut candidate in January 1990.

Also selected that year were civilian physicist Peter Jeffrey Kelsay Wisoff and Army aviator Nancy Jane Sherlock. Wisoff – who later married fellow astronaut Tammy Jernigan – was born in Norfolk, Virginia, on 16 August 1958, and recalled growing up as the first men set foot on the Moon. "I remember watching that live on TV and being very excited about it," he told the NASA oral historian, "and always thinking that would be a great thing to do. It wasn't really until the Shuttle programme came along that there were enough seats ... for someone with my background to participate, because I'm not a pilot." Wisoff's background was principally in laser physics. After high school in his hometown, he gained a degree in physics from the University of Virginia in 1980 and then a master's and doctorate in applied physics in 1982 and 1986, both from Stanford. His PhD encompassed laser physics and laser construction technologies. Wisoff next accepted a faculty position in the Department of Electrical and Computer Engineering at Rice University and worked on the development of new vacuum ultraviolet and high-intensity lasers, whose potential applications included the reconstruction of damaged nerves.

As for Nancy Sherlock, her career path toward the astronaut corps could not have been more different. Born on 29 December 1958 in Wilmington, Delaware, she admitted that a career in the space programme – or even in the military services – was unthinkable during her childhood. "Women *weren't* military pilots," she told the NASA oral historian. "Women *weren't* astronauts. It really wasn't a concrete goal of mine until much later in my life." Having said this, she considered it a surprisingly natural progression to enter the military and pursue an education which ultimately opened professional routes which might previously have been closed to her. Only at the end of her time in high school did the military begin to accept female aviators and only whilst she was at Ohio State University, studying the biological sciences, did NASA begin to accept female astronauts. Upon receipt of her degree in 1980, she entered the Army with a dream of becoming an aviator and perhaps eventually a military doctor. However, women were gradually being introduced into combat positions and after rotary-wing pilot training she served as an instructor pilot at the Army Aviation Center. Other positions followed – section leader, platoon leader and brigade flight-standardisation officer – and in 1985 Sherlock earned a master's degree in safety engineering from the University of Southern California.

Eventually, she became a Master Army Aviator and by the end of her military career in 2005 she had logged over 3,900 hours in the air aboard various rotary-wing and fixed-wing aircraft. Years later, she remembered her first day in a helicopter and the instant thought: *This is for me. This is what I want to do.* Her decision to pursue a master's in safety engineering is an interesting side note. "When I was in flight

3-21: STS-57's External Tank drifts away from Endeavour, towards a fiery destruction in the atmosphere, minutes after launch.

school," she explained, "we actually had an accident that killed my instructor pilots and two of the guys that I flew with every day. It was just kind of a strange coincidence that I wasn't in the aircraft. It was at that time that I decided to devote a portion of my career and my academic life to safety engineering." Later, during her NASA life, she went on to earn a doctorate in industrial engineering, with emphasis upon human factors. "Catastrophic things *can* happen," she reflected, sadly, "due to human error in the cockpit or human error combined with a malfunction in the aircraft."

Sherlock's arrival at the Johnson Space Center came in September 1987 as a flight simulation engineer on the Shuttle Training Aircraft. She had earlier applied unsuccessfully for admission into that year's astronaut candidate class. That same year, her daughter Stephanie was born. Three years later she was selected as an astronaut and for the first half of the 1990s until she met her future husband, Sherlock was a single parent. "It was somewhat difficult for a mom to have a lifestyle like that," she said of her early NASA career, "gone an awful lot on travel and studying a lot at home." That study was particularly intense in the months leading up to STS-57, for Sherlock served as the flight engineer, responsible for assisting Grabe and Duffy with monitoring Endeavour's systems during ascent and re-entry. By the end of her career, she had served as flight engineer on four Shuttle missions, more times than any other female astronaut.

All six members of the crew were responsible for aspects of the Spacehab-1 payload, which encompassed a wide range of experiments in the life and microgravity sciences. A dozen of these were sponsored by NASA Centers for the Commercial Development of Space and one by the agency's Langley Research Center of Hampton, Virginia, and many had flown aboard previous missions. The year 1993 was expected to mark a significant watershed for commercial access to space, with a total of 56 research payloads planned aboard Spacehab-1 and, in November, aboard Spacehab-2.

Aboard STS-57, materials science investigations included the Equipment for Controlled Liquid Phase Sintering Experiments (ECLiPSE), developed by the University of Alabama at Huntsville, which sought to explore 'sintering' – a process by which metallic powders are consolidated into metals at temperatures of only 50-75 percent of that needed to melt all of the constituent phases – of several metallic composites, with possible applications ranging from stronger, lighter and more durable materials for bearings, cutting tools, electrical brushes and contacts. Other areas of research included demonstrations of the feasibility of producing materials for high-speed digital circuits, as well as 'zeolites', commonly used in catalysts, molecular sieves, adsorbents and ion-exchange methods, and experiments in gas-permeable polymers and the processing of advanced polymer membranes in support of the commercial membrane industry.

In the life sciences, Astroculture – previously flown aboard the USML-1 mission a year earlier – continued a programme of testing the equipment for an operational plant-growth unit. BioServe Space Technologies, based at the University of Colorado in Boulder, provided a pilot laboratory to support various experiments, including the *Rhizobium trifolii* bacterium, which is known to form symbiotic relationships with Earth-plants, such as alfalfa, clover and soybean. Such research was expected to lead ultimately to new insights about manipulating the processes responsible for crop infections. Other experimental foci included observations of the growth and development of the *E. coli* bacterium and studies of gravitational influences upon frogs' kidney cells. BioServe's Commercial Generic Bioprocessing Apparatus, also aboard the Spacehab-1 module, supported a further 27 experiments into biomedical testing and drug development, controlled ecological life-support systems and agricultural development and manufacture of biological-based materials. These placed emphasis upon the processes by which cells destroy foreign materials and examined the ability of the immune system to respond to infections. Long-term implications for the work included an improved understanding of bone and developmental disorders, the healing of wounds and cancers and other cellular diseases. The development of wasps, seeds, algae and micro-organisms in the weightless environment was also monitored.

The Organic Separation (ORSEP) payload offered the commercial and scientific communities an opportunity to separate cells and particles for analysis, whilst next-generation protein crystal growth equipment employed vapour-diffusion and direct-control methods. Twelve adult male rats were also flown in a pair of Animal Enclosure Modules and were used to evaluate the performance of implanted growth factors. NASA provided a facility to accommodate cell and tissue cultures, a charged particle directional spectrometer, a three-dimensional accelerometer, the Space Acceleration Measurement System (SAMS) and an experimental life-support apparatus to test water-recycling methods for Space Station Freedom. A series of Human Factors Assessment studies, mainly performed by orbiter crew members Grabe, Duffy and Sherlock, involved the collection of data on the astronauts' ease of movement through the middeck-to-Spacehab tunnel, the Shuttle's acoustic and lighting environment and the usefulness of electronic procedures to fulfil tasks. The data was compared afterwards to similar experiments performed during a pair of Spacelab missions in June 1991 and September 1992. Endeavour's pilots also participated in neutral body posture investigations, sponsored by JSC's Space and Life Sciences Directorate, to observe changes in spinal lengthening and overall posture in the microgravity environment.

By the time the long-delayed STS-55 returned to Earth in early May 1993, the Spacehab-1 flight had slipped into early June. However, engineers discovered that one of two springs which keep ball bearings in place in the high-pressure oxygen turbopump of the No. 2 main engine had an improperly-placed inspection stamp, which it was thought might weaken it to the point of breakage. It was considered prudent to replace the turbopump and a firm launch date of the 20th – Brian Duffy's 40th birthday – was established. As the mission commander, Ron Grabe hoped to send his crewmate over the hill in style with an on-time launch, but it was not to be. The countdown proceeded normally until the standard 'hold' at T-9 minutes, when weather conditions at all three Transatlantic Abort Landing sites of Banjul in Gambia, Ben Guerir in Morocco and Moron in Spain, *and* the Shuttle Landing Facility itself at the Cape, were considered unacceptable to support Endeavour's flight. Nonetheless, a decision was taken to proceed down to T-5 minutes and wait out the weather, but to no avail. Eventually, the 71-minute launch window ran out and the attempt was scrubbed.

Next morning, conditions had improved markedly, but as the six astronauts boarded the orbiter a scratch was observed on the outer seal of the crew access hatch. After consultation with mission managers, it was decided to replace the seal and this was accomplished with little impact on the countdown. At T-5 minutes, the hold was extended by 22 seconds in order to shoo an unauthorised aircraft out of the launch danger zone and STS-57 rocketed off the pad at 9:07 am EDT. The ascent was electrifying, particularly for the three rookie members of the crew, with main engine cutoff and the onset of weightlessness both memorable. "The bear jumps off your chest," remembered Jeff Wisoff in an interview for the Smithsonian, "and you see your seatbelts float upward, which is kinda cool."

Despite initial difficulties, the RMS arm was initialised and checked out, in readiness for the retrieval of the EURECA free-flyer on the third day of the flight. A number of thruster firings were performed by Grabe and Duffy over the next 48 hours to effect closure with EURECA, whose orbit had already been lowered under ground command in recent weeks from its operational altitude of 515 km to more closely approximate Endeavour's altitude of some 300 km. With Grabe at the Shuttle's controls for the final phases of the rendezvous, the satellite was successfully grappled by the RMS, under the deft control of David Low, at 9:53 am EDT on 24 June. Endeavour's Ku-band radar satisfactorily tracked the satellite from a distance of 45 km to less than 30 m, with no loss of data. Although EURECA's solar arrays successfully folded up, its two antennas – which should have retracted and latched prior to grappling – failed to close. As a consequence, it was decided to use a portion of Low and Wisoff's planned EVA on the 25th to tend to the problem.

An EVA was not originally planned for STS-57, but in the aftermath of the Intelsat 603 repair mission NASA decided in late 1992 to add excursions to as many flights as possible in order to build up an experience base within the astronaut office. In mid-February 1993, an EVA was added to STS-57, during which Low and Wisoff would spend four hours outside "to refine training methods for spacewalks, expand the EVA experience levels of astronauts, flight controllers and instructors and aid in better understanding the differences between true microgravity and the ground simulations used in training". Since Endeavour carried an RMS for the EURECA retrieval, it was decided to make use of it to rehearse procedures ahead of the Hubble Space Telescope repair and servicing mission in December 1993. However, the criticality of Low and Wisoff's EVA was of a low level and it was stressed that only if STS-57's nominal mission duration was extended from seven to eight days would it be cleared to take place.

The decision was to be made shortly after launch and was based upon calculations of the propellant and energy expenditure and how well they matched predictions. Fortunately, these proved to be in line with expectations and on 23 June the Mission Management Team agreed to extend STS-57 to eight days. Plans for the EVA involved Low and Wisoff taking turns in a foot restraint on the end of the RMS – manipulated by Nancy Sherlock – and handling each other's mass to imitate the movement of large pieces of equipment. Moving large masses on the end of the mechanical arm was expected to be a critical requirement for both Hubble and Space Station Freedom missions. The spacewalkers also trained to investigate techniques for managing their safety tethers. The EVA, which lasted five hours and 50 minutes in total, ran exceptionally smoothly, with Sherlock positioning Low on the RMS to enable him to secure EURECA's antennas against their latching mechanisms. Payload controllers then remotely drove the latches to secure each antenna.

Since the EVA required the astronauts to depart Endeavour via the middeck-to-Spacehab tunnel adaptor, it was necessary for the pressurised module to be temporarily closed off. Subsequent investigation revealed no pressure loss from Spacehab during this time. It may, however, have induced an event which scared the daylights out of the four crew members inside the cabin. During the EVA, the Shuttle was positioned with her belly toward the Sun, in order to simulate the deep cold of space which would be experienced at phases of Hubble servicing and Space Station Freedom construction. "We were in this attitude and the four of us were on the flight deck," Brian Duffy remembered. "All of a sudden, it was as if somebody took the orbiter and *hit it* with a bulldozer! The whole vehicle *shook*. It got quiet on the flight deck and we thought maybe we were hit by something. We looked outside and didn't see any damage. Nothing came flying through the wings or through the payload bay." They informed Mission Control, who also saw nothing amiss in their data. The most likely explanation, it seemed, was that residual forces that had built up on the ground in the struts holding the Spacehab-1 tunnel adaptor in place had released "and it rang the whole vehicle". When the two spacewalkers returned to the cabin, they said that they had felt nothing. "They didn't have clue," Duffy chuckled, "but if they'd looked *inside* at that time, they would have seen eight *big* eyes!"

Low and Wisoff's EVA again demonstrated the importance and usefulness of having astronauts on hand to tend to unforeseen contingencies. It also provided EVA experience for another pair of astronauts, including the first spacewalk for a member of the 1990 astronaut class. Years later, Wisoff was asked about it by a NASA interviewer. "It's an incredible opportunity," he said, "an incredible experience. You're kind of in your own little spaceship in the space suit and you can't beat that view of the Earth while you're working. It's an incredible sight!" Despite the successful closure of EURECA's twin antennas, an attempt to activate the RMS arm's Special Purpose End Effector (SPEE) connector power relay, which was to provide electrical sustenance to the satellite failed. It subsequently became clear from on-orbit video footage and post-flight inspection that the SPEE had been incorrectly rotated by 180 degrees.

Spacehab-1, in particular, had proven an enormous success. The module's interior environment had a tendency to be quite cool at times, prompting a pair of in-flight maintenance procedures by the astronauts to successfully adjust the cold-water bypass valve and increase the temperature slightly. The cause was later attributed to lower electrical power requirements for Spacehab than had been predicted in pre-flight tests. In other words, it was not generating as much waste heat as predicted, but the cooling system was still working at full capacity. Overall, the module's experiments exceeded 90 percent of pre-mission plans. The mission had proven challenging for all six crew members, including Grabe, who – although on his fourth flight – had never done rendezvous or EVA before. "We had Ron kinda carrying the ball," Brian Duffy recalled in his NASA oral history. "Particularly in the training, he was just really good about knowing what was important for us, what we needed to know, what we should focus on."

Beyond Spacehab-1, in the payload bay, a bridge of ten Getaway Special canisters supported numerous other experiments from researchers as far afield as the United States, Japan, Canada and Europe. These gauged liquid behaviour and distribution in tanks, insulin-tagged brine shrimp, grew crystals of gallium arsenide and other semiconductors, observed bubble growth and movement in the boiling process, monitored high-frequency variations of the Sun and – part of an experiment sponsored by the Jet Propulsion Laboratory – characterised orbital debris and extraterrestrial particles in the payload bay as part of investigations into the development of a future comet-sample-return mission.

Two other major payloads – the Consortium for Materials Development in Space Complex Autonomous Payload (CONCAP)-IV and the Superfluid Helium On-Orbit Transfer (SHOOT) – were also carried. CONCAP-IV investigated the growth of 'non-linear organic' crystals by physical vapour transport. Such crystals were expected to yield applications in future optical computing techniques. SHOOT was designed to develop and demonstrate the technologies needed to resupply liquid helium dewars in space. Already, astronomical missions such as the Infrared Astronomy Satellite (IRAS) and Cosmic Background Explorer (COBE) had highlighted the importance of cryogenic cooling as an aid to their observing power, but the 'venting' of liquid helium during the cooling process meant that such instruments had finite lifespans. SHOOT contained a pair of Thermos-like dewars, between which liquid helium was pumped at a rate of 300-1,000 litres per hour, with each serving as the 'supplier' or the 'receiver'. The experiment and its associated instrumentation explored the process of liquid management in microgravity and filled a large gap in the knowledge of cryogenics behaviour in space.

With a one-day extension to the mission, STS-57 was scheduled to land on 29 June, but Endeavour's return actually came 48 hours later than planned. Unsatisfactory weather in the vicinity of the Shuttle Landing Facility at KSC forced a one-orbit wave-off, and when there was no improvement a 24-hour postponement was ordered. The next day, the 30th, offered no respite and *another* 24-hour delay was imposed. Finally, early on 1 July, Grabe and Duffy performed the de-orbit burn at 7:41 am EDT, committing the vehicle to its hypersonic dive through the atmosphere. Touchdown on Runway 33 came an hour later at 8:52 am, just 15 minutes short of ten full days since liftoff.

The re-entry, Duffy related, was nominal, but characterised by one unanticipated event. "This one was a flatter, lower trajectory" than his first flight, he said. "We were coming across northern Mexico and then across the Gulf, coming in almost due east, and it was ... after sunrise, maybe mid-morning." All at once, as they passed through Mach 21, a sudden *wham* echoed through the vehicle's airframe. About 30 seconds later, another one followed. Nancy Sherlock, sitting in the flight engineer's seat, looked up in surprise after the first event, but noticed that Grabe and Duffy – the veterans – seemed unperturbed and said nothing. When it happened again, Grabe turned to his pilot.

"Have you ever felt that before?"

"We didn't have a clue" 335

3-22: Spacewalkers David Low and Jeff Wisoff are pictured at work on EURECA in the rear portion of Endeavour's payload bay. This view – which also includes the flat-roofed Spacehab module in the foreground – was taken by one of their crewmates on the Shuttle's aft flight deck.

"No," replied Duffy, perplexed. It later became clear that Endeavour had been shaken by a 'density shear' – a difference in air masses at extremely high altitude, which Endeavour had transitioned at hypersonic speed. "Those things extend up at a high altitude," Duffy reflected later, "and when you're going very fast and you change from one to the other, you get an instantaneous notification that you're somewhere else. It's like somebody really picks up and shakes the whole orbiter. The whole vehicle rings, because [it's] not very sturdy. It looks sturdy, [but] it's about as stiff as a Twinky and, truth be known, during ascent you can actually *feel* it flexing."

With the exception of an ammonia cooling glitch on the runway, which required Grabe to work closely with ground crews to establish cooling and prevent an emergency power-down, Endeavour's return from her fourth flight was a spectacular success. It also cemented life-long friendships for the STS-57 crew and reinforced memories of their flight. "Nancy and I were on the flight deck one night," said Duffy, during the weather wave-off period, "where we were waiting to go to bed – basically hanging out – and she and I were upstairs, turned all the lights out on the flight deck and the two of us just sat there and floated in the window and watched the Earth going by in the dark. It was pretty neat." The down side of having to spend *two* additional days in orbit was that they had to budget their camera film and by 1 July it was all gone. "We didn't even have film to shoot" at the end, noted Duffy, "so we just got to *look* and just enjoy it. It was really cool."

ABORT!

"T-minus 30 seconds ..."

The words of the launch commentator at the Kennedy Space Center on the morning of 12 August 1993 were calm and measured, as all eyes focused upon Space Shuttle Discovery as she entered the final portion of the countdown to fly STS-51. The mission – a nine-day flight to deploy a NASA advanced communications satellite and release and retrieve an ultraviolet telescope on a Shuttle Pallet Satellite, as well as perform to a spacewalk – had already been postponed twice, with the astronauts aboard the vehicle. On 17 July, a flaw in a pyrotechnic initiator controller, needed to trigger the release of the Solid Rocket Boosters from the launch platform, had forced the first scrub. Troubleshooting later identified a thermal instability issue in a solid-state switch card in the ground support equipment and after replacement and testing the launch was rescheduled for the following week.

"Standing by to activate the sound suppression water system in five seconds ..."

The attempt to get Discovery into space on the 24th also proved fruitless, due to a turbine 'underspeed' condition in a hydraulic power unit in the right-hand Solid Rocket Booster. "The specification requires the turbine speed to be between 66,200 and 77,800 rpm," noted NASA's STS-51 Mission Report, "and the speed at that point was 65,000 rpm." As a result, the countdown was automatically halted at T-19 seconds and the attempt was scrubbed. Due to the July-August Perseid meteor shower associated with Comet Swift Tuttle, NASA decided not to make a third launch attempt until the second week in August. It was known that the Perseid debris field would pass closest to Earth on the 11th – producing a 1/1000 chance of an impact with the Shuttle – and so the launch was rescheduled for 9:10 am EDT the following day.

By 12 August, Discovery and her five-man crew, commanded by STS-38 veteran Frank Culbertson, were ready to go. The astronauts had closed and locked their visors and their eyes were focused intently on their instruments. Despite a two-and-a-half-minute hold in the countdown at T-5 minutes, due to lost synchronisation lock between the Mission Control in Houston and the Merritt Island Launch Area, everything proceeded smoothly. With five minutes remaining, pilot Bill Readdy reached over and switched on the Auxiliary Power Units. The trio of pumps hummed perfectly to life, bringing muscle and control to the Shuttle's hydraulics. The Shuttle's on-board computers automatically commanded a final series of checks on the three main engines and the elevons on the wings. Thirty-one seconds before launch, control of the countdown was handed off to Discovery's General Purpose Computers.

"*T-minus ten, nine, eight, seven …*"

The mission for which they had spent the last 18 months training was finally about to begin. Seated at the front of the flight deck, Culbertson and Readdy were the only veteran members of the crew; by contrast, none of the three mission specialists – uniquely for a flight in the post-Challenger era – had flown before. Sitting behind Readdy, Mission Specialist One was James Hansen Newman, a physicist by profession. Born in the Trust Territory of the Pacific Islands (today part of the Federated States of Micronesia) on 16 October 1956, he completed high school in San Diego and earned his degree in physics from Dartmouth College in New Hampshire in 1978. These were quickly followed by master's and doctoral degrees from Rice University in 1982 and 1984. Newman completed a year's post-doctoral work in atomic and molecular physics and was appointed as an adjunct professor in Rice's Department of Space Physics and Astronomy.

The space programme beckoned from an early age for Newman. "My uncle recommended that I go into the military and become a pilot and learn to fly and get into the astronaut office that way," he told a NASA interviewer, years later, "but I decided that my real love was for science and technology and that the right thing for me to do was to enter the space programme as a civilian scientist and to use those skills that I had developed in the laboratory to bring into NASA as part of the team that makes up a successful space flight." In 1985 he arrived at the Johnson Space Center to work on flight crew and flight control team training for Shuttle propulsion, guidance and control issues and was a simulator supervisor when selected as an astronaut candidate in January 1990.

Simulation Supervisors – 'Sim Sups' – have long gained a measure of notoriety among astronauts, since it was their role to introduce all manner of contingencies and failures into training exercises in order to keep a crew's skills sharp. According to fellow astronaut Tom Jones in his autobiography, *Skywalking*, Newman could not resist bringing a little of his Sim Sup expertise into astronaut candidate training. During one simulator session with Jones and legendary astronaut John Young, Newman mischievously pushed the main engine shutdown switches with his 'swizzle stick'. This obliged Young to execute an abort landing in a simulated Ben Guerir, Morocco. Jones wondered what Newman was playing at. "Why would you want to kill an engine on the most senior astronaut, the first man to command the Shuttle?" he wrote with incredulity. "Finally, Newman lost it; he burst out laughing as he confessed that he was the real culprit. Newman, the former instructor, had just been keeping his hand in, feeding Young one last malfunction. What surprised me was that John loved the practical joke."

"*Go for main engine start …*"

Now, on 12 August 1993, Newman was experiencing the final seconds of a countdown for real. Inside the cabin, the astronauts felt the immense vibration as turbopumps awoke, liquid oxygen and hydrogen flooded into the engines' combustion chambers and they roared to life ... and then, suddenly and shockingly, were arrested by the blaring sound of the master alarm. All three had been automatically shut down. Something had gone badly awry.

"... Three ... We have a cutoff ..."

Abutting Newman's left shoulder at the rear of Discovery's flight deck, in the Mission Specialist Two seat, was a small Navy engineer named Daniel Wheeler Bursch. His primary role was to serve as the flight engineer for ascent and re-entry ... and *his* eyes were as wide and saucers as he realised that his first mission had just suffered a potentially disastrous pad abort. (Nor was it the only time in Bursch's career that this happened; it occurred again on the first attempt to launch his *next* flight, STS-68, in August 1994.) He was born on 25 July 1957 in Bristol, Pennsylvania, and studied physics at the Naval Academy. Upon graduation in 1979, he became a naval flight officer and trained as a bombardier/navigator in the A-6E Intruder. Overseas deployments followed to the Mediterranean aboard the USS *John F. Kennedy* and to the North Atlantic and Indian Oceans aboard the USS *America* and upon his return to the United States he was selected for Naval Test Pilot School. Completion of the course in December 1984 was followed by A-6 test work and later duties as a flight instructor. He also served again overseas as Strike Operations Officer aboard the USS *Long Beach* and USS *Midway* and was in the process of completing his master's degree in engineering science at the Naval Postgraduate School when selected as an astronaut in January 1990.

"We have a main engine cutoff. Safing in work."

If Bursch and Newman and, up front in the pilots' seats, Culbertson and Readdy, could see the red lights on the instrument panel and catch a glimpse through Discovery's windows as she rocked to and fro in the first few seconds after engine shutdown, one man who had to rely upon his other senses was Mission Specialist Three. Astronaut Carl Erwin Walz was seated downstairs in the darkened middeck and although he could hear the roar of the engines – and the abrupt silence when they cut off – and feel the vibrations, he could see nothing else but a locker-studded wall in front of him. Walz had already gained a measure of fame as an Elvis Presley impersonator and this had earned him admission into the astronaut corps' rock band, 'Max Q', as their lead singer. On a survival training expedition in Spokane, Washington, shortly after his selection into the corps in the spring of 1990, Walz and fellow astronaut candidate Terry Wilcutt were conversing in a bar. Walz happened to mention his previous singing experience in a Cleveland band, called 'The Fabulous Blue Moons'. Wilcutt challenged Walz to sing, right there and then. According to a 2009 article by Michael Cassutt in *Air & Space* magazine, Walz "talked with the band, agreed on a couple of Elvis tunes, then rocked out". The rest was history.

Walz was born in Cleveland, Ohio, on 20 September 1955 and after high school entered Kent State University to study physics. With an upbringing in the 'Buckeye State', he grew up hero-worshipping fellow Ohioans John Glenn and Neil Armstrong. Walz got his degree in 1977, graduating *summa cum laude*, and followed up with a master's in solid state physics from John Carroll University in 1979. He was a Reserve Officers Training Corps student as an undergraduate, but took a two-year delay on 'inactive reserve' status to gain his master's degree before entering active service with the Air Force. After three years working on the analysis of radioactive samples from the Atomic Energy Detection System at the 1155th Technical Operations Squadron at McClellan Air Force Base in California, he trained as a flight test engineer at Edwards Air Force Base.

Walz was drawn to the astronaut business, since it offered a combination of his interests in engineering and science, and also drove his desire to enter the Air Force. He knew of other flight test engineers – including Ellison Onizuka and Jerry Ross – who had become astronauts and so he sought to follow in their footsteps. After qualification as Distinguished Graduate of the 1983 class, Walz spent three and a half years at Edwards, working as a flight test engineer to the F-16 Combined Test Force, during which time he participated in airframe avionics and armament development efforts. In July 1987, he was appointed Flight Test Manger at Detachment 3 of the Air Force Flight Test Center. Less than three years later, Walz was accepted into NASA's astronaut corps.

And three years after *that*, Walz found himself sitting through the *second* Redundant Set Launch Sequencer (RSLS) abort of 1993, the first case being Columbia's on-the-pad engine shutdown in March. The communications loop from the Launch Control Center provided a flurry of messages, verifying that the three main engines were in post-shutdown standby and requesting Readdy to shut down the orbiter's Auxiliary Power Units. No fire detectors on Pad 39B were tripped during the incident, which would later be traced to a faulty fuel-flow sensor in the No. 2 main engine. The engine had posted a 'major component failure', caused by the sensor glitch, about 0.6 seconds after ignition. "This condition," noted NASA's official STS-51 Mission Report, compiled and published after the flight, "caused a miscompare which violated the Launch Commit Criteria ... As a result of the failure, the engines were shut down and safing activities were initiated."

"Orbiter access arm now back in position ..."

Shortly thereafter, the five disappointed astronauts disembarked from Discovery, aware from previous RSLS events that their launch had been called off for several weeks at best. The main engines were replaced and an attempt was provisionally scheduled for 10 September, but this was itself slipped by two days, as a result of the failure, on 21 August, of NASA's Mars Observer, shortly before its arrival at the Red Planet. During the early investigation into the loss of Mars Observer, it was revealed that the spacecraft's Transfer Orbit Stage (TOS) – a near-identical booster to that of STS-51's primary payload, the $363 million Advanced Communications Technology Satellite (ACTS) – had exhibited a transistor failure. During the additional two days' delay, engineers and managers verified that there was no commonality between the Mars Observer TOS fault and the ACTS booster.

Liftoff of the long-delayed mission finally took place at 7:45 am EDT on the 12th, and a nominal ascent placed Discovery into the intended circular orbit of 300 km, inclined 28.45 degrees to the equator. Years later, Jim Newman recalled his first experience of flying in space. "When we first got to orbit, it was exhilarating," he told a NASA interviewer. "I can remember getting out of my seat and going to the windows in the aft flight deck. The orbiter was upside-down, so that we were able to look and see the Earth, 'beneath' us. If you've ever seen the IMAX movies, they *almost* capture it all, but to be *floating* – to be seeing the Earth with my own eyes – was really spectacular!"

Within two hours, the payload bay doors were open, exposing STS-51's twin-satellite cargo to the space environment for the first time. At the rear end of the bay, the 12,130 kg ACTS-TOS combination represented one of the most advanced communications satellites ever inserted into orbit. ACTS' purpose was to serve as a testbed for the development of high-risk advanced communications satellite technologies, employing sophisticated antenna beams and on-board switching and processing systems and bringing together government, academia and industry. Specifically, the satellite operated across three channels within the 30/20 GHz Ka-band, which boasted 2.5 GHz of available spectrum – some five times that available at lower-frequency bands – and very high-gain, multiple-hopping beam antennas which permitted smaller-aperture Earth stations. ACTS was begun in 1982 as a NASA project, which, as *Flight International* noted, had been "thwarted by politics, budget cuts and delays", producing the inevitable consequence that by 1993 precious few of its technologies were either 'advanced' or 'pathfinding'. Built by Martin Marietta, ACTS' switching systems allowed for interconnectivity between users at individual circuit levels and its microwave switch matrix provided for user-to-user communications rates of up to a gigabyte per second. "The development and flight validation of this advanced space communications technology by NASA's ACTS," read the pre-launch press kit, "will allow industry to adapt this technology to their individual commercial requirements at minimal risk." At the time of ACTS' development, the Ka-band was being adopted by Motorola for their Iridium global voice/data communications satellite system.

Physically, the 2,760 kg ACTS took the form of a three-axis-stabilised rectangular box, measuring 5 m high when stowed in the Shuttle's payload bay and expanding to 14 m across the span of its solar arrays and 9 m across the breadth of its main reception and transmission antennas when fully deployed in orbit. The satellite 'bus' also housed an apogee kick motor and equipment to provide attitude and thermal control, electrical power, command reception, telemetry transmission and ranging and propulsion. Contracts to develop the satellite were awarded by NASA's Lewis Research Center – today known as the Glenn Research Center – of Cleveland, Ohio, in August 1984 to an industry team, which included Lockheed for the spacecraft bus and systems integration, TRW for the communications payload, COMSAT Laboratories for the network control and master ground station, Motorola for the baseband processor and Electromagnetic Sciences for the spot-beam-forming networks. By 1988, as a result of a Congressionally-mandated funding cap, Lockheed Martin took over the development of the communications payload.

Attached to the base of ACTS was the 9,426 kg Transfer Orbit Stage, making its second flight after Mars Observer and its first and only flight aboard the Shuttle. On STS-51, the TOS was tasked with delivering the payload into an elliptical geosynchronous transfer orbit with its apogee at 34,620 km. Developed under the auspices of NASA's Marshall Space Flight Center in Huntsville, Alabama, the TOS

3-23: Carl Walz gathers tools in the forward portion of Discovery's payload bay, early in his EVA with Jim Newman. This unusual perspective highlights the outer airlock hatch from the middeck, which stands open, and the twin aft flight deck windows.

was built by Martin Marietta for Orbital Sciences and took the form of a single-stage, solid-propellant booster, measuring 3.3 m tall and 2.3 m wide. It carried navigation and guidance systems and reaction controls. An airborne support equipment cradle held the ACTS-TOS combination in the payload bay and positioned them for deployment.

In readiness for the release of the payload, Discovery's aft flight deck was a hive of activity on the first day of the mission, with deployment anticipated eight hours after launch, at 3:43 pm EDT, on the sixth orbit. The astronauts checked out the TOS' critical systems and unlatched and rotated the upper forward cradle into its 'open' configuration, after which the entire payload was elevated to an angle of 42 degrees. However, deployment was postponed by an additional orbit, lasting an additional 90 minutes, when orbiter S-band forward-link communications with Mission Control were lost. Flight controllers could receive telemetry and voice communications from Discovery, but not vice versa. The astronauts followed their malfunction protocols, waved off the planned deployment and changed the S-band to a lower frequency. This restored communications with the ground after 45 minutes. Post-flight analysis determined that the problem arose when the Payload Interrogator (PI) – part of the communications system between the Shuttle and the ACTS-TOS – was switched to its high-frequency mode. At this stage, according to the official Mission Report, "the PI and Tracking and Data Relay Satellite (TDRS)

S-band frequencies were close enough that the orbiter S-band forward link locked on to the PI, rather than the TDRS".

Deployment finally took place at 5:13 pm EDT, some nine hours and 28 minutes into the STS-51 mission. Under the direction of Walz and Newman, a 'Super*Zip' separation mechanism was fired and six springs on the aft cradle of the TOS pushed the payload away from the Shuttle. As the combination drifted into the inky blackness, Culbertson and Readdy fired Discovery's RCS thrusters to produce a separation rate of about half a metre per second, after which a larger, 28-second burn of the right-hand OMS pod was conducted to increase the gap between the pair. Finally, 45 minutes after deployment and with the Shuttle at a safe distance of some 19 km, the TOS' Orbus-21 solid-fuelled motor ignited for 110 seconds to accelerate the payload to transfer orbit. The Honeywell ring laser gyroscope of the guidance system had no moving parts, thus minimising the risk of in-flight malfunctions. Shortly after the engine shut down, ACTS was released and the TOS performed a perpendicular 'turn' to avoid placing itself in the satellite's path. NASA's advanced communications platform then employed its own manoeuvring assets to insert itself into its correct geostationary slot, high above the equator at 100 degrees West longitude. After a 12-week initial checkout, ACTS was ready to begin its two-year baseline programme of around 60 experiments.

It became clear from video footage of the deployment that extensive damage had been caused to the expanding tube assembly and doublers on the Super*Zip ring. Debris included sharp-edged metal and other non-metallic materials and well over half of the expanding tube assembly was recorded as being "no longer restrained to the airborne support equipment". Investigations revealed that the primary and backup separation 'cords' of the Super*Zip ring were fired simultaneously and despite concern about potential damage to the orbiter it was determined that the payload bay liners and thermal blankets could sustain impacts from the expanding tube assembly fragments if they came loose during re-entry. In mid-October, an investigative board was established by Jeremiah Pearson, NASA's Associate Administrator for Space Flight, to identify the cause of the Super*Zip incident. Headed by Robert Wingate of the Langley Research Center, the board found a total of 36 debris hits in Discovery's mid-body and aft bulkhead areas, resulting in tears, gouges, scratches and the deposition of residue on several surfaces. One area of penetration passed *through* the aft bulkhead itself. "None of the debris hits had any effect during the flight," noted NASA's post-flight summary, "and all damage sites will be repaired during turnaround operations."

Originally intended for a four-year lifespan, ACTS' equatorial station-keeping was halted in July 1998 so that it would oscillate slightly north and south of the equator on a daily basis. Two years later the satellite was moved to a new 'permanent' location at 105.2 degrees West longitude and in May 2001 the Ohio Consortium for Advanced Communications Technology assumed control of ACTS for educational research purposes. Its operations finally concluded in April 2004, when funding dried up, and it was placed into an orientation with its solar array edges facing the Sun to inert it. According to ACTS' operations manager at NASA-Glenn, Richard Krawczyk, the satellite will probably not re-enter the atmosphere for several thousand years.

Its 60 experiments were many and varied and encompassed communications technology, military activities and medical telescience. Several areas of study included a test in which Ohio University helped Huntington Bank to recover from a simulated natural disaster, thereby protecting it against total communications loss, with ACTS transmitting financial data – including deposits, account balances and funds transfers – in an effort to determine the reliability of the data links and ability to switch over to backup systems in a timely manner. The Jet Propulsion Laboratory evaluated a 4.8 kbps voice and data link between an aircraft and a fixed terminal to test ACTS' spot beam technology, the Public Broadcasting Service demonstrated high-definition television and Martin Marietta offered Ka-band volume to potential customers for evaluation. The satellite also supported medical research, with Georgetown University successfully transmitting magnetic resonance images and radiological images from Hawaii to Washington, DC, and performing 'tele-education' courses for radiologists. Attempts were made to conduct telemedicine and transfer high-resolution images and distance learning programmes were tested between the Florida A&M University's College of Pharmacy in Tallahassee and its Clinical Training Unit in Miami. Other experiments were supported by the US Army and encompassed videoconferencing, large-image transfer, logistics and medical databases, remote training and transmission of video, voice and data to battlefield locations.

With the deployment of ACTS thus behind them, the STS-51 astronauts pressed on with the remainder of their mission, which, although scheduled for nine days, was expected to be extended to a "highly desirable" ten days. The crew's second major payload was the German-built Shuttle Pallet Satellite, which had flown on two previous missions, but which was being carried for the first time in its new 'ASTRO-SPAS' configuration. Unlike its first-generation predecessor, ASTRO-SPAS had the capability to remain in autonomous free-flight for up to ten days, commanded by the mobile German SPAS Payload Operations Centre (SPOC). The power for the satellite and its payloads came from a new lithium-sulphate battery pack and precise attitude control was provided by a three-axis-stabilised cold-gas system, a star tracker and a space-borne global-positioning system (GPS) receiver. Weighing 3,150 kg, more than half of which was accounted for by operational science payloads, the carbon-fibre-composite ASTRO-SPAS measured 4.5 m across the width of the Shuttle's payload bay and was 2.5 m in depth from its front to rear faces.

The satellite's precise attitude-control capabilities enabled it to support sensitive astronomical and Earth-observation sensors, with several missions planned. Two of these would carry a set of infrared telescopes and spectrometers to examine the upper atmosphere, one was scheduled (but never flown) to demonstrate advanced automated rendezvous and capture technologies in support of Space Station Freedom and two others – including the ASTRO-SPAS aboard STS-51 – carried the Orbiting and Retrievable Far and Extreme Ultraviolet Spectrometer (ORFEUS). This instrument, with a large 1.2 m-diameter telescope, was designed to investigate very hot and very cold matter in the Universe, in conjunction with an Interstellar Medium Absorption Profile Spectrograph (IMAPS). Also affixed to the ASTRO-SPAS framework was a surface effects sample monitor to evaluate several future telescope material samples and a remote IMAX camera.

Of these, ORFEUS – which extended to a length of 2.4 m through the middle of the ASTRO-SPAS satellite – was by far the largest instrument. It was to observe the far ultraviolet (90-125 nm) and extreme ultraviolet (40-90 nm), a region of the electromagnetic spectrum obscured from ground-based astronomers by the atmosphere. Yet it bears the highest density of spectral lines – especially from various states of hydrogen and helium – which are emitted or absorbed by matter of very different temperatures. ORFEUS was expected to add a great deal to scientific understanding of the life-cycles of celestial objects by studying hot stellar atmospheres and white dwarfs, together with supernova remnants, active galactic nuclei and star-forming clouds of gas and dust. Operating alongside the telescope, IMAPS continued an earlier series of experiments aboard high-altitude sounding rockets to observe galactic objects at 95-114 nm and examine the fine structure of interstellar gas lines. During orbital operations, ORFEUS' two spectrometers – far and extreme – were operated alternately, by 'flipping' a mirror into the beam reflected off the instrument's primary mirror.

Deployment required Dan Bursch to operate the RMS, whilst Jim Newman monitored the health of the ORFEUS-SPAS systems from Discovery's aft flight deck. After powering up the payload by means of interfaces established by the arm's Special Purpose End Effector (SPEE), command and data links were established and the SPOC guided ORFEUS-SPAS through a lengthy pre-deployment checkout process. Its gyros were calibrated whilst still berthed, its data tape recorder was reset and Bursch prepared to raise the satellite above the payload bay. Preparations were slightly hampered by problems transmitting command files, which forced a one-orbit delay. When all was ready, Bursch released ORFEUS-SPAS from the arm at 11:05 am EDT on 13 September, a little more than a day into the mission, and the satellite drifted into the inky blackness. Frank Culbertson performed a separation manoeuvre to draw the Shuttle to a distance of about 20 km 'ahead' of ORFEUS-SPAS. Under SPOC control, the satellite performed an inertial 'attitude hold', then a second gyro calibration, and the IMAX camera began recording spectacular images of the departing Discovery.

For the next six days – far longer than earlier SPAS missions – the payload remained in free flight, with the Shuttle acting as a relay station to transmit ground commands from SPOC controllers to the satellite and vice versa. At length, on 19 September, the final manoeuvres to recapture ORFEUS-SPAS got underway, five and a half hours ahead of retrieval, and with Discovery then in a position about 45 km 'behind' its quarry. Closing at an approximate rate of around 15 km per orbit, Culbertson executed four mid-course correction burns and finally took manual control of the orbiter for the final moments of the rendezvous. Meanwhile, an automatic laser range-finder in the payload bay and a second, hand-held, device, operated by Bill Readdy, provided data on distances and rates of closure. Finally, at 9:49 am EDT, Bursch grappled the satellite and established an SPEE interface with the mechanical arm. Berthing and a full electrical 'power-down' of the satellite in the payload bay began at 10:05 am. The first flight of ORFEUS-SPAS had achieved more than 100 percent of its pre-launch scientific objectives. In addition to its astronomical goals, the satellite was employed as part of a GPS development test objective.

Between the deployment and retrieval of ORFEUS-SPAS, the third spacewalk of the year to build up NASA's EVA experience base in readiness for the construction of Space Station Freedom was conducted on 16 September by Carl Walz and Jim Newman. As the STS-51 press kit pointed out, one of the primary goals of the planned six-hour excursion (added to the flight in February 1993) was to "evaluate several tools that may be used during the servicing of the Hubble Space Telescope", including "a power socket wrench, a torque wrench, foot restraints, safety tethers and tool holders". Although the RMS arm was aboard, it was not be used as part of the EVA trials, since it was committed to the ORFEUS-SPAS retrieval.

Preparatory work began on the second day of the mission, when the cabin pressure was reduced to permit 'pre-breathing' protocols and Walz and Newman commenced standard checks of their space suits and tools. Early on the 16th, they donned the suits and underwent leak checks, a nitrogen purge and – a little earlier than intended – started their 40-minute period of pre-breathing. The airlock hatch into Discovery's payload bay opened at 4:39 am EDT and the first-time spacewalkers proceeded directly into their EVA timeline, with Walz examining debris from the Super*Zip malfunction. He verified Mission Control's consensus that the debris appeared sufficiently stable for landing, since it was securely held down in two places, and it was considered more prudent to leave it alone, rather than risk damaging their space suits from contact with sharp metal edges.

Working through their tasks, Walz and Newman found that operating in the microgravity environment, suited, was far easier than it had been in the Weightless Environment Training Facility (WET-F) water tank in Houston. An attempt to use the power tool in a high-torque evaluation was slightly delayed by a low-battery warning, which prompted Newman to return to the airlock for a spare, and the men opined that the mini work stations on the RMS offered "very little restraint for the torque operations". Later, Newman evaluated a portable foot restraint for the Hubble mission and found that it was much harder to egress the device than it had been in ground simulations. The astronauts returned to the airlock after seven hours and five minutes outside. In their post-flight debriefing, Walz and Newman stressed the importance of thermal vacuum chamber tests as part of EVA training.

Although the twin satellites and the EVA had been the most visible of STS-51's tasks, a variety of other investigations and objectives filled the remainder of their nine days in orbit. Two Getaway Special canisters in the payload bay supported a set of limited-duration experiments in material exposure, whilst Discovery's crew cabin housed studies of chromosome and plant cell division, polymer membrane processing and commercial protein crystal growth. Auroral photography and high-resolution imagery of the mysterious 'Shuttle glow' was also undertaken. Several military investigations monitored the on-board radiation environment and Discovery was used as a passive target for the Air Force's optical tracking site on the Hawaiian island of Maui.

Touchdown in Florida was planned for the morning of 21 September, but was postponed due to unacceptable weather forecasts in the vicinity of the Shuttle Landing Facility. Both opportunities for that day were thus called off, but 22 September proved more acceptable and Culbertson and Readdy performed the de-

3-24: The view from within Discovery's darkened flight deck, during the EVA, shows the astonishing panorama of Earth, the payload bay and spacewalkers Carl Walz and Jim Newman at work. At the rear of the bay is the gold-enshrouded bulk of the airborne support equipment cradle, which had earlier supported the deployment of ACTS.

orbit burn at 2:55 am EDT, committing Discovery to its hour-long hypersonic glide home. The orbiter landed safely on Runway 15 at 3:56 am and enjoyed a smooth drag chute deployment and rollout, punctuated by an earlier-than-intended APU shutdown when burning plumes were observed from the port-side exhaust ducts. Although seen on previous missions and not deemed abnormal, the plumes appeared more dramatic since Discovery landed at night. After more than two months of delays, the Shuttle fleet was back in action.

LONG ROAD TO SLS-2

"John, we're going to fly you one of these days," Launch Director Bob Sieck called over the communications loop on 15 October 1993. The disappointment of another scrubbed launch attempt was evident in his voice. "Just hang in there."

"Nice try," came the call from astronaut John Blaha on Columbia's flight deck, as he and his six crewmates prepared to disembark from the orbiter after two and a half hours on their backs in bulky, uncomfortable pressure suits, harnesses and parachute equipment. It was the second time that they had been through this routine in trying to get into space for what was to be NASA's longest Shuttle flight to date, lasting a little over two weeks. For now, however, Columbia was living up to her unenviable reputation: an immovable bear, difficult to get off the ground, but once in orbit imbued with the beauty and grace of a swan.

The delays in getting her previous mission, STS-55, off the ground had already pushed Columbia's STS-58 flight – the second Spacelab Life Sciences (SLS-2) voyage – from late August into mid-September 1993 and, following the STS-51 problems, eventually into the middle of October. Bad weather on the 14th forced a two-hour extension to the launch window and, when it finally cleared, the countdown clock ticked toward a scheduled 11:53 am EST liftoff. Then a scrub was ordered as a result of a failure in an Air Force range safety command message encoder verifier. An attempt on the 15th was called off when a problem with one of two S-band transponders meant that it could transmit data, but could not reliably receive. Although a replacement transponder was rushed to the pad, NASA managers were unhappy with the idea of exchanging it in the crew cabin in the final minutes before launch and then asking the astronauts to install it in orbit.

"Having been there myself," said Loren Shriver, a former astronaut and the new chair of the Mission Management Team, "there is no substitute for having a little help from the ground." Shriver was particularly concerned by the prognosis for a successful return to Earth in the event of a contingency, early in the mission. If the transponder failed, it might have left Columbia's crew out of contact with Mission Control for up to 15 critical minutes. It ultimately became a moot issue, for the weather closed in and the launch attempt was called off. The new transponder was installed and tested over the weekend of 16-17 October and on the morning of the 18th Blaha once again led his crew out of the Operations and Checkout Building, bound for the pad.

For Rhea Seddon, payload commander of STS-58, one of the worst aspects of a

Shuttle mission were the new partial-pressure suits that they were obliged to wear for ascent and re-entry. "It was crazy," she told the NASA oral historian. "They had technicians that got you into them prior to launch, but then you had to get yourself into them for landing ... and imagine the middeck, weightlessness, and seven suits floating around down there and 14 gloves and 14 boots and cooling garments." During training in the high heat of Florida, Houston and California, Seddon referred to the process as "suit-wrestling" and the difficulty increased for small astronauts, because they had the same weight of equipment, regardless of size or body weight. Yet it was about more than just bulk. "Those suits were built to be worn by high-altitude pilots," she said, "*regular-size guys*. If they had problems, they ejected. They didn't have to crawl out and run away. They didn't have to rappel down the side of their vehicle." Her concerns were realised in early May 1993, when she broke four metatarsal bones in her left foot, whilst practicing a fully-suited escape from the orbiter. "The STS-58 crew was practicing emergency egress," noted the NASA news release, dated 3 May. "As Dr Seddon was sliding down the slide, her left foot became pinned under her, causing four minor bones to break." It was fortunate that the injury was minor and that the training was of the refresher nature and therefore could be quickly caught up on after her recuperation.

Seddon was named as payload commander of SLS-2 in October 1991, with a projected launch two years later in July 1993. Her expertise from the SLS-1 mission was a crucial factor in the assignment ... and it was an assignment that Seddon had actively sought. "There was some controversy about my being on the next flight," she told the NASA oral historian, "because I was ... on the SLS-1 flight and they wanted four other subjects ... They wanted to get eight subjects altogether and if *I* flew, they were only going to get seven subjects, because I was a repeat. They already *had* data on me. They weighed the pros and cons of that, but I had been following SLS-2 as long as I'd been following SLS-1 and continued to follow it after the first flight." Coupled to that, Seddon had established good working relationships with the SLS investigators and had the benefit of having flown recently. "There were just so many things that came out of SLS-1," she said, "that they wanted to capture and I think they knew that if I wasn't on SLS-2, I would probably be busy with another flight and not be able to help them as much as they liked."

Two months later, in December 1991, veteran astronaut Shannon Lucid (a biochemist) and rookie Dave Wolf (a physician) were named as mission specialists. Seddon was relieved that a decision had been taken by NASA to assign a payload commander for SLS-2. Her previous mission, SLS-1, had worked well, because of the good working relationship she developed with crewmate Jim Bagian. However, having no one in overall authority to make the payload decisions proved "a little awkward". For Seddon, it meant that she could attend meetings, represent her mission's payload and make "reasonable decisions on our behalf".

Shortly after his assignment, in March 1992, David Alexander Wolf was jointly awarded the accolade of NASA Inventor of the Year. Together with Ray Schwarz and Tina Trinh, he had worked on the biotechnology team at JSC and had developed and designed a new class of horizontally-rotating tissue culture systems – known as a 'rotating-wall bioreactor' – to offer a ground-based means of simulating microgravity conditions for cell and tissue cultures. For Wolf, it represented the culmination of six years' work as the chief engineer and programme manager for the project.

Born in Indianapolis, Indiana, on 23 August 1956, Wolf was the son of a doctor and in his early childhood aspired to be a garbageman. "Mr Peacock, our garbageman, let me run the garbage truck," he told the NASA oral historian. "It was an old-time thing, with big levers." His parents encouraged him, as long as he worked hard to be the best garbageman, but in time Wolf's interests changed and focused upon the space programme. After high school he entered Purdue University to study electrical engineering. He received his degree in 1978 and was accepted into medical school at Indiana University, from which he graduated in 1982. Years later, he paid tribute to his family – "my father liked electronics, my mother liked sports, my uncle flew aerobatic airplanes" – for having fostered his interest in the sciences and engineering. Yet it was watching Ed White making America's first spacewalk in June 1965 which really captured Wolf's attention and changed his career aspiration from garbageman to spaceman. "I was nine years old," he later told a NASA interviewer. "It inspired me into technological fields, eventually medicine and engineering, and I always wanted to work at NASA from that point."

Shortly after receiving his medical degree, in 1983 Wolf did just that. After completing his medical internship at Methodist Hospital in Indianapolis, he trained as a flight surgeon at Brooks Air Force Base in San Antonio, Texas, and remained on reserve duty with the Air National Guard until 2004, rising to the rank of lieutenant-colonel. He joined NASA as a member of the Medical Sciences Division at JSC, working on the engineering development of the American Flight Echocardiograph and later the bioreactor. Even before entering the halls of the space agency, his experience was considerable: whilst working on his medical degree, Wolf had established himself as a pioneer in ultrasonic image-processing practices. He was accepted into NASA's astronaut corps in January 1990.

Also announced to the SLS-2 training complement in October 1991 were three candidates for a single payload specialist position on the flight: physician Jay Buckey, an assistant professor of medicine at the University of Texas Southwestern Medical Center, electrical engineer Larry Young, the director and professor of the Man-Vehicle Laboratory at Massachusetts Institute of Technology, and veterinarian Marty Fettman of the Department of Pathology at the College of Veterinary Medicine and Biomedical Sciences at Colorado State University. In October 1992, Fettman was announced as the primary payload specialist, making him the first professional veterinarian ever to travel into orbit ... and with a very specific purpose. "NASA's series of SLS missions play a central role in our programme of space biomedical research," explained Lennard Fisk, NASA's Associate Administrator for the Office of Space Science and Applications. "The experiments that Dr Fettman and his fellow SLS-2 crew members conduct will give us valuable information on how living and working in space affects the human body."

Martin Joseph Fettman came from New York, where he was born on 31 December 1956, and after high school in Brooklyn he studied animal nutrition at Cornell University. He earned his undergraduate degree in 1976 and remained at

Cornell to pursue a doctorate in veterinary medicine and a master's credential in nutrition. Fettman gained both degrees in 1980, followed by a second doctorate in physiology from Colorado State University in 1982. After being board-certified in veterinary clinical pathology, he was hired by the Department of Pathology at the College of Veterinary Medicine and Biomedical Sciences at Colorado State on an assistant professorship. In addition to teaching, Fettman's research encompassed the pathophysiology of nutritional and metabolic diseases, including the physiological biochemistry of energy, electrolyte and fluid metabolism. By 1988 he had risen to become section chief of clinical pathology at Colorado State's Veterinary Teaching Hospital. Sabbatical work at the University of Adelaide in Australia followed in 1989-90, where he focused upon the biochemical epidemiology of human colorectal cancer, and returned to a full professorship at Colorado State in 1992.

Several months after the announcement of the SLS-2 science crew, in August 1992 the three-man 'orbiter' team was named. John Blaha would command STS-58, having already flown three times and trained briefly for a spot on the SLS-1 mission. Joining him as Columbia's pilot was Richard Alan Searfoss, one of whose claims to fame in the 1990 astronaut class was that he designed their official 'Hairballs' logo and patch. Searfoss came from Mount Clemens, Michigan, where he was born on 5 June 1956, but received his high schooling in New Hampshire and entered the Air Force Academy. Whilst there, he joined the Church of Jesus Christ of Latter-Day Saints and would thus become one of only a handful of Mormons ever to enter NASA's astronaut corps. Searfoss emerged from the Air Force Academy in 1978 with a degree in aeronautical engineering and gained a master's qualification in aeronautics from Caltech on a National Science Foundation fellowship the following year. His formal Air Force career commenced shortly thereafter and he completed undergraduate pilot training at Williams Air Force Base in Arizona in 1980, then flew the F-111F Aardvark tactical strike aircraft at RAF Lakenheath in England, and at Mountain Home Air Force Base in Idaho. By 1987 – and having been selected as a finalist for Outstanding Young Men of America that year – he was serving as an Aardvark instructor pilot and weapons officer. His next assignment was Naval Test Pilot School at Patuxent River, Maryland, as an Air Force exchange student. Completion of the course in 1988 led to work as a flight instructor at the school and in January 1990 he was selected as a member of NASA's 13th class of astronauts.

Seated behind and between Blaha and Searfoss on Columbia's flight deck for STS-58 was William Surles McArthur Jr, the flight engineer. He was born in Laurinburg, North Carolina, on 26 July 1951, the son of an Army brigadier-general and Second World War veteran, and entered the Military Academy at West Point to study applied science and engineering. "In college, I'd really gotten fascinated with engineering," he told a NASA interviewer, years later. "And I liked aircraft; liked rockets. Engineering kind of put it all together." Upon receipt of his degree in 1973, McArthur was commissioned as a second lieutenant in the Army and after a tour with the 82nd Airborne Division at Fort Bragg, North Carolina, he entered the Army Aviation School in 1975 and graduated at the top of his class the following year. He served as an aeroscout team leader and brigade aviation section commander in the Republic of Korea and later assumed duties as a company commander, platoon leader and operations leader with the 24th Combat Aviation Battalion in Savannah, Georgia. In 1983, he gained a master's degree in aerospace engineering from Georgia Institute of Technology and returned to West Point as an assistant professor in the Department of Mechanics.

By this time, McArthur was aware that a fellow Army aviator, Bob Stewart, had been selected into NASA's astronaut corps. "And a little light came on," he remembered, "and I looked at it and, lo and behold, the goal all of a sudden became attainable." To McArthur, applying for NASA was like buying a lottery ticket and he submitted his application, knowing that "the chances might not be very good that you'll win, but they're a whole lot better than if you *never* buy the ticket!"

The ticket failed him – in a sense – in 1987, when he was unsuccessful in his bid to join NASA's 12th class of astronauts. Yet the cloud had a silver lining. McArthur completed Naval Test Pilot School that year, was designated as an experimental test pilot … and was accepted by NASA as an engineer on the Shuttle Vehicle Integration Test Team. A little more than two years later, he was selected for the astronaut corps. By the time he boarded Columbia for his first launch into orbit in October 1993, McArthur was 42 years old, one of the oldest members of his class. "Fortunately," he told an interviewer much later, "I haven't been forced to grow up just yet!"

Due to the life sciences bias of the SLS-2 flight, the crew timeline was planned as a single-shift operation and Blaha was clear that although he was responsible for the safety and success of the mission, it would be Seddon – as payload commander – who would take the lead for the biomedical research in the Spacelab module. Unlike many other Shuttle commanders, who viewed their role as little more than a truck driver, Blaha saw the mission differently: their goal was to obtain good science from SLS-2 and he willingly offered himself, Searfoss and McArthur as subjects for the non-invasive medical experiments. "In other words," said Seddon, "they wouldn't do anything that would make them sick or weak, because they might have to fly us home at any point in time."

The inclusion of veterinarian Marty Fettman on the crew had been on the cards since before the SLS-1 mission, since it would involve extensive physiological examinations with 48 male rats (*Rattus norvegicus*), caged in a pair of Research Animal Holding Facilities (RAHFs). It would also controversially feature the first-ever in-flight decapitation and dissection of six rats. As the payload commander, and a surgeon by training, Rhea Seddon assigned herself and Fettman to oversee the dissections. Not surprisingly, this had drawn much public criticism, but according to Fettman and NASA Associate Administrator for Life Sciences Harry Holloway it was an essential tool in measuring ongoing changes in the rats' body tissues during flight. "This is really a unique opportunity to collect biological specimens," said Fettman before launch. "We believe these tissues will provide some answers to questions that potentially will change our interpretation of past observations." It was rationalised that examinations of rats brought home from SLS-1 had been unable to conclusively differentiate between the effects of microgravity exposure and the effects of their readaptation to terrestrial conditions. The SLS-2 dissections would enable researchers to more precisely trace tissue changes.

Still, Holloway called for an unscheduled pre-launch assessment of the plans, led by Deputy Surgeon-General Robert Whitney of the Department of Health and Human Services. Holloway denied claims that the assessment was forced upon him by the White House or NASA Administrator Dan Goldin. Whitney's investigation described NASA's animal-care provisions as "superb" and commended the agency's use of the fewest number of rats as possible to satisfy the needs of more than a hundred investigators.

Another source of controversy, at least within the astronaut corps, surrounded Dave Wolf, although it would not enter the public consciousness until after the mission. The story was explored by Bryan Burrough in his book *Dragonfly*, but apparently involved an FBI 'sting', called 'Operation Lightning Strike', in which the unfortunate Wolf had become entangled. Although the astronaut himself was exonerated from blame and had not – as some journalists erroneously claimed – accepted bribes, the incident harmed his career for several years.

None of this had surfaced when Wolf accompanied his crewmates out to the launch pad on 18 October 1993 and roared aloft at 10:53 am EST. Shortly after entering orbit, Shannon Lucid and Marty Fettman began taking the first blood samples and Wolf took their blood pressures to acquire data on early adaptive processes to the microgravity environment. The findings correlated SLS-1 data by revealing a slightly lower central venous pressure than had been predicted in ground-based studies, coupled with a larger volume in the heart's left ventricle than would be expected with the lower pressure. This shed new light upon the basic physiology of the human heart in space. Immediately after the Spacelab module had been opened for business, Rhea Seddon took ultrasound measurements of Lucid's heart with a new echocardiograph imaging device. Both Lucid and Fettman had – like Drew Gaffney, two years earlier – ridden into orbit with catheters threaded into their arms, which ran to the tips of their hearts. Lucid's catheter was removed late on 18 October, followed by that of Fettman the following day. Data dropouts from the echocardiograph led to the crew resorting to the portable American Flight Echocardiograph, which Wolf had helped to design in his pre-astronaut days.

Within hours of activation, the SLS-2 payload was already shaping up to be a tremendous success. NASA had invested $175 million in the payload, which featured 14 major experiments, of which eight focused on the crew and six on the rats. Body tissues from the latter were to be preserved for distribution to US, French, Russian and Japanese medical scientists after the mission as part of an extensive biospecimen-sharing project. In fact, Russia had long been courted as a partner in space sciences research and in 1993 was being approached to play a leading role in the space station effort. Indeed, in August 1991, when President George H.W. Bush and Soviet Premier Mikhail Gorbachev met in Moscow to talk about potential co-operation in human space exploration, it was suggested that a physician-cosmonaut might fly SLS-2, in exchange for a NASA astronaut making a long-duration visit to Mir. "The missions will increase knowledge about life sciences and data will be shared by both countries," explained *Flight International* on 14 August. "They are also seen by some observers as the first step towards joint flights to Mars." Although there were no Russians on SLS-2, cosmonauts were included in two Spacehab flights, the first in

3-25: The STS-58 astronauts pose for a portrait during their pre-launch Terminal Countdown Demonstration Test at the launch pad. Kneeling from left are John Blaha, Marty Fettman and Dave Wolf, with Rick Searfoss, Bill McArthur, Shannon Lucid and Rhea Seddon standing, and the SRBs and External Tank barely visible in the background.

1994, and they subsequently flew regularly on Shuttle missions to Mir and the International Space Station.

Rhea Seddon and the members of the SLS-2 science team took frequent blood draws from the rats' tails during the early stages of the mission and performed additional radioisotope and hormone or placebo injections to measure plasma volumes and track protein metabolism. This was part of a study into how red blood cell masses changed in weightlessness. Ultimately, the six unlucky rats destined to meet their maker in orbit were decapitated by Fettman and Seddon on 30 October, using a modified laboratory dispatcher. Pre-flight studies had already concluded that it was best to decapitate, rather than anaesthetise, the rats, because the latter process would have degraded their neural tissues and impaired subsequent observations. "Things went pretty well," Marty Fettman recounted, after the six-hour procedure ended. "We're happy to accomplish this. It was a big day for us."

Despite the science, the mood aboard the Shuttle was sombre. At one stage, only Seddon and Fettman were at work inside the Spacelab module. John Blaha poked his head over Fettman's shoulder once or twice to check on their progress, and was quickly gone. After landing, the rat tissues were used as part of a series of neurovestibular and musculoskeletal investigations to explore changes in their gravity-sensing organs and the effect of microgravity upon their limb muscles and bones. "Gravity," said backup payload specialist Larry Young, "is as profound a factor on the evolution and development of biology on Earth as oxygen and water. Yet we know so little about its influence, because, until the Space Age, we simply couldn't get away from it."

It was originally intended to take a dozen or so organs from each rat and then dispose of the carcass, but NASA issued a Research Announcement for interested scientific parties to use the other body parts. "It became known as the Parts Programme," Seddon remembered, and the astronauts found that it was no more difficult to remove eyeballs and lungs and insert them into little bags of fixative to preserve them. "Some had to be frozen," she said, "and some of them had to be refrigerated and some of them just needed to be put in the fixatives." Significantly, the inner-ear mechanisms had to be placed into fixative within two minutes of dissection, and with the inner ear buried deep within the skull, this required immense skill from Seddon and Fettman. Limb muscles, too, had to be attached to muscle clamps and fixed within ten minutes. "It was just the choreography that was incredible," Seddon added, "and Marty was just terrific at this stuff."

In addition to the rat research, a series of joint US/Canadian experiments, originally carried aboard Spacelab-1 and SLS-1, were reflown to explore motion sickness and human vestibular changes. For SLS-2, the hardware included a rotating chair mounted in the centre aisle, which examined changes in the astronauts' reflexive eye motions. Seddon was the first to use the chair on 21 October, as part of studies of the vestibulo-ocular reflex in the eye, which enables us to see whilst we are in motion. Other experiments featured a rotating dome, placed over the astronauts' heads, whose interior face was coated with a pattern of dots which seemed to 'rotate' in the opposite direction. The subject used a joystick to indicate their perceived direction and velocity. Dave Wolf also donned a special skull-cap, the Acceleration Recording Unit, which was fitted with motion sensors and was used to record the time and severity of space sickness symptoms. Investigators hoped that if the science crew wore the cap throughout their working day it might enable them to correlate instances of sickness with periods of provocative head movement.

Studies of muscular atrophy included the ingestion of amino acids, labelled with non-radioactive isotopes of nitrogen, enabling the astronauts to track protein metabolism. Urine, saliva and blood samples were routinely acquired to determine the rates of protein synthesis and catabolism. Other experiments upon the rats looked at the performance of their hind limbs in microgravity, which showed an almost 40 percent reduction of muscle fibres at the end of the 14-day mission. The rats tended to rely more heavily on their forelimbs for bipedal locomotion and used their hind limbs only as grasping aids. After their return to Earth, they exhibited slow motions and an abnormally low body posture, all of which pointed clearly to a weakened muscular state, fatigue and co-ordination difficulties. Moreover, muscle protein 'turnover' in rats is much more rapid than in humans and two weeks of weightless exposure for them was roughly equivalent to two *months* for a human.

Other experiments focused on the astronauts' cardiovascular and regulatory systems. Data from SLS-1 highlighted increases in heart rate, size and output, which researchers attributed to the initial increase in central blood volume caused by fluid shifts within the body. Three SLS-2 studies assessed the functional capabilities of the system by monitoring the astronauts' cardiac outputs, heart rates, arterial and venous blood pressures, blood volume and the amount and distribution of blood and gases in the lungs. Cardiovascular 'deconditioning' had long been recognised as a problem after the return to Earth. Astronauts complained of light-headedness, an increased heart rate and decreased pulse pressure. Echocardiograph data, together with the catheters, an exercise bicycle and a Gas Analyser Mass Spectrometer supported much of this research.

Unusually for a Spacelab mission, STS-58 followed a single-shift system, although the whole crew typically put in 14-hour working days or more. In view of the long flight, and in line with Extended Duration Orbiter protocols, each astronaut received several periods of free time to relax. "The crew members are an important part of these investigations," stressed SLS-2 Mission Scientist Howard Schneider. "We want to assure ourselves that we continue to study the physiological effects of space flight and *not* the physiological effects of fatigue! If the crew is up there and is overly stressed, we don't get good science."

In aid of his own free time, Rick Searfoss took a huge atlas of the world in his personal effects. "Rick really wanted to focus on that," Seddon told the NASA oral historian. "He didn't have an awful lot of other things that he needed to do, other than managing the orbiter and when we asked he would come back and do some experiments for us. He was really our [photography] specialist and he got some really great pictures on that flight." Every so often, from Columbia's flight deck, Searfoss would call Seddon upstairs to the windows to take a look at the south-eastern United States, as her home state of Tennessee came into view. It was difficult to get a good glimpse of her hometown, Murfreesboro, but she managed to pick out the curvaceous lines of the Cumberland River in Nashville and the parallel lines of the I-24 and Nashville Highway. The 39-degree-inclination orbit had been designed in part to keep the astronauts' sleep-wake cycles approximately the same throughout the flight and to effect a landing at Edwards Air Force Base in California. STS-58 flew in a slightly higher inclination than her previous mission, STS-40, which meant she could see all the way up to Long Island and Cape Cod, as their orbital track carried them across the United States' eastern seaboard. She also saw the Himalayas on a number of occasions.

During the mission, John Blaha – who had offered during training to be the crew's on-board videographer – spoke for all of them when he declared that "we have a beautiful planet" and "we ought to take care of it and we ought to take care of ourselves". As STS-58 entered its final days, it was becoming clear that it would come close to – or even exceed – the 13 days and 19 hours record set by the first EDO mission in July 1992. In order to keep their flying skills sharp for the return home, Blaha and Searfoss took turns on a computer program known as the Portable In-flight Landing Operations Trainer (PILOT), which consisted of a high-resolution colour display and hand controller and offered them both the 'look' and 'feel' of the orbiter. Housed in a middeck locker when out of use, PILOT was assembled on the console in front of the pilot's seat and its joystick was attached to the top of Searfoss' own hand controller. The astronauts also participated in the customary Lower Body Negative Pressure suit runs to better prepare their bodies for the punishing onset of terrestrial gravity.

Early on 1 November, the final SLS-2 experiments were concluded and then Dave Wolf supervised the deactivation of the Spacelab module in time for the first landing opportunity at Edwards. Several hours later, Blaha fired Columbia's OMS engines to commence the hour-long glide back to Earth. He guided the vehicle perfectly onto concrete Runway 22 at 8:05 am local time (11:05 am EST) to end a mission of 14 days and 12 minutes, which established STS-58 as the longest Shuttle mission to date and the United States' fourth-longest human space flight at the time. For the astronauts, and particularly the science crew, it was the start of a week of post-mission medical experiments. NASA put them up in a resort called 'Silver Saddles' and despite the discomfort of frequent blood draws – "We began to look like drug addicts," joked Seddon, "because they kept drawing blood from us" – it was a pleasant time, being able to relax in the evenings and eat with their families.

"YOU AND THE REST!"

In the summer of 1992, astronaut Jeff Hoffman was in quarantine, preparing to launch aboard Space Shuttle Atlantis with the Tethered Satellite and the EURECA free-flying payload, when he fell into conversation with Don Puddy, the head of Flight Crew Operations at the Johnson Space Center. Puddy was interested in Hoffman's future plans. Several astronauts had already been approached about their willingness to be considered for flights to the Mir space station, but Hoffman's height ruled that out. Only one other mission captured his attention. The first servicing mission to the Hubble Space Telescope stood out like a jewel on the Shuttle manifest in December 1993 and as a professional astronomer, Hoffman found that it exerted an irresistible pull.

"What I'd really love," he told Puddy, "is to go on this Hubble mission."

Puddy laughed. "Oh, yeah. You and the *rest* of the office!"

Hoffman assumed that his chances of selection were minimal, but in addition to his flight experience he had one other credential which made him an attractive choice for the mission: he was one of few astronauts in the office, at that time, with EVA experience. Several years earlier, in April 1985, Hoffman had participated in the Shuttle programme's first contingency spacewalk in a fruitless attempt to activate a deployment switch on the malfunctioning Syncom 4-3 satellite. With several intricate and complex EVAs scheduled for the Hubble mission, NASA mandated that *all* members of the four-person spacewalking team must have prior EVA expertise. This obviously disappointed several unflown members of the office, including rookie astronaut Leroy Chiao.

"I was doing EVA training and showing some proficiency at it in the water tank," Chiao told the NASA oral historian. He had been approached by Dave Leestma, then serving as deputy chief of the office and later to serve as head of Flight Crew Operations, with what Chiao perceived to be a strong hint that he was in line to receive one of the EVA spots on the Hubble mission. "I was very excited about that," Chiao continued, "and my classmate Eileen Collins ... had heard through the grapevine that she was going to get assigned as the pilot on that flight. Then I started hearing rumours that the crew for that flight was going to be changed. They didn't want any rookies, at least on the EVA team. I have to say that was hard to swallow, because I had worked hard and I had shown proficiency and I had been *told* I was going to be on that flight and then, for political reasons or visibility reasons, they wanted to be able to say that it was an experienced crew if something had gone wrong." Certainly, in March 1994, *Flight International* noted that Collins' name had been proposed, "but was overruled because it was felt that an experienced pilot was needed". It was not simply a case of NASA being overly cautious. With the loss of Challenger still fresh and the embarrassing failure of Hubble's ability to resolve distant objects, due to a flaw in its primary optics, the mission to fix the $1.5 billion showpiece telescope was crucial. Congressional support for Space Station Freedom hung on the edge of a knife and any failure on NASA's part could spell its cancellation.

Today, Hubble has earned itself a well-deserved reputation as one of the most successful space-based observatories ever launched. Across more than two decades of operations, its instruments have peered deeper into the cosmos than ever before. It has acquired images of distant galaxies, made breakthroughs in physics and cosmology by accurately determining the Universe's rate of expansion, detected planets around far-off stars, witnessed the impact of a comet into Jupiter, tracked cloud movements in the atmospheres of Uranus and Neptune and created the best currently achievable 'map' of the surface of Pluto.

With the advent of the Space Age, it was hardly surprising that plans for a space-based telescope would become an important step forward and an attractive option for the fields of astronomy and astrophysics. Yet the ideas long pre-dated even the launch of Sputnik. Shortly after the Second World War, physicist Lyman Spitzer of Yale University had argued that an orbiting telescope would offer enormous advantages over ground-based instruments, its abilities unimpaired by the distorting effect of Earth's atmosphere and its sensors able to detect high-energy emissions, including X-rays, from distant celestial sources. Following the creation of NASA, the first real efforts to develop a space telescope got underway and in 1975 the agency tried to sell the project to the politicians. Funding was initially denied by the House Appropriations Subcommittee, who reasoned that it was too ambitious, too expensive at around $400 million and lacked the required support from the National Academy of Sciences. This prompted large-scale lobbying from NASA and leading astronomers *and* a supportive report from the National Academy of Sciences. International co-operation was directed by Congress and the newly-formed European Space Agency was invited to participate, with its role encompassing the creation of inexpensive solar panels for the telescope. The size of the mirror was reduced from 3 m to 2.4 m and together these measures halved the cost from $400 million to $200 million. There were other reasons for the reduction in mirror size. "The Shuttle could not lift a 3 m telescope to the required orbit," wrote Andrew Dunar and Stephen Waring in their book *Power to Explore*. "In addition, changing to a 2.4 m mirror would lessen fabrication costs by using manufacturing technologies developed for military spy satellites. The smaller mirror would also abbreviate polishing time from 3.5 years to 2.5 years."

In 1977 Congress granted approval for what was then known as the 'Large Space Telescope'. The primary candidates for the fabrication of the observatory's mirror were Perkin-Elmer Corporation, whose bid ran to $64.2 million, and Eastman Kodak, teamed with the defence contractor Itek, at almost $99.8 million. Despite being significantly higher, the Kodak-Itek joint bid included *two* independent tests of the grinding and polishing quality of the finished optics ... a 'double-checking' provision which Perkin-Elmer did not offer and which would not go unnoticed more than a decade later, when investigators dug into the cause of the telescope's unfortunate spherical aberration. Perkin-Elmer received approval from NASA to proceed with their bid in 1979. Meanwhile, Lockheed would build the spacecraft itself and the Europeans would make the solar arrays. In anticipation of the research bonanza, a Space Telescope Science Institute (STScI) was established at the Johns Hopkins University in Maryland in 1983 and the telescope itself was scheduled for launch by the Shuttle in 1985. By this time, it had been named in honour of the American astronomer Edwin P. Hubble, who, in the earlier part of the century, had not only conducted extensive research into the structure of stars and galaxies, but also made the surprising discovery that the Universe is expanding. The mirror was one of the most complex headaches of the project – both *before* and *after* launch. Optically, Hubble was a Cassegrain reflector and its two hyperbolic mirrors offered good imaging performance across what, for such a large telescope, was a wide field of view ... whilst also having shapes which were difficult to fabricate and test. Perkin-Elmer used custom polishing machines to precisely grind the mirror and, in case problems were encountered, NASA required the company to subcontract to Kodak to build a backup mirror using traditional polishing techniques. (The Kodak mirror is today on permanent display in the Smithsonian.)

In 1979, the construction of the Perkin-Elmer mirror began and was completed two years later, washed in hot, deionised water and coated with aluminium and protective magnesium fluoride. NASA remained sceptical about Perkin-Elmer's ability to competently grind and polish the mirror and the delays ultimately pushed

Hubble's launch back from April 1985 to first the summer and then the autumn of 1986. By this time, the total cost of the project had risen to a little more than $1 billion. At the time of its completion, Hubble housed five instruments: the Wide Field Planetary Camera (WFPC, nicknamed 'the whiffpick'), the Goddard High Resolution Spectrograph (GHRS), the High Speed Photometer (HSP), the Faint Object Camera (FOC) and the Faint Object Spectrograph (FOS). These devices gave the telescope a range which encompassed not only the visible area of the electromagnetic spectrum, but also the near-ultraviolet. Physically, Hubble was a cylindrical spacecraft, measuring over 13 m in length and weighing nearly 11,000 kg, which meant that it virtually filled the payload bay. It had been designed to be serviced by future Shuttle crews and, as such, was fitted with EVA-friendly hand holds, and would be deployed and retrieved using the RMS arm.

By the time that Challenger was lost, further processing delays had pushed Hubble's launch back to October 1986. The problems faced by Perkin-Elmer have already been mentioned, but the manufacturer of the telescope's bodywork, Lockheed, had also suffered difficulties. By the end of 1985, Hubble was over-budget by 30 percent and three months behind schedule, bringing it dangerously close to breaking the 'ceiling' which Congress had imposed on its budget. If

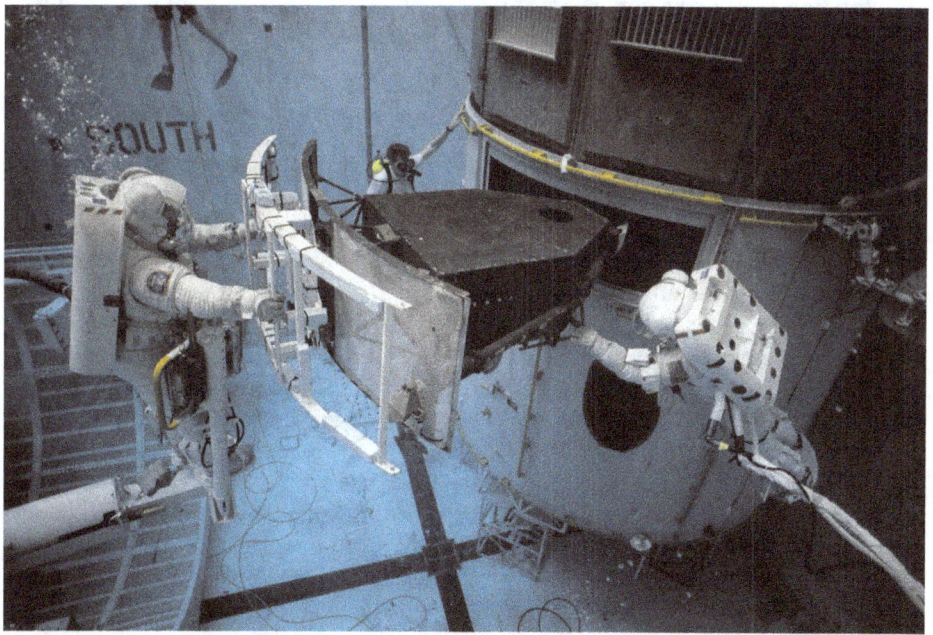

3-26: In the Weightless Environment Training Facility (WET-F) water tank at the Johnson Space Center, Kathy Thornton (left) and Tom Akers practice installing the Wide Field Planetary Camera (WFPC)-2 into the Hubble Space Telescope. The enormous size of the WET-F allowed for the emplacement of full-sized replicas of Hubble and the payload bay of the Shuttle itself.

Challenger smashed the dreams of so many within America's space programme, it also provided breathing room for payload development. Hubble came through a major thermal vacuum test with flying colours in June 1986 and the enforced down time was used to add more powerful solar arrays, enhance redundancy capabilities, improve the software, install better connectors and replace the nickel-cadmium batteries that were prone to failure with nickel-hydrogen ones. As a result, by the start of 1990 the Hubble team felt supremely confident that their observatory heralded a new dawn in the study of astronomy. As long ago as 1983, NASA Administrator Jim Beggs had encouraged his subordinates to treat Hubble in terms of significance on a par with the Shuttle itself and had even labelled it "the eighth wonder of the world".

In the first few weeks after Hubble's April 1990 launch, the problems seemed reasonably benign: a few communications glitches, drifting star trackers and snagged coaxial cables were part and parcel in the process of wringing out a new spacecraft. More serious concerns arose when temperature changes bent materials in the solar arrays' booms, the effect of which was magnified by the orientation mechanism in such a way that it 'bounced' the whole telescope. The result was a 'jittering' in Hubble's images and, since the booms only stabilised in the final few minutes of orbital daylight, the pointing system was only able to meet its design specifications for a fraction of its orbit. Engineers at the Marshall Space Flight Center worked with their counterparts at Lockheed to change the control program in the spacecraft's computer and successfully counteracted the vibrations. On 21 May, Hubble returned its first images of a double star in the constellation of Carina and these were lauded as being much clearer than were achievable with ground-based instruments.

Four weeks later, calamity befell the mission in a manner which could hardly have been anticipated. On 24 June, Hubble failed a *focusing* test. Its secondary mirror had been adjusted to focus the incoming light from a celestial source, but a fuzzy ring – like a halo – encircled even its best images, creating a blur. Additional tests revealed that the telescope was suffering from a 'spherical aberration' in its primary mirror; in essence, Perkin-Elmer had ground it to the *wrong* specification, removing too much glass and polishing it *too flat* ... by a mere *fiftieth* of the width of a human hair. The consequence was that Hubble was unable to acquire sharp images. With mounting horror, NASA realised that its attempts to sell its scientific showpiece on the basis of its ability to see further into the cosmos than ever before, with unprecedented clarity, now became very hollow indeed. The promised white knight of astronomy was now a white elephant. Even Hubble's chief scientist, Ed Weiler, admitted that it was comparable only to "a very good ground telescope on a very good night". Marshall staff were astounded and Senator Barbara Mikulski, a Democrat from Maryland, exploded that Hubble wasted taxpayers' money and was little more than "a techno-turkey". Meanwhile, Senator Al Gore – then a Democrat for Tennessee and later Vice President during the Clinton Administration – observed that, for the *second* time in less than *half* a decade, quality control shortcomings at NASA had been publicly exposed. The media had a field day. On 28 July 1990, the *New York Times* pointed out that – had Kodak-Itek's bid been accepted – then the mirror would have been subjected to *two* independent checks of its grinding and polishing accuracy, which certainly would have caught the error and enabled it to be rectified before launch. NASA responded to critics by asserting that, with 20-20 hindsight, it would have cost in excess of $100 million to incorporate additional testing and independent checking of the telescope optics into Perkin-Elmer's contract, but the effect on the general public was the same. The once-proud agency was rendered a laughing-stock on late-night TV talk shows. David Letterman compiled a pejorative list of Top Ten Hubble Excuses, whilst others criticised the Marshall Space Flight Center for having been in charge of *both* the Hubble development *and* the Shuttle's SRBs. Several analysts noted that NASA's attitude had changed from the 1960s, in which problems were anticipated and incorporated into planning, into one where there was apparently little effort to prepare for unforeseen obstacles. In the words of space policy analyst John Logsdon of George Washington University, "the agency was not being honest with itself or with anyone else".

In early July 1990, NASA established an investigating committee, chaired by Lew Allen, head of the Jet Propulsion Laboratory. His report – published in November – harshly criticised the incorrect assembly of the 'reflective null corrector', an optical device used to determine the figure of Hubble's mirror. The location of a lens in the device was improperly measured and the null corrector guided the polisher to shape a perfectly smooth mirror ... with the *wrong* curvature. Analysis revealed that the curvature flaw in the primary mirror exactly matched the flaw in the null corrector. A *second* null corrector, made only with lenses, was also built to measure the vertex radius of the finished mirror. It, too, clearly identified an error in the primary mirror. However, neither of these warning signs were heeded and Allen's report noted that "*both* indicators of error were discounted at the time as being *themselves* flawed". During the fabrication process, technicians had simply *assumed* the perfection of the mirror and of the reflective null corrector and therefore had rejected information from other independent tests, having convinced themselves that no problems existed. These errors were ultimately traced back to 1981-82, when Perkin-Elmer and the Marshall Space Flight Center had been distracted by serious cost and schedule difficulties. Allen's report was particularly critical of Perkin-Elmer quality control and communications failures, as well as the failure of the Marshall Space Flight Center to correct them.

In orbit, the spherical aberration was particularly obvious in its effect on Hubble's WFPC and FOC, both of which suffered in terms of spatial resolution and their ability to acquire images of individual celestial objects, including planets, star clusters and galaxies. Having said this, the aberration was well characterised and stable and, over time, astronomers were able to optimise the results obtained from Hubble by using sophisticated techniques, such as 'deconvolution', whereby software algorithms and microwave image processing methods were employed in an effort to remove many of the blurring effects of optical distortion. However, the results were still less than ideal. Spectroscopy by the FOS and GHRS instruments was less severely affected, because the instruments required less focused light, and by increasing exposure times it became possible to gather valuable images. Nevertheless, by the end of 1991, Hubble had made almost 2,000 quality observations of hundreds of astronomical objects, including storms on Saturn and images of Pluto's moon, Charon.

At the start of the following year, a quarter of all the papers presented before the American Astronomical Society's meeting drew on Hubble data. A repair was critical in order to restore the telescope to its pre-flight billing, although NASA announced in July 1991 that it was "not planning an early visit" to repair the telescope, preferring instead to stick with the previous schedule of an all-up servicing mission in late 1993. Even though Kodak-Itek had produced their mirror, this could not simply be inserted into the spacecraft, so the solution would have to overcome the problem with the existing primary mirror. A new device – the $50 million Corrective Optics Space Telescope Axial Replacement (COSTAR) – would need to be developed and manifested onto the first servicing mission to revitalise its vision. In October 1991, NASA awarded a $30.4 million contract to Ball Aerospace to fabricate COSTAR, whose ten small, coin-sized mirrors were expected to correct the spherical aberration and restore the potential of the affected instruments.

Interestingly, Bruce McCandless, one of the astronauts who carried Hubble into orbit on STS-31 in April 1990, was closely involved with the development of the optics, as related by his former crewmate Kathy Sullivan. She remembered McCandless telling her that all ideas of how to restore its capabilities were entertained, even the wildest and most outlandish ones, such as directly entering Hubble's telescope barrel. "The bad news is you did indeed screw up the mirror by an amount that is significant and should be avoidable in figuring astronomical mirrors," Sullivan told the NASA oral historian. "The *good* news is you screwed it up very precisely! This meant you knew the actual mirror shape, very precisely, and could do a precise difference calculation of the 'actual' versus the 'desired' and determine the needed correction out of that." This realisation was met with a determination that the correction could be accomplished through reflective lenses, instead of transmissive ones, and was intricate in its scope and brilliantly clever. As Sullivan put it: "I don't have to make all that light come through a lens. I can bounce it off several mirrors and through a couple of steps basically restore it to the focus that it should have had."

COSTAR's mandate was to optically correct the effects of spherical aberration upon the instruments ... at the expense of losing the phonebox-sized High Speed Photometer, which had been rendered virtually useless by the solar array jitters. COSTAR was invented by the Hubble Space Telescope Strategy Panel, a group of scientists and engineers convened by the STScI in late 1990, and after the removal of the 220 kg HSP would be installed in its stead. "Once in place," explained NASA's pre-mission press kit, "COSTAR will deploy a set of mechanical arms, no longer than a human hand, that will place corrective mirrors in front of the openings that admit light" into the affected instruments. In doing so, it would refocus light from Hubble's primary mirror *before* it reached those instruments and was expected to bring their overall optical performance "very close" to original specifications. Subsequent instruments for the telescope would be specifically designed with their own corrective optics. Dr John Wood of NASA's Goddard Space Flight Center was placed in charge of the COSTAR development effort. Its installation and that of a new whiffpick (WFPC-2) would come atop an already hefty pile of work for the First Servicing Mission (SM-1), which already involved the replacement of Hubble's twin solar arrays and drive electronics, two of the three Rate Sensing Units, one of two Electronic Control Units, one of two magnetometers and fuse plugs to correct wiring and sizing discrepancies.

"Whoever was going to be doing something on the Hubble mission had to have done it before," recalled Jeff Hoffman in his NASA oral history. "You had to have somebody who had already been a commander, somebody who had already been a pilot, four people who had already done EVAs and somebody who had already done a significant arm operation." In March 1992, Story Musgrave was assigned as payload commander for the flight, designated STS-61, and in August Hoffman and fellow astronauts Kathy Thornton and Tom Akers joined him to support at least five ambitious EVAs. The mission thus morphed into something which would represent nothing less than an opportunity for NASA to prove itself triumphantly ... or fail spectacularly. In its January 1990 manifest, the agency listed SM-1 as a five-day flight with a crew of five, suggesting a maximum of only two or three EVAs, but as 1991 wore into 1992 and onward into 1993 it became increasingly clear that the mission would run to as long as 11 days and evaluations of underwater simulations convinced managers that they should schedule as many as five back-to-back EVAs over five days. (The flight plan actually provided for a sixth and seventh EVA, and a mission duration of up to 13 days, although this did not become necessary.)

Such an enormous workload demanded a crew of seven, with two alternating teams of spacewalkers, to reduce fatigue and enhance the likelihood of mission success. Original plans called for all tools to be kept outside, in the Shuttle's payload bay, but the crew recognised at an early stage that EVA time was a critical limiting consumable and decided that the hour spent preparing equipment at the start of each excursion could be better spent starting the repair work. It was therefore decided that some tools would be kept inside Endeavour's crew cabin, enabling the spacewalkers to 'load-up' before opening the airlock and utilising their suits' consumables. "What we've done by going to five EVAs, rather than three, is to repackage our margin," said STS-61 Mission Director Randy Brinkley, "so that we have the capability to respond to the dynamics, or unknowns, of spacewalks. It improves the probabilities for mission success, while providing added flexibility and adaptability for reacting to real-time situations."

In December 1992, Dick Covey, Ken Bowersox and Swiss astronaut Claude Nicollier were named, respectively, as commander, pilot and operator of the RMS arm, which would pluck Hubble out of orbit, anchor it into the payload bay for repairs and subsequently deploy it back into space to begin its reinvigorated studies of the Universe. According to Tim Furniss, writing in *Flight International* in March 1994, Nicollier was specifically requested by NASA Administrator Dan Goldin for the RMS task, "to reflect the European involvement in the Hubble programme". Having received such a plum assignment, the STS-61 team bore the brunt of much good-natured ribbing from their fellow astronauts. On one occasion, Thornton quipped to her crewmates: "Well, guys, *everybody* is gonna *hate* us now!"

Covey's arrival on the crew was particularly interesting; in *Flight International*, Tim Furniss noted that he had been "strangely sidelined" since his last mission,

STS-38. Covey had flown the Syncom 4-3 retrieval and repair mission in August 1985 and thus was well-versed in the rendezvous, proximity operations and servicing associated with a large payload, but his previous stints as acting chief of the astronaut office *and* acting head of Flight Crew Operations certainly also factored into his selection. Late in 1992, he was called to a meeting by JSC Director Aaron Cohen and his deputy, Paul Weitz, and offered a choice: he could either be the 'permanent' chief of the astronaut office or he could command the Hubble flight. By his own admission, it took Covey "about two seconds" to make his choice.

"Doing five [EVAs] really pushed the bounds of what people thought we could do," Covey recalled in his oral history. "Even with four EVA crew members, even with an 11-day mission, it just started pushing the bounds. There was a lot of scrutiny on it and a lot of focus on it." The size of their quarry posed additional problems. Hubble was far larger than anything with which the Shuttle had previously rendezvoused in orbit and Claude Nicollier was faced with the unenviable challenge of manoeuvring his EVA crewmates, along with phonebox-sized pieces of hardware, into position with extreme delicacy and precision. "The integrated operations," said Covey, "of Shuttle manoeuvring, RMS activities and EVAs, although now commonplace, *wasn't* back then. So integrating all of those activities and the crew activities together was a big part of my role as the commander."

To be fair, the audacity of the mission filled NASA's top brass with dread and the memory of Challenger still loomed large in many minds. In May 1993, the Task Force on the Hubble Space Telescope Servicing Mission – established by NASA Administrator Dan Goldin the previous January and chaired by Dr Joe Shea – concluded that "the mission is achievable". This was primarily due to the fact that Hubble and most of its subsystems were originally designed for orbital maintenance. However, Shea's report noted that such missions were enormously complex in their scope and required more EVA time than had been achieved on any flight up to that time. With this stark reality in mind, the group advised that planning for the *next* Hubble servicing should enter the planning stage within 6-12 months of STS-61, in order to "handle tasks that might not be completed during the first mission or respond to failures that occur in the intervening months". Shea also found that the appointment of a Mission Director (Brinkley) was "necessary if the mission is to be carried out with confidence". However, the August 1993 loss of Mars Observer, only days before it was due to enter orbit around the Red Planet, added to the popular sense of NASA as an organisation of failure. An article in *Aviation Week* hinted that WFPC-2 might also be flawed and it so unnerved the Clinton Administration that Goldin was invited to attend an audience with the President to discuss the agency's level of preparedness for SM-1. It has been suggested that Clinton told Goldin that the agency had either to make the Hubble repair mission *work* or be drastically restructured. Mistakes could not be tolerated.

The heat was on the agency, on the STS-61 crew and on the thousands of engineers and technicians responsible for ensuring that COSTAR and WFPC-2 were ready for launch in December 1993. Nerves frequently became frayed in the final

3-27: One of the STS-61 spacewalkers captured this spectacular view from the end of the RMS mechanical arm, showing Endeavour backdropped against the blackness of space and the Sun displaying a 'rayed' effect.

months. Late in September, NASA announced that "further testing" of WFPC-2 "might be necessary", requiring it to be shipped back from the Kennedy Space Center to the Goddard Space Flight Center, although such a move was not expected to affect the scheduled launch in early December. According to Joseph Rothenberg, Associate Director of Flight Projects for Hubble Space Telescope at Goddard, results from earlier tests suggested that the focus point for HST might be outside the COSTAR adjustment range and that focusing both it and WFPC-2, simultaneously, might not be achievable. As circumstances transpired, such a move was unnecessary, but time constraints were critical. *Are we missing something?* was the introspective question often asked of themselves and each other. Others were irritated by constant claims that they were "testing things to death" and therein causing further delays, which prompted one manager to explode: "We wouldn't *be* here if you guys had tested this thing to death the *first* time!" Some engineers half-jokingly offered to spend the New Year in Acapulco, figuring that if SM-1 succeeded they might be able to return home to the United States and if it failed they would be forced to wait on restaurant tables in Mexico for the rest of their lives!

Training for the STS-61 astronauts was also difficult at times. "We were invited to come down to NASA Headquarters to meet with Dan Goldin," remembered Jeff Hoffman. "He told us quite frankly that NASA's future was in our hands. That was the time when we were waiting for Congress to approve the construction of the space station. Everybody recognised that assembling the space station was going to take a lot of sophisticated EVAs, of the sort that we were getting ready to do for Hubble, so if we went up thinking that we could fix Hubble and then it turned out that we *couldn't*, how could people trust NASA to build a space station? *That* was the attitude."

Story Musgrave had been assigned to lead the planning of the EVAs, as payload commander, and he had pushed very hard to commit to Endeavour as the orbiter which would carry out the mission. He particularly highlighted the vehicle's improved capabilities over her sisters, including the capacity for a long mission. He got his way: in NASA's August 1991 manifest the SM-1 mission was scheduled for Discovery and by January 1992 had moved to Atlantis and by December 1992 it had been shifted to Endeavour. All of the spacewalkers recognised the need to develop physical strength to handle the demands of their space suits and build the necessary stamina for six or seven hours outside. Kathy Thornton worked out in the gym, as did the others, although by Hoffman's admission most of the servicing tasks did not demand immense physical strength, but placed greater emphasis on "technical co-ordination", involving them "being very careful in how you moved things around and not messing anything up".

The delicacy involved in each of the spacewalking tasks was further complicated by the need for sunlight never to enter Hubble's interior, because it carried the potential to evaporate organic contaminants and potentially ruin the sensitive ultraviolet optics. In normal operation the telescope was never allowed to point anywhere near the Sun. Consequently, the mission was planned with the Shuttle's belly positioned to face the Sun – but this, in turn, meant that temperatures in the payload bay and upon the astronauts' suits would fall precipitously. Already, the critical nature of the EVAs had obliged NASA to assign veteran spacewalker Greg Harbaugh in March 1993 to train as a backup crewman and provide "some insurance in the event of the unavailability late in the training cycle of any of the four prime EVA crew members". Harbaugh would be ready to step in if circumstances dictated.

And they almost did.

On 28 May, Musgrave was performing an eight-hour suited equipment test in one of the thermal vacuum chambers at the Johnson Space Center, when he complained of intense coldness in his right hand. He persevered and then upon removing his suit at the end of the test, he noticed discolouration and numbness in his fingers. (They were "black and purple," according to Jeff Hoffman.) Flight surgeons quickly identified a case of severe frostbite and Musgrave was referred to a specialist in Alaska for treatment. At this stage, Hoffman was unsure what would happen. "I don't know exactly what they did to him," he told the oral historian, "but they managed to save his fingers, and he flew, but that definitely got management's attention."

That 'attention' was both positive and negative for Musgrave. On the positive side, it led to the elimination of the belly-to-Sun attitude, in favour of an attitude whereby Covey and Bowersox would execute attitude manoeuvres, during each orbit, to ensure that sunlight did not enter the telescope. In Dick Covey's recollections, the single-mindedness of Musgrave towards training for the mission caused some anger and even demands for his removal from the crew. "Story had been around for a long time," said Covey. "There was this concerted effort to use Story's injury as a reason to get him thrown off the crew by some people with the agency and [JSC] and I had to go fight that ... The reasons were political and personality-based, rather than technically based on his capabilities and whether he was going to recover from his injuries. That was hard to deal with." Harbaugh was also recovering from recent knee surgery and, for a while, struggled to even don a space suit, but both he and Musgrave were declared ready to support the mission. In fact, Harbaugh went on to serve as EVA Capcom for STS-61 and also flew aboard the second Hubble servicing mission in February 1997.

A windstorm on 30 October 1993 prompted the movement of Endeavour from Pad 39A to Pad 39B, following sand-blasting grit contamination of the payload changeout room, with much anticipation of a successful launch in the small hours of 1 December. The attempt was scrubbed due to high winds at the Shuttle Landing Facility, together with excessive cloud cover and the presence of a ship in the restricted waters, but, after a 24-hour delay, STS-61 blazed into the darkened Florida skies at 4:27 am EST on the 2nd.

Several burns of Endeavour's thrusters over the following two days closed their separation distance from Hubble by around 95 km per orbit, until, early on the 4th, Hoffman spotted the telescope through his binoculars ... and noted that one of its solar arrays – due to be replaced by Thornton and Akers during the second EVA – appeared to be bent in a 90-degree angle. "It was a beautiful sight when we finally could really see the Hubble," Dick Covey remembered in his NASA oral history, "and it is as bright as anything you can imagine, because of the silver-coloured insulation, and the gold of the solar arrays just made it spectacular when it first came into visual range and tracked on in." Two hours ahead of retrieval, at a distance of 12 km, Covey executed the Terminal Initiation burn, then performed several small mid-course correction firings to bring Endeavour to a position some 360 m 'below' and 150 m 'behind' Hubble. By this time, the telescope's twin high-gain antennas had been stowed against its main body. Closing in on their quarry just after orbital sunset, this approach was designed to minimise contamination from the Shuttle's thrusters pulses. As the orbiter drew closer, a ground-commanded manoeuvre of the telescope aligned its grapple fixture convenient for the RMS. Finally, at 3:48 am, Covey brought the orbiter to a position just ten metres from Hubble and Nicollier gingerly extended the arm and grappled the target. Endeavour was high above the South Pacific, to the east of Australia, at the moment of capture.

"Houston," radioed Covey, triumphantly, "Endeavour has a firm handshake with Mr Hubble's telescope."

"Roger, Covey," replied Capcom Susan Helms. "There's smiles galore down here."

The first major objective of the mission had been accomplished. Yet the real challenge of the five back-to-back EVAs and the uncertainty about whether the efforts of the astronauts would succeed remained to be seen.

With the enormous bulk of Hubble safely anchored to the Flight Support Structure – a rotatable and 'tiltable' turntable, a little like an oversized 'lazy susan' – in Endeavour's payload bay, the next step was EVA-1. The spacewalks would be performed daily, with Musgrave and Hoffman charged with the first, third and fifth and Thornton and Akers assigned to the second and fourth. Encased within their pressurised suits, the astronauts were identified by the presence (or absence) of markings on their legs: Hoffman (EV1) would have red stripes, Musgrave (EV2) would have no stripes, Thornton (EV3) would have dashed red stripes and Akers (EV4) would have diagonal broken red stripes. All four spacewalkers were extensively 'cross-trained' to allow them to perform any one of the mission's given EVA tasks using around 200 tools – from power ratchets and sockets to safety bars and articulating foot restraints and from portable work lights and locking connectors to instrument covers, handles and umbilical connectors – and a total of 7,200 kg of equipment would be transported into orbit for the repair.

Looking back on those adrenaline-charged days, Covey was filled with pride that his crew accomplished everything they set out to do. "There wasn't anybody that was chilling down on the middeck," he said. "Everybody was up top, working. There was concern about whether we could sustain that tempo. We went five straight days doing EVAs and that was the right answer. Everybody felt good about that. Nobody was getting excessively fatigued. The EVA crew members, because they were getting a day off in between were okay with that and so that facilitated us pressing on with five straight days of spacewalks."

It was testament to the planning of each of these six-hour-plus excursions that Musgrave and Hoffman were outside in the payload bay, an hour earlier than planned, at 10:46 pm EST on 4 December, to kick off EVA-1. Their main task was the replacement of Hubble's twin sets of Rate Sensing Units (RSU), which carried gyroscopes to effect tracking of the telescope. Three of a total complement of six gyroscopes had failed between December 1990 and November 1992 ... and *at least three* were required to maintain proper pointing. After setting up tools, safety tethers and equipment, Hoffman fitted a foot restraint onto the end of the RMS and was manoeuvred by Nicollier towards Hubble. In the meantime, Musgrave installed protective covers over the telescope's low-gain antenna and exposed voltage-bearing connectors. Next, the two men (whom Dick Covey had lightheartedly nicknamed "the odd couple" during training because they were assigned the 'odd-numbered' EVAs) opened Hubble's equipment bay doors and emplaced another foot restraint inside. Working inside the bowels of the largest space telescope yet placed into orbit, astronomer Hoffman had little time to ponder, and by 12:24 am EST on the 5th, less than two hours into the EVA, the first set of new RSUs were in place. The second set followed, as did the installation of eight fuse plugs to protect the electrical circuits. By now, Hubble had a full, healthy set of gyroscopes.

Then the spacewalkers hit their first major problem: the doors to the gyro compartment refused to close and seal properly. "The doors in the telescope gyro compartment are the biggest doors in the whole telescope," Hoffman explained. "In fact, they're asymmetrical; one of the doors is bigger than the other. We had opened and closed those doors a hundred times in the water. We knew how they worked. There were several latches, but there was one big handle. You turned that handle and that basically closed the latches; then you just had to throw a couple of bolts and tighten up the bolts and the door is secured." Upon closer inspection, it appeared that two door bolts did not reset correctly and Hoffman suspected the doors had somehow become 'warped', perhaps by uneven heating. If the doors did not close, Hubble would be lost, for its thermal control capability would be gone and light leaking into the telescope's innards would ruin its delicate instruments.

Whenever he tried to close the top of the door, the bottom would refuse to close, and its height precluded Hoffman from holding both ends at once. He asked Musgrave to help him. Unfortunately, Musgrave was tethered and floating freely and could only push the door with one hand, since he needed to steady himself with the other hand. "It was basically a five-handed job," said Hoffman, "and we only had *four* hands. We tried a few times." On one occasion, Musgrave even tried pushing with his helmeted *head*, to no avail. At length, the spacewalkers recommended to Mission Control the use of the payload restraint device – "kind of a webbing tool, with a ratchet" – to help bring the doors into closure. Lead Flight Director Milt Heflin agreed that the crew were in the best position to make the decision and gave them the go-ahead. It *worked* and the doors were successfully closed and latched ... but at the expense of EVA-1 turning into the second-longest Shuttle spacewalk to date: seven hours and 54 minutes.

With Kathy Thornton and Tom Akers slated to replace Hubble's twin solar arrays on the next day's EVA-2, the complexity of the back-to-back spacewalks became evident. "All four of us worked together," said Hoffman, "because when you get a space suit ready, you try to be each other's personal valet. There's just a lot of work to be done. When you're doing a spacewalk, it really takes over the *entire* Shuttle. You can't really be doing anything else while the EVAs are planned." Pre-flight training indicated that changing both arrays would require around five hours, thereby necessitating an entire EVA to be devoted to this objective. Although the arrays were designed to accommodate expansion and contraction caused by heating and cooling during orbital daytime and nighttime, actual experience demonstrated that this did not occur as smoothly as hoped. A temporary fix had been effected, whereby Hubble's pointing system compensated automatically for the jitter induced by the solar panels, but ESA was assigned the responsibility to produce a new set and reduce the effect to an acceptable level. The new arrays included thermal insulation sleeves on their 'bi-stem' supports to minimise heating and cooling and springs worked like shock absorbers to take up tension at the ends.

Thornton and Akers ("the even couple", according to Dick Covey) ventured into the payload bay at 10:35 pm on 5 December to the sight of a slightly different Hubble, for the roller-blind-like solar arrays had been commanded to fold up. The plan was to replace the arrays with new ones, carried aboard a Solar Array Carrier (SAC) in the payload bay, then load the old arrays onto the SAC for the return to Earth. One of them (the port-side array) folded up perfectly and could be stowed for return to Earth, but a bent bi-stem strut prevented the other one from doing likewise.

It could only be moved to a position about 30-percent-closed, because any attempt to fold it further risked breaking the bi-stem and creating a risk to the spacewalkers. This left Thornton and Akers with little alternative but to dump it overboard. Interestingly, Thornton lost voice communications with Endeavour or the ground until around three hours into the EVA, requiring Akers to serve as a 'relay'. (Earlier, she had also experienced lower-than-normal pressure in her vent garment, due to a temporary ice 'plug' in the suit's plumbing system.) An hour into the EVA, they dismounted the damaged array – which was 4.8 m long when folded up and weighed 160 kg – during orbital darkness, to minimise electrical activity, and Thornton held it until the next daylight pass. This would allow mission controllers to track its position and relative velocity. She threw it overboard at 11:52 pm, as Endeavour sailed high above Somalia, describing its departure as resembling a bird in flight.

"Then we had to fire our manoeuvring jets to get away from it," remembered Hoffman. "The solar array was just inert. That was really spectacular, because when the exhaust plume from the reaction control jets hit this solar array ... and it started to oscillate, up and down, it looked like the wings of a giant prehistoric bird, just flapping out in space." Watching from Endeavour, the crew of STS-61 was mesmerised by the spectacle, as the array somersaulted a few times in the vacuum. Those moments of silent awe were suddenly broken by a voice over the radio. It was Tom Akers. "Hey," he said, "isn't somebody supposed to be reading me the procedures? We have *work* to do!"

That broke the spell, recalled Hoffman, and they went back to work. EVA-2 ended after six hours and 36 minutes. Its relative brevity came as a relief, for the astronauts were at least able to enjoy dinner together and a proper night's sleep. In Hoffman's mind, if EVA-2 had lasted as long as EVA-1, it might have thrown them seriously behind on the timeline. He also had the fright of his life when he helped Thornton to remove her space suit; as he pulled off one of her gloves, Hoffman noticed that her fingertips were bright red. His first thought was that it was blood. As it turned out, a chunk of Thornton's red-coloured food bar had floated away from her mouth and somehow made its way down through her suit, into one of the arms and into the glove ...

"Not nearly as serious as it looked," Hoffman acquiesced, "but I got quite a shock when I pulled her glove off."

It was Musgrave and Hoffman's turn the following night, 6-7 December, with the primary task of installing WFPC-2 into the telescope. This had been developed by the Jet Propulsion Laboratory in 1985 as a 'spare' and after the discovery of the spherical aberration NASA and the WFPC team had installed an optical corrector. "The new design incorporates an optical correction by the refiguring of relay mirrors already in the optical train of the cameras," read NASA's pre-flight press kit. "Each relay mirror is polished to a new specification that will compensate for the incorrect figure on [Hubble's] primary mirror. Small actuators will fine-tune the positioning of these mirrors on-orbit, ensuring the very precise alignment that is required." The WFPC team also upgraded the instrument, by reducing the number of cameras from eight to four in order to develop an alignment system and adding improved charge-coupled devices to aid its ultraviolet sensitivity.

3-28: The repair of the Hubble Space Telescope was perhaps the singular mission, at the singular critical juncture in time, which saved NASA and enabled the future direction of the space agency to be assured. In the days following the mission, the intergovernmental agreements which produced the roadmap for today's International Space Station were signed and a new partnership with Russia was formalised.

An hour into the spacewalk, Hoffman crisply removed the old whiffpick from its housing in Hubble's bowels and inserted it into a storage container in the payload bay. A protective hood was then removed from the new device and it was installed perfectly at 1:05 am EST. Ground controllers ran an 'aliveness' test and verified that the 280 kg pie-wedge-shaped WFPC-2 was working correctly. The spacewalkers then replaced a pair of magnetometers, before returning inside Endeavour after six hours and 47 minutes. This proved to be an exceptionally good time, when one considers that training for the whiffpick replacement alone had typically taken four and a half hours in the water tank.

Thornton and Akers were next, on 7-8 December, with the long-awaited installation of the COSTAR optics package to fully restore Hubble's blurred vision. Before launch, Hoffman remembered being told not to worry if they did not accomplish *everything* on the manifest; as long as *either* the new whiffpick or the COSTAR was successfully installed, the scientists on the ground would be "deliriously happy". However, they were not fully appreciative of NASA's collective mindset of having a one-hundred-percent-successful mission. In the months prior to the mission, there was talk that STS-61 was *too* complex and that all of the tasks demanded far more than could be achieved by even five EVAs and a mammoth 11-day flight. Some managers considered splitting the mission into two halves. "But from a technical point of view," said Hoffman, "if you removed half the tasks from a mission, how do you know that you've not left the ones that you're going to *fail* at? At least if you have more things for us to do, we have a better chance of at least getting *some* of them done." To Hoffman and his crewmates, it made little sense to split SM-1 into two halves.

Exchanging the 220 kg High Speed Photometer for the 290 kg COSTAR involved Thornton and Akers opening the telescope's bay doors and loosening latches and removing electrical connectors in order to slide out the old instrument. The new corrective optics package was then fitted. In training on Earth, the operation had taken around three and a half hours. The intensity of the mission – an intensity which had impacted Story Musgrave for almost two years, to such an extent that he remarked, with the merest hint of jest, that the only peace and solace he could find from the mission was sitting in the *dentist's chair* – began to lessen somewhat when Thornton and Akers successfully removed the photometer and installed COSTAR in its place. By the end of their six hour and 50 minute EVA, both the new whiffpick *and* the corrective optics had been triumphantly fitted.

Musgrave and Hoffman's final EVA, lasting seven hours and 21 minutes on the night of 8-9 December, replaced the overheating solar array drive electronics on the telescope, installed magnetometer covers and an electrical connection box on the GHRS. All were listed as critical tasks. By the time the two men returned inside the Shuttle, STS-61 had accomplished five remarkably complex EVAs and a tally of more than 35 hours of spacewalking … in a *single* flight. Whilst this would be duplicated several times over the years, it must be borne in mind that STS-61 was the first Shuttle flight in which the bounds of accomplishment in terms of mission duration, complexity and the intricately linked EVA-RMS-orbiter operations were pushed to their absolute limits.

That night, the night after the final EVA, the crew of STS-61 celebrated their success above the roof of the world. "Of all of the programmes that I have been associated with," Dick Covey remembered, years later, "it's the one that was best planned and has been best executed, in terms of using astronauts and crewed vehicles to be able to support, enable and enhance the scientific mission of space." They did not yet know if the corrective optics would work, of course, but they had carried out their share of the repair. Prior to deploying Hubble back into space, its orbit was slightly boosted to around 595 km in order to overcome the drag experienced since its initial deployment in 1990.

Late on the evening of 12 December, preparations for STS-61's triumphant return to Earth got underway. Re-entering from almost twice the average altitude for a Shuttle mission, the de-orbit burn of Endeavour's OMS engines lasted almost five minutes, but was completed by 11:19 pm EST, committing the Shuttle to a 70-minute descent. Passing over Mexico City during their period of peak heating, Covey was convinced that Endeavour gave ground-based observers a great view. "The orbiter was fully enveloped in the ionisation plume," he said later, "and as we banked up into a left bank coming over Mexico City and the windows were *white* because of the plume, I could look out and still see all the lights. It was not washed out at all; it was very bright through that, so we had to be giving them a great show." Eventually, the crew of STS-61 returned to Florida airspace. At 12:25 am on the morning of the 13th, trailing double sonic booms in her wake, she swept like an enormous bird of prey into Florida and alighted perfectly onto the concrete surface of KSC's Runway 33.

STS-61 had done nothing less than save NASA itself. Few other human space missions since Apollo 11 had exerted such a positive influence on the agency's subsequent fortunes. Of course, we know today that fixing Hubble's optics was triumphantly successful and the telescope repair team received the prestigious Robert J. Collier Trophy in March 1994 for their work. The citation praised their "outstanding leadership, intrepidity and the renewal of public faith in America's space programme by the successful orbital recovery and repair of the Hubble Space Telescope".

Acapulco was once again a holiday destination, not a place for out-of-work NASA engineers ...

Over the following weeks, the Servicing Mission Orbital Verification (SMOV) got underway, encompassing the checkout of Hubble and a resumption of scientific activities as soon as possible. This included the optical alignment and focusing of WFPC-2 and the deployment of COSTAR's mechanical arms, as well as test observations. The announcement came on 13 January 1994, when NASA Administrator Dan Goldin revealed the first new images at a press conference at the Goddard Space Flight Center. Accompanied by Dr John Gibbons, Assistant to the President for Science and Technology, and Senator Barbara Mikulski of Maryland, Goldin told the gathered media that Hubble was "a true international treasure". Mikulski, who had earlier poured scorn and criticism upon the telescope after the discovery of its spherical aberration in mid-1990, now lauded the successful repair as "a wonderful victory for the Hubble team".

The astronauts, of course, knew of the success well ahead of the press conference.

And for one of them – the astronomer, Jeff Hoffman – it came as a particularly sweet gift. In the early hours of New Year's Day 1994, he and his English-born wife, Barbara, hosted friends to their Houston house. By the end of the evening, when everybody had left and Hoffman was cleaning up in the kitchen, the telephone rang. It was one of Hoffman's astronomer friends.

"Jeff, hi," came the greeting. "Do you have any champagne left?"

"Yeah. I still have a half bottle in the refrigerator. Why?"

"Well, crack it open, because we've just gotten the first pictures back from Hubble. *It works!*"

4

From foes to friends

"PINCH ME!"

On 8 February 1994, the ABC television programme *Good Morning America* broadcast a live link-up between three Russians aboard the Mir space station – cosmonauts Viktor Afanasyev, Yuri Usachev and Valeri Polyakov – and the six-member STS-60 crew of Space Shuttle Discovery. The time was 7:38 am EST and Mir was flying high above the southern United States, whilst the orbiter was somewhere over the Pacific. The political situation between the two nations had thawed substantially in recent years, but it was still a noteworthy event. Most noteworthy of all was the fact that in addition to five American members of the Shuttle crew ... was a *Russian* cosmonaut, 35-year-old Sergei Krikalev. The following day, Krikalev and STS-60 commander Charlie Bolden received a telephone call from Russian Prime Minister Viktor Chernomyrdin. During the exchange, Chernomyrdin invited the entire crew to come to Russia, as guests of President Boris Yeltsin, upon their return to Earth.

Discovery landed safely on 11 February and, shortly thereafter, the crew travelled to Moscow. One day, Bolden and his pilot, Ken Reightler, stood on the rampart of the Kremlin, gazing upon the grandeur of Red Square: from Lenin's mausoleum to the drab GUM department store and from the newly-reconsecrated Kazan Cathedral to the bright ice-cream domes of St Basil's. For Bolden, a colonel in the US Marine Corps, and for Reightler, a captain in the US Navy, the vista literally took their breath away. They had served long military careers to *attack* this place ... and *now* they were coming here as *partners* of the Russians. To be fair, it was not all champagne and roses. The Americans stayed in the old KGB headquarters, which had been converted into an ornate hotel, bristling with marble and hardwood floors and glittering chandeliers ... and bugging devices. On one occasion, Bolden remarked to his wife in their bedroom that he would *love* a Coke. As if by magic, a bellboy appeared at the door to offer them a Coke. "So we *knew*," said Bolden, "that there was still some semblance of the old Soviet Union left over."

But for the two US military officers, Bolden and Reightler, standing alone on the ancient walls of the Kremlin, against a chill wind, words were hard to find. Then one of them spoke. "Pinch me," he said to the other. "This *can't* be real."

AN AWKWARD START

Charlie Bolden was unimpressed by the Russians at first.

During his mission in command of STS-45, he had overseen the first ship-to-ship dialogue between the Shuttle and cosmonauts Alexander Volkov and Sergei Krikalev aboard the Mir space station, but could hardly have expected to be flying with one of them a little less than two years later. When he was approached to lead STS-60 – the first Shuttle flight to include a cosmonaut among its crew – Bolden opposed it. He had previously served aboard the Hubble Space Telescope deployment mission and made no secret of the fact that he wanted command of the coveted first servicing flight. More than that, as an active-duty Marine Corps officer, "it was just my upbringing" that "I did not have any desire to work with the Russians". For Bolden, it was as simple as that.

His mind was changed by the new NASA Administrator, Dan Goldin, appointed in April 1992 and with whom Bolden had worked at the space agency's Washington headquarters for several months that summer as Assistant Deputy Administrator. Goldin encouraged him to meet the two cosmonaut candidates for the flight, before making a final decision. Over dinner, Bolden was impressed by Vladimir Titov and Sergei Krikalev; the former was a veteran MiG-21 fighter pilot in the Russian Air Force and had commanded the world's first year-long space mission in 1987-88, whilst the latter was a civilian engineer and one of only two men to be off the planet during the tumultuous events of 1991 when the Soviet Union crumbled and was reborn as the Commonwealth of Independent States. Bolden had spoken to Krikalev in orbit during the March 1992 ship-to-ship radio ham session and knew that he spoke English. Titov, on the other hand, spoke none. "Zero, *nada*," said Bolden. The situation for the cosmonauts was equally difficult. "By the time I get through with all my studies for the classes," Krikalev once told a NASA manager, "I normally have ... from about one o'clock to two o'clock in the *morning* to do my English language training." Although the relationship was affable and professional, the political situation between the United States and the former Soviet Union remained mistrustful.

It was a far cry from the first co-operative manned mission between the two superpowers, the Apollo-Soyuz Test Project (ASTP), in July 1975, when a three-man Apollo craft docked in orbit with a two-man Soyuz and conducted joint experiments. In the wake of ASTP, the Americans and Soviets discussed the possibility of the Shuttle docking with the Salyut 7 space station and as early as August 1976 NASA manager Glynn Lunney, who directed the US half of ASTP, suggested flying a cosmonaut aboard one of the orbiters. It is one of the tragedies of history that no serious attempt to bring this idea to fruition was accomplished for almost two decades. Détente between the Soviet Union and the United States would veer sharply off course and by the end of the decade relations were once again at Cold War levels.

The Hensinki Accords, signed in the summer of 1975 under Gerald Ford's brief presidency, attempted to improve relations between East and West, including agreements on the inviolability of frontiers, the peaceful settlement of disputes and non-intervention in internal affairs, but superpower involvement in conflicts in the Middle East, Central Asia and the Americas served only to sour relations: the Yom Kippur War, the Chilean coup d'état, the Ogaden War in Ethiopia, the Angolan Civil War, the Nicaraguan Civil War, the Islamic Revolution in Iran ... and, perhaps most fundamentally, the Soviet invasion of Afghanistan.

By the spring of 1977, a new president was in the White House. Jimmy Carter, whose attitude towards the Soviets was much less sympathetic than Richard Nixon or Gerald Ford, very quickly signed into law Presidential Directive No. 18 on National Security to reassess the United States' position on détente. At the same time, in an effort to curtail the manufacturing of nuclear weapons and the development of new missile arsenals, Carter and Soviet General Secretary Leonid Brezhnev undertook the second round of Strategic Arms Limitation Talks (SALT II) and would lay down their signatures in Vienna in June 1979. Within six months, however, the Soviets had invaded Afghanistan, prompting an "open-mouthed" Carter and the CIA to begin to arm a native mujahideen insurgency. As the decade drew to an end, fears grew that the Soviets were seeking to expand their sphere of influence into Pakistan and Iran and even that they were positioning themselves for a takeover of oil in the Middle East. For his part, Carter refused to permit any outside forces to gain control of the Persian Gulf, terminated a 'wheat deal' with Russia and made the unpopular move of prohibiting American athletes from participating in the 1980 Summer Olympics in Moscow. Together with his national security advisor, Zbigniew Brzezinski, Carter started a $40 billion covert programme to train Pakistani and Afghan insurgents to foil the Soviet invasion. The 'hawkish' foreign policies of Brzezinski have even prompted mutterings over the years that he and Carter had begun arming the mujahideen *before* the invasion, as a way of drawing the Soviets into a protracted and gritty conflict; in essence, creating their own Vietnam.

When one views these important geopolitical events through the looking-glass of history, it is not difficult to understand why virtually no progress was made on the topic of co-operation in space after the return of Apollo-Soyuz to Earth. Today, in the era of the International Space Station and genuine co-operation in space between not merely Russia and America, but other nations, too, it is saddening to consider the possibility of what might have been. Certainly, ASTP Flight Director Neil Hutchinson once commented on how well he worked with his Soviet counterparts, to such an extent that he wished for another such mission. "It's like going to the Moon once and never going back," Hutchinson said. "Ninety percent of the battle is over with ... getting all the firsts done ... I could run another Apollo-Soyuz ... with a heck of a lot less fuss. Though some of the worry in both Houston and Moscow had been in vain, the two teams had confirmed that they could work together in analysing an unforeseen problem." Others saw it differently. Robert Hotz, then-editor-in-chief of *Aviation Week & Space Technology*, lamented that the fact that ASTP was a one-off stunt was the *real* tragedy ... that NASA and America had bet

4-01: Pictured during emergency bailout training, Sergei Krikalev was the first cosmonaut to fly aboard the Shuttle. At 35, he was the youngest member of the STS-60 crew, but had already accrued 15 months in space, spread across two previous Mir expeditions.

everything on a 'political fanfare', when it could have invested the money into a second Skylab orbital station – already built and waiting to launch – for greater long-term scientific return.

Still, to gain some idea of what may have been, it is necessary to return to the high-watermark of US-Soviet relations: the time in May 1972 when US President Richard Nixon and Soviet Premier Alexei Kosygin put their signatures to ASTP. At a press conference in Houston, NASA Administrator Jim Fletcher had responded to a journalist's question by stating that Apollo-Soyuz was merely "a first step in international co-operation" and, moreover, that co-operation in manned programmes "to save duplication of effort between the two countries" was his great hope. Only now, five decades since Gagarin and three and a half decades since ASTP, are we beginning to see the realisation of some of Fletcher's vision. Genuine co-operation was also the expectation of his deputy, George Low. During a visit to the Soviet Union in May 1975, Low spoke of the future with several of his NASA colleagues and his Russian counterparts, including Konstantin Bushuyev. A rendezvous and docking between a Salyut orbital station and the Shuttle was one possibility, as was Soviet participation in a Spacelab flight. Although the latter option was not seen in a particularly favourable light, the idea of a Salyut-Shuttle mission and also the joint development and construction of an 'international' space station were of interest to both sides, a possibility which *Time* told its readers in its summing-up of ASTP on 4 August 1975. Unfortunately, the enthusiasm of Fletcher, Bushuyev and Low and other like-minded colleagues lay at the mercy of the political climate ... and in the late 1970s and early 1980s, that mercy and that climate deteriorated dramatically.

It was kept alive for a time, however. Informal discussions continued between the Americans and the Soviets and culminated in a series of talks in October 1976 at NASA Headquarters in Washington, DC. These established "a meeting of minds" between the two sides on future manned co-operation, with two primary foci: a scientific venture involving the Shuttle and a Salyut space station or the development of "a space platform ... bilaterally or multilaterally". By May of the following year, this meeting of minds had crystallised further, when NASA Acting Administrator Alan Lovelace and Anatoli Alexandrov of the Soviet Academy of Sciences explored the topic of a Shuttle docking with a Salyut in greater depth. The result was a document, rather ponderously entitled 'Objectives, Feasibility and Means of Accomplishing Joint Experimental Flights of a Long-Duration Station of the Salyut Type and a Reusable Shuttle Spacecraft', which highlighted the benefits that both sides could bring to such a venture: the Soviet system could achieve long-term missions and the Americans could carry large scientific payloads into orbit.

On 18 May 1977, Soviet Foreign Minister Andrei Gromyko and US Secretary of State Cyrus Vance signed the space co-operation agreement, which took effect six days later ... exactly five years to the day since ASTP had been formalised. The new deal would run for a further five years. "This agreement," noted a December 1982 document, produced at the behest of Bob Packwood, then-chair of the US Senate's Committee on Commerce, Science and Transportation, "established the basis for Soviet-American space co-operation through the early 1980s. It was a very important political instrument, because it [ensured] continuity in Soviet-American space relations." The Soviet press, in particular, wrote glowingly of the plans and in November 1977 a meeting in Moscow began to discuss the technical aspects. By April 1978, when follow-on meetings were scheduled to take place in the United States, *Flight International* mentioned the joint Shuttle-Salyut venture, with a rendezvous scheduled for 1981 and a docking a few years later. At around the same time, in the spring of 1978, NASA's Associate Administrator for Space Science, Noel Hinners, testified before the House Science and Technology Committee that future US-Soviet co-operation was crucial, not least because "they have a station in Earth orbit now [Salyut 6] that may be capable of lasting 1.5 years to two years [and] *we* have nothing on the horizon approximating that staytime duration in space". Sadly, this glimmer of future co-operation on the horizon ultimately was nothing more than a glimmer. The plans did not come to pass.

The cause was chiefly political. Issues of human rights violations and the repression of political dissidents, including Anatoli Shcharanski, who was accused of treason and collusion with the CIA, had long bothered the Americans and the implementation of a new Soviet constitution – the 'Brezhnev Constitution' – in the summer of 1977 brought with it worrying signs that new guarantees of individual liberties were a mockery of justice. "Exercise by citizens of rights and freedoms *must not* injure the interests of society and the state and the rights of other citizens," read one proviso of the Brezhnev Constitution. "Obviously," *Time* told its readers on 13 June, "this statement gives legal sanction for the KGB to proceed, without having to manufacture pretexts, against dissidents exercising the right of free speech, assembly or religion." The situation steadily worsened. When the Carter administration re-established formal diplomatic ties with the Soviet Union's sworn enemy, China, in January 1979, the Brezhnev regime responded with undisguised anger, delaying the planned second round of Strategic Arms Limitation Talks until June. Within the year, the Soviets had invaded Afghanistan and relations had deteriorated still further. Reluctantly, the Americans agreed to abide by the SALT agreements, but were determined to exact punitive action in other areas. Space co-operation turned out to be one of them. Even before SALT and Afghanistan, in February 1979, the new NASA Administrator, Robert Frosch, spoke of a hypothetical joint Shuttle-Salyut venture in the distinctly more frosty language of *if*, rather than *when*. By December of that year, as Soviet tanks and troops invaded Afghanistan, the situation had scarcely moved. Frosch told Congress that the American and Soviet working groups had "been in abeyance for something over a year". A further meeting was scheduled for October 1980, but nothing was ever formalised and it never transpired. Senator James Exon of Nebraska described the relationship as having devolved into an "arm's length arrangement that we'll more or less continue" and noted, tellingly, that "the direct scientific activities may be affected, but not immediately, since there was no immediate action to be taken anyway".

In some minds, Jimmy Carter and Zbigniew Brzezinski have been seen as the key obstruction to the implementation of any joint plans. In a January 2002 oral history for NASA, Arnold Frutkin, deputy head of the agency's international affairs office until 1978, related that a breakthrough for more advanced co-operation with the Soviets may have been just around the corner. "It [seemed] so logical to continue ... because [ASTP] was so successful," he said. "It seemed to me the thing to do next would be to move [on] into a space station, but that was a huge undertaking at the height of the Cold War with the Soviet Union and it had to be done in such a way that we weren't transferring technology." The Soviets were interested in conceptual studies and developed a draft agreement with NASA, "to the point where they [actually] signed it! There was a *signed agreement* from them for a joint space station programme, but with this careful, limited, step-by-step [procedure, whereby] you would never proceed from one to the ... next ... unless there was complete comfort and satisfaction in the prior [phase]."

It would be another decade and a half before Russia would again approach the negotiating table with a view to a joint manned project with the Americans and opening up their space programme to a wider world. In the mid-1980s, a wind of change grasped the Soviet Union. It began blowing in the socialist states of Eastern Europe and gusted into a whirlwind towards the end of the decade as failed economic policies, flawed Five-Year Plans, detestable political surveillance and religious persecution and the actions of a string of unelected dictators intensified a popular cry for reform and democratic change. The Berlin Wall, once uniformly grey, now had virtually no grey left on its western face, so ubiquitous was the colourful, angry graffiti which coated it. West and East cried out for reunification.

The wind intensified in March 1985, when Mikhail Gorbachev – a long-time member of the Soviet Politburo and a lawyer by profession – was elected as General Secretary of the Communist Party. However, if Western leaders had reacted to the arrivals of post-Brezhnev General Secretaries Yuri Andropov and Konstantin Chernenko with concern, then their response to Gorbachev was far warmer; for here was a man who was considerably younger than his predecessors, a man who had travelled widely on Soviet business, a man who had seen how the West had prospered through freedom of speech, political transparency and the respect of human rights, and a man who had seen the policies of the Communist Party bring his once-proud nation to an economic standstill.

Within weeks of entering office, Gorbachev became the first Soviet leader to admit that the economy was virtually stagnant and reorganisation was crucial; a "vague programme of reform" was acutely needed to bring about rapid technological modernisation, increased industrial and agricultural productivity, and a sharpened, more efficient and less corrupt bureaucracy. His efforts to streamline the economy, maintain quality control, battle inferior manufacturing and combat rampant alcoholism necessitated broader reforms on the political level, but Gorbachev remained a staunch Communist. He was nowhere near willing to surrender the Soviet notion of a centrally-planned economy in favour of a free market. Speaking that summer to the economic secretaries of the Eastern Bloc states, he scoffed at the idea that 'the market' might prove to be their saviour. "Comrades," he told them, "you should not think about lifesavers, but about the *ship* ... and the ship is *socialism*!"

Yet the fact remained that the ship was immovable in some areas and downright leaky in others. *Perestroika* (political and economic 'restructuring') was a term which had become entrenched in Gorbachev's political ideology by the late spring of 1986. At face value, it called for the creation of more effective and dependable mechanisms "for accelerating economic and social progress", encouraging the initiative of the individual – a *real* Soviet first – and balancing a need for order and discipline with opportunities for criticism and self-criticism. By early 1987, the Central Committee proposed multi-candidate elections and the appointment of non-Party members to government posts, opponents of Stalin were rehabilitated and in 1988 *glasnost* ('openness') brought a measure of real freedom to the Soviet people for the first time. Thousands of political prisoners and dissidents were freed, private ownership of businesses was permitted and in March 1989 the Soviets cast their votes for the Congress of People's Deputies, marking the first open elections in the Union for more than seven decades. Today, many Western analysts see *perestroika* and *glasnost* as the two principal nails in the coffin of the Soviet Union, although the ultimate outcome could hardly have been further from Gorbachev's mind in 1985. Rather, he wanted to see socialism work more effectively for its people, with less 'cronyism' and corruption and greater transparency into government affairs.

As circumstances transpired, this snowballed much further than he could have anticipated: relaxed censorship on the state media led to the exposure of severe social and economic problems – previously denied or concealed – as well as food shortages, the terrible price paid to the effects of alcoholism, pollution and environmental destruction, dramatic mortality rates and the first evidence of the 'purges' imposed by Joseph Stalin on his own people, half a century earlier. Under Gorbachev, steps were taken to end a bloody war of attrition in the mountains of Afghanistan and, as time passed, his weakening of the authority of the Communist Party and the removal of its power to control the media emboldened and strengthened nationalists in the Eastern Bloc. Simmering discontent in Estonia, Latvia and Lithuania – illegally annexed by Stalin – would boil over, whilst other republics, including Ukraine, Azerbaijan and Georgia, would rise in a wave of nationalistic revolt. In the final year of the 1980s, the Eastern Bloc collapsed in dramatic and spectacular fashion. The Berlin Wall would topple, one dictator would be gunned down by firing squad and republic after republic would throw off the Communist yoke, announce elections and set foot on the thorny path to new nationhood. By the time Gorbachev was obliged to hand over the reins of power in Russia to Boris Yeltsin, late in 1991, the Soviet Union as a political entity would be over ... and years of economic turbulence, civil strife, crime and outright military conflict would ensue. It was at this tumultuous stage that the first steps on a new path of space co-operation were taken. In April 1989, President George H.W. Bush had re-established by executive order the National Space Council, chaired by Vice President Dan Quayle. With emphasis upon the construction of Space Station Freedom – a project already years overdue and mired in budgetary and technical issues. Given the need to reduce costs, accessing the Soviets' enormous experience in long-duration space flight, heavy-lift launch vehicles and reliable Soyuz-TM craft were hugely appealing. As the Soviet Union tottered on the road to collapse, there were more 'real-world' problems, such as helping to keep Russia as a nation together and preventing the haemorrhaging of technology to undesirable destinations, such as Iran. The first formal discussions of co-operation between Quayle and Gorbachev came in May 1990 and a little over a year later, in July 1991, President Bush himself met with the Soviet leader at a two-day summit in Moscow. At the summit, an agreement was reached on "flying a US astronaut on a long-duration ... Mir mission and a Soviet cosmonaut on a US Space Shuttle mission".

By the summer of 1992, the Soviet Union had collapsed. A bloodless coup the previous August – only three weeks after Bush and Gorbachev's summit – by hardline Communists had failed and in December Gorbachev stepped down from power. In his place at the head of the new 'Russian Federation' came President Boris Yeltsin and in June 1992 his fragile nation and the United States issued a 'Joint Statement on Co-operation in Space'. For the first time, it laid out the plans for cosmonauts to be flown aboard the Shuttle, a docking mission between the Shuttle and Mir in 1994-95 and at least one long-duration flight (of at least 90 days) by a US astronaut. After several other meetings during the summer, in October NASA Administrator Dan Goldin and his counterpart Yuri Koptev, Director-General of the new Russian Federal Space Agency, met in Moscow to sign the 'Implementing Agreement on Human Space Flight Co-operation', which explored the plan in further depth.

Late in October, Charlie Bolden's STS-60 crew was named, with the addendum that "an experienced Russian cosmonaut" would join them on an eight-day flight, scheduled for November 1993. In making the announcement, former astronaut Steve Hawley, then serving as Acting Director of Flight Crew Operations, noted that the flight was "a significant milestone in future space exploration" and a critical "first step in our co-operative agreements with our Russian partners". In addition to STS-60's primary payload of the second Spacehab commercial research module and the unique free-flying Wake Shield Facility, it was announced that "Russian Space Agency-sponsored life science activities" would be carried out. Within days of the naming of Bolden, pilot Ken Reightler and mission specialists Jan Davis, Ron Sega and Franklin Chang-Díaz, the two cosmonauts appeared before journalists for a photo opportunity at the Johnson Space Center (JSC) in Houston, Texas.

Their training was difficult in more ways than one. In addition to an introduction to basic Shuttle systems processes, the language and cultural problems remained a gulf to be bridged. "Even through an interpreter," said one NASA manager, "it was a different language ... [even] *after* we communicated!" As a case in point, Titov and Krikalev were given driving lessons, as it was feared they would be unable to cope with the aggressive Houston freeway traffic ... until a handful of NASA trainers visited Russia. "After a few hours in Moscow," one of them said, "I realised that we were totally silly for wasting our time trying to teach these guys how to drive in Texas. If they could survive *Moscow* traffic, then they had *no* problem whatsoever driving!" At length, early in April 1993, after "an intensive, three-month training programme", Sergei Krikalev was named as the prime candidate. By this time, further plans for the new partnership had evolved. In late November 1992, NASA revealed that it was actively investigating the usefulness of Russia's Soyuz-TM spacecraft as a possible emergency rescue vehicle for Space Station Freedom. Notably, this development emerged as US and Russian officials met at JSC for two weeks of working group meetings ... the first such meetings since the ASTP era. Then, in December 1993, only weeks before the STS-60 launch, Dan Goldin and Yuri Koptev agreed to stage several Shuttle docking missions to Mir, beginning in 1995, with US astronauts spending up to 21 months aboard the station, as 'Phase 1' of a three-phase programme to bring the Russians into the now-renamed 'International Space Station' effort. Under the terms of the agreement, Vladimir Titov would be assigned his own Shuttle flight, STS-63, which would also attempt a close-range rendezvous – but *not* a docking – with Mir in the spring of 1995.

Although it was not mandatory for them to do so, the STS-60 crew requested a measure of Russian language tuition. "While we didn't need it for flight," said Bolden, "it would be really a nice gesture if we could at least communicate some simple things in Russian from on-orbit". *Fluency* in such a short span of training time was impractical, but the ability to send greetings to Russian schoolchildren from space was a strong motivator. A newly-formed independent contractor, TechTrans International, Inc., arranged for a Russian language instructor to meet with the astronauts several times each week and teach them rudimentary terms, including a traditional children's lullaby called 'Tired Toys are Asleep'. "We were told that every Russian child would recognise the song," continued Bolden, "and if we did it right, we could sing it at 8 o'clock, Russian time, and they would be impressed."

Less impressive were the apartments into which Titov and Krikalev were placed, situated in a complex on the corner of NASA Road 1 and El Camino. Bolden had asked his pilot, Ken Reightler, to take the cosmonauts under his wing and within a few days he first became aware of problems. "Charlie, we've got to get them out of those apartments," said Reightler. There was nothing *wrong* with them, but Reightler felt that Titov and Krikalev would get a better feel for American life if their families were settled into individual family homes. "We took them around and put them with a realtor," said Bolden, "and let them pick the place where they wanted to live." NASA managers agreed, the costs were covered and the Krikalevs ended up in Middlebrook, a suburb of Clear Lake, whilst the Titovs settled in Friendswood. Bolden tried to organise happy hours, church activities and other social functions to bring the cosmonauts and their families into the fold.

All but one member of the STS-60 crew had prior flight experience. Sergei Krikalev had participated in two long-duration missions aboard Mir, whilst Bolden and Chang-Díaz had both flown three times and Reightler and Davis once apiece. Serving as Mission Specialist Two, the 'flight engineer' for ascent and re-entry, was Ronald Michael Sega, whose close professional involvement with the Wake Shield Facility was one primary reason for his assignment. Sega came from Cleveland, Ohio, where he was born on 4 December 1952. After graduation from high school, he entered the Air Force Academy and received a degree in mathematics and physics in 1974 and a master's credential in physics from Ohio State University the following year. Sega then entered active duty with the Air Force, learning to fly and serving as an instructor pilot at Williams Air Force Base in Arizona until 1979. He was then assigned to the physics faculty of the Air Force Academy, where he designed and constructed a laboratory facility to investigate microwave fields using infrared technologies. Concurrently with this research, Sega worked toward a doctorate in electrical engineering, which he received from the University of Colorado at Boulder in 1982. In the wake of his active military career, Sega remained an Air Force reservist.

With his PhD in hand, he joined the University of Colorado at Colorado Springs on an assistant professorship in the Department of Electrical and Computer Engineering and by 1990, when he was selected as an astronaut candidate by NASA, he had secured a full professorial post. During this period, Sega also worked as Technical Director for the Lasers and Aerospace Mechanics Directorate of the Frank J. Seiler Research Laboratory at the Air Force and, in 1989-90, he served as a research associate professor of physics at the University of Houston, affiliated with Alex Ingatiev, Director of the Space Vacuum Epitaxy Center (SVEC). Today known as the Texas Center for Superconductivity at The University of Houston (TcSUH), this 740 m^2 institution has for more than two decades explored thin-film deposition, processing and characterisation of semiconducting, high-temperature superconducting and ferroelectric oxide material systems. As long ago as the 1970s, NASA engineers published papers arguing that a satellite sailing through space would leave an 'ultra-vacuum' in its wake, but the absence of practical applications at the time meant that the idea was left unexplored until Ignatiev and his team revived it.

In 1986, they joined forces with nine other companies to form a Center for the

4-02: Discovery roars into orbit on 3 February 1994, beginning a new chapter in space exploration and a new era for the space programmes of both the United States and Russia.

Commercial Development of Space – one of several industry-academia partnerships sponsored by NASA's Office of Advanced Concepts and Technology – with the plan to build the Wake Shield Facility (WSF). It was conceived as a 3.6 m-wide stainless steel disk, to be deployed using the Shuttle's RMS mechanical arm for the purposes of generating an 'ultra-vacuum' within which to grow thin films for future advanced electronics applications. The purely functional shape of the WSF made it appear like a factory cast-off: of dull, silver-grey colour, it was a clumsy arrangement of boxes, rods, tubing and angular shapes. It was designed, built and managed by the SVEC, along with its industrial partner, Space Industries, Inc., of League City, Texas. In March 1989, SVEC and Space Industries, Inc., formally partnered with NASA and carried the payload from the drawing board to the launch pad in less than 60 months and around $12 million. From the outset, it was presented as a major 'first', since the generation and characterisation of an ultra-vacuum in space had never been attempted. The inclusion of Ron Sega as a member of the STS-60 crew was invaluable and the complex unberthing, atomic 'cleaning', deployment, rendezvous and capture operations were planned in the most optimum fashion to ensure the maximum possible scientific yield from the mission.

The primary objectives of WSF-1 (the first of up to four flights of the satellite) were to characterise the ultra-vacuum environment and employ the techniques of molecular and chemical beam epitaxy to grow a thin film of gallium arsenide (GaAs) on a prepared substrate. Upon this substrate, atomic or molecular beams of arsenic and gallium formed thin films in layers, to create a 'wafer', with an ultra-high-purity top region. The WSF was designed to grow epitaxial films on seven different substrate wafers and it was expected that, on STS-60, the free-flyer would produce at least one 'thick' GaAs film, measuring up to nine micrometres deep. The use of this material in digital cellphones, high-speed transistors, high-definition television and fibre-optic communications and optoelectronics was already known to present "a very promising economic advantage". Gallium arsenide was thought to yield electronic devices eight times faster than silicon, whilst at the same time requiring a mere tenth of the power demand ... but could not be easily produced on Earth of sufficiently high purity and quality. "If improved GaAs material were available," noted NASA's press kit for STS-60, "it could significantly impact the global semiconductor market", whose worldwide consumption in 1990 alone reached $56.8 billion and whose projection for 1994 had risen to $109 billion. "Within this giant market," continued the press kit, "GaAs currently holds only a 0.5 percent niche. It is predicted that the niche for GaAs should grow to 2 percent (or about $2.2 billion) by 1995."

Although a moderate, 'natural' vacuum was known to exist in low-Earth orbit, it was hoped that the forward-facing (or 'ram') side of the WSF would 'push' even the few atoms present out of the way, leaving a unique ultra-vacuum in its wake and producing conditions between a thousand and ten thousand times better than was attainable in the best ground-based vacuum laboratories. Despite ostensibly representing the relatively 'dirty' side of WSF, the ram could be used to support other experiments, including technology payloads, with a total 'real estate' of around six square metres. To make the most of this facility, four payload-attach points were placed on the ram side, capable of holding up to 90 kg of hardware. A fully-equipped spacecraft in its own right, the 1,800 kg free-flyer had cold-gas propulsion thrusters to effect a satisfactory separation distance from the Shuttle and silver-zinc batteries to provide 45 kilowatt-hours of power for the thin-film growth cells, heaters, process controllers and other instruments. Mounted on the 'wake' side of the spacecraft, in addition to the epitaxy process control equipment, were pressure gauges, mass spectrometers, potential analysers and a video camera, together with the attitude-control system, the batteries and solar panels. Meanwhile, attached to the ram face of the disk were the avionics hardware and associated support equipment.

In addition to the GaAs epitaxial growth hardware, a number of other co-operative experiments were aboard WSF-1. The University of Toronto's Institute for Aerospace Studies provided a space exposure package and Case Western Reserve University's Center for Materials for Space Structures offered a unique investigation to evaluate different spacecraft materials and coatings and their degradation in the environment of low-Earth orbit. Case's Materials Laboratory – or 'MatLab' – featured a number of leading industrial partners, including Martin Marietta, TRW and McDonnell Douglas, and was attached to the ram side of WSF-1. The materials were tested for thermal cycling, strain, micro-debris impacts, atomic oxygen erosion and the influence of harsh ultraviolet radiation, and the Materials Flight Experiment (MFLEX) carrier scanned sensors attached to each material and relayed data back to Earth, via the WSF-1 communications link. The US Air Force's Phillips Laboratory supplied the Charging Hazards and Wake Studies (CHAWS) experiment to develop a clearer understanding of interactions and inherent hazards between the space environment and space systems. Of specific interest to the Phillips investigators, from a civilian and military standpoint, were measurements of ambient, low-energy positively-charged particles on the wake and ram sides of WSF-1 and studies of the magnitude and directionality of current collected by negatively-charged objects. Several CHAWS objectives were to be achieved whilst the free-flyer was in the grasp of the RMS arm. Also aboard were a series of highly-sensitive accelerometers, the Shuttle Plume Impingement Experiment, a pair of 'SmartCans' housing materials science investigations and a pair of student-supplied plant-growth studies.

Since WSF-1 was mounted atop a bridge-like truss structure, it occupied roughly a quarter of Discovery's payload bay on STS-60 and plans called for it to be grappled and unberthed by the RMS arm, under the control of astronaut Jan Davis, on the third day of the mission. Not surprisingly, Ron Sega served as lead crew member for the payload itself and it is interesting that his then-wife, fellow astronaut Bonnie Dunbar, had also come into contact with the WSF in a professional capacity a few years earlier. In 1987-88, she was a member of the astronaut office's Science Support Group and worked extensively on future Shuttle payload proposals from industry and academia in the area of microgravity research ... one of which came from Alex Ignatiev at SVEC for the Wake Shield.

The original November 1993 launch date for STS-60 quickly moved into the spring of the following year as NASA's Shuttle manifest writhed and contorted in the wake of two pad aborts and their impact on downstream flight. The mission got off to a fine start, with liftoff from the Kennedy Space Center's Pad 39A at 7:10 am EST – exactly the time published in the press kit, several weeks earlier – on 3 February 1994. The only difficulty during ascent was a problem with Ken Reightler's portable headset, which was later swapped for an on-board spare. Shortly after orbital insertion, Franklin Chang-Díaz, in his role as the payload commander, and Jan Davis oversaw the opening of the Spacehab-2 module and the activation of its experiments. During this period, a procedure for removing a diffuser cap from the middeck floor air duct fitting and placing it on the floor of the tunnel adaptor was scheduled to be performed. However, due to the lack of sufficient clearance between the adaptor floor fitting and the adaptor floor itself, normal positioning was not possible. As a result, the diffuser cap was grey-taped into position and this posed no further difficulties.

Like its predecessor, Spacehab-1, in the summer of 1993, this second mission exploited the enlarged volume of the commercial middeck augmentation module, which almost tripled the amount of experiment space for middeck-sized lockers. In terms of systems performance, Spacehab-1 met 100 percent of its pre-mission success criteria and the NASA experiments achieved in excess of 90 percent of their requirements. The primary research focuses of Spacehab-2 were upon biotechnology (including the creation of improved pharmaceutical products and studying immune-system disorders) and microgravity research (including the sintering of metals and the production of important protein crystals). Twelve payloads within the module – and one atop its flat roof – supported a wide range of experiments, several of which were sponsored by a handful of NASA Centers for the Commercial Development of Space. The Equipment for Controlled Liquid Phase Sintering Experiments (ECLiPSE) had flown aboard Spacehab-1 and on STS-60 it was devoted to continuing the exploration of the sintering process for possible applications in the creation of tougher, more durable metals for bearings, cutting tools, electrical brushes, contact points and high-stress mechanical parts. The Space Experiment Furnace (SEF) supported a vapour transport growth facility with three sample-processing ovens, capable of providing temperature gradients. The Astroculture plant-growth investigation, already twice-flown on the Shuttle, had demonstrated the capacity for a system to supply water and nutrients and appropriate lighting; on STS-60, its focus was upon the demonstration of closed-air-loop temperature and humidity controls. Pennsylvania State University supplied a 'biomodule' to test the hypothesis that microgravity could alter microbial gene expression in commercially useful ways, such as to create an advanced and environmentally-friendly insecticide to tackle the ravages of the Colorado potato beetle.

BioServe – another of NASA's Centers for the Commercial Development of Space – was the Commercial Generic Bioprocessing Apparatus, which carried 32 experiments in biomedical testing and pharmaceutical development, the development of controlled ecological life-support systems and the agricultural manufacture of biological-based materials. Specimens studied included bone marrow cell cultures, brine shrimp and miniature wasps. Russia's first Shuttle crew member, Sergei Krikalev, took charge as the lead operator of another BioServe experiment in Discovery's middeck, called 'Immune-1'. This experiment involved a dozen rats in a pair of animal enclosure modules (whose design had been copied from a similar facility previously used by NASA's Ames Research Center) and sought to reduce or prevent immune-system changes after microgravity exposure. Half of the rats were treated before launch with an immune-system stimulant, whilst the others received a placebo. The stimulant, known as Polyethylene Glycol-Modified Recombinant Human Interleukin-2 (PEG-IL-2), was expected to yield Earthly benefits as an anti-viral and anti-bacterial agent and as a treatment for the elderly and for infectious diseases. In addition, the BioServe Pilot Laboratory again carried the *Rhizobium trifolii* and *E. coli* bacteria, with terrestrial applications ranging from improved control of crop infestation to better waste treatment and water reclamation.

The University of Alabama's Organic Separations (ORSEP) payload offered scientists the ability to separate cells and particles according to surface properties, using a process known as 'counter current phase partitioning', whose effectiveness on Earth is hampered by the influences of gravity and sedimentation. Samples carried aboard STS-60 included growth hormone vesicles, along with inert particles for diagnostics, together with lymphocytes and bone marrow cells. Protein crystal growth aboard Spacehab-1 had produced rave reviews from investigators, who lauded them as having produced "superior data when compared to the very best crystals ever obtained by Earth-grown methods" and this next mission was expected to advance this further. The Stirling Orbiter Refrigerator/Freezer evaluated the use of environmentally-benign helium as a coolant, in the hope that its technology might replace less reliable and efficient vapour-compression systems. Mounted on the roof of Spacehab-2 was the Sample Return Experiment, provided by the Jet Propulsion Laboratory, which captured intact cosmic dust particles as they came into contact with 160 transparent silica aerogel 'capture cells' for ground-based observation and analysis. Rounding out the payload for this second Spacehab flight were the Three-Dimensional Microgravity Accelerometer (3-DMA), provided by the University of Alabama at Huntsville, and the Space Acceleration Measurement System (SAMS) from the Lewis Research Center, which were to monitor and record accelerations during sensitive experiment runs.

As a crew, the astronauts and Krikalev divided themselves equally among the experiments. Since this was the first opportunity for NASA's Space and Life Sciences Directorate to work on a real manned mission with Moscow's Institute of Biomedical Problems, an additional thrust of the research was space medicine. Krikalev and Davis worked on a number of neural sensory investigations, inner-ear balance studies and the influence of the microgravity environment upon the eyes. Meanwhile, Bolden and Chang-Díaz were the guinea pigs for blood draws, urine samples and specimens of saliva. During their training cycle, the two men – neither of whose backgrounds had anything to do with medicine, for Bolden was a Marine Corps aviator and Chang-Díaz a space plasma physicist – trained and were certified as 'phlebotomists', qualified and capable of taking venous blood draws. Every couple of weeks, throughout 1993, at the flight surgeon's office they would be presented with a line of volunteers on whom to 'practice'. "Franklin and I both *hated* it," Bolden told the oral historian with a chuckle, "because we hated *sticking* people!"

4-03: Ron Sega had participated extensively in the development of the Wake Shield Facility – here pictured 'behind' him, at the end of Discovery's RMS arm – before joining the astronaut corps.

With the activation of Spacehab-2 completed and experiment operations underway, the major activity of the third day of the mission, 5 February, was the grapple, unberthing from its support structure and deployment of the Wake Shield Facility. Plans called for the disk-shaped craft to spend about 48 hours in free-flight, after which it would be recaptured by the RMS and returned to the payload bay for the return home. Under Jan Davis' expert control, the mechanical arm was checked out and WSF-1 grappled at 6:13 am EST and unberthed an hour later at 7:23 am. Davis firstly manoeuvred the payload into its so-called 'ram-cleaning' position, which was essential for the atomic oxygen of low-Earth orbit to 'scour' its ram side and atomically cleanse it in readiness for two days of semiconductor processing. It was during this three-hour process that the proper configuration of the payload was rendered uncertain, because its status lights were washed out by fierce sunlight. Attempts to troubleshoot the problem resulted in a loss of communications capability with the satellite and the crew was directed to berth it back in the payload bay overnight and await a decision from Mission Control on their next steps.

By 3:48 pm EST, the free-flyer was back on its support structure. Over the next few hours, engineers evaluated the situation and determined that the communications problem "was related to near-field radio frequency (RF) multi-path interference on the WSF or a payload RF communication problem". This problem was remembered clearly by Charlie Bolden, years later, for the Wake Shield Facility had cost a mere $12 million, as opposed to the hundreds of millions invested in most satellite projects. "One of the ways they saved money was limiting the amount of pre-flight testing," he said. "One of the crucial tests that they did *not* do was an Electromagnetic Interference Test, whereby you put the satellite together in its flight configuration and turn it up and see if there is interference among different components." WSF-1 was powered-up in the integration room at the Kennedy Space Center and appeared to function perfectly. "But when they bundled up all these feet of electrical cable," continued Bolden, "you get a *current* generated. When we powered up the Wake Shield Facility on-orbit and tried to turn on the attitude-control system, there was a 5 Hz signal that was generated by the power cables ... that shut down the attitude-control system and *wouldn't* let it run! Every time we tried to power it up, it would start spinning and cut itself off." The crew had no idea what was causing the problem, but they instinctively knew that they could not reliably deploy WSF-1 if they had no control over its motions.

It was reflective of NASA's 'faster, better, cheaper' policy, instituted by Administrator Dan Goldin. Bolden looked upon the matter with a measure of sadness. Although he admired the science and technology behind the Wake Shield, and dearly wanted it to succeed, he was disappointed that expense *had* been spared in its final months of testing. It was, he said scornfully, "manufactured in a storefront".

Early the following morning, Davis prepared for another bid to deploy the satellite. At 6:14 am EST on 6 February, she gingerly unberthed and extended WSF-1 on the end of the arm. Again, there were communications dropouts, but these were resolved with a hard reboot of the satellite's computer. However, all was still not well. During a standard checkout of the Attitude Determination and Control System (ADACS), data indicated that the satellite's attitude could not be properly ascertained and efforts to troubleshoot the problem by aiming its horizon sensor toward the Sun to warm it up were fruitless. A second deployment opportunity later that day, barely 50 minutes in length, was also missed because WSF-1 remained unready to go. Another attempt was possible on the 7th, but matters were hampered by the fact that there would be insufficient time to safely develop contingency procedures in the event that the satellite was unable to effect stable attitude control. Mission Control told the disappointed astronauts that WSF-1 would not be deployed and that all GaAs operations would take place whilst firmly attached to the RMS.

In the days to come, the mission would be disparagingly nicknamed 'The Wake Shield Fallacy' by some sections of the media, but even as it 'hung' on the end of the RMS arm, the satellite demonstrated its capabilities. Five thin semiconducting films were successfully grown and, in spite of the lack of continuous command capability, the CHAWS experiment collected its required data. The satellite was finally berthed in the payload bay at 7:18 am on 9 February. Although the results *were* demonstrative of the practicality of the Wake Shield concept, it was a far cry from what should have happened and it is important to consider the original flight plan for a fully-deployed WSF-1.

Had circumstances been different, after the ram-cleaning process, Davis would have put the satellite through an hour of vacuum measurements and released it over the starboard wall of Discovery's payload bay. Flying freely, WSF-1 would have employed its cold-gas thrusters to manoeuvre itself 'ahead' of the orbiter, thereby avoiding the risk of contamination from waste water dumps, fuel cell purges and engine bursts. It was expected that the two vehicles would remain at a distance of about 75 km for the next two days. With the Shuttle's pilots, Charlie Bolden and Ken Reightler, handling the orbiter, Franklin Chang-Díaz would have tracked the rate of separation distance with a laser range-finder, whilst Davis and Krikalev tended to the RMS arm and Ron Sega kept watch over the Wake Shield's systems. On 7 February, Bolden and Reightler would have manoeuvred Discovery from 75 km to a 'station-keeping' position of about 15 km, whereupon Krikalev would have powered-up the RMS for the retrieval. The pilots would then have performed a 'Terminal Initiation' burn, with the scope for four smaller course-correction firings, before Bolden took manual control for the final approach.

As the Shuttle drew nearer to its quarry, a series of plume impingement tests were planned. Between the distances of 120 m and 60 m 'ahead' of WSF-1, the pilots would have executed a four-hour series of thruster firings to characterise the behaviour of their exhaust and allow the satellite's instruments to measure the composition, accelerations and pressures. (Since the Wake Shield would have completed its ultra-vacuum work by this time, these tests would not have posed a contamination risk.) One of the key aims of this plume-impingement work was to assess the extent to which thruster firings could be executed in close proximity to other spacecraft, including Russia's Mir orbital station. Finally, Bolden would have guided Discovery to within 10 m of WSF-1 to enable Krikalev to grapple the payload. It would then be 'parked' above the payload bay, overnight, and the CHAWS experiment was scheduled for 8 February, followed by final berthing.

Although the Wake Shield would fly twice more after STS-60 – and both missions would involve successful deployments and retrievals and encouraging results – it was with a measure of angst that the crew returned home to KSC's Shuttle Landing Facility at 2:19 pm EST on 11 February, having been unable to run the new payload through its paces to completion.

For Bolden, it was a lost opportunity to fly a 'rendezvous', as none of his three previous flights had included a task of this nature. He had sorely desired command of the first Hubble servicing flight, but had not gotten it. "I had not flown a rendezvous mission," he told the NASA oral historian, "and I was really excited about being able to rendezvous with something, since I wasn't going to be around for the International Space Station." The list of astronauts waiting for their first flights was increasing and more were to be selected later in 1994, and Bolden had been approached to return to active duty in the Marine Corps as Deputy Commandant of Midshipmen at the Naval Academy. In August 1994, he retired from NASA and returned to the military. Of course, it was not to be the end of Bolden's involvement with the space agency and in July 2009 he was appointed as NASA's first African-American Administrator.

This was illustrated in light-hearted fashion by Bolden at the mission's post-flight press conference, when he offered the audience the tongue-in-cheek chance to ask questions about "what we *thought* we did ... in case we did something that you didn't *think* we did ... or we didn't do something that you *thought* we did". Yet the first flight of the Wake Shield Facility *had* shown what it could do and in September 1995 and November 1996 it would be reflown and its capabilities realised. A fourth flight, capable of producing up to 300 epitaxial thin-film wafers, was planned and its integration and test schedule was approved, but by the end of 1997 NASA funding ran out. Several months later, in May 1998, SVEC granted exclusive licence to Spacehab, Inc., to market, manage and operate the WSF and there were hopes that a five-year free-flyer might be launched. Ultimately, with the emphasis of the Shuttle upon building the International Space Station – which, in 2003, was punctuated by the loss of Columbia – this did not come to pass and the satellite never flew again.

Elsewhere aboard Discovery, beyond the Wake Shield and the Spacehab, were a multitude of other experiments. Early on 9 February, six Orbital Debris Radar Calibration Spheres (ODERACS) were ejected from a dustbin-sized Getaway Special (GAS) canister. These spheres, which were arranged in pairs and ranged in size from 5 cm to 15.2 cm, were designed to be tracked by radars and optical telescopes as part of efforts to calibrate the abilities of ground stations. Five hours after ODERACS, another GAS canister lid popped open and BREMSAT – a 63 kg satellite, built by students at Germany's University of Bremen – was deployed into space. Measuring 48 cm in diameter, BREMSAT carried a number of experiments to explore heat conductivity, acceleration and microgravity quality, micrometeoroid and cosmic dust densities, atomic oxygen prevalence and pressures and temperatures experienced during the onset of its own destructive re-entry. An additional cross-bay 'bridge' of GAS canisters supported experiments to investigate the behaviour of ball bearings, the vibration characteristics of the Shuttle itself, the process of boiling in microgravity and a two-phased capillary pumped loop.

As for Spacehab-2, many of its experiments produced exceptionally pleasing results. Astroculture performed to near-perfection, as did the BioServe Pilot Laboratory and the Commercial Generic Bioprocessing Apparatus. Early in the flight, the crew reported the beginning of the nucleation process – the first time this had ever been detected in space – within the crystal growth facility. The Immune-1 payload, under Krikalev's observation, was successful and all 12 rats survived the mission. Less satisfactory was ORSEP, both of whose sample-transfer disks suffered mechanical failures and none of its experiments could be activated. Perhaps the most obvious 'problems' associated with Spacehab-2 were a cross-talk glitch between the intercom and the module's caution and warning system, heard clearly on the space-to-ground communications link … and the partial collapse of the flexible rubber duct between Discovery's environmental control and life-support system supply line to the Spacehab floor fitting. In his post-flight press conference, Bolden referred to it as the "collapsed rubber duck". It was significantly 'crimped' and Chang-Díaz and Davis were sufficiently concerned that it might restrict airflow into the module.

Bolden called the ground and advised them of the crimped duct. A little investigation identified the cause as a blockage in the Spacehab Atmospheric Revitalisation System (ARS) fan inlet debris screen in the fan inlet muffler. The blockage increased the suction in the environmental control and life-support system line and caused the duct to collapse. Whilst Mission Control mulled over the issue, Sergei Krikalev floated over to find out what was happening. As a cosmonaut with two long-duration Mir missions behind him, Krikalev had nine times more flight experience than the rest of the Shuttle crew, put together. He was perplexed at the Americans' stance. His question was simple: "Why don't we *fix* it?" As logical as the question seemed, it did not fit in with the normal Shuttle protocols, which required support and consensus from the ground. "The stuff you guys do just doesn't make any sense," he said, as paraphrased by Bolden in his oral history. "On Mir, I wouldn't even have *called* them. I'd have fixed it and then I'd let them know what we had done."

Several days later, Chang-Díaz happened to be running a televised guided tour of the orbiter, speaking in Spanish, for a Costa Rican audience. As he passed through the tunnel from Discovery's middeck into the Spacehab module, he happened upon the crimped duct. He mentioned it in passing to his audience. Not long afterwards, Mission Control called Bolden in surprise: "You guys didn't tell us it was *that* bad!" Proposals to fix the duct were quickly sent up to the crew, but the astronauts found them unsatisfactory. Then the orbiter passed out of direct communications with the ground and Krikalev suggested an impromptu repair to clean the screen and insert the stiff plastic cover of the on-board atlas to reinforce the collapsed duct. "If you … rolled it up and stuck it inside the hose," explained Bolden, "then it expands. It wants to go back to being flat … and makes the hose stand out." Upon seeing videotaped footage of the fix, Mission Control made no comment, but after the flight someone approached Bolden. "We wish you hadn't just gone ahead and done it," he said. "We *wouldn't* have disagreed with it if you'd told us that's what you were planning to do!" In the minds of both Bolden and Krikalev, it was a lesson in the different working practices of the Americans and the Russians.

COOL CREW

When astronauts John Casper and Andy Allen took their seats on the flight deck of Columbia, early on 4 March 1994, they did so with the knowledge that they were the coolest, most chilled-out astronauts around. "We're as cool as a chilled martini at sunset," Casper joked before launch. He was referring to a new set of water-cooled underwear, fitted with more than 320 km of tubing, which he and Allen wore under their pumpkin-orange pressure suits to keep them comfortable during the eight-and-a-half-minute climb to orbit.

Not only was STS-62's ascent comfortable, but the entire countdown was exceptionally smooth. In fact, the only reason that Columbia did not fly on her first attempt, at 8:54 am EST on the 3rd, was due to weather forecasters' predictions that conditions would be unacceptable. Citing "predictable weather patterns", it was noted that surface winds of 30-50 km/h, well in excess of the maximum-allowable 24 km/h rule specified by the Mission Management Team, might hamper the crew if they needed to make an emergency return to KSC's Shuttle Landing Facility. Their ability to make such predictions (with upwards of 90-percent confidence) meant that the Mission Management Team could postpone the launch *before* engineers began loading the External Tank with cryogenic oxygen and hydrogen propellants. The countdown clock was held at T-11 hours on the 3rd and kept at that mark until conditions improved. Eventually, late that evening, it began ticking again, the giant tank was fuelled without incident and Columbia thundered aloft at 8:53 am EST the following day. By now, the Shuttle's ability to set off a cacophony of car alarms with the roar of its main engines and twin Solid Rocket Boosters had become the stuff of lore. STS-62 was no exception and the noise and vibration were easily sufficient to affect vehicles more than a dozen kilometres from Pad 39B.

Still, *arriving* at the Kennedy Space Center was always a difficult time for Andy Allen. "The worst thing," he explained in an interview for the Smithsonian, years later, "is saying goodbye to your kids. My process in the quarantine period before launch was to get my will all squared away and write notes and letters to my kids." Before each of his three Shuttle missions – of which STS-62 was his second – Allen always tried to do something special for his children; not only notes and letters, but perhaps also a recording of songs. "Having been in the Marine Corps and been on aircraft carriers and had gazillions of close calls as a fighter pilot," he continued, "nothing is as stretched out as getting ready for a space flight. Partly it was because my kids ... were at an age where they understood what was going on. *Daddy might blow up* and he might *not* come back!"

Thankfully, STS-62's climb into space was outstanding. The smooth countdown and perfect liftoff – the only deviation in procedures being a slight delay in despatching recovery ships to retrieve the SRBs, due to high seas – set the stage for what would be one of the quietest and problem-free missions in Shuttle history. At

396 From foes to friends

4-04: Marsha Ivins and Pierre Thuot at work in Columbia's cramped flight deck during the 14 days of STS-62.

the risk of being dubbed 'boring', STS-62 was a 14-day flight with a payload bay and middeck literally packed to the rafters with materials science, space technology, medical, solar physics and robotics experiments, most of which had flown before. All five members of the crew – Casper in command, Allen as pilot and mission specialists Pierre Thuot, Charles 'Sam' Gemar and Marsha Ivins – had flown before.

Most visible in the payload bay of Columbia was the second United States Microgravity Payload (USMP-2), continuing the research begun by STS-52 a little over a year earlier with the joint US-French MEPHISTO furnace and the SAMS accelerometer, as well as three new experiments: the Advanced Automated Directional Solidification Furnace (AADSF), the Isothermal Dendritic Growth Experiment (IDGE) and the Critical Fluid Light Scattering Experiment (nicknamed 'Zeno' by its research team). Like USMP-1, the experiments were mounted atop a pair of bridge-like Multi-Purpose Equipment Support Structures (MPESS) at the midpoint of Columbia's payload bay and were operated via 'telescience' by ground controllers. "There is the potential for a number of important benefits and applications to come out of this flight," Casper explained in a pre-launch press conference, "but the payoff is not always right away, and that's the way it is with this flight." Some of these benefits were improved semiconductors for computers, calculators and infrared detectors, as well as materials to fabricate stronger turbine blades for aircraft and powerful electronic components. USMP-2's twin furnaces – MEPHISTO and AADSF, both affixed to the 'front' MPESS – played a pivotal role in this research.

Building on experience from STS-52, MEPHISTO conducted studies of the actual process of 'directional solidification', whereby materials were melted, then solidified, from end to end. The USMP-1 investigations had traced the motion of the 'interface' between the solid part of a sample and the liquid part; on USMP-2, this was expanded to look more closely at the location, shape and behaviour of the interface. Three rod-shaped samples of a bismuth-tin alloy were again flown and, despite a temperature sensor glitch on 5 March, the furnace was returning analysable data within a couple of days. During USMP-1, regular cellular patterns had been detected on the alloy's structure, close to the point at which solidification became unstable. This offered a chance to make adjustments and better manipulate and control the mechanical and electrical properties of materials. The use of telescience also enabled MEPHISTO scientists to make real-time tweaks to their operating procedures and thus improve their data. In fact, cross-sectional analyses of the samples after STS-62 landed provided the best-ever information on the shape of the solid-liquid interface front. Meanwhile, the AADSF employed directional solidification to process a cylindrical sample of mercury-cadmium-telluride, which has seen terrestrial uses in remote-sensing and astronomical instruments. To achieve this, the furnace moved the sample through three separate temperature 'zones' – from a 'hot' region at 870 degrees Celsius to a 'cool' region at 340 degrees Celsius – to slowly cool and solidify them. The result was a 'flatter' solidification front and crystals which could be studied with greater clarity. In total, the AADSF processing runs exceeded 240 hours.

Telescience had already been used with great fanfare on STS-52 and, on this second mission, investigators transmitted hundreds of commands from the ground to adjust settings and parameters and make changes as new and unexpected data emerged. The technique was also receiving intense consideration for autonomously operating payloads on the International Space Station, which, at the time, was expected to operate for lengthy periods of time without the presence of a permanent crew. "Telescience," said USMP-2 Assistant Mission Scientist Don Reiss, "is the closest thing a scientist can get without actually being there." After STS-62 landed, the AADSF sample was polished and etched to permit investigators to better determine the position and shape of the solid-liquid front.

As the two furnaces continued their work, the IDGE and Zeno payloads mounted on the aft MPESS were operating. The former grew strange, tree-like crystalline structures, known as 'dendrites', whilst the latter explored the behaviour of xenon as it approached the 'critical point' at which it temporarily, and simultaneously, exhibited the characteristics of both a liquid *and* a gas. Dendrites are known to evolve as materials solidify under certain conditions and by improving scientists' understanding of them, it was hoped to improve the industrial production of a wide range of materials, including steels and super-alloys, used in applications ranging from making tin foil to cars and jet engines. Each of these materials is formed under conditions which yield dendrites. During USMP-2, dendrites were grown and photographed. As each growth 'cycle' ended, the tiny tree-like structures were re-melted and another dendrite was produced at a different temperature. Cameras recorded the entire process and the resultant data revealed that dendrites grew much faster and to larger sizes in the absence of gravity than their terrestrial counterparts.

Zeno, meanwhile, studied the behaviour of xenon at its critical point; a point which is very difficult to attain in ground-based experiments, because the fluid becomes highly compressible or 'elastic'. In such cases, even if the critical point could be reached, its stability could not be maintained for long enough to complete meaningful observations. In space, with a virtual absence of gravity, these complications were dramatically reduced. A slower-than-expected cooling of Zeno, soon after reaching orbit, put the experiment slightly behind on its timeline, but by 5 March investigators reported that they were closing in to within a 20 millionth of a degrees Celsius of xenon's critical point. (By the time the study ended, it had reached a remarkable a 500 millionth of a degree.) During one televised demonstration, Andy Allen explained to his audience how gases reached their critical points, with the help of a freon-filled vial, and added that knowing the critical point of water led to the developments of new methods to decaffeinate coffee.

Unlike previous missions, which typically operated payloads side-by-side, STS-62 was virtually *two* Shuttle flights in one. The USMP-2 research filled the first ten days in orbit, after which they were shut down and operations with a payload sponsored by NASA's Office of Aeronautics and Space Technology – dubbed 'OAST-2' – took precedence. For both payloads, Pierre Thuot served as the lead crew member. This transition was also highlighted by the lowering of Columbia's orbital altitude, early on 14 March, to enhance the data-gathering capabilities of the OAST-2 experiments. By the time USMP-2 research ended, late on the 13th, virtually all of the scientists involved with its experiments were raving over what Mission Scientist Peter Curreri called "a fantastically rich mission". Operations with the $30 million OAST-2 began almost immediately and came on the heels of the OAST-1 mission, which flew several years earlier, in August 1984. Since then, the office had actually been renamed as the 'Office of Advanced Concepts and Technology', but the original nomenclature stuck. It consisted of six experiments mounted on a cross-bay Hitchhiker bridge and was devoted to conducting advanced technology research, including evaluations of new solar cells and energy-storage hardware and investigations of the influence of atomic oxygen and plasma upon the Shuttle.

All but one of those experiments – the Thermal Energy Storage (TES), housed in a pair of GAS canisters – were controlled directly from an avionics box aboard the Hitchhiker. TES was designed to investigate the behaviour of two thermal energy storage salts – lithium fluoride and lithium fluoride with calcium difluoride eutectic – in microgravity and was operated by the STS-62 crew from a laptop on Columbia's aft flight deck. It collected and stored solar energy, which was then converted into electricity whilst in Earth's shadow, and was being run as part of efforts to develop future solar dynamic power receivers for the International Space Station.

Elsewhere on OAST-2 was the Solar Array Module Plasma Interaction Experiment (SAMPIE), which exposed samples of four different solar cell materials to the harsh environment of low-Earth orbit to evaluate their performance and degradation over time. It was recognised that conducting surfaces, whose electrical potential is highly negative with regard to the plasmas of this environment, would be prone to 'arcing' and potentially destructive to solar cells. Before STS-62, engineers used ground-based plasma chambers to simulate temperatures and conditions in low-Earth orbit, but could not accurately replicate them, owing to differences in pressure, plasma flow, electron temperatures and ion species. Generation and storage of heat energy were also expected to yield benefits for the upcoming station programme. The Cryogenic Two Phase (CRYOTP) experiment evaluated the responses of a nitrogen 'heat pipe', whilst the Brilliant Eyes Thermal Storage Unit (BETSU) advanced the process a step further by absorbing thermal energy from a heater, using methyl-pentane, a 'phase-change' material, in a pair of cryocoolers. After activation, the device cooled the material down to minus 150 degrees Celsius and the heater was switched on to effect the melting process. Several freezing and melting cycles were run to better quantify ground-based models.

Shielding astronauts from the harsh radiation environment of low-Earth orbit was also considered a crucial prerequisite for International Space Station operations and the Emulsion Chamber Technology (ECT) experiment on OAST-2 sought to gather data on its effects using a series of photographic plates. Highly sensitive X-ray films were interleaved with sheets of lead and the 'tracks' left by radiation particles passing through them were examined after Columbia's landing. Particles of atomic oxygen and nitrogen in the near-Earth environment had long been known to have an adverse influence on the Shuttle, producing auras of light around the wing tips and vertical stabiliser and threatening measurements taken by optical or other scientific instruments. The Experimental Investigation of Spacecraft Glow (EISG) and the Spacecraft Kinetic Infrared Test (SKIRT) were designed to study these impacts in more detail. A set of pressurised nitrogen canisters were mounted beneath the EISG sample plate and used to produce ionising atoms which generated the 'glow' patterns around the Orbital Manoeuvring System (OMS) pods and vertical stabiliser. Far-ultraviolet and visible imaging spectrometers then gathered data on a series of samples coated with paints of differing thermal properties. The Shuttle pilots, John Casper and Andy Allen, manoeuvred their craft into seven different elliptical orbits and four circular orbits, one of which took the crew, on 16 March, to their lowest altitude of just 170 km.

Lowering their orbit in this manner increased the yield of glow-related data. Additional measurements were taken during thruster firings and also from Earth's own atmospheric 'airglow'. Meanwhile, SKIRT complemented EISG with an infrared imaging capability. During the course of the mission, the astronauts performed six manoeuvres: four involving 'nose-down', 360-degree rolls and two using the Moon for calibration. Overall, the study was highly successful, with the exception that EISG's ultraviolet sensitivity declined later in the flight. Marsha Ivins used the Shuttle's RMS camera at one stage to look at the device for troubleshooting purposes, but flight controllers suspected that it was either operating at a lower-gain or was partially blocked. Still, operations with both glow instruments ran smoothly and Pierre Thuot remarked that the phenomenon was much more pronounced when in the lower orbit than in a higher one.

If preparation for the International Space Station was one key goal of STS-62, then the work continued inside the orbiter's cabin. Squirrelled away in the middeck was the Middeck Zero-Gravity Dynamics Experiment (MODE), lovingly tended by Sam Gemar, which took the form of a set of miniaturised station modules and girder-like trusses. Like a previous experiment performed by Gemar on STS-48, this sought to understand the dynamic behaviour of large deployable structures in microgravity. Outside, a nifty extension to the RMS arm – known as the Dexterous End Effector (DEE) – demonstrated a set of powerful, U-shaped electromagnets to generate an attraction force of 1,450 kg. This was an alternate to the traditional RMS method of grappling payloads by means of a wire 'snare' which closed around a capture 'pin' on the target. From the crew's perspective, DEE also offered a sense of 'touch' with the mechanical arm and yielded vital benefits when working on the development of smaller and more compact grapple fixtures for new spacecraft. It allowed several new components to be tested, including the magnetic end effector, a targeting and reflective alignment camera, a capture latch assembly with two dummy 'payloads' and a force torque sensor. Overall, the system was a reliable means of maintaining a solid grip on a payload and a safer means of releasing it the event of difficulties with the RMS. All three STS-62 mission specialists took turns on the DEE evaluations and praised its precision.

Also aboard Columbia was the frequently flown Shuttle Solar Backscatter Ultraviolet (SSBUV) instrument, continuing a year-on-year effort to calibrate ozone measurements from the Total Ozone Mapping Spectrometer (TOMS) and Upper Atmosphere Research Satellite (UARS). Already, it had successfully helped to confirm a ten-percent level of ozone depletion in the northern hemisphere at mid-latitudes, resulting from the combined effects of residual aerosols distributed in the upper atmosphere originating from the eruption of Mount Pinatubo in the Philippines in 1991 and cold stratospheric temperatures during the winter of 1992-93. A final key payload was the Limited Duration Space Environment Candidate Materials Exposure (LDCE), which, as its name implied, exposed a series of material samples to the conditions of low-Earth orbit.

By 17 March, the day before the scheduled landing, STS-62 had become virtually a miniature space station mission in its own right. Mounted at the back of the payload bay, the Extended Duration Orbiter (EDO) pallet provided sufficient cryogenic supplies to enable Columbia to comfortably remain aloft for two full weeks. To assist with readaptation to terrestrial gravity, the astronauts followed exercise protocols with the Lower Body Negative Pressure (LBNP) apparatus – a cylindrical device which sealed around their waists and drew bodily fluids into their legs – as part of countermeasures to better prepare them for the punishing return to Earth. Throughout their time aloft, all five crew members undertook several 45-minute-long 'ramp' sessions with the LBNP unit and on the 17th Gemar wore it continuously for four hours. Elsewhere, Casper and Allen spent time practicing their flying skills with the Pilot In-Flight Landing Operations Trainer simulator. In view of the long-duration nature of their mission, each astronaut was given two half-days off, which allowed them to soak up Earth's spectacular beauty. Allen also received some good news on 12 March, when Mission Control advised him that his parent service, the Marine Corps, had promoted him from major to lieutenant-colonel.

The good times came to an end on 18 March, however, and at 7:16 am EST Casper fired Columbia's OMS engines to begin the hour-long glide back home. He knew that he would come close to – but would not exceed – the 14-day record established by the STS-58 crew the previous November. "It would have been nice to get the record," he said after landing, "but I think we did a lot of good things." Casper brought the orbiter down onto KSC's Runway 33 at 8:09 am, less than an hour shy of 14 full days since launch. However, the landing was *not* entirely perfect, for infrared images revealed four fragments of debris falling from the Shuttle's underside as her nose gear deployed. Two sources were later identified: one from around the nose gear's starboard door and another from close to its thermal barrier. Nine years later, during the inquiry into the loss of Columbia at the end of STS-107, Admiral Harold Gehman's investigative board would focus on the debris liberation on Casper's mission in greater depth. No evidence of plasma had entered the wheel well during re-entry and the thermal barrier performed without incident, but it was serious enough for NASA to spend two full days after STS-62's landing searching for debris on the runway. It was another worrying sign that the Shuttle was not, and never could be, a truly safe machine.

RADAR LOVE

The Empty Quarter, or *Rub' al Khali*, in southern Oman, is a barren and desolate place. Rich in scorpions and rodent life, it is the largest sand desert in the world, covering much of the bottom third of the Arabian Peninsula, with a total extent of some 650,000 km^2, and measures around 1,000 km from its south-western to north-eastern extremes and about 500 km from north to south. A seemingly endless sea of 'hyper-arid' sand dunes – hued a fiery reddish-orange colour by the ubiquitous presence of feldspar – together with plains of gravel and gypsum and brackish salt flats render this landscape one of the most inhospitable on Earth. Thousands of years ago, several shallow lakes existed here, particularly in the south-western corner of the Empty Quarter, supporting a plethora of plant and animal life ... and some evidence of human habitation, too.

Human habitation, that is, with the notable exception that no human remains have been found here, only the traces of their chipped flint tools, more than 2,000 years old. Nowadays, the Empty Quarter is the most oil-rich site in the world, but in remote antiquity it served as a caravan route for the frankincense trade through the interior of Arabia. One biblical city, supposedly involved in this trade, was Ubar, whose inhabitants became so corrupted with their new-found wealth that the Holy Koran says God destroyed them. The legend of the lost city endured for centuries and many explorers and adventurers – including T.E. Lawrence – vowed unsuccessfully to find it. Then, in 1983, a Los Angeles filmmaker, Nicholas Clapp, drew the possibility of using a space-borne science instrument – the Shuttle Imaging Radar (SIR) – to look for traces of Ubar. The radar first flew aboard the second Shuttle mission, STS-2, in November 1981, and had mapped more than 40 million km^2 of terrain at resolutions of less than 40 m. SIR-A had penetrated dry sand dunes

4-05: The Space Radar Laboratory (SRL), containing the Shuttle Imaging Radar (SIR) and the X-band Synthetic Aperture Radar (X-SAR), flew twice during the course of 1994.

in the Sahara Desert of northern Sudan and revealed to investigators the location of long-dried-up river channels, ancient oases and Palaeolithic settlements.

And *this* whetted many archaeologists' appetites.

The radar's full capabilities were left undemonstrated on STS-2, because the planned five-day flight was cut short by a fuel cell problem and returned to Earth after just 54 hours. Three years later, however, it flew again as 'SIR-B' and its success provoked astonishment. Over an eight-day period in October 1984, it identified ancient caravan trails, allowed geologists to construct three-dimensional maps of subtle features on California's Mount Shasta, permitted contour modelling of parts of eastern and southern Africa and examined intricate structural features, including fault-lines, folds, fractures, dunes and rock layers. The lost city of Ubar, though, was magnetic in its appeal. "I was surprised to find that we were able to readily detect ancient caravan tracks in the enhanced Shuttle images," admitted geologist Ronald Blom of NASA's Jet Propulsion Laboratory. "One can easily separate many modern and ancient tracks on the computer enhanced images, because older tracks often go directly under very large sand dunes. We could never have surveyed the vast area where Ubar may have been, nor could we be confident of its location without the advantage of computer enhanced images from space."

Excavation began in the summer of 1990 and found a remote well, together with towers, rooms and artefacts from at least 2000 BC. The first Gulf War, which erupted early the following year, prevented further exploration, but in November 1991 archaeologists returned to the Omani desert, this time with ground-penetrating radar. Suspicion was strong that the lost city – which the Holy Koran described as having been swallowed up by the desert – may have collapsed into an underground cavity. As digging progressed, the remains of the site closely paralleled the Koran's description: an octagonal fortification, with towers some ten metres tall and thick walls, within which lay a variety of structures, storage rooms, frankincense burners and pottery sherds. In February 1992, the *New York Times* exulted that archaeologists were "virtually sure" that the site *was* Ubar.

By this time, at least two more missions of the Shuttle-ferried radar were planned, but had been extensively delayed. On the eve of the Challenger disaster, a SIR-C radar was scheduled to form part of the Space Radar Laboratory (SRL) and scheduled for launch on Mission 72A in March 1987 from Vandenberg Air Force Base in California. The payload would have operated from a near-polar inclination of 88 degrees, offering its radar instruments a broad imaging swathe across most of Earth's landmasses. In the aftermath of Challenger, all Vandenberg missions were suspended (and ultimately cancelled) and the launch site for SRL-1 changed to the Kennedy Space Center in Florida. This meant that the maximum achievable orbital inclination for the mission was 57 degrees. When NASA issued its January 1990 manifest, the launch of SRL-1 was not anticipated before the summer of 1992 and Shuttle delays forced an inevitable slip into the latter quarter of the following year. By the time that astronaut Linda Godwin – a veteran of STS-37 – was named as the SRL-1 payload commander in August 1991, the launch of her mission had moved forward slightly to the late summer of 1993. Six months later, 'rookie' astronaut Tom Jones joined her as a mission specialist and the pair began a lengthy period of preparatory work on SRL-1, ahead of the assignment of the other four crew members.

In his memoir, *Skywalking*, Jones recounted the electrifying experience of receiving this first assignment to a space mission. One February morning in 1992, he was summoned to the JSC office of the Director of Flight Crew Operations, Don Puddy. Sat at the conference table with Puddy was Godwin. "I vaguely remembered that Linda had been assigned earlier to work on long-range planning for one of next year's Shuttle missions," Jones wrote, "but I couldn't recall which one." Quickly, Puddy asked Jones if he would be prepared to serve as Godwin's 'science deputy' on SRL-1? Naturally, it was a question with only one answer, but the background of Jones, with a doctorate in planetary science, would prove invaluable for one of the most dramatic research missions of the decade.

Thomas David Jones came from Baltimore, Maryland, where he was born on 22 January 1955, and grew up with the early astronauts of Mercury and Gemini as his heroes. "I knew all of these guys by what missions they flew," he later told a NASA interviewer, "and I was really hoping to follow directly in their footsteps." Although no one in his family had served in the military, or had been a scientist, Jones was guided by his teacher father and by the influence of the early astronauts, all of whom showed him "how important it was to excel in academics and school". One day, as a Cub Scout, Jones visited the Martin Aircraft production plant at one of their 'Open House' days, where the Titan II boosters for NASA's Gemini VII and VIII missions were being prepared. Gazing upon those two enormous machines, Jones was entranced, enthralled ... and hooked for life.

After leaving high school, he entered the Air Force Academy and earned a degree in basic sciences in 1977, then began his military career. He underwent flight training and spent six years flying strategic bombers at Carswell Air Force Base in Texas and served as a pilot and commander of the B-52D Stratofortress, leading a six-strong combat crew. By the time he resigned from active duty in 1983, Jones had attained the rank of captain and was recipient of the Air Force's Commendation Medal. As a civilian, he was admitted into the University of Arizona in Tucson and began working towards his doctorate in planetary sciences, which he completed in 1988. Jones' research focuses included the remote sensing of asteroids, meteorite spectroscopy and the applications of space resources. During this period, he tried twice for admission into NASA's astronaut corps, but was unsuccessful. Until he had received his PhD, Jones wrote in his memoir, *Skywalking*, "I just wasn't competitive". Shortly thereafter, he was hired by the CIA as a programme management engineer at the Office of Development and Engineering in Washington, DC, and sent off another application for the space agency's scheduled 1990 astronaut intake.

"I tried to keep my expectations low," Jones wrote. "About five percent of all qualified applicants would be interviewed; fewer than one percent would be hired." Nonetheless, he completed a week-long series of interviews and medical and psychological evaluations in October 1989. At around the same time, he was selected as a senior scientist with Science Applications International Corporation (SAIC) in Washington, DC, supporting advanced mission planning for NASA's Solar System Exploration Division. Jones was to assume his new role on 22 January 1990 and took a week-long vacation with his family before his start date. During the course of that week, he received a telephone call, inviting him to become an astronaut, and that led to an awkward, but gracious, conversation with his SAIC boss, Harvey Feingold. Jones explained that he *could* accept the post with SAIC ... but he could *only* accept it for six months, as NASA required him to be in Houston in July. "For someone who had just been told his company had, in effect, finished in second place," Jones wrote, "Harvey was gracious to a fault. He said that although he was sorry to lose me, he was happy that my dream was moving closer to reality."

That dream moved even closer, two years later, in February 1992, when Jones sat in Don Puddy's conference room, alongside Linda Godwin, and was introduced to the scientific goals of the SRL-1 mission. Operating from an orbital inclination of 57 degrees, the payload's imaging radar would map a substantial portion of Earth, including regions as far north as Juneau in Alaska and a little further south than Tierra del Fuego at the tip of Argentina. When the remaining four members of the SRL-1 crew – which had by now received the mission designation of 'STS-59' – were announced in March 1993, only two of them had previous experience of such a high-inclination orbit; one of the highest ever attained by the Shuttle. Mission specialists Jay Apt and Rich Clifford had flown 57-degree orbits on STS-47 and STS-53 and were keenly aware of its usefulness for Earth observations. Apt wrote extensively about the experience in his book *Orbit*, co-authored with Justin Wilkinson and Michael Helfert. "On my first space flight," Apt wrote, referring to STS-37, which operated at an inclination of 28.5 degrees, "I flew no farther north than the glorious Himalaya. I shot photo after photo of Tibet, with the Sun low in the sky and the shadows long. Central and northern Asia were a mystery to me. On my second and third flights, I was on the flight deck for a 12-hour shift, when it was 'night' in Houston and 'day' in Asia. We flew over almost the entire continent."

From 57 degrees, Apt saw plumes of smoke from cellulose plants at the south shore of Lake Baikal, irrigation channels cutting across the Taklimakan Desert, multi-coloured soil tones in Kamchatka – the latter of which, he wrote, was "worth the trip up the north" – and the puzzling landscape of the Korean peninsula. Thousands of kilometres away, in the northern Americas, Apt beheld drifting volcanic ash from Mount Spurr in Alaska, together with the breathtaking grandeur of Yellowstone National Park and the geological variety of the western states. Elsewhere, northern Europe seemed to be almost completely covered by farmland and during one pass over Australia, on STS-59, a woman on the ground relayed Apt's ham radio signals to enable him to speak directly to cosmonaut Valeri Polyakov aboard the Mir space station. South America was most memorable for Apt, as he was able to see smoke and fires in Tierra del Fuego and Patagonia, presented to his eyes as far-off points of light. Even though he had flown to 57 degrees on STS-47, Apt found that lighting conditions were not good – "Our windows were pointed into the Sun most of the time" – but SRL-1 benefitted from much better conditions. In preparation for the immense amount of photography which would be conducted, alongside the radar observations, all six crew members became proficient in the use of appropriate camera lenses with consistent shutter speeds and light meters to determine proper exposure times.

To understand how the SRL-1 payload evolved, it is important to comprehend the accomplishments of the SIR-A and SIR-B radars. The first mission was restricted to recording the ground-track directly 'beneath' the orbiter, but in preparation for SIR-B the radar was engineered to 'tilt' at angles of between 15 and 57 degrees to the side and its imaging resolution was enhanced from 40 m to 25 m. By varying its 'look' angle in this fashion, it became possible to assemble 'mosaics' of adjacent surface features, collected over periods of several days. The third mission, SIR-C, offered multi-frequency, multi-polarisation imagery and represented the first spaceborne radar with the ability to transmit and receive horizontally and vertically polarised waves at both the L-band (23 cm) and C-band (6 cm) wavelengths. Measuring 12 m long and 4 m wide and weighing 10,500 kg, SIR-C was the most massive piece of flight hardware ever built at the Jet Propulsion Laboratory (JPL) in Pasadena, California. It was a 'synthetic-aperture' radar and, like a huge dining table, consumed almost half of Endeavour's payload bay for STS-59. "Synthetic-aperture radar technology," explained Tom Jones in his memoir, "uses the motion of a spacecraft or aircraft to electronically synthesise an antenna much larger than its physical size, yielding higher imaging resolution." Originally assembled from spares left over from NASA's 1978 Seasat mission, SIR-C was affixed to its own truss support structure, which, in turn, was mounted onto a Spacelab pallet, providing a 'side-looking' viewing angle, some 47 degrees from the nadir. As a result, during data-gathering operations, the Shuttle was required to reorient itself to precisely direct SIR-C at its ground targets. Unlike the previous SIR missions, its radar beam was formed from hundreds of transmitters, embedded within the surface of the antenna. By properly adjusting the energy from these transmitters, the beam could be electronically 'steered' and, when combined with Shuttle manoeuvres, offered the scope to acquire images from various directions.

However, SIR-C was not the only radar aboard SRL-1. Running like a strip along the uppermost edge of SIR-C was 'X-SAR' – a 12 m × 0.4 m synthetic-aperture device, built by a partnership of organisations, including the German Dornier and Italian Aleniz Spazio companies, together with the German (DARA) and Italian (ASI) space agencies – which offered a single-polarisation radar, operating at X-band (3 cm) wavelengths. X-SAR was a follow-on project from Germany's Microwave Remote Sensing Instrument, flown aboard Spacelab-1 in late 1983. Its 'slotted waveguide antenna' was finely tuned to produce a narrow energy beam and X-SAR was designed to be mechanically aligned with the L-band and C-band beams of SIR-C. Before launch, NASA's SRL-1 press kit noted that resolutions as high as 10 m were a possibility. Throughout the projected nine-day mission, the combined SIR-C/X-SAR imaging suite was expected to gather around 50 hours of observations, covering more than 50 million km^2, acquiring 32 terabytes of raw data and storing it all on 180 cassettes, using three high-density, digital, rotary-head tape recorders. Both radars would work together to acquire the best possible data. For instance, the shorter-wavelength X-SAR data was expected to be particularly useful for the determination of snow types, with the L-band and C-band capabilities of SIR-C estimating snow volumes. Although the United States, Germany and Italy were involved in the development of the payload, more than 50 investigators from 13 nations (Australia, Austria, Brazil, Canada, China, the United Kingdom, France, Germany, Italy, Japan, Mexico, Saudi Arabia and the United States) provided experiments and investigations as part of the science team. Also operating as part of the payload was an instrument called the Measurement of Air Pollution from Satellite (MAPS), which had also flown as part of the SIR-A and SIR-B missions. It sought to measure global distributions of carbon monoxide in the troposphere, between the altitudes of 5-15 km, in which the 'weather system' is most active, and it had already pointed to a worrisome trend that 'greenhouse gases' had become increasingly severe throughout the early part of the 1980s. MAPS also highlighted disturbing levels of pollution – particularly in the tropics – caused by the seasonal burning of biomass.

SRL-1 would not be scanning the planet in an indiscriminate fashion. Rather, a series of 19 'super-sites' of particular geological, hydrological or ecological interest were selected and would be observed during daily radar passes. Experiments were set up on every continent in the world, save Antarctica, and particular regions of focus included Mammoth Mountain in California, for remote-sensing of the water content of the Sierra Nevada ice pack, Chichkasha in Oklahoma for soil-moisture studies, Michigan's Upper Peninsula to investigate forest biomass, together with temperate forests in North America and Central Europe and tropical forests in South America's Amazon basin, surface and internal waves within the Mid-Atlantic Gulf Stream and Hawaii's Volcanoes National Park. During the SRL-1 mission, 'ground truth' teams at various sites undertook measurements of vegetation, soil moisture, sea state, snow cover and weather conditions to correlate with the Shuttle data.

With such an enormous workload of preparation for the mission, it came as something of a relief when in mid-1992 NASA took the decision to shift SRL-1 into the spring of 1994 and move the critical Hubble Space Telescope servicing mission to the left. In the meantime, at the beginning of March 1993, the names of the four astronauts who would accompany Godwin and Jones were announced. Mission specialists Jay Apt and Rich Clifford would be joined by commander Sid Gutierrez and pilot Kevin Chilton; all four had flown before. (In *Skywalking*, Jones recalled a telephone call from Clifford, with a simple message of unbridled joy: "We're flying together on STS-59, T.J.") Gutierrez soon broke the crew into two shifts – 'red' and 'blue' – to supervise orbiter systems and the SRL-1 payload around the clock. He would lead the red team, with Godwin and Chilton, whilst Apt would lead the blue team, joined by Jones and Clifford. "He needed the two pilots on the same sleep cycle," Jones explained, "both fresh and wide awake for launch and landing days. Linda would join them, enabling Sid to get the payload commander's input on any experiment or orbiter problems without having to wake the opposite shift."

By the end of March 1994, Endeavour was 'hard-down' on the surface of Pad 39A, with launch anticipated early the following month. Flying for the sixth time, she carried six new-specification thermal protection tile, known as 'Toughened Uni-Piece Fibrous Insulation' (TUFI), which offered a varying density level from high at the outer surface to low at the interior surface and carried a greater capability to withstand debris impacts. Produced by the Shuttle's prime contractor, Rockwell International, the new tiles were affixed to a section between her three main engines,

4-06: Surrounded by free-floating objects on Endeavour's flight deck, STS-59 pilot Kevin Chilton was making his second space voyage. During an 11-year career with NASA, he flew three Shuttle missions, before returning to the Air Force. By the time Chilton retired from active military duty in February 2011, as a four-star general, he had earned the highest rank of any military astronaut.

on the base heat shield, which had historically been susceptible to damage from launch pad debris. Their presence was expected to reduce the post-flight maintenance process. Extant tile systems utilised a reaction-cured glass (RCG) coating, but as Daniel Leiser, of the NASA Ames Research Center's Thermal Protection Materials Branch, explained, they tended to receive little support from the underlying tile and this caused cracking or chipping when an impact occurred. TUFI, on the other hand, permeated the pores closer to the surface of the insulator, which supported and reinforced the outer surface and rendered it less susceptible to impact damage. Rather than creating an expanding crack, it would yield only a small dent which should be easier to repair after landing.

According to the STS-59 press kit, launch was scheduled for 7 April, but an inspection of a Shuttle main engine at Rocketdyne's facility in Canoga Park, California, had found that the critical dimensions of nickel-alloy liquid oxygen guide vanes in the turbopump preburner were 'out of tolerance'. Inspectors found that two components had sharp, rather than rounded tips, presented an increased possibility of fragmentation … and *that* enhanced the likelihood of a premature engine shutdown. "A thinned or deformed vane," explained Tom Jones, "could break off in the oxidiser flow, shattering the turbopump blades downstream." A 24-hour delay was enforced to offer technicians the opportunity to examine Endeavour's vanes, which turned out to be in the proper configuration. The delay proved fortuitous, for Kevin Chilton contracted a virus and had to be placed under observation, 'isolated' even from his already-isolated crewmates in quarantine. "It was a race," wrote Jones, "between Chili's recuperative powers and the ticking countdown clock." By the 7th, Chilton had recovered sufficiently to rejoin the rest of the crew. High winds the following morning threatened the launch. Overcast conditions led the Mission Management Team to lengthen the 'hold' in the countdown at T-9 minutes and, eventually, late in the 2.5-hour 'window', the clouds began to clear. From the flight deck, Gutierrez, Chilton, Apt and Clifford described a beautifully clear blue sky through their windows. Downstairs on the middeck, Godwin and Jones could see nothing, but it seemed a positive change. Unfortunately, this break was accompanied by an increase in wind speed, which exceeded the maximum allowable limit for a Return to Launch Site abort at the Shuttle Landing Facility. Jones tried to invoke St. Theresa, patron saint of aviators ... and St. Joseph of Cupertino, the *patron saint of astronauts*, for their help.

Suddenly, the cabin fell silent. Then Gutierrez spoke. "What's St. Joseph of Cupertino got to do with *astronauts*?" he asked.

"Well," Jones replied, "he used to levitate over the altar of his monastery chapel. He discovered zero-G *three hundred years ago!*"

They gave it a try, but alas it was not to be and the weather gods were just too strong on 8 April 1994. With the winds continuing to gust out-of-limits, Launch Director Bob Sieck ordered a scrub. Liftoff was rescheduled for 7:05 am EDT on the 9th and it was six anxious astronauts who climbed aboard Endeavour that morning, as *another* scrub would produce a conflict with other scheduled launches on the Eastern Range and effect a delay until at least the 23rd. Lying uncomfortably on their backs, they listened to the milestones over the intercom. By 6:56 am, as Endeavour emerged from the last scheduled 'hold' in the countdown at T-9 minutes, they tightened their harnesses and steeled themselves for the controlled explosion that would soon come. At T-5 minutes, Kevin Chilton leaned over from the pilot's seat and flipped a trio of switches to activate the orbiter's Auxiliary Power Units. Endeavour now had hydraulic muscle.

In the acronym-laden minutes which followed, the astronauts were instructed to close and lock their helmet visors and activate the flow of oxygen into their suits. The Albuquerque-born Sid Gutierrez, the only Shuttle pilot of Hispanic ancestry, led his team in their final 'comm' checks and was wished "*Vaya con Dios*" – "Godspeed" – by launch controllers as a last send-off. With 31 seconds to go, they received the 'Go' for autosequence start, as Endeavour's computers assumed primary command of vehicle critical functions. Fifteen seconds later, navigational indicators on the instrument panel snapped smartly into their correct launch-ready positions. "Nav init," confirmed Gutierrez.

At seven seconds, the swirling sparks of the hydrogen burn igniters gave way to a familiar rumble and sheet of translucent orange flame, as the three main engines roared to life, quickly reaching full power and producing three dancing Mach diamonds. "*Three at a hundred!*" yelled Chilton over the growing crescendo. Finally, at T-zero, the twin boosters flared plumes of brilliant orange flame and the STS-59 crew received a smart punch in the back as they departed Planet Earth. Writing almost a decade later, Tom Jones remembered lucidly the adrenaline-charged minutes of his first climb into orbit. A "nasty shaking" was accompanied by a peculiar sensation of the entire cabin whipsawing around him, as the computer-controlled 'Roll Program' manoeuvre, ten seconds after liftoff, oriented Endeavour for her 57-degree orbit. Jones thought of his father, who had died a little more than a year earlier. Two minutes into the ascent, the SRBs were safely jettisoned and the Shuttle continued to power her way into orbit under the thrust of her main engines. At length, Gutierrez and Chilton congratulated Jones as Endeavour crossed the 100 km Kármán line and passed the official boundary between the 'sensible' atmosphere and the edge of space. Suddenly, the G-imposed pressure on his chest was gone, he felt light under his harness straps ... and was hit by the instant realisation: "*This must be weightlessness!*"

In *Skywalking*, Jones noted that the sensation of perpetual free-fall was perplexing, but that he encountered no difficulties deciding which way was 'up' and which was 'down'. Then, as his body adapted to microgravity, fluids normally pulled into his legs and lower abdomen instead migrated toward his chest and his inner-ear balance organs sent confusing messages to his brain ... and produced a wave of nausea. It came, he wrote, as a sudden doubled-over spasm, eyes closed, feeling miserable, and was gone within minutes. A shot of the anti-nausea drug Phenergan from crewmate Rich Clifford provided instant relief, but the malaise would return to haunt Jones a couple of times during the early stages of the flight. Yet the view of Earth was glorious. "As Endeavour rose toward sunrise, I gasped," he wrote. "Between heaven and Earth was a vision of pure beauty, the robin's-egg-blue of the atmosphere backlighting the darkened horizon." For an instant, his eyes filled with tears.

But the clock was forever their enemy. As time ticked down toward the first sleep period for the blue team, Linda Godwin's red shift led the activation of SRL-1. Despite initial problems with the power-up of the X-SAR amplifier – caused by an overly sensitive protection circuit – the German-built radar entered full operations on 10 April. Although the mission had always been baselined for nine days, it was expected that with appropriate use of consumables, a tenth day could be squeezed out of Endeavour. The decision to extend STS-59 did not drag its heels and came late on Day One, setting the mission on its path of science-gathering discovery in fine fashion.

The astonishing experience of 16 sunrises and sunsets in each 24-hour period was best illustrated through one of Tom Jones' written recollections. Orbiting at twice the average Shuttle inclination, and a much lower altitude of just 220 km, he and his crewmates were provided with an astonishing vista of the Home Planet. "Now the eggshell-blue light of the sunrise is coating the horizon," he wrote at one point, late in the mission. "The payload bay is now going bluish-white as we come up out of the darkness. Across Nova Scotia now, and Labrador, and still no sunshine visible. I can still see the stars. No, not for long. Here comes the orange of the Sun. *Boom!*

Sunrise! Now the payload bay is pink-orange, yellow, going to white, and it will soon be brilliant. *Fantastic!*" Amidst all the high technology and intensive science workload, it was a profoundly spiritual experience. On the second Sunday after Easter, Jones, Chilton and Gutierrez – all Catholics, and Chilton a Eucharistic minister – gathered on the flight deck for a short service of Communion. They also had the opportunity, on 16 April, to speak via the Shuttle Amateur Radio Experiment with fellow astronauts Norm Thagard, Bonnie Dunbar and Ken Cameron, who had recently moved to Russia to support the early Shuttle-Mir effort.

Overall, SRL-1 was a remarkable scientific triumph. In its post-flight mission report, NASA announced that the SIR-C/X-SAR observations had accomplished 97 percent of their required data takes from 400 primary science targets and 99 percent from the 19 critical super-sites. The crew also handled additional requests, including imaging Germany's Rugen Island, in the Baltic Sea, and examining Japanese rice fields. Ninety-four hours of radar data, taken over 44 discrete nations and covering an area in excess of 70 million km^2, were stored on 165 of the digital tapes. Elsewhere, the MAPS experiment performed flawlessly, acquiring data on the regrowth of forests in a fire-scarred area of China. On one occasion, Tom Jones relayed a verbal report on thunderstorms over Taiwan, the Philippines and New Guinea to augment the MAPS data. On the middeck, a Visual Function Tester assessed the astronauts' visual acuity and the National Institutes of Health-supplied space tissue loss investigation explored the effects of microgravity on muscles, bones and endothelial cells. Outside in the payload bay, mounted in a GAS canister, was the Consortium for Materials Development in Space's fourth Complex Autonomous Payload (CONCAP-4), which produced crystals and thin films through physical vapour transport. Other GAS canisters supported a study of water-ice crystals from New Mexico State University, a French experiment to explore the thermal conductivity of silicone oils and a Japanese investigation to examine the behaviour of the cellular slime mould, *Dictysotelium discoideum*.

On 19 April, the payload bay doors were closed and sealed on time, but cloud conditions at the Shuttle Landing Facility caused the first scheduled landing attempt of the day to be waved off, just half an hour before the anticipated de-orbit burn. "Unfavourable and dynamic" weather later that afternoon also put paid to a second attempt to bring STS-59 home. This forced mission managers to reschedule the landing for the following day, the 20th. However, weather at KSC remained 'No-Go' and Gutierrez and his crew were diverted to Edwards Air Force Base in California, touching down without incident at 9:54 am PDT (12:54 pm EDT) on concrete Runway 22, after a mission lasting a little more than 11 days. STS-59 had proven a spectacular success, gathering sufficient data to fill an estimated 20,000 encyclopedias, taking more than 15,000 photographs and requiring in excess of 400 manoeuvres to position the Shuttle for the radar observations.

Two weeks later, with the ink barely dry on their crew flight report, Tom Jones returned to his Houston office to begin his next assignment. More than two years of his professional life at NASA had been devoted to the Space Radar Laboratory and in August 1993 he was assigned to serve as payload commander for the second flight, SRL-2. The Jet Propulsion Laboratory, which developed the radar, wanted two of its own experts to serve as payload specialists, but had been told that NASA career mission specialists could handle the tasks. Next, JPL insisted that at least one crew member should fly both missions. This followed typical NASA practice of 'carrying over' an experienced crew member from one payload to the next on important science flights ... but Jones' transition from SRL-1 to SRL-2 offered something a little different. Originally, the two radar flights were supposed to fly at least a year apart, as shown by NASA's February 1991 and January 1992 manifests, which anticipated a 15-month gap between them. However, when the decision was taken, in mid-1992, to advance Endeavour's Hubble repair flight *ahead* of SRL-1, the two missions drew much closer – within four months of each other – on the manifest.

By the time Jones was named as the SRL-2 payload commander, he was still almost eight months away from flying SRL-1. "I recalled my surprise," he wrote in *Skywalking*, "when I met with chief astronaut Robert 'Hoot' Gibson ... He had just invited me to fly again." By then, the manifest envisaged SRL-2 – aboard STS-68 – flying aboard Endeavour in mid-August 1994, a mere four months after SRL-1. "Hoot laughed at my startled reaction," Jones continued, "but he *wasn't* kidding. What else could I say but *yes*?" Two months later, in October 1993, NASA announced the names of the remainder of the STS-68 crew: Mike Baker in command, joined by pilot Terry Wilcutt and mission specialists Steve Smith, Dan Bursch and Jeff Wisoff. The five men had pestered Jones mercilessly during his SRL-1 training and even whilst he was in orbit.

"*Don't forget you start sims with us next week*," read one note, authored by Baker. It was signed off with simplicity: "*Your STS-68 Associates*".

Little did any of them realise that SRL-2 would only get off the ground after a particularly hair-raising on-the-pad engine abort, less than *two seconds* ahead of liftoff.

THE MAJESTY OF 'COLUMBIA'

The name 'Columbia' has long been associated with the people and culture of the United States, originating from the earliest arrival of European settlers, under the command of the Italian-born explorer Christopher Columbus, in the Americas in the autumn of 1492. Since then, adjectives such as 'pre-Columbian' and 'post-Columbian' have been routinely applied to the epochs before and after his historic arrival. Ships have been named in his honour, including two which profoundly altered humanity's fortunes in space. The accomplishments of both of these 'Columbias' were appropriately hailed by Capcom Mario Runco on the morning of 16 July 1994, as seven astronauts – the crew of STS-65 – orbited Earth aboard Space Shuttle Columbia. By the time they landed, a week hence, they would have smashed the record for the longest Shuttle mission so far.

Yet Runco's message paid tribute to *another* Columbia, too. "On this day, at this moment, 25 years ago," he began, "three of your predecessors began an epic journey that would change the way we viewed our world. Columbia's journey today, as her namesake did back then, is pushing the frontiers of knowledge and science for all mankind." Runco was, of course, referring to Apollo 11, whose command and service module – also named 'Columbia' – had ferried them to the Moon in support of humanity's first piloted lunar landing. A quarter of a century, to the very day, later, the astronauts of STS-65, commanded by Bob Cabana, were halfway through a two-week flight which was not destined to travel so far, but whose contribution to the space sciences was no less important.

That flight was to be Columbia's last, before she was scheduled for a year-long period of refurbishment and maintenance, and STS-65 again carried the Extended Duration Orbiter to support the second International Microgravity Laboratory (IML-2). The mission encompassed more than 80 experiments, provided by 200 scientists from 13 countries, including the multi-national European Space Agency, together with the space organisations of Canada, France, Germany, Japan and the United States. Unlike IML-1, flown two years earlier, the second flight was twice as long and carried twice as many facilities in its Spacelab long module. Several of these were controlled remotely by ground-based investigators, with US and European research teams connected by intercontinental voice, video and data links. Such 'telescience' was seen as useful practice for the International Space Station. In another indication of the importance of those experiments, the STS-65 crew was divided into two 12-hour shifts to run the laboratory around the clock. The red team comprised Bob Cabana, together with pilot Jim Halsell, payload commander Rick Hieb and Japanese payload specialist Chiaki Mukai, whilst the blue team consisted of mission specialists Carl Walz, Leroy Chiao and Don Thomas.

The shift patterning caused some consternation at first, which Chiao attributed to the decisions taken by Rick Hieb. "We were not the most cohesive crew," he explained in his NASA oral history. "I really don't want to blame anyone, but maybe it was the payload commander set up the rivalry from the beginning – the red shift and the blue shift – because the mission specialists and the payload specialist were assigned first. We went through this training with this dynamic of a competition between the two." Years later, Chiao was convinced that Hieb's competitiveness was good-natured, but it could be perceived that during training 'The Reds' were choosing the 'good' shifts for themselves. In simulations, for example, it seemed to Chiao that the blue team always worked the undesirable 6:00 pm-6:00 am night shifts in Houston. Body clocks quickly became confused, as the astronauts travelled around the world, spending a couple of weeks in Houston, a couple of weeks in Europe, a couple of weeks back in Houston, a couple of weeks in Japan, and so on, working with experiment facilities and investigators. (It did not help that NASA Administrator Dan Goldin decided that his personnel could no longer fly in business class. "All those trips, except where we could beg our way up to the front, were done in economy class," Chiao lamented.) Still, working in shifts certainly engendered a measure of team bonding, particularly as there were four members of the Group 13 astronaut class aboard ... and *three* of them were on the blue shift. Fellow 'Hairball' Jim Halsell was on the red shift. Every so often, Walz, Chiao and Thomas would offer him some advice. "You know, you *really* should be blue," they told him. "You *really* don't belong with those red guys!"

The inclusion of Mukai, who became the first female Japanese spacefarer and

4-07: Commander Bob Cabana (left) and pilot Jim Halsell listen attentively to instructions, ahead of an STS-65 ascent simulation.

who represented her country's National Space Development Agency (NASDA), was illustrative of Japan's enormous contribution to the IML-2 mission. Backing her up was French astronaut Jean-Jacques Favier. "To go into space became my dream when I was 32 years old," Mukai later told a NASA interviewer, "because when I was a child Japan didn't have a space programme, so my first dream was to be a medical doctor." She was born in Yayebayashi, in the far-western part of Gunma Province, on the central Japanese island of Honshu, on 6 May 1952. After completing senior high school in Tokyo, she entered medical school at Keio University – Japan's oldest extant university – and received her doctorate in 1977. This was followed, in 1988, by a PhD in physiology, also from Keio. Between the receipt of her two doctorates, she was board-certified in medicine, served a residency in General Surgery and worked variously as a general surgeon at Keio University Hospital and as an emergency surgeon at Saiseikai Kanagawa Hospital.

By the beginning of 1985, Mukai was an assistant professor in Keio University's Department of Cardiovascular Surgery. She then spotted a NASDA advertisement, calling for payload specialist candidates for a joint US-Japanese Spacelab-J mission on the Shuttle. "I thought the space programme is such a wonderful area to use my medical expertise," she told the NASA interviewer, years later. "I applied and was lucky enough to be selected." Admission into the first group of NASDA astronauts in June 1985, she was joined by chemist Mamoru Mohri (later to become the first 'professional' Japanese astronaut) and physicist Takao Doi (later to become the first

Japanese spacewalker), and the trio began work on Spacelab-J. Mukai continued her work as a cardiovascular physiologist and as a visiting surgeon at Keio University and in 1989 she was board-certified as a cardiovascular surgeon.

Although she did not fly on Spacelab-J, Mukai was assigned in October 1992 as the prime payload specialist for IML-2. The announcement of crew members for this important life and microgravity sciences mission got underway a month earlier, at the beginning of September 1992, when NASA selected veteran astronaut Rick Hieb as the payload commander. Eight weeks later came the naming of mission specialists Chiao and Thomas, as well as Mukai. Finally, in late August 1993, the 'orbiter' crew of Cabana, Walz – then a few weeks away from his first flight, STS-51 – and Halsell were assigned to round out the seven-strong team. "Both Don and Leroy bring strong materials science backgrounds to the IML-2 payload crew," explained Acting Director of Flight Crew Operations, former astronaut Steve Hawley, at the time. "Their strengths will complement the previously assigned crew members in achieving the multi-science objectives of this important international mission." In time, both men would be assigned to long-duration expeditions aboard the International Space Station – the very end-game for which IML-2 served amply as a test model – and in Chiao's case his career would culminate in commanding the multi-national research laboratory for six months in 2004-05. For Thomas, however, the fortune of a lengthy ISS mission would not smile upon him.

Donald Alan Thomas was born in Cleveland, Ohio, on 6 May 1955, and attended high school in his home state, before entering Case Western Reserve University to study physics. He earned his degree in 1977 and went on to gain a master's and a doctorate in materials science from Cornell University, in 1980 and 1982, with an emphasis upon the evaluation of crystalline defects and sample purity in the superconducting properties of niobium. Clutching his PhD in hand, Thomas joined AT&T Bell Laboratories as a senior member of the technical staff, helping to develop advanced materials and processes for high-density interconnections of semiconductors. In 1987, he moved to Lockheed to review materials used in Shuttle payloads and, a year later, he joined NASA's Johnson Space Center in Houston, Texas, for the first time, as a materials engineer. Thomas' particular emphasis was upon the projected lifetime of advanced composites being considered for use aboard Space Station Freedom. Following an unsuccessful attempt to join the 12th astronaut class in June 1987, Thomas was selected as a member of the 13th intake in January 1990.

Leroy Chiao came from Milwaukee, Wisconsin, the child of Chinese parents who fled the Communist takeover in the aftermath of the Second World War by first settling in Taiwan and then emigrating to the United States in the 1950s. (In his later astronaut career, Chiao selected 'Shandong' for his radio callsign, to honour the coastal province in eastern China where both his mother and his father were born.) Chiao himself was born on 28 August 1960 and became interested in the space programme as a child, when he saw Neil Armstrong and Buzz Aldrin walking on the Moon. "I remember being in my parents' home a hot summer day," he recalled in a NASA interview, "and then, later that evening, actually take the first steps on the Moon. To me, that was *really* a big event!" However, unlike most children of his generation, who sought after the astronaut goal for a while, then gave it up, Chiao kept it at the back of his mind through high school and his university education.

"Mechanical things and electrical things and chemical things" intrigued him as a youngster, together with a fascination for mathematics and the sciences, which led Chiao to select chemical engineering for his degree choice at the University of California at Berkeley. Receiving his undergraduate award in 1983, he proceeded to the University of California at Santa Barbara and earned his master's and doctoral degrees in 1985 and 1987. Chiao initially worked at Hexcel Corporation – "a composite materials supplier" – and at Lawrence Livermore National Laboratory, working on the fabrication of filament-wound and thick-section aerospace composites. In September 1989, after submitting an application to join NASA's astronaut corps, he was invited to Houston for a week of interviews and physical and physiological tests. Four months later, he was selected. It was not easy. "The thing I tell people who ask me how do you become an astronaut," he explained later, "there really is no one way to do it ... If you're a person who puts all your eggs in one basket and says, well, I'm going to study something I'm not really interested in and hope to become an astronaut, you're in for a lot of disappointment. My advice ... always is to study something that you're interested in and work in a field that interests you. *Then* apply."

As the IML-2 payload crew, Hieb, Chiao, Thomas and Mukai supported a wide range of the facilities in the Spacelab module. Germany's Biostack sandwiched biological specimens between 'plates' of radiation detectors as part of a project to determine the impact of high-energy cosmic rays passing through the module's hull. The specimens would be examined after Columbia landed to identify the paths and entry points of heavy ions and assess physical changes or damage. The impact of such radiation, it was recognised, could prove detrimental to astronauts on future long-duration missions. Three sealed Biostack containers were filled with shrimp eggs and salad seeds to study their behaviour. Other radiation-measuring experiments were conducted using a Japanese monitor. More significantly, IML-2 marked the first time that such radiation data had been transmitted to Earth in real time. This immeasurably aided the research in 'Biorack', a device built by the European Space Agency (ESA) to specifically investigate the impact of microgravity and cosmic radiation upon genetically modified rapeseed roots, cress seedlings, fruit flies and even human skin cells. Elsewhere, the Aquatic Animal Experiment Unit (AAEU) housed a complement of Japanese red-bellied newts, goldfish and Medaka fish. Obviously, the 'perishable' nature of these living specimens meant that loading them aboard the Spacelab module had to occur a few hours before launch. Fortunately, the processing of IML-2 – and Columbia herself – was exceptionally smooth and was even described as "fairly simple plug-in work" by NASA spokesman George Diller.

The summer of 1994 was one of the few occasions on which all four Shuttle orbiters were present in Florida at the same time. This was particularly interesting, as the Kennedy Space Center possessed only *three* Orbiter Processing Facility (OPF) hangars for them. Moreover, continuous early-summer rain complicated the issue further, obliging technicians to 'shuttle' the vehicles backwards and forwards between the three OPF bays and a 'transfer aisle' in the cavernous Vehicle Assembly Building (VAB) to keep them under cover. At length, Columbia entered that VAB on 8 June for stacking onto her External Tank and twin boosters and was 'hard-down' on Pad 39A by mid-morning on the 15th. In spite of the need to replace a faulty main engine controller, the main problem was the weather on the early afternoon of Launch Day, 8 July, which forecasters predicted to be unacceptable.

Chiao was sceptical that they would actually get to fly on the 8th. The night before, there had been reports of thunderstorms and lightning in the area and the Transoceanic Abort Landing sites were forecasted to also be poor, should an emergency during ascent demand their use. Still, fuelling of Columbia's External Tank went ahead and the astronauts suited-up, rode the bus out to the pad and were strapped into their seats for a scheduled launch at 1:11 pm EDT. By this time, Air Force meteorologists were expecting a 40 percent chance of early-afternoon thunderstorms, which might impair Cabana and Halsell's visibility in the dire eventuality that they had to perform an abort landing at the Cape. "We were totally in the mindset of 'Okay, we're going to get suited-up and go out, strap in and then probably come back, eat a hamburger and sit around for another day or two'," Chiao recalled. Yet the countdown progressed, halting, as planned, at T-9 minutes, then emerging without issues to continue down to T-5 minutes.

Then, just before Halsell was instructed to activate Columbia's Auxiliary Power Units, the weather started to clear and was declared 'Green'. Not surprisingly, that decision "got *everybody's* attention", said Chiao, and STS-65 rocketed safely into orbit – watched by dozens of Japanese journalists and dignitaries – at 12:43 pm EDT. The launch created a blast of such intensity that a corrugated iron roof was torn from a building, close to the pad. Columbia's rousing rise to the heavens also lifted the roof for the astronauts themselves, four of whom had never flown into space before. Chiao was sitting in the so-called 'MS1' seat, directly behind Halsell, for ascent, and for him one of the most lasting experiences was the separation of the Solid Rocket Boosters, two minutes into the climb. The event was accompanied by a loud bang as the separation rockets ignited, followed by a period of unearthly silence and smoothness, which momentarily convinced him that the main engines had stopped. It gave Halsell, watching the data tapes, a start, too. Both men quickly verified that the engine tapes were still 'Green', and the engines still running, but the change in acceleration and sound was so dramatic that they were sure something was awry.

Minutes later, the sensation of three times a 'normal' gravitational load was replaced, in the blink of an eye, by the purity and unalloyed freedom of *weightlessness*. "For me, it felt like a forward tumble," Chiao remembered, "and then the full-headedness." But there was little time to think, only time to *do*. Helmets had to be removed and stowed, communications headgear had to be removed and stowed, cameras had to be *un*stowed and passed up by their crewmates on the middeck to photograph the just-jettisoned External Tank. Those first moments in perpetual free-fall were puzzling and brought with them a little headache, a queer dizziness ... but no disorientating nausea. Space sickness seemed not to affect him. Chiao was one of the lucky ones, for historically upwards of 25 percent of space travellers have reported an element of the debilitating malaise upon arrival in this strange new realm.

Later that evening, as the payload crew readied the IML-2 module for science operations, Bob Cabana showed Mission Control a video recorded from a tiny, lipstick-sized camera inside the cockpit during the bone-jarring climb to space. "As you can see," he remarked, matter-of-factly, "it vibrates *pretty good*!" His words were echoed by James Donald Halsell Jr, the STS-65 pilot, who was embarking on the first of what would become a five-mission astronaut career. Halsell was born in West Monroe, Louisiana, on 29 September 1956, the son of parents who believed strongly in setting ambitious personal goals and pursuing his interests. "I also had an uncle, Tommy Thompson, who was an airline pilot," he once told a NASA interviewer. "When I was growing up, I considered him one of my heroes, because he was flying airplanes and they were *paying* him to do it. *That* seemed like a good way to spend your life!" By the time Halsell finished high school, he had decided on the Air Force as a career path and that began with the prestigious Air Force Academy, from which he earned an engineering degree in 1978. Undergraduate pilot training followed and Halsell flew the F-4 Phantom aircraft from Nellis Air Force Base in Nevada and Moody Air Force Base in Georgia, gaining qualifications in conventional and nuclear weapons delivery. Two master's degrees followed – one in management from Troy University in 1983 and a second in space operations from the Air Force Institute of Technology in 1985 – and Halsell was selected to attend test pilot school. *Air Force Academy, fighter pilot, test pilot* had long been his goals and in 1986 he graduated first in his Test Pilot School class and was awarded the coveted Liethen/Tittle Trophy for performance and academic excellence. He served as a test pilot in the F-4, the F-16 Falcon and the SR-71 Blackbird, before entering NASA's astronaut corps, alongside Leroy Chiao and Don Thomas, in January 1990. "I think it might be wrong to paint myself as a super-charged 16-year-old, thinking I was going to be an astronaut," Halsell said later, reflecting upon his youth, "but certainly I had it in the back of my mind that could be a path that I would like to explore."

That exploration got underway, for Halsell, as soon as Columbia reached orbit, for the red team proceeded with the activation of the IML-2 research facilities and initiation of its first experiments. Meanwhile, the blue shift – led by Carl Walz – bedded down for an abbreviated sleep period, ahead of their first 12 hours on duty. The starboard side of the middeck was equipped with phone-booth-sized 'sleep stations', but going to sleep after four hours in space was like asking a child to sleep on Christmas morning. Chiao took a couple of Restoril, "which was the sleeping medication of choice, back then", but found himself floating, zipped up in his sleeping bag, for some period of time, before the pills kicked in. But he still was not comfortable. At length, he convinced himself to lie on his side, as he would in his earthly bed. Peculiar as that seemed, in an area with no up, down or sideways, it *worked* and Chiao fell quickly to sleep.

"We're looking forward to a super two weeks up here," Cabana exulted, as within three hours of launch Chiaki Mukai activated the Advanced Protein Crystallisation Facility (APCF) in a pair of middeck lockers. This device employed vapour-diffusion, liquid-to-liquid diffusion and dialysis to grow high-quality protein crystals and during the course of STS-65 more than 7,000 video images of the slowly growing crystals would be obtained. In the meantime, Leroy Chiao spent part of his first blue shift initiating the Biorack payload. Mukai had already transferred many of the perishable specimens from middeck storage lockers into the facility, allowing Chiao to kick off a five-day experiment to examine calcium loss from human bones. Similar work had been undertaken on the IML-1 mission and had suggested that bones typically did not lose a sizeable amount of calcium, as long as they were exposed to periods of 'compression', often through vigorous exercise. However, it was recognised that more data was needed to counteract the effect of microgravity on the human skeletal structure. Biorack housed 19 experiments, including a Norwegian investigation into the growth of genetically modified rapeseed roots and cress seeds and a study of premature aging in fruit flies, possibly induced by their difficulty in coping with weightlessness. Indeed, at the start of the mission, Rick Hieb reported that the flies were "buzzing around with excellent vitality", but that after ten days or so their bull-at-a-gate response had slowed substantially and tailed off and they were acting more like their ground-based 'control' counterparts. Other Biorack work included observations of cultures of human skin fibroblasts – cell-producing connective tissues – and bacterial cells.

Germany's Slow Rotating Centrifuge Microscope, nicknamed 'NIZEMI', focused upon the response of living and non-living specimens to various levels of gravitational influence. The centrifuge applied loads ranging from a thousandth of terrestrial gravity to around 1.5 G on slime mould, Moon jellyfish (*Aurelia aurita*), cress roots, lymphocytes, green algae, immune system T- and B-cells and a transparent material known as 'succinonitrile-acetone'. All of the researchers involved with the NIZEMI research praised its "great success", which mirrored the comments from other investigators. Canada again flew its back-pain experiment (previously carried aboard IML-1), which sought to explore the possibility that problems caused by the lengthening of the human spine in weightlessness might be induced by changes in the function of the spinal cord or spinal nerve 'roots'. IML-1 found that astronauts 'grew' on average between 5-7 cm during a mission and experienced a 'flattening' of the normal spinal contour. During IML-2, the spacing between disks in the astronauts' spinal vertebrae was measured and, each day, the entire crew filled in questionnaires to describe occurrences of back pain or symptoms of spinal column dysfunction. They also took stereo photographs of themselves in differing postures, subjected themselves to electrical impulses to stimulate their sensory nerves, squeezed hand grips as electrodes gathered data on their blood pressures and heart rates and would surrender to MRI scans after landing.

Japanese scientists were particularly impressed with the work in the Animal Aquatic Experiment Unit, which offered the opportunity to examine the spawning, fertilisation, embryonic stage, vestibular function and general behaviour of live fish and small amphibians. IML-2 carried almost 150 pre-fertilised newt eggs, as well as four adult newts, in cassette-sized containers. Unfortunately, two of the newts died, but the others seemed to develop at approximately the same rate as their ground-

4-08: Bob Cabana (left) and Don Thomas pose with Columbia on the runway after concluding their 15-day space voyage.

based 'controls'. Other AAEU passengers included goldfish and Medaka fish, both of which were shown to respond well to light stimulation within their tank.

Although life sciences consumed a sizeable portion of IML-2 crew time, a significant thrust of the mission's research surrounded materials science, processing and fluid physics. One of these was ESA's Critical Point Facility (CPF), which employed five high-precision thermostats to investigate the behaviour of substances at the peculiar stage in which they are technically neither a liquid or a gas. (More accurately, the material's properties fluctuate to and fro from a liqueous to a gaseous state, such that its 'bulk' state is indistinguishable.) During the mission, the CPF was used to measure the wave motion of heat within sulphur hexafluoride and at one stage a wire was charged inside the test cell to 500 volts in order to simulate the pressures induced by terrestrial gravity. "The effect," said investigator Richard Farrell, "was like turning the *gravity* on and off."

Elsewhere in the Spacelab module was the Electromagnetic Containerless Processing Facility, known by its German acronym of 'TEMPUS', which provided a levitation melting device for processing metals. Containerless processing was also considered hugely advantageous, because Earth-based processes involving liquids can cause them to be affected by the properties of their holding vessel; in microgravity, on the other hand, positioning and control can be effected with greater precision. This, in turn, reduced motions within the sample liquid and was less intrusive upon the physical phenomena under study. Moreover, containerless processing was known to eliminate contamination of the sample caused by the material in the container's walls. For IML-2, TEMPUS carried 22 spherical specimens, each measuring up to 10 mm in diameter, which were housed on a storage disk. This rotated until the desired sample was positioned over a mechanism that transferred it to the processing area, where it could be placed into either a vacuum or an ultra-pure helium-argon 'atmosphere'. Most of the experiments were run by computer command, with the members of the payload crew keeping a close eye on their progress. At one stage, Rick Hieb watched over a zirconium-cobalt alloy as the TEMPUS ground team sent telescience commands to levitate, and then melt, the small metal sphere inside the processing chamber. Such materials were expected to aid the production of near-perfect metallic 'glasses' with unique mechanical and physical properties. Others samples included alloys of niobium-nickel, aluminium-copper-cobalt and nickel-tin and on one occasion all thruster firings were temporarily suspended to provide the experiments with an ultra-pure acceleration environment.

European involvement in IML-2 was represented also by RAMSES – a French acronym for 'Applied Research on Separation Methods Using Space Electrophoresis' – which separated and collected ultra-pure components of biological samples, including haemoglobin and bovine serum albumin, according to their electrical charges. Meanwhile, the Bubble, Drop and Particle Unit explored bubble growth, evaporation, condensation and temperature-induced thermocapillary flows in microgravity. Video and still cameras mounted in close proximity to the device enabled scientists to carefully scrutinise the behaviour of fluids under various conditions, including the injection of vapour bubbles into a test cell filled with an alcohol-water solution. Liquid refrigerants were also boiled and the coalescence of large bubbles, apparently unaffected by gravity-induced buoyancy, were captured by Spacelab's video cameras. With a Japanese crew member aboard, it seemed inevitable that the involvement of her parent nation in IML-2 was strong. In addition to the aquarium, a Large Isothermal Furnace was used to heat and rapidly cool materials – including ceramic-metallic composites and semiconductor alloys – in various temperature ranges to identify relationships between their structures, processing and physical properties. Usually, the furnace followed a pre-programmed heating and cooling cycle, reaching a maximum temperature of around 1,600 degrees Celsius.

Working around the clock, the payload crew closely monitored the IML-2 experiments, whilst their counterparts on the 'orbiter' team – Cabana, Halsell and Walz – kept watch over Columbia's systems, which performed with near-perfection. A jammed solid waste compactor piston on the toilet was quickly fixed by Halsell and the pilots spent time with the Pilot In-Flight Landing Operations Trainer, practicing simulated approaches, whilst the entire crew participated in Lower Body Negative Pressure runs. As was now customary on long-duration flights, each astronaut received two half-days of off-duty time, which were variously spent exercising, conducting photography or gazing at the majesty of Earth. Their return home was originally scheduled for 22 July, after a mission of 13 days and 17 hours, which would have fallen slightly short of the 14-day record set by the STS-58 crew the previous November. However, both landing opportunities at KSC on the 22nd were waved-off, due to unacceptable cloud cover at the east end of the Shuttle

Landing Facility, and although weather conditions were fine at Edwards Air Force Base in California mission managers opted to wait and try their chances in Florida. When the mission exceeded the record later that afternoon, the crew were congratulated by STS-58 astronaut Bill McArthur, who was serving as the Capcom in Mission Control.

Seventeen hours later, at 6:38 am EDT on 23 July, Columbia settled gracefully onto Runway 33 at KSC, wrapping up a spectacular voyage of almost 15 days. A subsequent sweep-down of the runway revealed a flattened fish, apparently dropped by an osprey or an eagle in terror as it heard the Shuttle's double sonic booms. The enormous success of IML-2 received official acknowledgement in a review meeting at the European Space Operations Centre in Darmstadt, Germany, in early November 1994. The consensus from all of the scientists and investigators was hugely positive. "This is the important part of the mission, where the scientists get their samples," said IML-2 Mission Scientist Bob Snyder. "In some cases, this is going to take many months, up to possibly several years, in the cases where this huge amount of data has to be analysed."

"SAFING IN PROGRESS"

In the pre-dawn gloom of 18 August 1994, Space Shuttle Endeavour sat on Pad 39A, bathed in million-candlepower xenon floodlights, as the final seconds of the countdown to her next voyage into orbit evaporated. Although STS-68 would be her seventh flight, it was in many ways a repeat of her sixth, for the orbiter carried the massive Shuttle Imaging Radar (SIR-C) and Germany's X-band Synthetic Aperture Radar (X-SAR) to support the second Space Radar Laboratory (SRL-2) mission. Barely four months earlier, on STS-59, Endeavour had flown for 11 days on SRL-1, gathering 94 hours of radar data over 44 nations and covering an area in excess of 70 million km^2. On STS-68, she would repeat that mission to monitor changes between late spring and late summer. "JPL had always planned to fly the new instrument at least two and preferably three times," explained SRL-2 payload commander Tom Jones, "testing the radars' ability to monitor seasonal variations in rain forests, croplands, wetlands, ocean currents, sea ice, soil moisture, glaciers and snow cover." Similarly, the MAPS air-pollution experiment sought to track ongoing changes in carbon monoxide production. "JPL would have preferred six months between missions," wrote Jones, "to capture a full seasonal swing, but competing demands on the Shuttle schedule moved SRL-2 up to late summer."

For Jones, seated alongside STS-68 mission specialist Jeff Wisoff in Endeavour's darkened middeck, the flight also offered the chance to seize for himself a new record in the annals of human space exploration. Between June and October 1985, Steve Nagel had flown two space missions within 128 days of each other and Jones was expected to smash that achievement, with a gap of only 120 days since his STS-59 landing.

Alas, for Jones, today would bring him cruel fortune ... and for his crewmate Dan Bursch, it would be crueller still.

Liftoff of STS-68 was scheduled for 6:54 am EDT. After strapping in and checking their equipment, Jones and Wisoff killed a little time by playing rock, scissors, paper. As the countdown resumed ticking from the T-9 minute hold, the astronauts had been heartened to learn that Range Operations had given them a green light to go. Weather was good, as were the systems and payloads aboard the orbiter herself. Launch Director Bob Sieck wished the crew good luck, to which Endeavour's commander, Mike Baker, responded with a heartfelt thanks for getting them ready to fly this important 'Mission to Planet Earth'. A few minutes later, pilot Terry Wilcutt reached over and activated the ship's Auxiliary Power Units. "Here comes the vibration of the vehicle," Jones wrote in his memoir, *Skywalking*, "flight control surfaces moving, engines are cycling now with the hydraulics."

Endeavour was primed and ready to go. Or so it seemed.

"Go for Autosequence Start. Endeavour's on-board computers now have primary control of all the vehicle's critical functions ..."

As the countdown ticked to the all-important T-31 seconds, control of the final stages was handed off from the Launch Control Center to Endeavour's on-board General Purpose Computers. It was they, and they alone, which would monitor hundreds of separate sensors and execute decisions as to whether the mission would fly today. The disembodied voice of the launch announcer echoed from loudspeakers, crisply acknowledging the seconds as they worked their way backwards toward the ignition of the three main engines, the ignition of the twin Solid Rocket Boosters and a date with a high-inclination (though relatively low-altitude) orbit to radar-map the Home Planet for the next ten days or so.

"Twelve, eleven, ten, nine, eight, seven ..."

At ten seconds, the darkness was punctuated by the bright cascade of sparks from the swirling hydrogen burn igniters.

"We have a Go for main engine start ..."

With a familiar rumble and a sheet of translucent orange flame, Endeavour's three main engines thundered to life. From the roof of the Launch Control Center, several kilometres away, Jones' wife, Liz, together with their two children, the other crew families and a handful of astronaut escorts – including STS-59 veteran Rich Clifford – braced themselves for the upcoming crescendo of sound and vibration. "In the growing light of dawn," Jones recalled in *Skywalking*, "she saw the gout of orange exhaust flare beneath the orbiter and saw the steam billow from the flame trench as the engines spooled up to full power."

"We have three main engines running ..."

All seemed normal. Then, with shocking abruptness, something went wrong.

"Three, two, one ... and ... we have main engine cutoff. GLS safing is in progress."

As the Ground Launch Sequencer automatically kicked in to safe the vehicle, the three blazing engine bells suddenly fell dark and silent. For the third time in less than 18 months (and only the *fifth* occasion in the Shuttle's 13-year operational history), a Redundant Set Launch Sequencer (RSLS) abort had been called, after engine start, necessitating a highly hazardous on-the-pad shutdown. However, whereas previous aborts had occurred at around T-3 seconds, STS-68 had gotten down to a mere 1.9 seconds ahead of Solid Rocket Booster ignition. The attention of everyone in the Launch Control Center was riveted upon the Main Propulsion System (MPS) fire detectors; if any of them had tripped, the abort carried the prospects of turning into a bad day, for an invisible hydrogen fire at the base of the main engines could easily spread across the launch pad and trigger an explosion. And *that* would necessitate a 'Mode One Egress': a hairy evacuation of the astronauts from the orbiter.

"We have a cutoff of the main engines. The countdown clock has stopped."

In the seconds which followed, an urgent flurry of acronym-laden communications passed between engineers, controllers and managers in the Launch Control Center, and with the astronauts themselves aboard Endeavour. *"We have main engine cutoff ... RSLS safing is in progress ... All three main engines are in post-shutdown standby ... GLS is Go for orbiter APU shutdown ..."* The gathered spectators at the Cape watched in alarm as the famous countdown clock starkly read T-00:00:00, yet no Shuttle ascended into the heavens and only a large smudge of grey cloud rose ominously above Pad 39A. Then came the call which brought a measure of calm to the proceedings: *"No MPS fire detectors tripped."* There was no evidence of fire on the pad, meaning a 'Mode One Egress' of the vehicle would probably be unneeded. The 'white room' was moved back into position alongside Endeavour's crew access hatch, to facilitate the departure of Baker and his men. Pilot Terry Wilcutt shut down the three APUs. Through Endeavour's tiny side hatch window, Tom Jones could clearly see the Pad 39A gantry visibly *swaying* backwards and forwards; the vehicle was still rocking from the 'twang' effect induced by the ignition of her main engines.

Had the engine shutdown been triggered a couple of seconds later – *after* SRB ignition – it would have placed the crew in an unenviable situation of having to perform the Shuttle programme's first Return to Launch Site (RTLS) abort, riding the boosters for two minutes until their expiry, then separating from them and the External Tank to perform an emergency landing, back at the Kennedy Space Center. "The RTLS is a daunting prospect for the crew," wrote Tom Jones. "We would have to fly the orbiter and attached ET through half an outside loop, then ride *backward* through our exhaust plume at Mach 5. Ditching the empty tank, we would then try to make it back to the Kennedy runway. No Shuttle crew had *ever* flown such an emergency approach. None wanted to be the first to try."

It was a dramatic introduction for both Wilcutt and fellow STS-68 rookie, mission specialist Steve Smith, raising the curtain on what would ultimately turn into four-flight Shuttle careers for them both. (Interestingly, *all six* members of the crew would go on to log four missions apiece.) In the case of Terrence Wade Wilcutt, the urge to someday become a spacefarer was neither all-consuming, nor did it originate at an early age. "I didn't have a desire to be an astronaut until much later in my Marine Corps career, as I was finishing up my test pilot career," he later told a NASA interviewer, "but the desire to fly airplanes ... was always there." As a child, whilst on a Little League Baseball team, Wilcutt saw (and heard) jets breaking the sound barrier overhead and convinced himself that flying "as high and as fast as you possibly could" was the most exciting job in the world.

Born in Russellville, Kentucky, on 31 October 1949, Wilcutt was raised and received his early schooling in Louisville, before entering Western Kentucky University to study mathematics. Whilst there, he was a member of the Lambda Chi Alpha fraternity, whose motto *Naught without Labour* guided his steps. Upon receipt of his degree in 1974, Wilcutt taught mathematics in high school for two years, then entered the Marine Corps. He earned his aviator's wings in 1978, flew the F-4 Phantom fighter, attended the Navy's Fighter Weapons School ('Top Gun') at Miramar, California, and was deployed overseas to Japan, Korea and the Philippines. His career took him through F/A-18 Hornet conversion training and in 1986 he graduated from the Naval Test Pilot School, after which he worked on numerous classified aircraft projects and was selected as a NASA astronaut candidate in January 1990. Thinking back over his life, Wilcutt was philosophical about the chances he had been given and how these ultimately put him in a position to apply for astronaut training. "Some of it's kind of happenstance," he said later, "or I just got lucky."

For Steven Lee Smith, the mission of STS-68 brought both euphoria and disappointment, for when he was named to the crew he became the first of his astronaut class – Group 14, selected in March 1992, who dubbed themselves 'The Hogs' – to draw a flight assignment. By the time he eventually got to fly, however, he would fall behind STS-64 crewman Jerry Linenger for having reached orbit first. Although his route to NASA followed a distinctly civilian path, Smith was, like Terry Wilcutt, intrigued by aviation from childhood. "When I was growing up in San Jose, California, my father and mother took me to the local airport to watch the airplanes take off and land," he recalled. "We'd sit at the end of the runway and watch those airplanes and *that* really first grabbed my aviation interest." Exploration was another powerful magnet for the young Smith and the images of America's first spacewalkers, floating in the void, prompted him to draw them as a boy. Remarkably, his parents kept all of his childhood drawings of rockets and astronauts. "That dream ... carried itself through high school and on to college," he told a NASA interviewer, years later. Little could he have foreseen that, by the time he completed his fourth and final Shuttle mission in April 2002, Smith would have established himself as the world's third most experienced spacewalker ... and although others have since surpassed him, even at the end of 2012, he remained within the worldwide 'Top Ten'.

Smith came from Phoenix, Arizona, born on 30 December 1958. His interest in science and mathematics started early and he would invite his former high school calculus teacher, Mr Lanborn, to the STS-68 launch. He "pushed us very hard," Smith recalled, reflecting on "being very inspired to understand math and technology, just by his incredible interest". Smith studied extensively at Stanford University, receiving his undergraduate and master's degrees in electrical engineering in 1981 and 1982, respectively, *and* a second master's credential in business administration in 1987. A keen sportsman, he was a seven-time high school and collegiate All-American in swimming and water polo and a two-time National Collegiate Athletic Association (NCAA) Champion at Stanford in water polo. He also captained the 1980 NCAA Championship team and learned to fly and scuba-dive. Betwixt his two advanced degrees, he worked for IBM as a technical group lead within the Large Scale Integration (semiconductor) Technology Group and, until 1989, served as a product manager with the Hardware and Systems Management Group. That same year, Smith joined NASA as a payload officer within the Mission Operations Directorate at the Johnson Space Center and was selected as an astronaut (on his *fifth* try) in March 1992. "Why do we fly in space?" he once rhetorically asked a NASA interviewer. "We fly in space to make people's lives better. *That's* the bottom line." On STS-68, with its powerful radar, Smith would be part of a team which would scour much of Earth's surface ... and potentially make millions of lives better.

In the wake of an engine abort, on the morning of 18 August 1994, it was necessary that Endeavour be rolled back to the VAB for repairs. Since the Mission Management Team regarded an RSLS contingency as equivalent to an actual launch this mandated a complete re-inspection of all main engines in the VAB engine shop, creating a delay of approximately six weeks. Tom Jones must have groaned as he lost his chance to eclipse Steve Nagel's record. Meanwhile, Dan Bursch, sitting in the flight engineer's seat behind Mike Baker and Terry Wilcutt, was even more unlucky, having also sat through the STS-51 pad abort, almost exactly a year earlier. Now, his STS-68 crewmates teased him without mercy and, according to Jones, "he lamented that no one would be coming to another of his launch attempts". In fact, Bursch would gain an unenviable reputation as the *only* astronaut to sit through as many as two RSLS aborts in his career.

In the hours and days which followed, the attention of technicians focused upon a problem with the No. 3 main engine's High Pressure Oxidiser Turbine (HPOT). One of its sensors detected a dangerously high discharge temperature, which exceeded Launch Commit Criteria rules, and Endeavour's computers halted the countdown after the Engine Start Command (ESC) had been issued. "From ESC+2.3 seconds through ESC+5.8 seconds," explained NASA's STS-68 post-flight report, "the HPOT discharge temperature must not exceed 1,560 degrees R[ankine]. The SSME HPOT discharged temperature Channel A attained 1,576 degrees R. The Channel B measurement attained 1,530 degrees R and that was also higher than predicted." (Normal HPOT discharge temperatures were around 1,400 degrees R.) As a result, the No. 3 engine was commanded to shut down at 4.72 seconds after ignition, followed, within the next couple of seconds, by its No. 1 and 2 counterparts. Although the No. 3 engine had been used on two previous Shuttle missions, its HPOT was new and was undertaking its first flight. By 24 August, the STS-68 stack was back inside the VAB, where all three main engines were removed and replaced with a flight-qualified trio earmarked for Atlantis' forthcoming STS-66 mission in November. Endeavour returned to the launch pad on 13 September, with a new flight date scheduled no earlier than the 30th. In the meantime, the troublesome No. 3 engine was transported to NASA's Stennis Space Center in Mississippi for an extensive period of evaluation on the test stand. It was fired for 340 seconds without difficulty on 4 September. "The liftoff could have taken place and the engine could have functioned normally," *Flight International* noted, "but conservative launch rules had been built into the on-board launch computer."

For the astronauts, it seemed their best bet of getting off the ground was to convince Endeavour that the unlucky Dan Bursch was *not* aboard. As a result, when the STS-68 crew arrived at the Kennedy Space Center, a couple of days before launch, Bursch ensured that he climbed out of his T-38 jet in an appropriate 'Groucho Marx' disguise. Things did not seem to be going well, for all of them, save Jeff Wisoff, had colds. In spite of the jinxed nature of their flight, 30 September turned out to be charmed and Endeavour rose perfectly at 7:16 am EDT, just as the 2.5-hour launch window opened. Upon arrival in orbit, Jones expressed satisfaction that he had taken anti-nausea medication before launch, but poor Steve Smith fell victim to the malaise within a couple of hours. Dan Bursch, the designated medical officer, offered a Phenergan shot, which seemed to do the trick. It was perhaps fortuitous that Smith, Bursch and Jones – the STS-68 blue shift – bedded down for their first abbreviated sleep period, shortly after reaching space. Activation of the SRL-2 payload and its inaugural observations were conducted under the auspices of the red shift, with Jeff Wisoff taking charge of the radar instruments.

Endeavour flew over many of the same sites that SRL-1 did, enabling the SIR-C and X-SAR research teams to examine seasonal changes. In addition to its enormous scientific yield, the first radar mission had demonstrated that it could acquire high-resolution data *and* could endure adjustments to its timeline to cater for new events on the ground. For example, SRL-1 observed severe floodwaters in the mid-western United States and in Thoringen, Germany, as well as taking three different views of Tropical Cyclone Odille as it formed and contorted in the Pacific Ocean. Snow and ice classification maps had been assembled from data over a supersite at Oetztal in Austria and the late-summer flight of STS-68 offered a chance to observe the Patagonian district of southern Chile, home to the largest glaciers and ice fields in South America.

Volcanic sites were a key objective. Already, on SRL-1, the radars had observed Mount Pinatubo in the Philippines – which erupted in mid-1991, affecting global stratospheric temperatures and aerosol levels – and several locations in the Galapagos Islands. Imagery of Pinatubo during the summer monsoon season, when new mud flows were predicted to occur, was high on SRL-2's list of priorities... but, early in the flight, a serendipitous observation of another eruption, in the Russian Far East, was made. Klyuchevskaya Sopka is the highest mountain on the Kamchatka Peninsula and the highest active volcano in the whole of Eurasia; it burst into fiery life shortly after Endeavour entered orbit, giving the STS-68 crew a surprise. They saw its tremendous black plume on the horizon, 'ahead' of their flight path. At first, it looked like a vast thunderhead. Mike Baker was the first to recognise it as a volcano. Jeff Wisoff was amazed. "It shows how much change nature can produce in a short period of time," he recalled in a Smithsonian interview. "You had this huge eruption, but then by the end of the flight it had largely stopped erupting. There was still a small smoke trail, but it had re-snowed on top of the soot. In the span of ten days, it was almost *white* again!"

Tom Jones, too, remembered the event lucidly in his book, *Skywalking*. In his recollection, it was the red team of Baker, Wilcutt and Wisoff who called the blues up to the flight deck to see the eruption. Like Wisoff, Jones thought at first that it

4-09: In an event of pure serendipity, Klyuchevskaya Sopka – the highest mountain on Russia's Kamchatka Peninsula and the highest volcano in Eurasia – erupted a few hours after the STS-68 crew entered orbit. The violence of the eruption left a vast ash trail in the lower atmosphere.

was an anvil-shaped thunderstorm, or a clump of dust lofted by high winds, but the smoke-free nature of the surrounding terrain (and the location, Kamchatka) quickly assured them that it was a volcano. Jones had seen it before, during the SRL-1 radar passes, but less than six months earlier Klyuchevskaya had been silent and still under a blanket of April snow. Now, at the end of September, it was in full fury. "We soon had every camera ... zeroed-in on the eruption," Jones wrote, "as Endeavour gave us a dramatic, down-the-throat view of this impressive geology lesson." In another example of responding to unforeseen events on the ground, the SRL-2 investigators reprogrammed several radar passes over the coming week to scan the eruption site as many as three times *per day* and Jones hoped that their work might lead to the implementation of permanent, Earth-orbiting satellites to watch for volcanic events.

By the sixth day of STS-68, as expected, consumables remained at a level sufficient to enable the Mission Management Team to formally extend the flight an additional 24 hours. Endeavour would now land on 11 October. (She would actually touch down at Edwards Air Force in California at 10:02 am PDT, after thick cloud at the Kennedy Space Center forced NASA to switch landing sites.) The radar's filing-cabinet-sized payload recorders performed well, although one had to be removed and replaced after it failed to play back properly. Rerouting the data stream between the remaining machines, the repair by Smith and Wisoff was scheduled over a comparatively 'empty' Pacific orbital pass. "With the radar inactive over the ocean," wrote Jones, "Steve and Jeff were well into the repair before we made landfall again. An hour later, the swap was complete and the pair had stowed their wrenches and screwdrivers with almost no loss in science data."

Overall, more than 110 hours of radar observations were acquired in 950 data takes, recorded on 199 digital tapes, and covered an area of over 83 million km^2. One of the bonuses of flying the mission a second time was the ability to use 'interferometry' during the final three days of STS-68 and SIR-C/X-SAR data was used to record topographical changes between April and October in California's Long Valley Caldera and Hawaii's Kilauea. Interferometry is analogous to stereo photography, although to achieve it between STS-59 and STS-68 demanded that their repeat orbital paths be separated by no more than a hundred metres or so. "Mission Control and our crew combined to perform the most precise orbital manoeuvres ever seen in the Shuttle programme," explained Jones, "putting Endeavour in an orbit for the first six days that nearly matched our SRL-1 flight path." At times, the respective flight paths differed by less than ten metres.

Also flawless were operations with the Measurement of Air Pollution from Satellite (MAPS), which gathered 256 hours of data and achieved a 100-percent success rate. However, the MAPS results reinforced the STS-59 consensus that although carbon monoxide concentrations in the southern hemisphere were relatively low, with exceptionally clean air, the situation worsened in the northern hemisphere, with the highest concentrations to the north of the 40-degree latitudinal band. MAPS also observed intentionally-set fires, monitored by scientists from the University of Iowa and the Canadian Forest Service, to assess their wind fields, thermal evolution and carbon monoxide emissions and calibrate SRL-2 infrared data. "These fires," noted NASA in one of its STS-68 news releases, "were planned in advance of the mission and would have been set for forest-management purposes, even if the Shuttle mission were not in progress." Other 'controlled' observations included an experimental 'spill' of 477 litres of diesel oil and 117 litres of algae products in the North Sea to test SRL-2's capacity to differentiate between the spill and naturally-produced film caused by the products of fish and plankton.

Aside from SRL-2 operations, a trio of GAS canisters also flew aboard Endeavour. North Carolina A&T State University supplied experiments which sought to understand the behaviour and mating characteristics of the milkweed bug in microgravity and grow crystals of rochelle salt, whilst another from the University of Alabama at Huntsville supported research into algae development, the mixing of concrete ("for future Moon Base applications"), roots and metals, and a Swedish team provided a germanium-gallium crystal growth investigation. Elsewhere, half a million Apollo 11 commemorative stamps were flown by the US Postal Service and, inside the Shuttle's cabin, commercial protein crystal growth and a series of biological research experiments focused on the early development of gypsy moths in weightlessness.

Orbital life aboard Endeavour herself was exceptionally smooth ... and humorous for the irreverent STS-68 sextet. During training, at reviews of the food menus, Mike Baker had expressed his fondness for smoked turkey, which he enjoyed folding into tortillas. Now, in space, this love would return to haunt him. "When Bakes asked what was for dinner, the answer nearly always was ... *smoked turkey*," wrote Tom Jones. "Noticing our escalating laughter, by the third day our commander was convinced he was the victim of a practical joke instigated either by us or some of his 1985 astronaut classmates. We maintained our innocence and post-flight investigation showed the on-board turkey surplus was his *own* doing!" To be fair, an entire pantry of other goodies awaited them – shrimp cocktails, horseradish, beef tips and mushrooms, spaghetti and meat sauce, buttered asparagus, strawberries, chocolate puddings – but Jones could not help but wonder what turkey-loving Baker ate for his Thanksgiving meal ...

EYES ON THE EARTH, ON THE STARS, ON THE FUTURE

One hundred and twenty-three times before 16 September 1994, astronauts and cosmonauts had clambered outside their ships and manoeuvred themselves around in the harsh vacuum of space. They had used handholds, they had used tethers and they had used specialised manoeuvring units to prevent them from losing contact with their craft and floating away. When STS-64 astronauts Mark Lee and Carl Meade ventured outside Discovery's airlock on the morning of that September day in 1994, they wore something quite different and entirely new over their space suits ... a $7 million device known as the 'Simplified Aid for EVA Rescue', or 'SAFER', whose descendents continue to be utilised aboard today's International Space Station. Unlike the Manned Manoeuvring Unit (MMU), trialled and used operationally in 1984, SAFER was not intended for satellite repairs, but as a design solution to the Shuttle programme's requirement to offer a means of 'self-rescue' in the event that an EVA crew member became untethered during the course of a spacewalk.

True to the sentiment behind its acronym, SAFER's design centred on the fact that in the event of a tether failure during an EVA, it would be hugely difficult to undock the Shuttle from the space station, rendezvous and retrieve the lost crew member and redock in a safe manner. SAFER quite literally provided a *safer* means of going about EVA and, on STS-64, supported the United States' first untethered spacewalk in almost ten years. Developed by the Automation and Robotics Division at NASA's Johnson Space Center, it weighed 37.7 kg (less than a third of the

MMU), carried 24 fixed-position compressed nitrogen thrusters and was attached to six 'hard-points' at the base of the life-sustaining backpack. Powered by a 28-volt battery pack, SAFER's attitude-control system included an automatic attitude-hold and six degrees of freedom. Unlike the MMU, it had no bulky arms for hand controllers, instead being equipped with thruster 'towers' extending up the sides of the backpack, and on STS-64 Lee and Meade operated it from controls hard-secured to their suit torsos. (In the current ISS configuration, the SAFER's controller is embedded within one of the thruster towers and is swung out by the simple pulling of a lanyard.)

When the STS-64 crew was announced in November 1993, it originally comprised five veteran astronauts: Dick Richards in command, Blaine Hammond as pilot and mission specialists Susan Helms, Carl Meade and Mark Lee. Unusually, in February 1994, a *sixth* crew member – Navy physician Jerry Linenger – joined the crew as a mission specialist. "This assignment," read the space agency's press release, "was made to more efficiently distribute the crew workload for this complex flight." However, there existed a secondary reason for the arrival of Linenger, whose experience was expected to be "of great value in ongoing human physiology investigations". Specifically, those 'investigations' were the need for medically-qualified astronauts to fly long-duration (and medically focused) expeditions to the Mir space station ... and, coupled with Russia's insistence that *all* of the NASA fliers must have prior flight experience, Linenger was slotted into the earliest available Shuttle mission. Selected as an astronaut candidate in March 1992, with STS-64 he became the first of his class – nicknamed 'The Hogs' – to fly into space and established a Shuttle-era record of less than 30 months between admission into the astronaut corps and first liftoff.

Jerry Michael Linenger was born on 16 January 1955 and raised in Eastpointe, Michigan. After leaving high school, he entered the Naval Academy to study bioscience and earned his degree in 1977. Additional academic credentials followed rapidly: a doctorate in medicine from Wayne State University School of Medicine in 1981, a master's degree in systems management from the University of Southern California in 1988, a master's in public health policy from the University of North Carolina at Chapel Hill and a PhD in epidemiology from the University of North Carolina, both in 1989. "Whenever anyone teases me about the number of academic degrees I have accrued," Linenger wrote in his autobiography, *Off the Planet*, "I generally respond that I was a rather slow learner – I had to keep going to school until I got it right!" As a physician, he completed his surgical internship, his aerospace medicine training, his flight surgeon training and served as medical adviser to Admiral Jim Service, commander of the Navy's air assets in the Pacific and Indian Oceans.

One of Linenger's key responsibilities on STS-64 was to provide 'intravehicular' support for Mark Lee and Carl Meade during their EVA. Although he would go on to make his own EVA, later in his astronaut career, for Linenger the sight of his crewmates floating against the serene backdrop of Earth and the ethereal of blackness of space was unforgettable. "The one thing I'll always picture from STS-64 is watching Carl Meade and Mark Lee out there testing the new jetpacks," he told a

432 From foes to friends

4-10: Carl Meade flies freely away from Discovery during America's first untethered EVA in ten years. During his spacewalk with STS-64 crewmate Mark Lee, he evaluated the effectiveness of the Simplified Aid for EVA Rescue (SAFER). The LITE instrument is visible at bottom right, as is the gold-enshrouded Spartan-201 satellite and, at top right, the extended RMS arm.

Smithsonian interviewer, "spinning on the end of the arm, totally free of the Shuttle and the Earth below them. It's one of those pictures that's pretty much imprinted in my brain, even though I wasn't doing the spacewalk."

During training, Lee and Meade had spent considerable time working with the SAFER engineers and performed evaluations on the ground. Their four crewmates supported the EVA from Discovery's aft flight deck, intricately guiding the orbiter itself, controlling the mechanical arm and overseeing the timeline for their 6.5 hours outside. The spacewalkers were to venture no more 8.2 m from the Shuttle, operating within a 'box' of space at the forward end of the payload bay and their tasks encompassed familiarisation with SAFER and a programmed series of nitrogen-jet tests to gather engineering data, together with 'tumbling' tests and demonstrations of precision manoeuvring. With Helms operating the RMS, the men took turns standing in mobile foot restraints on its end effector and 'tumbling' his colleague. The 'tumbled' astronaut then activated SAFER's automatic attitude-hold system, stabilised himself and manoeuvred toward the arm, which Helms pulled away to simulate a separation rate of about 0.06 m/sec. The astronauts quickly determined that the device used less nitrogen than predicted and on one occasion Meade 'rolled' Lee at 2 revolutions per minute, somewhat quicker than planned, but his SAFER successfully stabilised him without incident. They evaluated their capability to replenish nitrogen from a recharging unit in the payload bay and flew the unit precisely along the length of the RMS.

Overall, SAFER's first trial was a huge success, with the only problem of note arising from glitches with a battery-powered Electronic Cuff Checklist, which was capable of storing 500 pages of information, including graphics and photographs on its small screen. On a number of occasions, it did not respond to commands when the upper-middle sextant was depressed and efforts to update its contents resulted in an error message being displayed. However, it remained usable and Lee and Meade spoke glowingly of its performance later. Unlike previous printed notebooks attached to the space suit cuff, which held 25-50 pages of notes, usually detailing critical functions, such as emergency operations, the two-megabyte Electronic Cuff Checklist and its touch screen enabled its wear to scroll through much more information in support of more complex EVAs, such as Hubble Space Telescope servicing missions or space station construction tasks. It could also be updated electronically from one of the orbiter's laptop computers. For this assessment, Lee and Meade wore an Electronic Cuff Checklist on the left arm of their suits and a standard, printed checklist on the right arm. The EVA lasted six hours and 51 minutes and formed the capstone of a multi-faceted mission which showcased virtually all of the Shuttle's capabilities: from spacewalking to scientific research and from satellite deployment to rendezvous, proximity operations and payload retrieval.

The primary objectives were the deployment and recovery of the Shuttle-Pointed Autonomous Research Tool for Astronomy (Spartan)-201, with its battery of solar physics instruments – previously flown aboard STS-56 in April 1993 – and the operation of the Lidar In-Space Technology Experiment (LITE). The latter was originally slated for two Shuttle missions in the 1994-95 timeframe, but ultimately flew just once, as several 'stand-alone' primary payloads haemorrhaged from the manifest in response to the growing Shuttle-Mir effort. Developed by NASA's Office of Advanced Concepts and Technology and Mission to Planet Earth, the 2,680 kg LITE instrument took the form of a light detection and ranging ('lidar') device to transmit narrow pulses of laser light into Earth's atmosphere and used a telescope to measure the reflectance qualities from clouds, suspended aerosol particles and the surface itself. The usefulness of lidar as a remote-sensing tool derived from the fact that it could obtain very high vertical and horizontal resolutions and LITE marked the first occasion on which lidar had ever been employed in space for atmospheric observations. More than 43 hours of data were gathered and Discovery's relatively low-altitude (260 km) and high-inclination (57-degree) orbit enabled the instrument to cover a broad swathe of the Home Planet. Simultaneously, on the ground and in the air, investigators at more than 50 sites in 20 discrete countries around the world made their own complementary observations. Five international aircraft flew directly 'beneath' portions of Discovery's flight path, covering parts of Europe, the south-western United States, the Caribbean, South America and the South Atlantic. Mounted on a Spacelab pallet, LITE gathered its data during a series of ten sessions, each lasting 4.5 hours, as well as performing a handful of 15-minute 'snapshots' of selected target sites. (These snapshot sites yielded 'ground-truth' comparison points and included a lidar situated at NASA's Langley Research Center in Hampton, Virginia.) The returning lidar signals were converted to digital data and stored on tapes aboard Discovery for transmission to the ground. "Lasers ... produce a tight, coherent beam that spreads very little as it travels from its source, compared to ordinary light," noted NASA's pre-mission press kit. "From its orbital altitude, LITE's laser beam would spread to only about 300 m wide at the surface. This allows the LITE instrument to measure a very small, narrowly-defined column of the atmosphere with each pulse." With pulses transmitted ten times per second, and each lasting less than 30 *billionths* of a second, at precisely-known wavelengths corresponding to ultraviolet, infrared and visible green light, it was possible to achieve a resolution of about 15 m. By the end of the first day of the mission, the LITE ground team was reporting "terrific-looking returns" from the instrument.

Operating throughout the mission – with operations temporarily suspended whilst Lee and Meade performed their EVA – the LITE instrument investigated the organisation of cloud structures in the Western Pacific Ocean, together with cloud decks off the coasts of California and Peru, smoke plumes from biomass fires in South America and Africa and the transport of dust from the Sahara Desert. Additionally, low-atmosphere aerosols were examined over the Amazon rainforest and gravity waves (a mechanism that transfers momentum from the troposphere to the stratosphere) over the Andes mountain chain and the reflection characteristics of desert surfaces in the United States, Africa and China. LITE also dramatically outlined the structure of the Super-Typhoon Melissa, providing unprecedented views of the storm's eye. It was expected that the technology demonstration could lead to future dedicated satellites to perform year-round lidar observations of clouds, urban smog or dust storms. The instrument took the form of a one-metre-diameter telescope and optics package with photomultiplier tubes to provide its visible green and ultraviolet sensitivity and a silicon avalanche photodiode for its infrared capability. Interestingly, the telescope was actually an engineering model from NASA's 1960s-era Orbiting Astronomical Observatory programme and its use for LITE reportedly saved the space agency an estimated $8 million. In spite of periodic laser shutdowns, caused by coolant-loop problems, NASA noted in its post-mission report that the system "performed more effectively than originally expected, demonstrating the ability of a lidar ... to penetrate multiple clouds and aerosol layers down to the Earth's surface".

Although the STS-64 crew members looked visibly identical to their predecessors when they left the Operations and Checkout Building, bound for the launch pad, on

9 September 1994, they were in fact clad in upgraded versions of the pressure suits worn in the wake of the Challenger accident. The new Model S1035 Advanced Crew Escape Suit (ACES) – popularly nicknamed 'the pumpkin suit', owing to its bright 'international orange' colour – was a direct descendent of Air Force high-altitude garments worn by SR-71 Blackbird pilots and, latterly, of the Launch and Entry Suits (LES) worn by each Shuttle crew since STS-26. Manufactured by the David Clark Company of Worcester, Massachusetts, the new suit differed fundamentally from the LES in that it operated at full (rather than partial) pressure. "It was always realised," wrote Dennis Jenkins in his monumental history of the Shuttle, "that the LES was not the optimum suit and in 1990 a new development effort was initiated that resulted in the Model S1035 ACES ... designed to be simplified, lightweight, low-bulk, full-pressure ... that facilitated self-donning/doffing and provided enhanced overall performance." The first ACES units were delivered to NASA in May 1994. They boasted gloves on disconnecting wrist-rings, better liquid cooling and ventilation and extra insulation. Operating at 24.1 kPa, oxygen was fed into the suit through a connector at the left thigh and transmitted to the helmet via the base of the neck ring. Textured glove palms offered a measure of dexterity and flexion, particularly for the commander and pilot when operating the stick, whilst heavy leather 'paratrooper' boots with zippers, a survival backpack with life raft and emergency light sticks completed the fully-equipped 41.7 kg ensemble.

Liftoff of STS-64 occurred at 6:22 pm EDT on 9 September 1994, very close to the end of that day's 'window', which had opened at 4:30 pm and was due to close at 7:00 pm. A late-afternoon launch was required in order to permit 'night' operations of LITE early in the mission, but weather violations close to Pad 39B required the six astronauts to wait uncomfortably on their backs for almost two hours later than planned. Upon arrival in orbit, however, the entire crew moved crisply about their tasks. By the time of Lee and Meade's EVA on the 16th, the LITE operations were well underway and the second primary objective of the mission – the deployment and retrieval of Spartan-201 – had also been successfully concluded. On 13 September, Susan Helms lifted the boxy satellite out of Discovery's payload bay and released it into space at 5:30 pm EDT; it spent the next two days performing observations of the solar corona with its twin instruments, a White Light Coronagraph and an Ultraviolet Coronal Spectrometer.

During the deployment, however, data from Discovery's Ku-band rendezvous radar exhibited "questionable readings", which Mission Control attributed to a late acquisition of Spartan. An hour after the satellite's release, the radar passed several self-tests, without obvious problems, but according to NASA's post-mission report, "multiple searches were performed in both the General Purpose Computer and Automatic modes ... all without success". Finally, an hour after the deployment, the Ku-band successfully acquired the Spartan and remained locked onto its target until the Communications Mode was selected. However, as a contingency measure, and in case the radar was unavailable for retrieval, plans were developed for a 'workaround' rendezvous profile on 15 September, possibly incorporating ground-based navigation data and the Shuttle's own star trackers. "These procedures are not as precise and would require slightly more propellant than normal," noted a Mission Control

Center Status Report on the evening of the 14th, but acquiesced that "the propellant margins are adequate to support a 'no-radar' rendezvous and the crew and flight control teams are trained for such a scenario." As the situation evolved over the following hours, the radar posed no further difficulties and from a distance of 39 km to just 25 m it served to guide Dick Richards perfectly into position for him to complete proximity operations and for Helms to grapple the satellite.

Elsewhere in the payload bay, aboard two GAS canisters, the Robot-Operated Materials Processing System, which formed the rather questionable acronym of 'ROMPS', was another 'first' of STS-64, in that it represented the first robotic processing of semiconductor materials in space. The experiment employed a robot to transfer 100 sample 'pellets' from storage racks into small halogen-lamp furnaces, where their crystalline structures were re-formed in heating and cooling cycles. A 10 m instrumented extension boom for the RMS arm, known as the Shuttle Plume Impingement Flight Experiment (SPIFEX), was tested to gather data about the effect of 86 thruster firings upon large structures, whilst ten GAS canisters on a cross-bay bridge housed experiments in ozone-monitoring, the growth of seeds, the behaviour of insects, the formation of superconductors, crystal growth, solar observations and materials science. These GAS payloads represented a wide range of investigators and a wide range of age groups, from young schoolchildren to higher academic institutions, and came from as far afield as Canada, China, Japan and The Netherlands, as well as the United States.

The mission had already been extended from nine to ten days, but the scheduled touchdown at the Kennedy Space Center's Shuttle Landing Facility on 19 September dawned gloomy, with thunderstorms and low, thick cloud cover. Two opportunities existed to land in Florida on the 20th, together with two more at Edwards Air Force Base, and although Florida was the preferred option the weather finally turned against STS-64. Rain showers along the Space Coast made even the 20th a 'No Go' in terms of Mother Nature and Dick Richards and his crew were diverted to California instead. Under picture-perfect skies, Discovery swooped onto Runway 04 at Edwards at 2:12 pm PDT (5:12 pm EDT), concluding a voyage of a little less than 11 full days. The safe return of STS-64 heralded the start of a remarkable phase in which no fewer than *three* Shuttle missions took place within an eight-week period. Ten days after Discovery made landfall in California, Endeavour thundered into orbit from Florida on STS-68, and was followed, three weeks after her own Edwards Air Force Base touchdown, by the launch of the final mission of 1994. That mission, STS-66, marked the return of Atlantis, following a lengthy period of refurbishment and modification at Rockwell International's Palmdale facility on the West Coast ... a modification period which had been longer than anticipated as the details of the agreements between the United States and Russia evolved into a concrete series of Shuttle-Mir flights. Most of those flights would be conducted by Atlantis.

Upon her return from STS-46 – the Tethered Satellite and EURECA deployment flight – in the summer of 1992, Atlantis was transferred to California in October of that year for what would turn out to be 19 months. Whilst at Palmdale, she was outfitted with improved nose-wheel steering, the drag chute in her aft compartment and internal plumbing and electrical connections to support the Extended Duration Orbiter pallet. Although the EDO hardware would never be fully incorporated or used, it might have made Atlantis capable of 28-day missions. Preparatory work was also undertaken to equip her with an Orbiter Docking System (ODS) to render her compatible with Mir. The full installation of the ODS hardware was scheduled to take place at the Kennedy Space Center, after her return to the East Coast on 19 May 1994. She was originally scheduled to fly STS-66 on 27 October, but the need to refurbish three more main engines to replace those of Endeavour enforced a week-long delay. Other minor processing headaches included an overhead flight deck window, 'borrowed' from Columbia to replace a scratched one on Atlantis, and attention to her plumbing.

ATLAS-3, the third flight of the Atmospheric Laboratory for Applications and Science, would be the final mission in a series which was originally supposed to fly annually throughout the Sun's 11-year cycle of activity. As NASA and Russia drew closer and plans for a single Shuttle-Mir docking mission crystallised into an entire series of perhaps nine or even ten flights, between 1995 and 1998, in order to exchange long-duration NASA resident astronauts, several key scientific payloads had to be dropped from consideration. These included a second LITE, the German Spacelab-D3, a third Spacelab Life Sciences mission, the pan-European Spacelab-E1, multiple Spacehabs, a fourth Wake Shield Facility, a reflight of the EURECA payload ... and at least two additional missions of ATLAS. As late as January 1992, an ATLAS-4 flight was pencilled-in for the end of 1995, joined by a deployable Spartan satellite equipped with an experiment to investigate waves in space plasma. By the April 1994 manifest, all such missions had vanished, replaced – for Atlantis, at least – by a series of back-to-back Mir-docking flights. Although Shuttle-Mir would be critical to the success of today's International Space Station, from an international-relationships perspective, it has been lamented over the years that a great deal of potential science was lost by deleting so many promising research missions from consideration. It was clear when Ellen Ochoa, veteran of ATLAS-2, was named as the payload commander of STS-66 in August 1993 that ATLAS-3 would be the end of an era. Five months later, she was joined by her five crewmates: Don McMonagle in command, Curt Brown as pilot and a trio of rookie mission specialists, all of whom would go on to carve their own niches in both the Shuttle-Mir and International Space Station eras.

Joseph Richard Tanner, born in Danville, Illinois, on 21 January 1950, served as Atlantis' flight engineer during ascent and re-entry and his knowledge of Shuttle systems, even *before* admission into NASA's astronaut corps, was profound, for he had been an instructor in the Shuttle Training Aircraft ... and his whole aviation career seemed decidedly unusual. By his own admission, his entire family were "adventurers at heart". Growing up, he drew excitement from the unfolding effort to plant American boots on the Moon in Project Apollo. A competitive swimmer at school and in college, Tanner earned a degree in mechanical engineering from the University of Illinois in 1973 and joined the Navy, won his wings and served as an A-7 Corsair attack pilot aboard the USS *Coral Sea*. He wrapped up his active military career as an advanced jet instructor in Pensacola, Florida, and entered

4-11: CRISTA-SPAS is readied for deployment, early in the STS-66 mission.

NASA's Johnson Space Center in 1984 as an aerospace engineer and research pilot. During the next several years, Tanner taught approach and landing techniques to astronaut candidates in the Shuttle Training Aircraft and instructed both pilots and mission specialists on the T-38 Talon. He also rose to become deputy chief of the Aircraft Operations Division. With almost 9,000 hours in military and NASA aircraft, Tanner had significantly more flight time than several Shuttle pilots and commanders. An unsuccessful application for admission into the astronaut corps in 1987 was followed by success in March 1992, but Tanner is distinct amongst NASA astronauts in that he had neither a background in military flight-test or an advanced degree. "My career path is a little bit different," he told a NASA interviewer, years later. "Academically, I'm not quite as qualified as most of the other people in the office. I guess I have to rely on my job experience!"

With a substantial European involvement in the ATLAS-3 project (including the French-built SOLSPEC instrument), it is perhaps unsurprising that a Frenchman, Jean-François André Clervoy, should have been selected as a member of the STS-66 crew. Born on 19 November 1958 in Longeville-lès-Metz, north-eastern France, Clervoy grew up with a desire and a willingness to someday fly into space. "When I was in second grade," he explained, "my teacher used to tell us that when we would be grown up we would be able to fly in space, the same way he was able to buy a ticket to go to the United States." As his education advanced, Clervoy realised it would not be so straightforward, but with a fighter-pilot father and a family yearning for the romance of adventure, the urge to explore space remained alive. He read and was inspired by the adventures of the French volcanologist Haroun Tazieff and the Atlantic explorer Alain Bombard. Clervoy gained his baccalauréat from the Collège Militaire de Saint Cyr l'Ecole in 1976 and passed Math. Sup. and Math. Spé. M' at Prytanée Militaire, La Flèche, in 1978. He graduated from Ecole Polytechnique, Paris, in 1981, Ecole Nationale Supérieure de l'Aéronautique et de l'Espace, Toulouse, in 1983, and qualified as a flight test engineer from Ecole du Personnel Navigant d'Essais et de Réception, Istres, in 1987. By this stage in his career, Clervoy had already begun to gravitate towards becoming an astronaut. He was selected with the second group of French astronauts – or *spationautes* – in September 1985 and entered the European Space Agency (ESA) in May 1992. The eventual paths of this seven-strong group are fascinating in their breadth: two never made it into space, whilst Jean-Jacques Favier flew aboard the Shuttle, Clervoy and Michel Tognini flew aboard both the Shuttle and Mir and Claudie André-Deshays and Jean-Pierre Haigneré (who subsequently married) participated in lengthy Mir missions.

If there was a crewman aboard STS-66 who appeared – in theory – at odds with the mission's primary objectives, it was Scott Edward Parazynski. A physician by training, on paper he looked out-of-place amongst physicists and test pilots and aviators and engineers, yet he went on to fly with John Glenn, helped to build the International Space Station, volunteered at the Houston Astrodome for two weeks in the summer of 2004 to support the victims of Hurricane Katrina and became the first veteran spacefarer to summit Mount Everest in May 2009. As will be explored in the final volume of this series, he also missed out on a long-duration mission to Mir, on account of his height. Yet Parazynski's assignment on STS-66 was beneficial, for in addition to the primary ATLAS-3 payload Atlantis carried a large number of medical and biological experiments. These explored the effects of microgravity upon the muscles and nerves of five pregnant rats in a pair of middeck Animal Enclosure Modules, as well as the development of their optic nerves, their immune responses and their brain and heart behaviour.

Born on 28 July 1961 in Little Rock, Arkansas, Parazynski was well-travelled as a youth, since his father was a Boeing engineer during the Apollo era. "I sort of had a nomadic upbringing," Parazynski admitted to a NASA interviewer. "There are no camels involved, but a *lot* of travel." An understatement indeed: he received his early schooling in Dakar, Senegal, and Beirut, Lebanon, and attended high schools in Tehran and Athens. "I think the fundamental lesson that I learned through all that experience," he continued, "is that some of life's greatest lessons come *outside* of the classroom. Travel opened up for me new experiences. Being adventurous, meeting new people, travelling to new places, really broadens your horizons and it gave me the motivation to do some of the things I did later in life." Returning to the United States, he entered Stanford University, graduated in biology in 1983, and remained to study for his medical degree, which he received in 1989. At approximately the same time, he competed on the US Development Luge Team and was ranked among the Top Ten competitors in the nation during the 1988 Olympic Trials. Parazynski served his internship at the Brigham and Women's Hospital in Boston, Massachusetts, and was two years into a residency programme in emergency medicine when he was selected by NASA in March 1992 as an astronaut candidate.

Like the previous ATLAS missions, the crew of STS-66 was divided into two 12-hour shifts for orbital operations, with Don McMonagle leading the red team of Ochoa and Tanner and Curt Brown leading the 'blue' team of Clervoy and Parazynski. Atlantis roared into orbit at 11:59 am EST on 3 November 1994, following a short delay, as mission managers discussed wind speeds at one of the Transoceanic Abort Landing sites. Weather at Zaragoza and Moron, both in Spain, was initially unacceptable, with Ben Guerir in Morocco also described as "marginal" for a time, in terms of cross winds. Fortunately, the situation improved to an acceptable level by T-5 minutes and the launch attempt proceeded. This was fortuitous, for if Atlantis had succumbed to a scrub on the 3rd her launch team would have had to stand down until at least the 14th in order for the cryogenic dewars on the CRISTA-SPAS satellite – one of her primary payloads – to be replenished.

Within three hours, ATLAS-3 had been powered-up and its seven pallet-mounted instruments – the Atmospheric Trace Molecule Spectroscopy (ATMOS), the Millimetre-Wave Atmospheric Sounder (MAS), the Shuttle Solar Backscatter Ultraviolet (SSBUV), the Active Cavity Radiometer Irradiance Monitor (ACRIM), the Measurement of the Solar Constant (SOLCON) and the Solar Spectrum (SOLSPEC) and the Solar Ultraviolet Spectral Irradiance Monitor (SUSIM) – had begun their initial calibration. This was the first occasion on which the payload had flown in the autumn/winter period and consequently was tasked with taking measurements of the northern hemisphere during the time at which processes in the atmosphere began to 'shift' from relatively quiescent summertime conditions to more active wintertime conditions.

It was known that the chemical processes associated with the much-publicised Antarctic 'ozone hole' tended to peak in early October, when increased springtime sunlight struck air cooled during the southern hemisphere's winter season and solar ultraviolet radiation triggered chemical reactions which were both creative and destructive of ozone. By late November, ozone-rich air from mid-latitudes mixed with the Antarctic air to fill the lost ozone and chemicals such as nitrogen oxides began to consume free chlorine; this served to 'repair' the lost ozone. ATLAS-3 took place at an 'intermediate' period of the year, during which time the ozone hole had begun the process of recovery, but had not yet dissipated. As a result, ozone science was a key part of the ATLAS-3 research. "Observations of both areas," noted

NASA's pre-flight press kit, "should provide valuable data for comparison with the spring data of ATLAS-1 and 2." After STS-66, preliminary data analysis in the post-mission report indicated that the Antarctic hole was a self-contained region. Although it had seen a large increase in the quantity of freon-22 – a chemical which was "used as a replacement for chlorofluorocarbons (CFCs)" – in the stratosphere, it was not as great a threat as CFCs. "It is, however, still a growing source of stratospheric chlorine," cautioned the report. "Also of note is the lack of a direct link between the Antarctic ozone hole and ozone depletion in the mid-latitudes, indicating that there are atmospheric processes that are still not well understood."

In a similar vein to the two missions which preceded it, ATLAS-3 sought to measure global temperatures in the middle atmosphere, together with trace-gas concentrations, and to provide this to the scientific community for comparison with data from other spacecraft. "The ATLAS-3 mission is the most complete global health check on the atmosphere that has ever been done," explained Mission Scientist Tim Miller, "measuring more trace gases that are important in ozone chemistry than any previous research effort." All of its instruments had flown before, but several had been extensively modified in readiness for STS-66 and their precise objectives shifted slightly. ATMOS, for example, benefitted from an improved recorder controller to provide ground-based scientists with more data about its status and performance. The solar science experiments were also prepared for quite different results, compared to their predecessors; for during ATLAS-1 and 2 the Sun had been at a period of near-maximum activity in 1992-93 and on ATLAS-3 the crew directed the orbiter's payload bay toward our parent star on no fewer than four occasions to investigate its behaviour at a time when its cycle of activity was wearing down toward a 'minimum' in 1996-97.

The instruments in Atlantis' payload bay were complemented by two others attached to a free-flying atmospheric observatory, which was itself mounted atop the German-built Shuttle Pallet Satellite (SPAS). Like the ORFEUS mission a year earlier, the observatory employed the uprated 'ASTRO-SPAS' platform, capable of extended-duration operations, and it was intended that it should spend up to eight days in free flight. However, whereas ORFEUS was devoted to ultraviolet astronomy, the two instruments on the STS-66 payload were devoted to infrared observations of the middle and upper atmosphere. The first instrument was a German one, known as the Cryogenic Infrared Spectrometers and Telescopes for the Atmosphere (CRISTA), and sought to gather the first global information about medium- and small-scale disturbances in middle-atmosphere trace gases.

Provided by the University of Wuppertal, and first conceived in 1985, it was capable of scanning the horizon simultaneously in three directions, promising to provide new data about disturbances induced by winds, wave interactions, turbulence and other physical processes. "While the actual instrument was manufactured by industry," explained Professor Dirk Offermann, who co-ordinated CRISTA science from NASA's Marshall Space Flight Center in Huntsville, Alabama, "students did the calculations, constructed the cryostat, designed the optics, then integrated the equipment with the help of university technicians." In fact, around 20 students from undergraduate to doctoral level had been involved in the project between 1985 and 1994. CRISTA consisted of three telescopes with four spectrometers to measure emissions in the near-infrared and far-infrared portions of the electromagnetic spectrum. It had the capacity to acquire complete vertical profiles of trace gases within a minute as the lines-of-sight were scanned through the atmosphere. Its measurement speed and high sensitivity derived from a cryogenic liquid helium vacuum container, which cooled its optics and detectors. Abutting CRISTA was the Middle Atmosphere High Resolution Spectrograph Investigation (MAHRSI). Developed by the US Naval Research Laboratory, this instrument was designed to scan the horizon to observe ultraviolet emissions from nitric oxide and hydroxyl in the middle atmosphere and lower thermosphere, between the altitudes of 40-120 km. This region of the atmosphere is too high for high-altitude balloons to access and sounding rockets pass through it too quickly to gather useful data, making the capabilities of MAHRSI hugely important. Hydroxyl was known to contribute directly to the destruction of atmospheric ozone. "This information," explained NASA's press kit, "will be used to test many theories that have been based on assumed values and will provide the first global vertical measurements of hydroxyl in the stratosphere."

Preparations to deploy CRISTA-SPAS began a few hours after launch. Atlantis' RMS mechanical arm was checked out satisfactorily and at 4:45 pm EST on 3 November it was employed to grapple the payload. The arm was left 'parked' overnight, with the satellite still firmly latched into the payload bay, until it was unberthed and released into space by Jean-François Clervoy at 7:50 am on the 4th. Fittingly, perhaps, the deployment took place whilst Atlantis orbited high above Germany. Now physically separated from, and thus independent of, orbiter manoeuvres, the 3,400 kg CRISTA-SPAS operated for the next eight days between 40-70 km 'behind' Atlantis, communicating through the Shuttle with its control station at the Kennedy Space Center every ten hours or so. The two instruments worked in conjunction for several tasks, including measurements of temperature and nitric oxide quantities at an altitude of 100 km – the coldest part of the atmosphere – in order to better understand the thermal balance and its role in ozone chemistry. At the same time, high-precision ozone-measuring balloons launched from the Hohenpeissenberg station in Germany and instrumented aircraft despatched from the west coast of Scotland, formed part of a vigorous ground-based campaign in conjunction with the satellite. In total, CRISTA gathered 180 hours of data, eclipsed slightly by MAHRSI with 200 hours, and it was remarked in NASA's post-mission report that "current-generation orbital satellites would require almost six months to collect a data set as large as [was] gathered on this flight".

During the rendezvous phase on 11 November, two exercises of note were performed. One was dramatically described as 'The MAHRSI Football', due to the elliptical shape of the pattern flown by Atlantis around the satellite. It involved MAHRSI making ultraviolet observations of the Shuttle and its environment, including the enigmatic 'glow' phenomena around its flight surfaces. CRISTA-SPAS was finally grappled by the RMS arm at 8:05 am EST, as Atlantis flew to the southeast of New Zealand, and was berthed in the payload bay at 11:50 am. The second rendezvous exercise was labelled 'Detailed Test Objective 835' – innocuous enough, it seemed, were it not for its subtitle: *Mir Approach Demonstration*. Atlantis' next flight, in mid-1995, was destined to undertake the first Shuttle docking with Russia's Mir station and DTO 835 sought to evaluate a new rendezvous profile. "Rather than approaching for the final approximately [3 km] to CRISTA-SPAS from a point directly *ahead* of the satellite," explained NASA's press kit, "Atlantis will approach from *beneath* the satellite." The orbiter would effectively trace and imaginary line, extending 'upwards' from the centre of Earth to the mass of CRISTA-SPAS. Known as 'R-Bar' (or 'Earth Radius Vector'), the profile differed fundamentally from the 'V-Bar' (or 'Velocity Vector') technique employed on all previous Shuttle rendezvous missions. In approaching the target from 'beneath', rather than 'ahead', it was possible for Atlantis to exploit the gravitational gradient to brake her approach. In fact, because she would have to thrust *against* gravity to maintain the approach, this would also provide a margin of safety in the event of an RCS thruster failure. Moreover, it was recognised that plume impingement would have to be minimised during proximity operations with Mir, lest the thrusters cause damage or deposit contaminants upon the station's surfaces and, in particular, its solar panels. If such firings *were* necessary, using the R-Bar approach, they could be executed in a so-called 'Low-Z' mode, whereby Atlantis would use highly offset thrusters on her nose and tail, such that the plumes were not aimed toward the station. Earlier in the mission, during the CRISTA-SPAS retrieval effort, Joe Tanner had demonstrated a hand-held laser-ranging device which would be used on the Shuttle-Mir missions to gather precise data throughout the rendezvous process.

With the successful retrieval of CRISTA-SPAS, the mission of STS-66 was drawing inexorably toward its conclusion. Early on 14 November, the ATLAS-3 payload was deactivated and the crew prepared for their return to Earth. That return was hampered, in Florida, at least, by the ravages of Tropical Storm Gordon, which repeatedly battered the East Coast and the Caribbean with high winds, rain and cloud from the 8th until the 21st. Both scheduled landing opportunities at the Kennedy Space Center were thus called off and Don McMonagle guided his ship smoothly to a touchdown on concrete Runway 22 at Edwards Air Force Base in California at 7:33 am PST (10:33 am EST), concluding the third 11-day Shuttle mission in less than three months. Atlantis' first deployment of the drag chute was uneventful and the attention of NASA shifted sharply to her next mission, STS-71, scheduled for May-June 1995, which would attempt the long-awaited first docking with Mir.

Although Shuttle-Mir, which formed 'Phase 1' of the new partnership between the United States and Russia, took centre stage from 1995 until 1998, the reusable orbiters continued to undertake their 'solo' missions to release and retrieve satellite payloads, operate scientific instruments and support increasingly ambitious EVA tasks. Yet all of these missions had their sights fixed on the future; none were divorced from it and all played their own part in establishing the groundwork for today's International Space Station, whether through establishing research and operations protocols, EVA protocols or staging extended-duration flights. The 'payload commander' concept, for example, was originally designed with the intent that it would mirror the duties of a future station commander ... and several

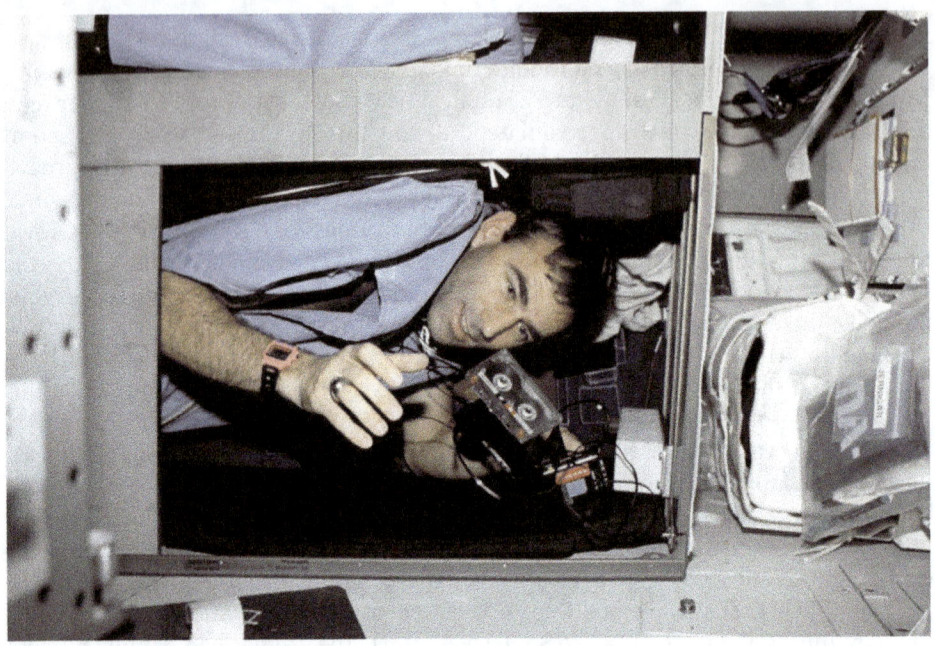

4-12: STS-67 pilot Bill Gregory ejects a cassette and prepares to bail out of one of the phonebox-sized sleep stations on Endeavour's middeck, ready to begin his 12-hour shift in support of the ASTRO-2 payload.

payload commanders – including Tammy Jernigan – had been named in August 1993 to support a series of complex research missions.

For Jernigan, one of the few professional astronomers on active duty within the astronaut corps, it was little surprise that she was assigned as payload commander for the ASTRO-2 mission, a reflight of the three ultraviolet telescopes first carried aboard Columbia on STS-35 in December 1990. In spite of technical difficulties related to the telescopes' pointing system, ASTRO-1 had been an enormous success and in May 1991 the first steps were taken to manifest another flight "as soon as the NASA flight schedule can accommodate it". Particular note was paid to the mission's "high-quality science data", which provided the driving force behind the decision and even its problems did not prevent NASA from claiming an 80 percent success rate, with 135 of 250 planned celestial targets acquired. By the time NASA's January 1992 manifest was published, ASTRO-2 appeared as STS-67 in September 1994, although its complexity (and flight duration) would steadily increase to such an extent that when it finally flew in March 1995 it would seize the record for the longest Shuttle mission to date. Originally scheduled to fly aboard Discovery for just seven days, by the middle of 1993 ASTRO-2 had been shifted over to Columbia, whose Extended Duration Orbiter capabilities permitted a longer flight of 13 days or more. As the flight accelerated towards its launch date, another astronomer-astronaut, John Grunsfeld, was named as a mission specialist in October 1993, followed by the announcement of the rest of the crew three months later.

Unsurprisingly, the two payload specialist positions would be taken by ASTRO-1 veterans Sam Durrance and Ron Parise. They were selected, alongside a third candidate, electrical engineer Scott Vangen, for training in May 1993. The success of ASTRO-1 was amply demonstrated by the fact that in 1994 – a full *four years* after the mission – more than one hundred scientific articles were published in academic journals, detailing results from investigators.

For John Mace Grunsfeld, it would be the first space voyage in a stellar, five-mission career which earned him the record for the only human being to visit the Hubble Space Telescope as many as three times. Like fellow astronaut Scott Parazynski, he was an avid mountaineer, and as Parazynski became the first astronaut to summit Mount Everest, Grunsfeld became the first astronaut to summit Mount McKinley in Alaska in June 2004. Unlike many astronauts, whose pre-NASA profession became more 'generalised' after admission into the corps, Grunsfeld remained a devoted astronomer, to such an extent that he served as the agency's Chief Scientist in 2003-04 and, at the time of writing, in early 2013, he serves as Associate Administrator of the Science Mission Directorate at NASA Headquarters in Washington, DC. He was born in Chicago, Illinois, on 10 October 1958, the son of the distinguished architect Ernest 'Tony' Grunsfeld III and grandson of Ernest Grunsfeld Jr, who designed the Adler Planetarium, the first of its kind ever built in the western hemisphere.

By his own admission, Grunsfeld yearned to be an astronaut from early childhood, although his interests later shifted towards physics and cosmology. "I had a vacuum cleaner with a hose that became my liquid-cooling unit," he told a NASA interviewer, years later, "walking out to the space launch pad. I turned those big ice-cream tins, by cutting a hole and putting cellophane over it, into a space helmet. It seemed very natural at the time, because here was a nation starting the exploration of space." Exploring Highland Park, on the north side of Chicago, wading through knee-high snowdrifts and forests and into ravines, Grunsfeld pretended that he was an astronaut on another world. It was a love of nature and science which would remain with him. After high school, he entered the Massachusetts Institute of Technology to study physics and earned his degree in 1980. Whilst there, he met a post-doctoral student named Jeff Hoffman, who was an astronomer. It was Hoffman's subsequent path in NASA which convinced Grunsfeld that it was accessible to him. To make ends meet as a student, he worked the graveyard shift in the control room for the Small Astronomy Satellite (SAS)-3, changing data tapes and monitoring strip-chart recorders. This led to a year-long post as a visiting scientist at the University of Tokyo, working with X-ray astronomer Minoru Oda. Grunsfeld returned to the United States and entered the University of Chicago, gaining master's and doctoral degrees in physics in 1984 and 1988, with a research focus upon energetic cosmic rays. After graduation, he worked as a senior research fellow at the California Institute of Technology, where he and his wife both learned to fly; in late 1991, when NASA invited Grunsfeld to attend an astronaut interview, he flew his own aircraft to Houston.

Until he received the call from the space agency, he thought his job at Caltech was the best in the world. "I was working with the Compton Gamma Ray Observatory," he said. "I had observatories, like Palomar, that I went to use big telescopes." It was electrifying to consider that *now* was the Golden Age of Astronomy and *he* was part of it, but that did not stop him from applying, unsuccessfully, to join NASA in January 1990. "I was studying neutron stars and black holes," he continued, "which are incredibly interesting and exotic objects. I had wonderful students. I thought life couldn't be better ... and then I got a call from Houston." He chatted it over with a colleague, Bruce Margon. The consensus was that whilst Grunsfeld could *always* be an astronomer, he might *not* always have the chance to become an astronaut. In March 1992, he was selected by NASA.

Grunsfeld was not the only first-timer aboard STS-67. He was joined by Air Force test pilot William George Gregory – nicknamed 'Borneo' – and Navy helicopter pilot Wendy Barrien Lawrence. When their names, and that of the mission commander, Steve Oswald, were announced by NASA in January 1994, ASTRO-2 was scheduled for the following December, with an envisaged flight duration of at least 13 days. However, as the year wore on it was decided to remove Columbia from service for a lengthy period of refurbishment and maintenance at prime contractor Rockwell's facility in Palmdale, California. Her sister ship, Endeavour, had been outfitted with EDO capability during construction and was retasked to fly STS-67, necessitating a slight delay into the early spring of 1995. By the beginning of that year, it was clear that the mission would secure a new duration record; according to the pre-launch press kit, it was scheduled to run for 15 days and 13 hours, easily eclipsing the 14 days and 17 hours achieved by STS-65. As circumstances transpired, Endeavour would remain in orbit for longer still and would seize the single-flight-duration record from Columbia for a little over a year.

Bill Gregory, the STS-67 pilot, came from Lockport, New York, where he was born on 14 May 1957. Of Albanian ancestry, his family originated from the city of Korçë, in the mountainous south-eastern portion of the country. Gregory graduated with distinction in engineering sciences from the Air Force Academy in 1979, followed by master's credentials in engineering mechanics from Columbia University in 1980 and in management from Troy University in 1984. As he laboured towards his advanced degrees, he served as an operational fighter pilot for five years, flying the F-111 Aardvark aircraft as an instructor at RAF Lakenheath in England and at Cannon Air Force Base in New Mexico. In 1987, he was sent to Edwards Air Force Base for test pilot training and later flew the F-4 Phantom, A-7D Corsair and all five models of the F-15 Eagle. He was selected as an astronaut in January 1990.

Seated behind and between Gregory and Oswald on Endeavour's flight deck was petite Navy helicopter pilot Wendy Lawrence. Aged 35, she was the youngest member of the crew. Born in Jacksonville, Florida, on 2 July 1959, it seemed that both naval and astronaut careers were more than simply options, for her father was Vice-Admiral William P. Lawrence, a finalist for Project Mercury, former Vietnam prisoner of war and former Superintendent of the US Naval Academy. Seeing Neil Armstrong walk on the Moon, and watching *Star Trek*, imbued her with excitement, and her father further guided her steps. "His very wise advice to me was follow in the footsteps of the first several groups of astronauts," she told a NASA interviewer. And the first stage on that route was the Naval Academy. Lawrence left the academy in 1981 with a degree in ocean engineering and when she flew STS-67 she became its first female graduate ever to journey into space. She completed flight training with distinction and became a naval aviator in July 1982, ultimately flying six different types of helicopter, accruing 1,500 hours and more than 800 shipboard landings. At one stage, attached to Helicopter Combat Support Squadron Six, she was one of the first two female helicopter pilots to make a long deployment to the Indian Ocean as part of a carrier battle group. In 1988, she gained a master's degree in ocean engineering from the Massachusetts Institute of Technology and Woods Hole Oceanographic Institute and later worked as a physics instructor at the Naval Academy from October 1990 until her selection as an astronaut in March 1992. Flying her first mission on STS-67 was the pinnacle achievement of her life, but for Lawrence the entire experience at NASA had been filled with excitement. "How many people," she once rhetorically asked, "can say that they're *living* their dream? To me, that in itself is a huge motivation. It's really refreshing to get up in the morning and look forward to going to work."

As darkness fell on the evening of 1 March 1995, Endeavour sat silently under the glow of xenon floodlights at Pad 39A, awaiting her eighth launch into space. It seemed unlikely that she would fly at all, with Air Force meteorologists predicting only a 40 percent chance of acceptable weather conditions; nevertheless, Steve Oswald led his crew out of the Operations and Checkout Building and into the glare of television and camera lights at 10:20 pm EST. They boarded the 'Astrovan' and arrived at the base of the launch pad within 25 minutes, to begin the laborious process of being strapped into their seats on Endeavour. By midnight, all seven astronauts were aboard and communications checks were concluded. As the clock ticked over into 2 March, the Shuttle's middeck hatch was sealed and all remaining technicians and pad personnel evacuated the launch area. With liftoff scheduled for 1:37 am, at the start of a 2.5-hour 'window', everything proceeded normally until 1:26 am, when a problem was detected with the B-supply secondary heater of the Flash Evaporator System (FES), one of the devices responsible for cooling the orbiter's electronics. This seemed to indicate that the heater was approaching a 'redline' condition. The countdown proceeded and was briefly held at T-5 minutes, during which time the FES was verified as being healthy. The resolution of the problem consumed a mere 73 seconds of the launch window: at 1:34 am, Bill Gregory activated the orbiter's Auxiliary Power Units and at 1:38 am Endeavour thundered gloriously into the black sky. "*Instant daylight*," was Bill Gregory's description in the aftermath of the event. Using a hand-held mirror, Wendy Lawrence – in the flight engineer's seat – and John Grunsfeld, to her immediate right side, were able to momentarily glimpse the launch pad, the smoke plume and the steadily expanding Florida coastline.

Almost as soon as he entered space for the first time, Grunsfeld convinced himself that he really *didn't* want to return to Earth. "I had this real feeling of peace," he said later, "that I never had here on Planet Earth." There was little time to acclimatise, for the STS-67 mission was an around-the-clock operation, involving red and blue shifts. Oswald, Gregory, Grunsfeld and Parise formed the red team, with Jernigan, Lawrence and Durrance on the blue team. And for Steve Oswald, the first few hours of the flight afforded him precious little time to check on his three 'rookie' crewmates. "It never ceases to amaze me," he told an interviewer from the Smithsonian, "the *fire drill* that goes on down in the middeck in those first couple of hours after we reach orbit." Seven bodies floated hither and thither, removing pumpkin-orange pressure suits, watching as someone's sock comically floated past, grabbing a vomit bag to cope with an unexpected attack of motion sickness. "As the commander, you try to figure out how much extra time you need to add to the schedule, based on how many rookies you've got," Oswald continued. "Sometimes guys are semi-Velcroed to the wall, throwing up, while the folks you least expected to be heroes are just chugging along, executing the plan." That plan was enormous and complex in its scope; so enormous, in fact, that all seven astronauts would receive two half-days of free time at various points during the long mission.

Outside, in Endeavour's cavernous payload bay, sat 7,885 kg of equipment, comprising the Hopkins Ultraviolet Telescope (HUT) from Johns Hopkins University, the Ultraviolet Imaging Telescope (UIT) from NASA's Goddard Space Flight Center and the Wisconsin Ultraviolet Photopolarimeter Experiment (WUPPE) from the University of Wisconsin at Madison. These were mounted atop an Instrument Pointing System (IPS), which itself stood on a pair of Spacelab pallets. Attached to the forward pallet was a pressurised igloo to hold electronics and instrumentation, whilst at the rear end of the bay was the large wafer of the Extended Duration Pallet, enshrouded within white insulation.

Within four hours of reaching orbit, under the supervision of Tammy Jernigan, the IPS had been rotated into its upright orientation, whereupon Sam Durrance applied power to the three telescopes, preparatory to the first ASTRO-2 observations. Early activities proceeded normally, despite an RCS thruster leak which twice forced the closure of the telescopes' aperture doors to safeguard their optics from contamination. The mission suffered from none of the IPS pointing troubles which plagued its predecessor, ASTRO-1, in December 1990. According to NASA's STS-67 press kit, a special test team was assembled at the Marshall Space Flight Center, and this "extensively modified and tested the IPS software and made other improvements to ensure the IPS works properly for ASTRO-2". Specifically, an image motion compensation system – designed to eliminate the effects of 'jitter', induced by crew movements and thrusters firings – helped to refine instrument pointing and stability for the telescopes. This was particularly vital in the case of UIT, whose images were recorded on film with the individual exposures lasting as long as 30 minutes. WUPPE experienced a somewhat slow start, however, since the activation and verification of its detector was delayed by problems keeping it aligned with a test target. Twelve hours into the mission, controllers at the Johnson Space Center, declared the IPS fully operational and transferred control of its equipment to the ASTRO-2 payload team at Marshall. With this transfer, John Grunsfeld and Ron Parise began a lengthy procedure known as 'Joint Focus and Alignment', to ensure that all three instruments were capable of pointing in precisely the same direction.

Despite the day-long process of calibration, astronomical observations of the ultraviolet sky got underway with pace and gusto. Early on 3 March, the HUT and

UIT science teams had locked their instruments onto the Cygnus Loop, an ancient supernova remnant, with the former instrument gathering temperature, density and chemical data and the latter imaging 'filaments' of excited gas and energising shockwaves. WUPPE demonstrated that its optics were in perfect working order by observing a calibration star, Beta Cassiopeiae, followed by a study of the hypergiant luminous blue variable star P Cygni. Meanwhile, HUT examined EG Andromedae, a 'symbiotic' system of a relatively cool, orange giant star and a tiny, exceptionally hot blue star. White dwarfs, globular clusters – some of which appeared to be more than 16 billion years old, far older than the Hubble Space Telescope data suggested the Universe *itself* to be, according to UIT astronomer Steve Maran – and 'Wolf-Rayet' stars formed the centre of attention over the following days. The latter include EZ Canis Majoris and are thought to represent one of the final phases in the evolution of supermassive stars, whose luminosities vary between 100,000 and a million times as bright as our Sun. Their powerful ionised-gas emissions, or 'stellar winds', were believed to accelerate their aging process. Extremely bright-centred Seyfert galaxies, distant quasars and interactive binary star systems also received attention. In the latter case, WUPPE was directed to observe an X-ray binary, known as Vela X-1, with astronomers speculating that a neutron star was gravitationally 'stripping' material from its companion star, causing a large oval disk to form in orbit. Polarisation measurements by WUPPE enabled measurements to be made of the size and shape of this disk, as well as calculations as to the quantities of mass transferred between the two companions.

Closer to home – and illustrated on the STS-67 crew's mission patch – the largest planet in the Solar System, Jupiter, also came under ASTRO-2 scrutiny. HUT investigators paid particular attention to its immense magnetosphere, as well as the planet's curious, volcanically active moon, Io. A recent eruption on Io had deposited material onto the surface and into its tenuous atmosphere, prompting HUT co-investigator Paul Feldman to seek evidence of changes in the number of sulphur and oxygen ions in its environment. "As Io orbits Jupiter once every 42 hours," noted one of NASA's news summaries, early in the mission, "some of this material is left behind, forming a doughnut-shaped torus of sulphur and oxygen plasma around Io's orbit."

The sheer 'strangeness' of the Universe was illustrated by the peculiarities of so many ASTRO-2 targets. Phi Persei, a hot, rapidly spinning star, exhibited an unusual ultraviolet spectrum, possibly due to a 'shell' of gas which may have been an outer layer shed by its fast rotation. Two active Seyfert galaxies, both strong emitters of very bright ultraviolet radiation and thought to have supermassive black holes at their hearts, were studied; one of them, NGC 4151, was five times brighter in March 1995 than it had been when ASTRO-1 observed it, more than four years earlier. Indeed, the galaxy exhibited a 10-percent luminosity increase in a matter of *days* during ASTRO-2. Ancient stars and young stars, stellar graveyards and stellar nurseries, came under the observatory's ultraviolet gaze, including an open cluster, called N4, whose youthful occupants were believed to be less than ten million years old. Elsewhere, M104 – a distinctive spiral galaxy, nicknamed 'The Sombrero Galaxy', due to its likeness to the wide-brimmed Mexican hat – was observed, with

4-13: Unlike its predecessor in December 1990, which suffered multiple hardware and software glitches, the 16-day voyage of ASTRO-2 operated with exceptional smoothness.

star-forming regions thought to reside within its 'brim' and older stars (and maybe a black hole) within its 'crown'.

Periodically, the professional astronomer in Tammy Jernigan was overtaken by a sense of childlike wonder at the environment in which she found herself. "There was a lot of time when the cockpit was darkened," she remembered, in an interview with the Smithsonian, years later. "You would monitor observation of an object for maybe 20 minutes before you had to regroup, repoint the Instrument Pointing System and set up the instruments again. There were whole blocks of time where you could just look out and reflect, talk to the other crew members who were awake on your shift and really have a sense of what a beautiful Universe we inhabit."

In a sharp contrast with the technical difficulties experienced by its predecessor, the ASTRO-2 observatory was also a beautiful payload, from the standpoint of systems performance, with the hitherto-troublesome IPS and its Image Motion

Compensation System (IMCS) performing "in an outstanding manner", according to NASA's official post-mission report. "The IPS and IMCS for the first time achieved operational capacity," exulted ASTRO-2 Mission Manager Robert Jayroe, then quipped: "In *my* estimation, the IMCS and IPS teams have done everything but make the hardware *stand up* and do a *tap dance!*" The WUPPE team gathered more than three times as much data as had been gathered during ASTRO-1, whilst the UIT investigators reported that all planned celestial targets had been acquired and the HUT scientists announced more than 100 successful observations.

Aside from the instruments in the payload bay, Endeavour's crew undertook a handful of other experiments in the cabin. The Middeck Active Control Experiment (MACE), tended by Steve Oswald and Bill Gregory, evaluated a closed-loop control mechanism to compensate for around 200 discrete motion disturbance situations, whilst protein crystal growth and materials dispersion investigations were monitored by Lawrence. In the payload bay, two GAS canisters housed the 100 mm aperture 'Endeavour' ultraviolet binocular reflecting telescope, provided by the Australian Space Office at the Department of Industry, Science and Technology. Within days of launch, the STS-67 crew received word from the investigators that the small telescope had already surpassed its 100-percent mark of desired observations.

Already scheduled for more than 15 days aloft, the ASTRO-2 crew quietly eclipsed the previous Shuttle duration record on the evening of 16 March, with the expectation that Endeavour would make landfall at the Kennedy Space Center at 3:09 pm EST on the 17th. However, it was not to be. Unacceptable weather conditions in Florida forced the Mission Management Team to scrub the attempt and reschedule it for the 18th. With the situation on the East Coast showing little evidence of improvement, it was decided to divert to Edwards Air Force Base instead. Already, STS-67 had passed the 16-day mark, which was the maximum 'standard' length for an Extended Duration Orbiter mission, and managers were anxious not to press Endeavour's consumables any further. Sweeping across the Pacific, the orbiter entered US airspace at the California coastline and alighted on Runway 22 at Edwards at 1:47 pm PST (4:47 pm EST), concluding a remarkable mission of 16 days, 15 hours and eight minutes, with 262 circuits of Earth.

For Steve Oswald, the commander, it would be a nice record on which to end his astronaut career. Yet the *length* of the mission, and the months of preparation for it, had taken their toll. "The training is structured such that it trains to the lowest common denominator," he explained in a NASA Tacit Knowledge Capture interview, "and it just takes forever. You're going through all the stuff again for those that haven't flown before. It got to be kind of a long, drawn-out deal. It was a great flight, great crew; I had a great time, but afterwards, I was just done." He was not the only astronaut to be physically and mentally burned out by the intensity of training for a space mission. For his crewmates, other adventures would await: Tammy Jernigan would fly longer still, on STS-80, late in 1996, whilst John Grunsfeld would fly *higher* still, by becoming the only human being to visit the Hubble Space Telescope's 600 km altitude as many as three times, and Wendy Lawrence would be a member of the crew which resumed the Shuttle programme in the wake of its second disaster.

Disaster, indeed, and glory would travel together, virtually hand-in-hand, in the months and years after STS-67. *Nineteen-ninety-five* was the year in which 'The Partnership' between Russia and the United States reached its most exalted height since the Apollo-Soyuz Test Project, a full two decades earlier. Few could have foreseen the dramatic demise of the Soviet Union and still fewer could have envisaged the rapid pace with which partners were formed from old enemies. For the first time, the spacegoing aspirations of both nations were inextricably intertwined; the Shuttle manifest had morphed from one of 'solo' flights into one which embraced the Russians as partners, whilst the Mir space station would serve as home to English-speaking Americans, on a near-continuous basis, for almost three years. And *that* experiment in international relations, as much as science and technology, would be merely the prelude to one of the most remarkable endeavours in human history: the International Space Station, in which more than a dozen nations – some of them on the brink of outright conflict only a few decades earlier – collaborated on the grandest scientific and engineering accomplishment ever conceived.

As will be explored in more depth in the sixth and final volume of this series, the partnership would be one of the rockiest roads ever travelled in the history of human space exploration; tempers would fray, technologies would fray and – some said – lives would be placed at risk in a rickety old space station, no longer fit for purpose. Certainly, lives would be risked, and lost, aboard the Space Shuttle, whose veneer of being 'operational' was paper-thin. Yet the remarkable period from 1995-98 marked a true watershed in US-Russian relations and many astronauts, cosmonauts, managers and politicians are united in their belief that the 'international' partnerships which function routinely aboard today's International Space Station could not have been achieved without the trials, troubles, tribulations, tragedies and triumphs of Shuttle-Mir and the tumultuous years which followed.

5

Dawn of a new era

A GLOOMY ROAD TO FREEDOM

Early on 16 March 1995, a historic radio call took place between the Space Shuttle Endeavour and the Mir space station. NASA astronauts Steve Oswald and Norm Thagard spoke briefly to one another. Three years earlier, the pair had flown in space together, but now both were fulfilling distinctly separate missions: Oswald in command of STS-67, the longest Shuttle flight to date, and Thagard resident aboard Mir, participating in the longest American flight to date. They extended mutual congratulations on the success of their respective flights, but the conversation was laced with symbolism, for it marked a true passage from 'The Old' to 'The New'. For more than two decades, the longest period a US crew had remained in orbit was 84 days – a record established by the final team of Skylab astronauts in February 1974 – and it had long been recognised that experience of lengthy space voyages was acutely necessary ... not only for the International Space Station (ISS) programme, but for subsequent missions further afield, to the Moon, Mars and beyond.

Early in 1990, a handful of NASA managers and astronauts – including Dan Brandenstein and Jerry Ross – visited Tyuratam in the Soviet Union to watch the launch of cosmonauts Anatoli Solovyov and Alexander Balandin aboard Soyuz TM-9. Shortly thereafter, the first meetings at a political level to discuss co-operation in their human space programmes began. In July 1991, at a summit in Moscow, US President George H.W. Bush and Soviet General Secretary Mikhail Gorbachev agreed to fly a cosmonaut aboard the Shuttle and an astronaut for several months to Mir. "The purpose of the exchange of flights," explained a NASA news release, "is to conduct life sciences research of mutual interest. It would advance current efforts to standardise in-flight medical procedures, which would improve comparability of data taken by each side." As noted in Chapter 4, these efforts developed further in the wake of the collapse of the Soviet Union and by the end of 1992 cosmonauts Sergei Krikalev and Vladimir Titov had begun training at the Johnson Space Center in Houston, Texas, for STS-60. The NASA astronaut, on the other hand, was to spend 150-180 days aboard Mir. At around the same time, in November-December

1992, US and Russian working groups began exploring the possibility of utilising the Soyuz-TM spacecraft as an "interim" Assured Crew Return Vehicle for Space Station Freedom, for use in an emergency whilst the Shuttle was not present.

To put it mildly, the first year of the Commonwealth of Independent States (CIS) was the most difficult in the post-Soviet era. Almost seven decades of iron-fisted single-party rule over an area of more than 22 million km^2, including virtually all of Northern Asia, and control of more than 293 million people from various ethnicities, religious faiths and cultures, had finally collapsed in the aftermath of the unsuccessful coup in Moscow in August 1991 by Communist hardliners opposed to liberal reform. By then, several former socialist republics had already broken away from the Soviet sphere. The Berlin Wall had fallen, many Baltic and Eastern European states had progressed towards democratic elections and attempts to crush Lithuania with military force in January 1991 had prompted intense criticism of the Soviet Union, both domestically and internationally, to intensify. Sixty-nine years after its creation, the Union of Soviet Socialist Republics officially ceased to exist at the end of December 1991 ... and the new year brought small comfort or solace to a new cluster of economically crippled, ideologically adrift and ethnically volatile nations.

In space, this was perhaps best illustrated by the decision to keep Soyuz TM-12 cosmonaut Sergei Krikalev aboard Mir for an additional five months, from October 1991 until March 1992. As described in Chapter 2, Krikalev was supposed to be relieved by Soyuz TM-13's Alexander Kaleri, but in order to pacify Kazakhstan – which had declared its sovereignty in October 1990 and within whose territory lay the Tyuratam launch site – the Soviets had little choice but to drop Kaleri in favour of including an ethnic Kazakh cosmonaut, Toktar Aubakirov, on the crew. Media reports labelled Krikalev 'The Last Soviet Citizen', because he launched under the banner of the Hammer and Sickle and returning to the tri-colour red, white and blue horizontal bars of the revived Russian national flag. Yet the end of the Soviet era was also indicated by a marked change in stance of the space programme. In June 1991, amidst the crisis of dissolution and ahead of the final collapse, it was reported that the Soviets were planning "a space spectacular" in 1992, involving Mir, Soyuz-TM and their own Space Shuttle. The latter had undertaken an unpiloted maiden voyage several years earlier, with the expectation that it would eventually transport cosmonauts into orbit. According to Yuri Semenov, Chief Designer of the Energia bureau in Moscow, quoted by *Flight International*, the plan was for Buran to be launched, without a crew, late in 1992, and dock automatically at Mir's Kristall module. The station's Soyuz TM-15 cosmonauts would board Buran, test its remote-manipulator arm, and then return to Mir, whereupon the Shuttle would undock and commence a period of independent flight. Another crew, aboard Soyuz TM-16, would then rendezvous and link up with Buran, "to demonstrate the spacecraft's ability to operate independently from Mir", before separating and docking at the space station itself. According to the plans, the Shuttle – which closely mirrored NASA's own winged orbiters in physical appearance – would return to Earth under automatic control and touch down on the runway at Tyuratam.

In terms of lost opportunities and failed dreams, Buran has all the hallmarks of a

Homeric tragedy, from its conception until its ignominious end. It was created under the auspices of the Central Aerohydrodynamic Institute in Moscow in 1974, along with a mammoth booster known as 'Energia' ('Energy'), in response to the United States' Shuttle programme. The payload capacity of the latter convinced Soviet planners that the reusable orbiters might be used for the launch of reconnaissance and intelligence satellites, and possibly space-based weapons, too, and by the early 1980s the Buran effort had gained considerable ground. As part of its development process, five suborbital flights of small-scale models were executed from the Kasputin Yar launch site in southern Russia in 1983-86. In the meantime, between July 1977 and September 1985, four classes of test pilots were selected by the Gromov Flight Research Institute in Zhukovsky, near Moscow, for Shuttle training. It was mandated that at least one member of a given crew (both prime and backup) needed space flight experience, but of the five pilots chosen in July 1977 only two actually flew Soyuz familiarisation missions. Igor Volk spent a week aboard Salyut 7 in July 1984 and Anatoli Levchenko spent a week aboard Mir in December 1988; the others – Oleg Kononenko, Alexander Shchukin and Rimantas Stankyavichus – were all killed in aircraft accidents before they could get their chance. Unfortunately, Levchenko died of a brain tumour, a few months after his Mir flight, leaving Volk as the sole space-experienced member of the group. It was intended that the Buran trainees would take the controls of a Tu-154 aircraft, shortly after touching down aboard the Soyuz, and fly from Tyuratam to Artyubinsk, also in Kazakhstan, then return to Tyuratam at the controls of a MiG-25 fighter. The Tu-154's instrument suite was outfitted to virtually mimic that of Buran, thereby giving Soviet engineers a realistic gauge of how well a cosmonaut could land a winged vehicle on a runway after several days in weightlessness.

On 15 November 1988, in an unmanned capacity, Buran roared into orbit for what would turn out to be its first and only space mission. Measuring 36.4 m long and 23.9 m across its wingspan, the vehicle weighed 42,000 kg and was virtually identical to NASA's Shuttle, with the notable exception that its main propulsion system was integrated into the Energia booster, upon which it rode, 'piggyback' style, into space. Buran was equipped with its own tail-mounted orbital manoeuvring thrusters, fed by gaseous oxygen and kerosene, which yielded lower toxicity than the Shuttle's monomethyl hydrazine/nitrogen tetroxide mixture, as well as improved performance. After Energia achieved a low orbit, Buran separated and employed these thrusters to raise its apogee, before returning to a newly-built runway at Tyuratam after 206 minutes and two orbits. (For more than two decades, until the US Air Force launched its X-37B classified spaceplane in April 2010, the Buran mission marked the only remote-controlled runway landing of a winged spacecraft.) After the first flight, it was intended that Buran would fly again, for 15-20 days, in 1993, but most popular interest focused on the second orbiter in the fleet, informally nicknamed 'Ptichka' ('Little Bird'). In the final months of the Soviet Union, Ptichka – with its distinctive red framework atop its payload bay doors – was the most likely contender for the 1992 "space spectacular". After a 24-hour inaugural flight, it was to launch for an eight-day mission to support the Mir docking, the independent unpiloted mission, the Soyuz TM-16 docking and an automated return to Earth.

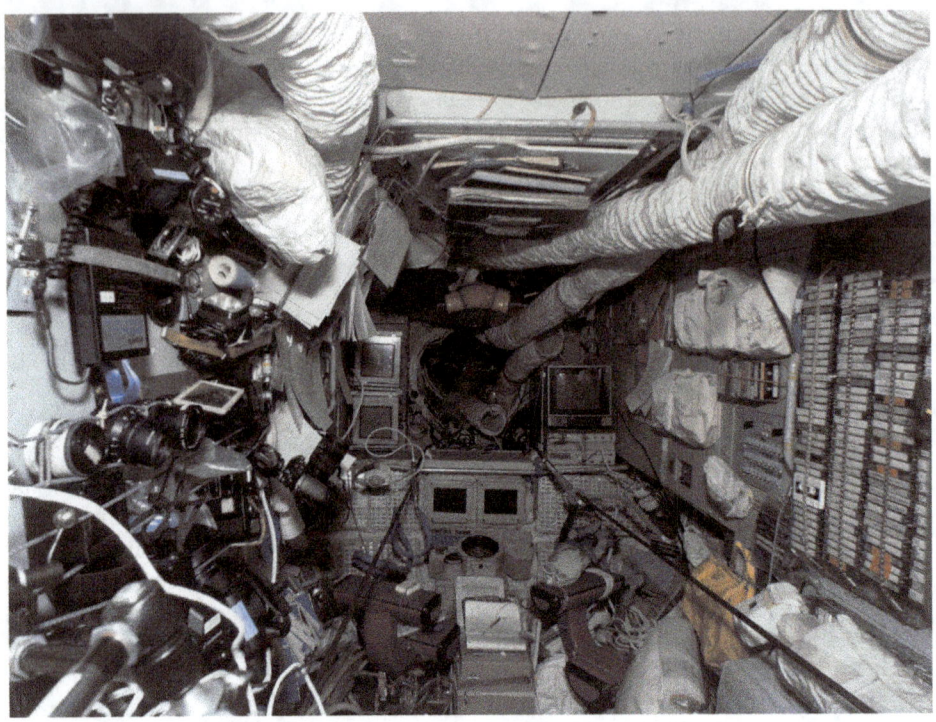

5-01: The cluttered interior of Mir's base block, looking toward the main control panel and hatch into the multiple docking adaptor.

None of this talk was backed by anything of substance in the dying, cash-strapped days of the Soviet Union and in June 1993 Russian President Boris Yeltsin cancelled the Energia/Buran programme, after almost two decades of work and around 70 billion roubles. The final indignity, however, came in May 2002, when the roof of the dilapidated Tyuratam hangar, housing the mothballed Buran, collapsed through negligence, killing several workers, together with an Energia mockup ... and the Shuttle itself.

In addition to the Buran-Soyuz-Mir spectacular, the Soviets revealed in mid-1991 that the station's final two modules – Priroda ('Nature') and Spektr ('Spectrum') – would also be launched in the following 12 months. Both would be positioned on Mir's forward multiple docking adaptor. However, at the start of the following year, it was revealed that the modules had been delayed until at least 1993 and then as the political and economic situation worsened they were postponed indefinitely. With so many projects facing the axe, Russia was quick to assure Germany and France that their commercial missions to Mir in 1992, involving Klaus-Dietrich Flade aboard Soyuz TM-14 in March and Michel Tognini aboard Soyuz TM-15 in July, were safe from cancellation. In the meantime, nine republics of the new Commonwealth – Azerbaijan, Belarus, Kyrgyzstan, Tajikistan, Uzbekistan, Armenia, Kazakhstan, Russia and Turkmenistan – established an Inter-State Space Council to better consolidate the management of the space programme in the post-Soviet era. "Although each republic can have its own programme," explained Tim Furniss in *Flight International* in January 1992, "decisions by the Council will be endorsed by heads of government and will be financed out of contributions from each republic." The republics' joint armed forces were made responsible for the provision of civilian and military space services, and although pledges were given "not to interrupt normal functioning", it was cautioned that "the use of space centres ... will be subject to separate agreements". In a practical sense, Tyuratam in Kazakhstan fared the worst. On the eve of the launch of Soyuz TM-14, it was reportedly suffering a "catastrophic cash shortage" of such severity that military officers threatened to strike "unless their economic demands" were met. At the same time, plans to replace Mir's core module were delayed until "at least 1997", placing more pressure upon cosmonaut crews to replace or repair vital equipment in order to enable the continuous habitation of the station. Media reports indicated that only five of Mir's 12 solar arrays were fully functional and – barely two years since its launch – four of Kvant-2's six gyrodynes were also dead. This latter problem required TM-14 cosmonauts Alexander Viktorenko and Alexander Kaleri to perform a hazardous, two-hour EVA on 8 July 1992 to inspect the gyrodynes and evaluate the immense difficulty of reaching and repairing them. The men used large-handled shears to cut through thermal insulation on Kvant-2 to access the electrically-driven devices and televised their work for controllers on the ground.

In the midst of these problems, the first murmurings of the Shuttle-Mir effort with the United States and increased European interest in flying long-duration missions to the station surfaced. Late in May 1992, it was reported that the European Space Agency (ESA) sought to negotiate a series of flights to Mir and had that very month selected a new cadre of astronaut candidates: Maurizio Cheli of Italy, Jean-François Clervoy of France, Pedro Duque of Spain, Christer Fuglesang of Sweden, Marianne Merchez of Belgium and Thomas Reiter of Germany. "Individual European nationals have flown on Mir," noted *Flight International*, "but ESA itself has never sponsored a flight to the station." Frenchman Michel Tognini's two-week mission was expected to net a profit of $12 million for Russia and hopes were high that an Israeli cosmonaut might join the Soyuz TM-16 crew for a launch as early as November 1992. Other interested parties included South Korea and Chile.

In the meantime, life and work aboard Mir progressed. From March until July, Viktorenko and Kaleri worked alone to maintain the rickety old station, supporting the arrival of two Progress cargo freighters in April and June and performing their EVA. It was stressed that the men were dividing their time between experiments and the need to repair or replace life-support equipment and their work included the use of the 'Liva' Earth-resources camera. (At one point, early in the summer of 1992, it was stated by a high-ranking representative of the Energia design bureau that Mir was in such a condition that it could no longer be 'mothballed' and left without a caretaker crew; it would "doze off forever", within two months, he noted, if left crewless.) On Earth, attempts on the part of the Russian government to tend to the interests of remote communities were illustrated when representatives of the administration, civil defence and media from Novosibirsk – upon whom overflying rockets had frequently dropped their spent stages, jettisoned hatches and other hardware – were invited to view the Progress M-13 launch. Four weeks later, at 9:08 am Moscow Time on 27 July, the next long-duration crew of Anatoli Solovyov and Sergei Avdeyev, joined by Michel Tognini of France, blasted off from Tyuratam aboard Soyuz TM-15.

For Solovyov, it was his third mission, but for Sergei Vasilyevich Avdeyev it represented the first flight in a career which would establish him from 1999-2005 as the acknowledged record-holder for the longest cumulative time spent in space. In his three missions to Mir, Avdeyev spent a total of 749 days – a little more than two full years – away from the Home Planet and this accomplishment received much popular attention when it was announced that he had also secured the record for the greatest 'time dilation' experienced by a human being. Travelling at an orbital velocity of 27,360 km/h, Avdeyev 'aged' roughly 20 milliseconds (0.02 seconds) less than an Earthbound person, due to the special relativistic effect of time dilation. Avdeyev came from Chapayevsk in Samara Oblast, where he was born on 1 January 1956, and attended the Moscow Physics-Engineering Institute. Upon graduation in 1979, he worked as an engineer-physicist for eight years and was selected as a cosmonaut candidate in March 1987.

Also making his first space mission aboard Soyuz TM-15 was Michel Ange-Charles Tognini, France's third man in space. Born on 30 September 1949 in Vincennes, within the eastern suburbs of Paris, Tognini grew up with aviation in his blood. "I wanted to be a pilot," he recalled in a NASA interview, "because my family was in the Air France company and they were dealing with planes and aircraft." His initial aspiration was to fly military transport aircraft and he earned a mathematics degree from Military School Grenoble in 1970, but after graduating in engineering from the Ecole de l'Air (French Air Force Academy) in 1973 Tognini broadened his scope into fighter and test piloting. He was posted for advanced fighter training at the Normandie-Niemen squadron and from 1974-81 served in the French Air Force as an operational fighter pilot, flying Mirage F-1 aircraft out of Cambrai Air Base. By 1979, he had risen to become a flight commander. Three years later, Tognini was selected for the Empire Test Pilot School at Boscombe Down in Wiltshire, England. At the end of his studies in 1982, Tognini was awarded the Patuxent Shield Trophy – an award won by two other future French astronauts, including Patrick Baudry (in 1978) and Jean-Pierre Haigneré (in 1981). After graduation, Tognini returned to France as a chief test pilot at the Cazaux Flight Test Centre. His work included weapons system testing for the Mirage and Jaguar aircraft, as well as responsibilities associated with the safety of pilots, experimenters and flight engineers.

By his own admission, Tognini's interest in the space programme was driven by France's short-lived Hermes mini-shuttle project in the mid-1980s, which proved attractive to his test-piloting and engineering mind. "They needed pilots," he said of the project, "and I knew that I had better chances to apply to be an astronaut at that time. If the Hermes programme had not been going on, they would not have paid any attention to the pilots that applied. I did not want to apply for nothing. The only reason I applied was because Hermes was there." Tognini was selected in September

1985 as part of the second astronaut candidate group of the Centre Nationale d'Études Spatiales (CNES) – the French national space agency – and from November 1986 he found himself assigned as backup to fellow Frenchman Jean-Loup Chrétien for the Soyuz TM-7 mission to Mir. This three-week flight took place in November-December 1988 and required Tognini to become proficient in EVA tasks. Subsequent work on Hermes was interrupted in 1991 when he was selected for Soyuz TM-15, which would involve a two-week visit to Mir on a mission designated 'Antares'.

Preparing for the mission was, in Tognini's words, very similar to Shuttle payload specialist training, with a focus on the scientific demands and limited participation in the operation of the spacecraft itself. After Soyuz TM-15, Tognini would join the ranks of NASA's astronaut corps, based in Houston, Texas, and would thus gain experience of training in both the Russian and US systems. Years later, he remembered that the Russian syllabus seemed "twice as long", overwhelmingly dominated by classroom instruction. "Before you fly, you have to go through anywhere from 50-70 different exams," he recalled. "If you don't go through them all, you can't fly. The last simulation you do is also an exam. You get either a good mark or a bad mark. Russian training is more skill-oriented and American training is more task-oriented." However, he noted that since Russian missions were (at the time) considerably longer than American ones, the need to accomplish discrete tasks in very short periods of time was unnecessary. "Although the cosmonauts are not prepared to do a particular task," he said, "they possess the skills to fulfil a task from instructions given to them while on the mission."

Two days after launching from Tyuratam, Soyuz TM-15 docked successfully at Mir, inaugurating an extended period of joint work with Alexander Viktorenko and Alexander Kaleri. During his time aboard the space station, Tognini undertook ten experiments in medicine, technology and physics, using 300 kg of scientific equipment delivered to Mir aboard Progress freighters. His work included the *Orthostatisme* investigation into blood pressure and cardiovascular regulation, and the *Illusion* experiment, which explored the adaptation of human sensory organs to the microgravity environment. At length, on the evening of 9 August 1992, he joined Viktorenko and Kaleri for the return to Earth. Their landing aboard Soyuz TM-14 was harsher than anticipated, about 130 km south-east of Jezkazgan, at 4:07 am Moscow Time on the 10th. "The ground was very hard," Tognini remembered of their summertime touchdown on the hard steppe of Kazakhstan. "We had a little bit of wind ... but nobody was hurt. After we landed, the capsule started to turn around. Everything became very unorganised, because we did not pack it up very well. The craft ended up on its side. I was in the highest part, Viktorenko was in the middle and Kaleri was on the bottom. We had some trouble getting Kaleri out of the capsule. It took close to 45 minutes." In total, Tognini had spent 14 days in space, compared to 145 days for Viktorenko and Kaleri.

Aboard Mir, Soyuz TM-15 crewmen Anatoli Solovyov and Sergei Avdeyev settled down for an extended mission, which they already knew would entail multiple EVAs. The stage was set for this spacewalking extravaganza in mid-August, when Progress M-14 arrived with a rather unusual payload: a 700 kg thruster package, known as the *Vynosnaya Dvigatelnaya Ustanovka* (VDU, or 'External Engine Unit'). It was to be mounted atop Kvant-1's Sofora girder in order to provide better roll control of the station. Two weeks later, on 2 September, automatic commands were transmitted from the ground to 'unload' the VDU and on the 3rd Solovyov and Avdeyev floated outside to begin work. Their first EVA lasted a little less than four hours and saw the two men install a locking device on Sofora to hold the truss securely whilst it was bent backwards. Next, on 7 September, they began the installation in earnest. Firstly, they bent back Sofora on its hinge a third of the way along its length and locked it into position to receive the VDU. Solovyov and Avdeyev laid a 14 m power cable and attached metal braces to secure the thruster package onto the girder. The 'real' world was not far away in the ethereal blackness, as the cosmonauts worked by helmet lamps in orbital darkness, for one of their tasks was to remove a Soviet flag – placed there in 1991 and since shredded by ultraviolet degradation and micrometeoroid impacts – from the top of Sofora. Part of their space-to-ground communications link was also severed when Ukrainian ground stations temporarily suspended their service to the Russian government. Neither of the cosmonauts was placed in any danger during this event and the EVA concluded after five hours. Finally, on 11 September, they spent a further six hours attaching the VDU to Sofora, straightening the girder and completing electrical connections. The VDU installation effort was originally scheduled to require four EVAs, but Solovyov and Avdeyev completed all of the tasks in three excursions. (Little was revealed at the time about how effectively the VDU supported Mir. One Russian specialist reported that it was "functioning normally", but it subsequently became clear that a software glitch had rendered it impossible to use.) A fourth spacewalk was performed, on 15 September, but for other activities: the men spent three and a half hours retrieving solar-cell samples from Mir's hull and transferred the Kurs rendezvous device from the 'front' to the 'rear' end of Kristall, ahead of its intended use by the US Space Shuttle. Mir had a T-bar configuration, with the Kristall module aligned perpendicular to the main axis of the station. At its far tip were two docking mechanisms, one 'axial' and the other 'radial', and it was intended that the axial port would be tested by Soyuz TM-16 crewmen Gennadi Manakov and Alexander Poleshchuk in January 1993.

In the meantime, with the EVAs behind them, Solovyov and Avdeyev settled down to an extended phase of research work aboard Mir. This encompassed semiconductor processing of barium and copper samples in the ovens of Kristall, geophysical observations of selected targets on Earth and photographic and spectrographic analyses of surface features in several of the former Soviet republics across Central Asia. It was announced that such imagery would enable the collection of more information pertaining to the ecological situation of agricultural areas and water basins. On 20 November, they also released the 16.5 kg MAK-2 subsatellite from Mir's scientific airlock to study Earth's ionosphere.

SEEDS OF FRIENDSHIP

The first year of the Commonwealth of Independent States had been fraught with drama, and on the space stage, little 'common' ground was seemingly found, with threats of strikes at the Kazakhstan-owned Tyuratam launch site and, in October 1992, efforts by Ukraine to privatise the Yevpatoria tracking and control station in its newly-established sovereign territory on the Crimea peninsula. Already, earlier in the year, it had been noted by at least one high-ranking Russian space official that Mir was in such a precarious state that it *could not* be mothballed; it needed a permanent crew in attendance to keep it alive. Then, in February 1993, only days after the launch of the Soyuz TM-16 crew, *Flight International* cited reports from Russia's Channel 1 TV that no engines were available for the next six boosters under assembly at Tyuratam. It seemed to many observers in the West that Mir's days were numbered ... and the entire Russian space programme along with it, for Buran's planned second flight had been scrapped, the new Spektr and Priroda modules had been repeatedly delayed and hopes of constructing a much larger 'Mir-2' seemed optimistic at best in such a dire economic and political situation. The saving grace for Mir was, ironically, the Soviet Union's old arch-enemy, the United States, with whom dialogue had already been established about conducting several joint crewed space missions. By the end of 1992, cosmonauts Sergei Krikalev and Vladimir Titov had begun training in Houston for a Shuttle mission and agreements had been signed for a NASA astronaut to fly to Mir in March 1995 and for the Shuttle to dock with the station in May-June 1995.

The latter mission was made possible by the Androgynous Peripheral Attach System (APAS)-89, situated on the end of the Kristall module, which was originally intended to support Buran. As a concept, APAS originated in designs by Soviet space scientist Vladimir Syromyatnikov for the Apollo-Soyuz Test Project and its basic premise was that, unlike a standard probe-and-drogue docking mechanism, its androgynous nature meant that its APAS attachment ring could mate with any other APAS attachment ring. In other words, each side of the mechanism could act in an 'active' or 'passive' capacity. During ASTP in July 1975, a series of shovel-shaped guide 'petals' were extended on the 'active' side and retracted on the 'passive' side to interact for gross alignment of the two coupling vehicles. The ring holding the guides shifted to align the active unit latches with those of the passive unit, after which shock absorbers and mechanical actuators dissipated residual impact energy on both sides and the active unit 'retracted' to bring both docking collars together. Undocking was achieved with a system of four spring-push rods. The subsequent Buran (and Shuttle)-class APAS on Kristall was reduced in external diameter from 2.03 m to 1.55 m, the internal pressurised tunnel was some 80 cm wide (like the other collars of the space station), the 'petals' were angled inward, rather than outward, and there were mechanical, fluid and electrical systems within the docking mechanism.

Work to prepare the APAS-89 on Kristall began during Anatoli Solovyov and Sergei Avdeyev's fourth EVA on 15 September 1992, when they repositioned the Kurs rendezvous antenna from the 'front' end to the 'rear' end of Kristall, close to

5-02: In a scene which should have featured Buran, this view of Atlantis docked at the end of Mir's Kristall module in June-July 1995 represented something which could hardly have been anticipated in the darkest days of the Cold War. Extending 'upwards' from the Shuttle's payload bay are Kristall, the multiple docking adaptor, the Mir base block and the Kvant-1 module. Extending from the radial ports of the multiple adaptor are Kvant-2 (left) and Spektr (right).

the APAS-89. Rather than a standard probe-and-drogue docking mechanism, Soyuz TM-16 was equipped with the APAS-89 hardware for the first time, in the first major test of the Kristall docking hardware. Soyuz TM-16 blasted off from Tyuratam at 8:58 am Moscow Time on 24 January 1993. Hopes of an Israeli guest cosmonaut joining the flight had come to nothing and the mission became the first in more than two years to have a basic two-person crew. Two days after launch, Gennadi Manakov and Alexander Poleshchuk prepared to become the first Soyuz to dock at a location other than Mir's forward or aft longitudinal ports. At a distance of 150 m, as planned, they disengaged the automated rendezvous system, closed to within 70 m of Kristall and completed a perfect docking shortly thereafter. According to Rex Hall and Dave Shayler, in *Soyuz: A Universal Spacecraft*, it provided data on docking a spacecraft off the longitudinal axis of a target vehicle. During the next few days, as the new crew worked jointly with Solovyov and Avdeyev, a series of 'Resonance' experiments were carried out to evaluate the dynamics and structural characteristics of the complex, which now boasted seven discrete components – the Mir base block, Kvant-1, Kvant-2, Kristall, Progress M-15, Soyuz TM-15 and Soyuz TM-16 – and exceeded 90,000 kg. Although no further Mir-bound and APAS-equipped Soyuz craft would dock with Kristall, the success of Manakov and Poleshchuk highlighted discussions in Houston, Texas, late in 1992, which envisaged the Soyuz as an Assured Crew Return Vehicle for Space Station Freedom.

For Manakov, the commander of Soyuz TM-16, it was his second mission, having flown for four months in 1990, but for Alexander Fyodorovich Poleshchuk it was his first voyage into space. He came from Cheremkhovo, in the Irkutsk region of Siberia, where he was born on 30 October 1953, and studied mechanical engineering at the Sergo Ordzhonikidze Moscow Aviation Institute. He also achieved the rank of captain in the Soviet (later Russian) Air Force Reserves. Poleshchuk initially worked for the Energia design bureau from 1977, focusing on repair and assembly techniques for space missions, and in January 1989 he was enrolled as a civilian cosmonaut candidate. By February 1991 he had qualified as a 'test-cosmonaut' and in March 1992 he was teamed with Manakov and Frenchman Jean-Pierre Haigneré as the backup crew for the Soyuz TM-15 'Antares' mission.

With the new team of cosmonauts now aboard Mir, the mission of Solovyov and Avdeyev drew inexorably toward its conclusion. Early on 1 February 1993, Soyuz TM-15 undocked from the station and underwent an uneventful re-entry ... and a decidedly precarious landing. Touching down near Jezkazgan at 6:47 am Moscow Time, after almost 189 days in orbit, this marked the longest mission by a Soyuz capsule to date, but after descending through low clouds it rolled partway down a hillock and came to rest just 150 m from a frozen marsh. Solovyov and Avdeyev's replacements in space had a busy six months ahead of them. One of the first tasks for Manakov and Poleshchuk was an activity with the unmanned Progress M-15 freighter, which undocked from Mir on 4 February and deployed a 20 m diameter foil reflector, known as 'Znamya' ('Banner') from its orbital module. This was a test for a future solar sail to illuminate and warm areas of Earth's surface at high latitudes when not in direct sunlight. Although the Znamya experiment lasted barely six minutes, it was visible in the skies above central Europe, from Lyon in southern France, through Prague in the Czech Republic, to Gomel in eastern Belarus. Observations continued for a period of 24 hours, with skywatchers in Canada catching a glimpse of the reflector, before it eventually burned up in the dense atmosphere. A new Progress arrived in late February, laden with supplies, including a new air-conditioning unit, replacement parts for Mir's computers and components for the communications system which linked the station with the Cosmos 2054 tracking and data-relay satellite in geosynchronous orbit. Manakov and Poleshchuk also installed new gyrodynes in Kvant-2 and successfully tested them.

In early March, *Flight International* outlined the details of three planned EVAs for the Soyuz TM-16 mission. The first excursion began on 19 April, when the cosmonauts opened the Kvant-2 airlock hatch and ventured outside to install solar array electric drives on the sides of Kvant-1. It was intended that these drives would later receive the retractable Kristall arrays. For the first time, the spacewalkers worked on a 'contractual' basis, with one source reporting that they were being paid a million roubles for three EVAs. Poleshchuk – who experienced problems with his suit's ventilation system during the spacewalk – manoeuvred himself along the full length of Kvant-2 to the base of the Strela boom, attached to the base block, whilst Manakov fixed himself to the end of the boom. Poleshchuk then swung the boom to move his colleague (and a container holding one of the electric drives) over to the installation site. The men experienced some difficulty attaching the drive onto its framework and Poleshchuk noticed that one of the two Strela control handles had become disconnected and floated away into space. It was realised with disappointment that a new handle would need to be sent from Earth, before the installation work could continue, and the cosmonauts returned to Mir after five hours and 25 minutes. One high-ranking Russian source reflected that "we will be sure to screw the handle on tighter next time".

This meant that the second EVA, originally planned for 23 April, had to be suspended until at least the end of May, but Manakov and Poleshchuk had much to occupy their time: routine maintenance on Mir's water-regeneration apparatus, its electricity supply and its computers. By this point, it was becoming clear that the cosmonauts were spending all of their work time for technical maintenance on the aging space station. On 24 May, Progress M-18 arrived with the replacement handle and other parts, including new pumps for the thermal-control system. This set the stage for the resumption of EVA work and on 18 June Manakov and Poleshchuk spent four and a half hours outside Mir and installed the second solar array drive mechanism, this time with few problems. So successful was the work that the cosmonauts were able to spend a few minutes videotaping the exterior of the station for downlink to mission controllers.

The expedition of Manakov and Poleshchuk was entering its homestretch and the next crew, together with Frenchman Jean-Pierre Haigneré – who had served as Michel Tognini's backup on the July 1992 flight – were shortly scheduled for launch aboard Soyuz TM-17. Yet the political and economic realities of life in post-Soviet Russia and its newly-independent republics continued. Shortly before the 5:33 pm Moscow Time liftoff on 1 July 1993, there was a temporary power blackout at the launch pad and the electricity supply in the nearby city of Leninsk failed completely. Aboard Soyuz TM-17 with Haigneré were two Russian cosmonauts: Alexander Serebrov, the flight engineer, was making his fourth space flight, but for Vasili Vasilyevich Tsibliyev, the commander, it was his first voyage. According to Bryan Burrough in his book *Dragonfly*, the relationship between the two men during training was as close as if they were family. Tsibliyev was born into a poor collective farming family in Orehovka, in the Crimean region of today's Ukraine, on 20 February 1954. Excited by the mission of Yuri Gagarin, he finished school and entered the army in 1971, later training as a Soviet Air Force fighter pilot. Tsibliyev graduated from the Kharkov Military School of Aviation in 1975 and later flew MiG jets along the Inner German border. His first attempt to enter the cosmonaut corps in 1976 was rejected, but after a deployment to an Odessa squadron on the Black Sea and eventual (after five applications) admission into test-pilot school he was successful in March 1987. In time, Tsibliyev would fly two long-duration Mir missions – the results of which were tarnished by two particularly unfortunate collisions with the station – and after his retirement from active duty rose to become chief of the cosmonaut detachment in Star City.

The presence of Jean-Pierre Haigneré on the Soyuz TM-17 crew made him the third Frenchman to board Mir ... and, interestingly, his backup on the mission, biologist Claudie André-Deshays, would later become his wife in 1998. Both had been selected into the CNES astronaut corps in September 1985 and, years later, both would be honoured by having Main Belt Asteroid 135268 named *Haigneré* for them. Born in Paris on 19 February 1948, Haigneré graduated in engineering from the French Air Force Academy at Salon de Provence and later qualified as a fighter pilot at Tours. He rose to become squadron leader on the Mirage V and Mirage IIIE aircraft. In 1981, he completed the Empire Test Pilots School at Boscombe Down in Wiltshire, England, winning the Hawker Hunter and Patuxent Shield awards, and subsequently served as project test pilot (and later chief test pilot) for the Mirage 2000N. As a pilot, Haigneré's accomplishments and qualifications were impressive: a colonel in the French Air Force, he flew 105 different types of aircraft during his career, with test-piloting and air-transport professional licences, Airbus A300 and A320 credentials, a helicopter private licence and mountain and seaplane ratings. Upon admission into the CNES astronaut corps, he headed the Manned Flight Division of the Hermes and Manned Flight Directorate, developing a parabolic flight programme, and in December 1990 was assigned as Michel Tognini's backup on the Franco-Soviet Soyuz TM-15 'Antares' mission. In the wake of this flight, Haigneré was assigned to the TM-17 flight, designated 'Altair'.

Arriving at Mir on 3 July 1993, the space station's population was temporarily increased to a five-man crew for almost three weeks. In fact, uniquely, the Progress M-18 supply craft undocked from the front port of Mir at 8:25 pm Moscow Time, bound for a destructive re-entry, and Soyuz TM-17 arrived to dock at the *same* port ... just 20 minutes later, at 8:45 pm. (It was noted at the time by Russian sources that Progress M-18 had remained attached to the station for longer than intended, as part of "a systems longevity test" to provide data which would help to determine the maximum time that a Soyuz could safely remain in space.) The complex programme of French experiments soon got underway and Haigneré reflected on 15 July that everything was going well, although the workload was excessive. The Frenchman experienced no symptoms of space sickness or nausea and reportedly slept and ate normally, although a space-to-ground conversation with President François Mitterand was cancelled due to a lack of available communications channels. A keen amateur radio user, Haigneré was unsurprised by a marked increase in French users contacting Mir during his mission ... but on one occasion, whilst talking to his parents, he had to ask one operator to leave the frequency so that he could hear his mother's voice. On another occasion, he expressed some annoyance with an Italian amateur, who used a very strong signal which often blocked the uplink channel for too long and limited Haigneré's ability to make contact with French and other radio operators.

On 22 July, Haigneré, Poleshchuk and Manakov – whose style as commander of Mir had reportedly seen him grasp the microphone tightly during each and every 'comm pass' over ground stations, sidelining his understudy, Tsibliyev – boarded Soyuz TM-16 and undocked from Kristall, bound for a touchdown in Kazakhstan. Viewing their departure through Mir's windows, Tsibliyev and Serebrov were able to see the penetration of the descent module into the 'sensible' atmosphere, together with the intense trail of plasma which enveloped it. Soyuz TM-16 thumped onto *terra firma*, near Jezkazgan, at 9:42 am Moscow Time, concluding the second-longest French space mission to date and a 179-day flight for Manakov and Poleshchuk.

By that summer, the plan was for Tsibliyev and Serebrov to spend four months in space, returning to Earth in November, after welcoming their own replacements, the Soyuz TM-18 crew of Viktor Afanasyev and Yuri Usachev. And *that* crew would itself be replaced, in May 1994, by Soyuz TM-19's Yuri Malenchenko and Gennadi Strekalov, joining a physician-cosmonaut, who would spend 18 months aboard Mir. Candidates for the last position included Valeri Polyakov (who had already completed a 241-day mission aboard Mir several years earlier), Gherman Arzamazov and Boris Morukov. According to *Flight International* in April 1993, the 18-month mission was expected to begin as early as December of that year. It was also hoped that Mir's next pressurised module, Spektr, would be ready for launch and incorporated into the gigantic complex before the end of 1993.

August was relatively quiet for the TM-17 crew, although the Perseids meteor shower, which peaked on the 11th and 12th, produced a spectacular display for them. In readiness for a possible emergency return to Earth, Russian aircraft and rescue forces were placed on alert, and Tsibliyev and Serebrov watched, around-the-clock, from Mir's windows as a total of 240 meteoroids burned up in the atmosphere. Several impacts were observed on the space station's windows, creating pit-like craters from 1.5 to 4 mm in diameter, and particle fluxes were 2,000 times higher than normal. Tsibliyev referred to them as "battle wounds" and noted that they had caused minor damage to solar panels on the base block and Kristall. Although Mir sustained no obvious structural damage, it was decided to stage an EVA in September to inspect the exterior.

The two cosmonauts spent more than four hours outside on 16 September, followed by another three hours on the 20th, primarily to assemble 'Rapana' – a 26 kg cylindrical girder, extendible to some 5 m – atop Kvant-1, which had design implications for Russia's planned Mir-2 station. It was expected that the latter would feature a large truss structure to support solar dynamic power systems and antennas and although the demise of Buran and a lack of funding threatened its own cancellation, there was hope that at least some elements of Mir-2 would be incorporated into Space Station Freedom as a joint project. Then, on the 28th, they carried out a two-hour inspection, known as 'Panorama', in which a 5 mm hole was spotted in one of Mir's solar arrays. The damaged area was surrounded by cracks, running for several centimetres, but the cosmonauts were unable to determine if a Perseid strike was responsible. This EVA was scheduled for four hours, but ended earlier than planned when a cooling issue was experienced with Tsibliyev's suit; he was forced to remain close to the Kvant-2 airlock, whilst Serebrov completed the photography of Mir and collected detector plates from a NASA-provided exposure experiment. A fourth and fifth EVA on 22 and 29 October concluded the Panorama inspections and enabled them to examine the entire outer skin of Mir. In the excursion on the 22nd, Serebrov suffered a problem in the oxygen flow system of his Orlan-DMA suit, which had been worn 13 times by previous cosmonauts and had exceeded its recommended operational lifetime. As a consequence, the spacewalk was curtailed and the cosmonauts returned to the Kvant-2 airlock after just 38 minutes. The final EVA on the 29th experienced no such problems, however, and Serebrov established a world record for the most spacewalks by one person, with ten in total. They ended the excursion by tossing overboard the Orlan suit which had caused problems on 22 October ... after rigging it so that it appeared to be saluting, "like a soldier". The results of the Panorama observations indicated that the station had suffered many small impact events, but none had penetrated the hull. Tsibliyev and Serebrov considered the micrometeoroid damage to be benign, although their primary concern was the state of the interior of Mir, whose life-support systems required near-constant attention. Still, the inspections provided a measure of assurance for US and European engineers, both of whom would be despatching their astronauts to Mir in 1994-95, that the aging station could adequately support their experiments.

At around this time, in October 1993, it was announced by the Russians that the Soyuz TM-17 crew would not return to Earth in November, but were to remain aboard Mir until the following January, due to problems obtaining engines for the powerful Soyuz-U variant of the venerable R-7 rocket which would loft their replacements. Budget cuts were acknowledged to have delayed the manufacturing of the engines at the factory in Samara, whose managers refused to deliver them until it had received payment from the Russian government. Tsibliyev and Serebrov agreed "reluctantly" to this extension of their mission. It was further added that the long-duration flight by the physician-cosmonaut – with Polyakov identified as the prime candidate, backed-up by Arzamazov – would also begin on Soyuz TM-18, whose launch was postponed from 16 November until 8 January. As a result, Polyakov's mission was shortened slightly to 14 months. The relationship between the prime and backup physicians became decidedly frosty. It was reported in January 1994 that Arzamazov accused Polyakov of being unqualified to fly. "Arzamazov says that Polyakov has not practiced medicine for a long time," *Flight International* noted, whilst quoting Russia's deputy health minister as saying that the disgruntled backup had "psychological problems". In any event, Arzamazov was discharged from the cosmonaut corps in late 1995, following disciplinary procedures.

THE PARTNERSHIP EVOLVES

If 1992 was the year in which the initial contracts for US-Russian co-operation in the human space exploration arena were signed, then 1993 saw those agreements accelerate at a rapid clip. Attention was already being devoted to including Russia in the International Space Station project – the new name for what had been Space Station Freedom – and the evolution of this partnership will be explored in greater depth in the final volume of this series. Yet there were other potential missions under consideration. NASA was considering diverting a future Shuttle flight, STS-63 in May 1994, from a standard 28.5-degree-inclination orbit into a 51.6-degree path, in order that it could rendezvous with Mir, 'station-keep' at a distance of 200 m and evaluate radio communications links, as a preliminary to the actual docking mission. This coincided with a summit in Vancouver, Canada, that same April, between US President Bill Clinton and Russian President Boris Yeltsin, which spurred discussions to stage not one, but *several*, long-duration missions to Mir by NASA astronauts. Initially, the talk was of a pair of three-month flights and as many as four others, lasting perhaps six months apiece, and it was agreed that the new Spektr and Priroda modules would be utilised (and subsidised) for US research equipment. The Shuttle-Mir docking mission itself, scheduled for May-June 1995, was also the subject of some controversy, particularly when former astronaut Guy Gardner, in his capacity as head of NASA's Russian programmes, recommended using the mission to transport two cosmonauts to the station for their expedition and bring home the long-duration US crewman. The Russians were concerned that such a move might "interfere with training regimes", although the potential benefits of a Shuttle visit were enormous: including the capacity to transport 500 kg of much-needed equipment to the space station and return 1,000 kg of failed gyrodynes for engineering inspections. Joint EVAs were also considered and it was explained by Clay Morgan, in his seminal work *Shuttle-Mir: The United States and Russia Share History's Highest Stage*, that such collaborative work would serve "to extend Mir's useful lifetime through the end of 1997".

As will be explored in the final volume of this series, the US-led Space Station Freedom project had long been teetering on the brink of collapse and was forced through an expensive – and bitterly divisive – process of redesign in early 1993, under the direction of the new president, Bill Clinton. The United States had partnered with Japan, Canada and the European Space Agency, all three of whom were providing pressurised and unpressurised components for Freedom ... and the Europeans were becoming increasingly frustrated that their home-grown space station module, Columbus, faced the axe of cancellation. To hedge its chances of maintaining a viable human space flight agenda, in May 1993 ESA opted to proceed with a pair of lengthy 'EuroMir' missions at a total cost of $53 million. The first, scheduled for September 1994 on Soyuz TM-20, would feature a European cosmonaut spending a month aboard Mir, and the second, due to begin in August 1995 on Soyuz TM-22, would last 135 days ... by far the longest flight by a non-Russian at that time. The missions were considered to be precursors for Europe's work aboard the International Space Station's Columbus module, although funding

5-03: Pictured during STS-42, his second Shuttle flight, Germany's Ulf Merbold became the first non-Russian and non-US spacefarer to fly aboard both Russian and US spacecraft. His 30-day stay aboard Mir in October-November 1994 marked the beginning of the ambitious 'EuroMir' programme.

limitations would require the EuroMir-94 cosmonaut to rely heavily on equipment left aboard the station by previous European residents Franz Viehböck, Klaus-Dietrich Flade, Michel Tognini and Jean-Pierre Haigneré to perform his 23 experiments in life and materials science and technology. A total of 140 kg of science hardware, lofted aboard a Progress freighter, would precede the launch of EuroMir-

94, with a further 10 kg accompanying the cosmonaut himself aboard the Soyuz-TM spacecraft. The limited volume available inside the descent module meant that only around 16 kg of experiment samples could be brought back to Earth, with the remainder staying aboard Mir for return by the Shuttle, several months later in mid-1995.

Candidates for the two flights included veteran German astronaut Ulf Merbold and Spain's Pedro Duque for 'EuroMir-94' and Thomas Reiter and Christer Fuglesang, of Germany and Sweden, respectively, for 'EuroMir-95'. Moreover, an EVA was manifested into the second mission, offering Germany or Sweden the chance to become the fourth discrete nation (after Russia, the United States and France) to have its own spacewalker. In August 1993, the four men arrived at Star City, near Moscow, and their gruelling syllabus emphasised technical preparation, biomedical activities and more than 530 hours of crew-specific mission training. "Studying the Mir systems is interesting," admitted Merbold, "but I would have liked to spend more time preparing for the experiments which we shall be carrying out on the mission." His three comrades added that the cultural differences were a challenge – "The Russian way of thinking," explained Duque, "is not the same as ours" – as well as the need to learn the complex Cyrillic alphabet and language. Alongside EuroMir, France had also agreed to pay $25 million for another two-week flight in mid-1996. At the same time, ESA's unmanned projects suffered. The European Retrievable Carrier (EURECA) – a free-flying microgravity platform, deployed and recovered by the Shuttle – had been earmarked for a second mission, but EuroMir and the possibility of ESA involvement in a future Spacelab flight led to its cancellation. "A EURECA-2 launch and retrieval by the Shuttle," lamented *Flight International* in May 1993, "would have cost about $250 million, but would have allowed a nine-month flight with an array of autonomous experiments, compared with a possible ten-day Spacelab mission, costing about the same." It was a price to be paid for the 'realpolitik' of space exploration.

Several months later, in September 1993, US Vice President Al Gore and Russian Prime Minister Viktor Chernomyrdin chaired the first meeting of the US-Russian Joint Commission on Energy and Space. The two political leaders agreed to begin 'Phase 1' of International Space Station co-operation with the Shuttle-Mir project, encompassing two years of total US astronaut time aboard Mir. Within weeks, an addendum had merged the American, European, Japanese and Canadian components from the former Space Station Freedom with Russia's planned Mir-2. Under its provisions, 'Phase 2' would see the construction of the International Space Station and 'Phase 3' would see full operations as an international scientific research facility. Gore visited the Russian mission control centre in Kaliningrad in December and it was highlighted at the time that American financial support was the only lifeline left which might save the decrepit Mir. On the 16th of that month, NASA Administrator Dan Goldin and his Russian Space Agency counterpart, Yuri Koptev, signed the $400 million 'Contract for Human Space Flight Activities' and Shuttle-Mir was open for business. As the ink on the contract dried, it became clear that ten Shuttle-Mir missions would fly, including the STS-63 rendezvous, with a joint solar dynamics power system being assembled in 1996 and at least four US astronauts spending a total of 24 months aboard the space station. The ultimate goal of the new partnership would be the beginning of International Space Station construction in late 1997 and a fully-operational, six-person outpost by 2001.

Almost immediately, the grandiose plan slipped. NASA announced that STS-63 would be delayed from mid-1994 until at least January 1995, due to a paucity of bookings to conduct experiments in its primary payload, the commercial Spacehab-3 pressurised research module. In the months to come, experiments already assigned to the third and fourth planned missions were combined and produced a full plate of mainly-NASA payloads for STS-63. "Spacehab's position reflects the lack of interest in commercial microgravity processing," explained *Flight International*, "which has been caused mainly by the high cost of experiments and the long-term nature of the research to develop viable processes."

Riding the coattails of this new effort of collaboration in space, Soyuz TM-18 roared into orbit at 1:05 pm Moscow Time on 8 January 1994, carrying cosmonauts Viktor Afanasyev in the commander's seat, Yuri Usachev as flight engineer and Valeri Polyakov, the physician, on the first stage of his marathon mission. Two days later, at 1:50 pm Moscow Time on the 10th, Afanasyev guided his ship to a smooth docking at Mir and radio communications picked up by amateur listeners established that the mood and health of the new arrivals was good. By now, the delayed availability of the Soyuz-U rocket, together with plans for flying the first NASA astronaut to the station in March 1995 and the schedule of the first Shuttle docking in May-June, had truncated Polyakov's voyage from 18 to 16 to around 14 months, although this still represented a 16-percent empirical duration increase over the 366-day record set by cosmonauts Vladimir Titov and Musa Manarov in December 1988.

For almost two decades, Valeri Vladmirovich Polyakov has held the world record for the longest single space mission, spending a total of 437 days in orbit aboard Mir from January 1994 until March 1995. When added to the 241 days he also spent aboard the space station in 1988-89, Polyakov also ranks within the top five on the list of the world's most experienced astronauts and cosmonauts, with a personal achievement of 678 days (more than 22 months). Yet the name which today continues to grace the Guinness Book of Records is not even his birth name. He was actually born Valeri Ivanovich Korshunov on 27 April 1942 in the industrial city of Tula, a couple of hundred kilometres south of Moscow, but changed his middle patronymic and his family name to 'Vladimirovich' and 'Polyakov' at the age of 15, when he was adopted by his stepfather. Polyakov completed his secondary schooling in Tula in 1959 and entered the I.M. Sechenov First Moscow Medical Institute, from where he gained his doctorate. He then specialised in astronautical medicine at the Institute of Medical and Biological Problems at Moscow's Ministry of Public Health, his fascination with how the human organism adapts to the microgravity environment having been inspired by the flight of the first medical specialist, Boris Yegorov, aboard Voskhod 1 in October 1964. A little over seven years later, in March 1972, Polyakov was selected, along with two other physicians, Georgi Machinsky and Lev Smirenny.

However, due to his ongoing postgraduate research towards a candidate of medical sciences degree, Polyakov did not begin his cosmonaut training until October 1972. He received his degree in 1976 and joined another cosmonaut selection, two years later, to complete the remainder of his training. He was an early candidate for a dedicated medical research mission to Salyut 6 in late 1980 and served as physician Oleg Atkov's backup on the long-duration Soyuz T-10B flight in 1984. He might also have flown aboard the Soyuz T-13 mission, but was replaced by military engineer Alexander Volkov.

Several days after the arrival of Polyakov, Afanasyev and Usachev, the time came for the departure of Soyuz TM-17 and the outgoing resident crew of Vasili Tsibliyev and Alexander Serebrov. Undocking from Mir's forward port occurred without incident at 7:37 am Moscow Time on 14 January and Tsibliyev commenced his assignment of a short inspection flight around the station, prior to departing. This task required the commander to assume manual control, withdraw to a distance of 45 m, then steer his spacecraft to within 15 m of Kristall to photograph the APAS-89 system for NASA. Nine minutes after undocking, at 7:46 am, Tsibliyev radioed his first complaint that something was amiss. The controls handled "sluggishly" and Serebrov voiced his own concern that they were drawing dangerously close to Mir's solar arrays.

"For some reason," wrote Bryan Burrough in *Dragonfly*, "his thruster control button momentarily froze. Unable to control the ship, Tsibliyev watched in amazement as it floated slowly toward the station." Aboard Mir, Afanasyev urgently instructed Usachev and Polyakov to prepare their vehicle, Soyuz TM-18, for an evacuation, certain that a collision was about to occur. The loss of control occurred at a distance of about 30 m and at 7:47 am Soyuz TM-17 hit Mir, not once, but *twice*, with a two-second gap between each impact. The collision was very slight, and almost imperceptible for the cosmonauts, as Soyuz TM-17 quickly rebounded away. Aboard the station, Afanasyev, Usachev and Polyakov also felt nothing, but Mir's attitude-control system registered the angular velocity and automatically switched to free-flying mode. Immediately after the impact, radio communications with ground controllers were lost for a period of about ten minutes. These were only fully restored, after a period of spotty comm, at 8:02 am. Subsequent discussions revealed no damage to either craft and Soyuz TM-17 began its descent to Earth at 10:15 am Moscow Time. An hour later, at 11:24 am, the descent module hit the steppe of Kazakhstan, 130 km west of Karaganda, and Tsibliyev and Serebrov were safely home after 197 days in space.

In the aftermath, it was considered prudent for Afanasyev and Usachev to execute an EVA to inspect the impact site – thought to lie close to the junction between Kristall and Mir's base block – and verify its integrity. Post-flight investigations would blame a switch error: the hand controller aboard Soyuz TM-17's orbital module which handled acceleration and braking was switched on, which effectively disabled the equivalent controller in Tsibliyev's hands in the descent module. Although the commander was able to use his other hand controller to avoid the solar arrays and docking ports, he could not react rapidly enough to avoid hitting Mir. His expert piloting actually prevented more serious damage. In *Dragonfly*, Burrough added more intrigue to the story, writing that TM-17 touched down more than 100 km from its predicted point, which was blamed upon the excessive weight of returned cargo and other 'personal' items from Mir. On 24 January, whilst relocating Soyuz TM-18 from the aft port to the front port, Afanasyev, Usachev and Polyakov reported seeing only a handful of scratches on the hull where their predecessors' craft had collided.

Launched in January 1994, with the Soyuz TM-18 crew, Polyakov knew that he would return to Earth with different comrades, for Afanasyev and Usachev were assigned a standard six-month mission and would land in July. Another crew – Soyuz TM-19's Yuri Malenchenko and Kazakh cosmonaut Talgat Musabayev – would then occupy Mir from July-November 1994 and a third, that of Soyuz TM-20, with Alexander Viktorenko and the first female long-duration resident, Yelena Kondakova (the second wife of veteran cosmonaut Valeri Ryumin), until March 1995. As a consequence, Polyakov's extra-long flight would span no fewer than *three* full Mir expeditions. During his 14 months aboard the old space station, Polyakov and his transient crewmates would oversee 25 experiments, mainly in the life sciences. Specific focuses included the diet, the function of the human muscular system, the lungs and the immune system in the peculiar microgravity environment. Changes in the blood and central nervous system were examined, together with problems with metabolism, alterations in blood volume and the function of the sense of balance in the inner ear. One investigation utilised the German 'Video OkluGraphie' equipment, used by Klaus-Dietrich Flade in March 1992, and again by Soyuz TM-20's 'EuroMir-94' astronaut Ulf Merbold in October 1994. Still other investigations explored calcium depletion in the bones, with the KARKAS 'advanced vacuum trousers' used to draw blood into the abdomen to simulate the higher volumes in this area due to terrestrial gravity. Whilst Afanasyev and Usachev were aboard, Polyakov took measurements every third day of variations in the circumference of their legs, their blood pressure, their cardiac output and the changing position of their hearts. The cosmonauts' sleep patterns were also observed, as was their psychological behaviour, including reaction rates, short-term memory and manual skills. Materials science, Earth observation, astrophysics and biotechnology assignments also consumed a large portion of their time.

With both Polyakov and Afanasyev making their second space missions, the only 'rookie' member of the Soyuz TM-18 crew was Yuri Vladimirovich Usachev. Born in Donetsk in eastern Ukraine on 9 October 1957, he grew up wanting to become a pilot and later studied mechanical engineering at the Moscow Aviation Institute. Following his graduation in 1985, Usachev worked for the Energia design bureau and it was whilst there that he first came into contact with a group of real cosmonauts. "I thought this could be a good idea to try to work like them, like they did," he told a NASA interviewer, "and I tried to pass some exams, some medical tests, and there you go." Of course, his route into the cosmonaut corps was arduous, but he was selected as a civilian engineer in January 1989. Following two years of general training, he was assigned to the backup crews for Soyuz TM-16 and, later, Soyuz TM-17, before securing a position on Soyuz TM-18 for his first mission. Asked years later to reflect on the most important attributes a cosmonaut requires for long-duration flight, Usachev was philosophical and paid tribute to his mother.

"I think my mother may have more influence," he said, "because she is very patient. It's very useful for me to have enough patience to live six months in space!"

Patience was most definitely a virtue for the three cosmonauts, as they settled down for a protracted period aboard Mir. One of their earliest tasks was the relocation of Soyuz TM-18, originally scheduled for 21 January, but delayed until the 24th to permit a longer inspection of Kristall and the impact point of Tsibliyev and Serebrov's craft. Their observations and photographs convinced ground controllers that an EVA inspection would be unnecessary. In the meantime, the first of three Progress visitors arrived at the end of the month, but the second – Progress M-22, originally scheduled for launch on 16 March – was postponed due to heavy blizzards at Tyuratam. The rails used to transport the rocket the short distance from the assembly building to the pad were covering with snow drifts, some 7 m deep. (A fire also broke out in one of the assembly and test complexes on 7 March and spread to the maintenance unit headquarters.) The launch eventually took place on the 22nd and the Progress successfully docked with Mir two days later.

Activities aboard the station were described as "routine" during this period, with a multitude of experiments and observations of fires along the borders of Siberia and Mongolia, but the situation on Earth remained in flux. Following the agreements to include Russia in the new International Space Station project, Tyuratam received its first visit by a US government official in March 1994, when senior defence secretary

5-04: Pictured in Russian Sokol launch and entry suits, astronauts Norm Thagard (left) and Bonnie Dunbar participate in Soyuz-TM training in September 1994.

William Perry toured the complex. Three weeks later, contracts were signed between Russia and Kazakhstan, in which the former leased Tyuratam from the latter for launch services at a cost of $115 million per annum. The agreement came at the end of two years of bitter dispute, in which the Kazakhs initially demanded a fee as high as $7 billion per year and the Russians angrily reacted by threatening to cut their losses and build an alternate launch site at Svobodny in the Amur region of southern Siberia. "The deal lifts the tension between the two countries," noted Tim Furniss in *Flight International*, "which has affected the operational state of the space centre." Yet even this deal did little to detract from the fact that Tyuratam was in a pitiful state: the buildings and infrastructure were decaying, staff morale had hit rock-bottom – having had half of its 100,000-strong workforce laid off – and the 7 March fire had caused an estimated $1.75 million in damage. The visit of William Perry highlighted the importance of renewed investment at Tyuratam, ahead of the launch of the first US astronaut – identified in February 1994 as four-flight veteran Norm Thagard, backed-up by Bonnie Dunbar – to begin NASA's series of Mir missions in March of the following year.

At around the same time, the General Accounting Office in the United States was highly critical of the increased costs associated with collaborating with Russia, including the need to upgrade the Shuttle with rendezvous and docking hardware. These criticisms were rebuffed by NASA Administrator Dan Goldin, who described Shuttle-Mir as a "win-win" situation, both in terms of the partnership and the prospects for the International Space Station. By early June 1994, five astronauts were assigned to STS-71, the first Shuttle docking mission, and a cadre of Russian cosmonauts arrived in Houston, Texas, to begin training on US systems. Amongst their number was 53-year-old Gennadi Strekalov, who had already flown four times into space, and was scheduled to launch aboard Soyuz TM-21 in March 1995, shoulder-to-shoulder with Thagard and fellow cosmonaut Vladimir Dezhurov. Strekalov's involvement in the programme is an interesting one, for in January 1994 he had been identified as the third crewman aboard Soyuz TM-19. His retention made him one of only a few cosmonauts with much to smile about in 1994, for in the middle of the year more than half of Russia's corps was axed, ostensibly due to budget cuts.

Strekalov's removal from TM-19 – whose roster also included 'rookie' cosmonauts Malenchenko and Musabayev – left this mission as the first Russian flight since 1977 to launch into orbit with no experienced crew member aboard. It was claimed that the decision to remove Strekalov was purely "economical" and had been taken because "his seat will be used for an extra amount of cargo", amounting to around 85 kg ... but it was left unsaid as to why *he* lost his seat and not Musabayev. In the days following the announcement, Russia stressed that Musabayev's place on the mission was secure because he was "a good cosmonaut" and it was not a cynical piece of politics to placate Kazakhstan. However, looking back with the benefit of hindsight, it seems a little too convenient that the decision to retain Musabayev, at the expense of the vastly experienced Strekalov, was made in such close temporal proximity to the signing of the leasing contracts over Tyuratam. These suspicions about the political undercurrents behind the flight seemed vindicated on 25 June, when Kazakhstan insisted that Musabayev was flying "as a representative of Kazakhstan" and Russia replied that he would indeed be aboard Soyuz TM-19 "as a foreign guest-cosmonaut". It was also revealed that, although Russia paid for Musabayev's training, the government of Kazakhstan would contribute to the cost of his work in orbit. Musabayev, though, had few problems with the political status of his flight; whether Russian, Russian-Kazakh, CIS or international, he regarded himself as a representative of humanity and emphasised that nothing could hamper the "good friendship" between the Russian and Kazakh people.

As for Mir itself, cosmonauts Afanasyev, Usachev and Polyakov continued their work, receiving the third Progress of their expedition in late May and a load of supplies which totalled around 2,000 kg. Concurrently, the Russians announced a slight delay to the Soyuz TM-19 launch, from 26 June to the early part of July, due to problems with the production of the payload fairing for the rocket. The liftoff eventually occurred at 3:24 pm Moscow Time on 1 July, carrying the two-man crew of Malenchenko and Musabayev towards Mir. Soyuz TM-19 reached the vicinity of the station early on the morning of the 3rd and docked at the aft port of Kvant-1 without incident at 4:55 pm Moscow Time, expanding the crew of Mir to five members.

If the inclusion of Kazakhstan's Musabayev seemed political, then ongoing wrangling between Russia and Ukraine over the future of the Yevpatoria control centre in Crimea made the presence of Ukrainian-born Yuri Ivanovich Malenchenko similarly notable. Today, Malenchenko – born on 22 December 1961 in Svetlovodsk, on the banks of the Dniepr River, in central Ukraine – has established himself as one of the top ten most experienced spacefarers of all time, with 641 days away from the Home Planet under his belt, spread across five missions between 1994 and 2012. Malenchenko attended local schools and later admitted to a NASA interviewer that his original career goal was "to have a profession which would not be boring ... something dynamic, something ever-changing and challenging". Early interests included sailing and, later, flying. In terms of education, he majored in mathematics, but changed paths and graduated in 1983 from the Kharkov Higher Military Aviation School as a pilot-engineer. In his military career, Malenchenko served as a pilot, senior pilot and multi-ship flight lead in the Soviet Air Force, before being accepted into the cosmonaut corps in 1987. He later served on the Soyuz TM-18 backup crew and, whilst training for that mission, he graduated from the Zhukovsky Air Force Engineering Academy. "I never regretted it," Malenchenko said years later of his decision to become a cosmonaut. "It requires a lot from a person. However, at the same time, it is a very rewarding profession, because the tasks you need to complete are very important, complicated and require a lot of responsibility from different points of view."

Alongside Malenchenko was Talgat Amangeldyevich Musabayev, the second ethnic Kazakh from an independent Kazakhstan to venture into space. He came from the north-western district of Kargaly, where he was born on 7 January 1951. As a citizen of the Soviet Union, Musabayev graduated from the Engineering Institute of Civil Aviation in Riga, within the territory of today's Latvia, in 1974, and later completed Higher Military Aviation School in Akhtubinsk in 1983, from where he obtained an engineering diploma. He was selected for cosmonaut training in May 1990, then promoted to the rank of major and transferred to the Soviet Air Force cosmonaut group in March of the following year. Musabayev served as backup to his fellow Kazakh countryman, Toktar Aubakirov, on Soyuz TM-13, after which he entered training for a long-duration mission to Mir.

That long-duration mission really began on 9 July 1994, when the Soyuz TM-18 spacecraft undocked from Mir, carrying Viktor Afanasyev and Yuri Usachev, and touched down safely at 1:33 pm Moscow Time, about 110 km north-east of Arkalyk in Kazakhstan. Their mission had lasted 182 days. Original plans called for Progress M-24 to arrive at the space station in mid-July, but the launch was postponed until late August. Part of its cargo was to be equipment for ESA's EuroMir-94 mission – whose prime cosmonaut, German astronaut Ulf Merbold, the first person to fly aboard both Russian and US manned spacecraft – had been formally announced in June. For the three men aboard Mir in the summer of 1994, an equally important part of its cargo was their own much-needed supplies. "The cancellation of the M-24 has brought complaints from the crew," explained *Flight International* on 10 August. "Having to drink recycled water and eat food past its sell-by date is making the crew unhappy." Already, an earlier Progress had arrived at Mir without most of its food supply, having been raided and the food stolen by launch-pad technicians. Power shortages at the nearby workers' city of Leninsk, coupled with spartan living conditions and economic cuts, had left the workforce truly "desperate", their morale decimated. This incident offered yet another indicator that space exploration in post-Soviet Russia and the troubles of the real world, back on Earth, were intrinsically linked together.

After roaring into orbit on 25 August, Progress M-24 reached the vicinity of Mir on the 27th. Then, during the final proximity operations, ahead of docking, it was recognised that a problem existed with the cargo craft and flight controllers opted to maintain Progress in an autonomous state whilst troubleshooting got underway. A second approach on the 30th went ahead, but failed when the ship actually impacted Mir on a handful of occasions, producing slight shocks which were felt by the three cosmonauts. It was noted at the time that further problems would have dire consequences, as the cargo ship included not only 140 kg of EuroMir-94 equipment, but also US equipment and other supplies for Malenchenko, Musabayev and Polyakov. On 2 September, it was decided to proceed with the docking, using Mir's TORU system, which enabled a cosmonaut to remotely control the Progress. The system comprised a pair of hand-controllers in Mir's base block and the cosmonaut was aided by televised views provided by a camera in the Progress docking mechanism. Under Malenchenko's expert control, the cargo craft was eventually brought in for a smooth docking at the forward port of the multiple adaptor. Years later, Malenchenko reflected on his actions for a NASA interviewer, blaming the initial failure upon a problem with Progress M-24's docking system. "Then we tried the second attempt," he said of the effort on 30 August, "and everything went nominally up to 10-15 metres. At that point, the Kurs system ... did not work correctly and it was giving us the wrong messages. Everything was *right*, except that the data that we were receiving from the Kurs system was *wrong*." The Progress seemed to veer to the left of its intended course and Malenchenko remembered that flight controllers were none too keen on using TORU for the first time under such circumstances, but they looked into his experience and asked him if he felt comfortable with the task. For Malenchenko, it was a question with only one answer and the success brought him much-deserved praise.

In spite of the flawless docking, Progress M-24 had impacted Mir on at least two – and perhaps as many as four – occasions, only lightly, but sufficient for an inspection to become necessary. Four EVAs were manifested for Malenchenko and Musabayev, for the purpose of transferring solar arrays from the Kristall module over to Kvant-1, but by late August this assignment had been truncated to just two excursions. On 9 September, the spacewalkers ventured outside for a little over five hours and made a detailed analysis of the glancing damage caused by Soyuz TM-17 in January and, more recently, by Progress M-24. Inside Mir, Polyakov watched their progress. They reported that Tsibliyev and Serebrov had caused light scuffing to the junction between Kristall and Mir's base block, knocking off a 30 cm × 40 cm thermal blanket and leaving a few minor scratches, and found no appreciable damage from the Progress collision. Malenchenko and Musabayev returned outside on 14 September for six hours, during which they set to work on the effort to reconfigure Mir, ahead of the first Shuttle docking in mid-1995. The men inspected mounting brackets and other equipment on Kristall's movable arrays, which were scheduled to be retracted and transferred to Kvant-1, thereby removing a potential obstacle for the Shuttle.

Obstacles remained elsewhere, however. Mir's next additional module, Spektr, already long-delayed, had no chance of making it into orbit before the end of 1994 and it was not expected to be ready until early or even mid-1995. (By October 1994, Spektr's launch had slipped still further from February to no earlier than 10 May.) Since the module would carry 1,130 kg of US research equipment to be used by Norm Thagard during his 90-day mission, the delay – for which NASA admitted partial responsibility – proved a decidedly bitter pill for the Americans to swallow on what would be their first long-duration mission in more than two decades. Nevertheless, the Americans opted to retain Thagard's original flight schedule and his launch aboard Soyuz TM-21 was slipped only by a matter of days from 3 March until 14 March 1995. Rumours of a delay to the EuroMir-94 flight proved unfounded and at 1:42 am Moscow Time on 4 October Soyuz TM-20 launched with Alexander Viktorenko in command, joined by female flight engineer Yelena Kondakova and Germany's Ulf Merbold. An expansion of Mir's crew from three to six members, for the relatively lengthy period of 30 days, required attention to the life-support systems, including the waste collection system, which had to be adapted to accommodate its first long-duration female resident. In fact, Kondakova became only the second woman – after Britain's Helen Sharman – to board the aging space station.

Yelena Vladimirovna Kondakova, born on 30 March 1957 in Mitischi, within the Moscow region, actually became only the third Soviet or Russian woman to venture into orbit, after Valentina Tereshkova and Svetlana Savitskaya. At the time of writing, in early 2013, she also remains the most recent, although Yelena Serova is scheduled for a six-month mission to the International Space Station in 2014-15. Kondakova graduated from Moscow Bauman High Technical College in 1980 and worked for several years for the Energia design bureau, before being selected into the cosmonaut corps in January 1989. By contrast, her commander on Soyuz TM-20, Alexander Viktorenko, was making his fourth flight, and Ulf Merbold his third, after two previous Shuttle missions.

The three new arrivals reached Mir at 3:28 am Moscow Time on 6 October 1994 and successfully docked at the station's forward port. During the final approach, at a distance of 130 m, a malfunction of Soyuz TM-20's computer caused the craft to yaw unexpectedly and prompted Viktorenko to assume manual control. It was reported that observers at ESA's facility in Cologne were "amazed" at how quickly and skilfully he manoeuvred into position and completed the picture-perfect docking. Soyuz TM-20 had suffered the same sort of anomaly during its final approach as did Progress M-24: the computerised flight control system, common to both vehicles, had been improperly programmed with the centre of mass of the spacecraft. The result was that as they made their final alignments, the Soyuz and Progress experienced unexpected rotational motions.

5-05: Mir's Spektr ('Spectrum') module, launched in mid-1995, would form the quarters of the NASA astronauts and the location for much of their research activities.

Events aboard Mir itself over the following days proved anything but picture-perfect, for on 13 October the crew were unable to activate a video camera and television lights whilst recharging Soyuz TM-20's batteries. A short circuit had disabled the computer which controlled Mir's solar arrays, forcing the station to drain its batteries. The consequence was that all systems in the base block were shut down, including attitude-control hardware and the crew's ability to direct their antenna for communications via the geosynchronous-orbiting Luch satellite was impaired. Reports from Russian sources were mixed: some blamed the aging nature of Mir as having caused the situation, whilst others emphatically declared that the incident would have no adverse impact on EuroMir-94, and Viktorenko and Kondakova even joked at one stage that at least the exchange rate of the US dollar remained stable. For a while, it was the only thing that *did* remain stable, as the batteries were recharged and engineers sought to adjust the station's attitude to permit the solar arrays to function as efficiently as possible. By 15 October, normal conditions had been restored and, speaking in December, Deputy Flight Director Yuri Antoshechkin explained that only the core module had been affected. Many of Merbold's experiments which required substantial electrical power loads were shifted to the tail end of the mission.

The flight of EuroMir-94 proceeded, notwithstanding these problems, and was actually extended by 24 hours when a decision was taken to duplicate the Progress M-24 approach anomaly with the departing Soyuz TM-19. Merbold reported that all of his medical and biological tasks ran smoothly and that none of the samples suffered any damage, even when the ESA-provided freezer – capable of holding around 100 blood, urine and saliva specimens for return to Earth in a frozen state – temporarily shut down as a result of the power outage. At the opposite extreme, EuroMir-94's Czech-provided CSK-1 materials-processing furnace malfunctioned and its five planned experiments were delayed. It was revealed that spare parts for the oven would be delivered by a subsequent Progress and the work would be resumed at a later date. On 2 November, with Malenchenko, Musabayev and Merbold aboard, Soyuz TM-19 undocked from Mir for a period of 36 minutes in a demonstration of the automatic approach system. No anomalies were detected and, two days later, on the morning of the 4th, the three cosmonauts departed the station for their return to Earth. They touched down on *terra firma*, 170 km north-east of Arkalyk, at 2:18 pm Moscow Time. Reflecting on his first space mission, Yuri Malenchenko later expressed awareness of its "challenging" nature, but remembered with greater fondness the "good memories about the time I spent there and the crew members I worked with".

FOOTSTEPS TO THE FUTURE

As 1994 drew to a close, Mir experienced something of a quiet spell, ahead of the most dramatic year of its operational life: 1995 and the dawn of Shuttle-Mir. Spacewalks were urgently required to remove and relocate the collapsible solar arrays from Kristall to Kvant-1, but *Flight International* noted in December that plans for Viktorenko and Polyakov to complete the work had been cancelled and rescheduled for the crew of the next expedition, Soyuz TM-21's Vladimir Dezhurov and Gennadi Strekalov, in May 1995. The delayed launch of Spektr, and Russia's insistence that it required a month-long period of activation and checkout, also conspired to push STS-71 – the first Shuttle-Mir docking flight, commanded by veteran astronaut Robert 'Hoot' Gibson – into June at the earliest. It would be a year filled with drama, in which the capabilities of the Shuttle and Mir, and their astronauts and cosmonauts, would be pushed to their very limits. It would also be a year of records, for on 8 January 1995 Valeri Polyakov quietly eclipsed the 366-day single-flight achievement, set by Soyuz TM-4 cosmonauts Vladimir Titov and Musa Manarov several years earlier. Since no woman had previously attempted a long-duration mission, and none from either the United States or Russia had flown a single flight for longer than 14 days, Yelena Kondakova had snatched a personal record for the books within the first quarter of her Soyuz TM-20 expedition. Ulf Merbold had spent 30 days in space, which, combined with his 18-day aggregate from two previous Shuttle missions, easily made him the most experienced non-Russian and non-US spacefarer at that time. And, not to be ignored, Alexander Viktorenko became the first person to fly as many as four times to Mir.

Yet for Russia itself, the omens for the future were bad. Even as it launched NASA's Norm Thagard into space on 14 March 1995, its civilian space programme was described in several quarters as "doomed", with its budgets declining to barely ten percent of their 1989 levels and the financing of military and rocket-production efforts reduced by a factor of ten. Only half of Russia's planned launches in 1994 were accomplished, the new Russian Space Agency – still less than three years into its existence – had received a meagre 12 percent of its required fiscal allocation, barely enough to keep it alive, and around half of all planned long-term programmes had been either cancelled or severely curtailed. Senior managers felt that, even with American dollars, Mir could probably not survive for much longer and might have to be evacuated. Even the booster for a Progress freighter, scheduled for April 1995 to carry some of Thagard's research equipment into orbit, had to be extracted from the Russian military arsenal. Thousands of skilled workers were either being laid off or were departing the space industry in droves, to such an extent than in February 1995 the Russian parliament adopted a new resolution to finance the space programme at a level of no lower than one percent of the gross national product and to implement a new national space policy. The early post-Soviet years were among the most desperate ever experienced by Russia's once-proud human space effort.

For America, 1995 would see the nation finally rise beyond the 84-day national space endurance record set in February 1974 by its final Skylab crew, for aboard Mir astronaut Norm Thagard would spend almost four months in orbit. The Shuttle would dock with the space station, not once, but *twice*, and the crew of STS-63 in February would accomplish a rendezvous exercise, closing to within several metres of Mir. The plan for long-duration American residents on the aging outpost would expand with new plans and new assignments and the following three years – from 1995 until 1998 – would prove a rollercoaster of emotion for both the United States and Russia, and for their international partners, too, as they learned to work together with old foes, setting aside fundamental ideological differences, overcoming stark challenges of culture and language, and steadily forging a true partnership which continues to this very day. The sixth and final volume of this series will explore the evolution of this 'Partnership' and its implications, not just for the two former superpowers, but for the Europeans, Canadians and Japanese participants in the International Space Station. All three groups of nations had been involved, with NASA, in the project from the early days of Space Station Freedom and all three had dreamed big and suffered badly as that project writhed and contorted its way to the doldrums of political neglect and finally teetered on the brink of outright cancellation by early 1993.

In the final volume of this series, the fate of Freedom will be explored, from its earliest designs, its grandiose scientific goals and the brutal and harsh realities of international politics which decimated its capabilities in the post-Cold War era. The very name 'Freedom' was originally conjured in 1988 by the Reagan Administration for political and ideological reasons, in order to counter a perceived threat from a closeted Communist group of republics. By the strangest of ironies, Freedom would transform into the International Space Station which, today, in the second decade of the 21st century, is large enough to be seen as a bright star in the night sky. The evolution of the new station, through the trials and tribulations of Shuttle-Mir, collisions and depressurisations, public and political scorn and the calamitous loss of Columbia in February 2003 and its aftermath will be followed as this most unlikely of partnerships began to tread a painful common path by which our species will continue its journey in space.

Bibliography

'Hispanics are Encouraged to Apply for Astronaut Program.' NASA Lyndon B. Johnson Space Center, Houston, Texas, 12 September 1979
'Space Transportation System Cargo Projects Milestones, Schedules and Status Summary.' NASA John F. Kennedy Space Center, Merritt Island, Florida, 15 March 1982
'Spas forms Eureca's backbone.' *Flight International*, 1 January 1983
'National Space Plan gets underway.' *Flight International*, 25 June 1983
'51D, 61D Crew Announcements.' NASA Lyndon B. Johnson Space Center, Houston, Texas, 2 February 1984
'Shuttle hopefuls named.' *Flight International*, 24 March 1984
'NASA Names Crews to Deploy Satellites in Year-End Flights.' NASA Lyndon B. Johnson Space Center, Houston, Texas, 29 January 1985
'Shuttle Shortcut for Senator.' *Flight International*, 16 February 1985
'RAF wins space race.' *Flight International*, 4 May 1985
'NASA Names Astronaut Crews for Ulysses, Galileo Missions.' NASA Lyndon B. Johnson Space Center, Houston, Texas, 31 May 1985
'UK astronaut experiments approved.' *Flight International*, 6 July 1985
'Eureca aims for 1988.' *Flight International*, 20 July 1985
'UARS borrows from Solar Max.' *Flight International*, 27 July 1985
'Eureca: European free-flyer.' *Flight International*, 31 August 1985
'Spacehab launch date set.' *Flight International*, 14 December 1985
'Hubble Delays Benefits EOM.' *Flight International*, 1 February 1986
'Italy fires Iris upper stage.' *Flight International*, 9 August 1986
'Spacehab moves ahead.' *Flight International*, 25 October 1986
'Orders in for Spacehab.' *Flight International*, 29 November 1986
'Shuttle commercial payloads rebuffed.' *Flight International*, 3 October 1987
'NASA assigns Shuttle payloads.' *Flight International*, 9 April 1988
'USAF prepares space-based sensor.' *Flight International*, 7 May 1988
'Lageos-2 delivered.' *Flight International*, 23 July 1988
'Spacehab to sign with NASA.' *Flight International*, 6 August 1988
'Space Shuttle Crew Members Named to DoD, Life Sciences Missions.' NASA Lyndon B. Johnson Space Center, Houston, Texas, 24 February 1989

'Astronauts Named to Space Science Missions (STS-37, STS-40).' NASA Lyndon B. Johnson Space Center, Houston, Texas, 5 April 1989
'Mir comes down to Earth.' *Flight International*, 15 April 1989
'McBride to Leave NASA; Brand Named Commander of STS-35.' NASA Lyndon B. Johnson Space Center, Houston, Texas, 24 April 1989
'Astronauts Named to DoD Missions in 1990 (STS-38, STS-39).' NASA Lyndon B. Johnson Space Center, Houston, Texas, 11 May 1989
'New Shuttle named.' *Flight International*, 27 May 1989
'Partial Shuttle Crew Assignments Announced.' NASA Lyndon B. Johnson Space Center, Houston, Texas, 29 June 1989
'Mir programme cost $2.7 billion.' *Flight International*, 29 July 1989
'Germany books third Spacelab.' *Flight International*, 5 August 1989
'Taiwan joins Spacehab.' *Flight International*, 23 September 1989
'Astronauts Named to Shuttle Crews: STS-39 (IBSS), STS-41 (Ulysses), STS-45 (ATLAS-01), STS-46 (TSS-1), STS-47 (SL-J).' NASA Lyndon B. Johnson Space Center, Houston, Texas, 29 September 1989
'Progress M-6 will carry recoverable capsule.' *Flight International*, 7 October 1989
'Walk for Freedom.' *Flight International*, 6 December 1989
'Astronaut Crew Named to International Microgravity Mission.' NASA Lyndon B. Johnson Space Center, Houston, Texas, 2 January 1990
'NASA Announces Payload Specialists for Spacelab IML-1 Mission.' NASA Headquarters, Washington, DC, 19 January 1990
'Science Payload Commanders Named; Carter Replaces Cleave on IML-1.' NASA Lyndon B. Johnson Space Center, Houston, Texas, 25 January 1990
'Gamma Ray Observatory Set for Shipment to Florida Launch Site.' NASA Lyndon B. Johnson Space Center, Houston, Texas, 2 February 1990
'US Air Force cancels November Shuttle mission.' *Flight International*, 28 February 1990
'Soyuz TM-9 crew starts Mir work.' *Flight International*, 28 February 1990
'NASA's First Spacewalk in Over Five Years is Set for November.' NASA Lyndon B. Johnson Space Center, Houston, Texas, 7 March 1990
'Intelsat faces loss of $150 million satellite.' *Flight International*, 21 March 1990
'Tethered Satellite completed.' *Flight International*, 21 March 1990
'NASA studies Intelsat-VI rescue.' *Flight International*, 28 March 1990
'NASA and Rockwell Sign Agreement for EDO Pallet.' NASA Headquarters, Washington, DC, 2 April 1990
'Telescope is set to peer at space and time.' *New York Times*, 9 April 1990
'Samuel T. Durrance Medically Qualified for ASTRO-1 Flight.' NASA Lyndon B. Johnson Space Center, Houston, Texas, 3 May 1990
'Shuttle and Mir to talk.' *Flight International*, 16 May 1990
'Columbia delays threaten five Space Shuttle missions.' *Flight International*, 23 May 1990
'Shuttle Crews Named for 1991 Missions (STS-43, STS-44, STS-45).' NASA Lyndon B. Johnson Space Center, Houston, Texas, 24 May 1990
'White Elephant, White Knight.' *Flight International*, 27 June 1990

'Hydrogen Leak Discovered on Shuttle Atlantis.' NASA Headquarters, Washington, DC, 29 June 1990
'Soviets to attempt Soyuz repair.' *Flight International*, 4 July 1990
'Space Shuttle Endeavour Powered-On.' NASA Lyndon B. Johnson Space Center, Houston, Texas, 6 July 1990
'Shuttle Crew Commanders Reassigned.' NASA Lyndon B. Johnson Space Center, Houston, Texas, 9 July 1990
'Space Shuttle Drag Chute Tests Set to Begin at Ames-Dryden.' NASA Lyndon B. Johnson Space Center, Houston, Texas, 18 July 1990
'Losing bid offered 2 tests on Hubble.' *New York Times*, 28 July 1990
'NASA B-52 tests Shuttle parachute.' *Flight International*, 1 August 1990
'NASA Selects Microgravity Mission Payload Specialist Candidates.' NASA Lyndon B. Johnson Space Center, Houston, Texas, 6 August 1990
'Shuttle leakages "not connected".' *Flight International*, 8 August 1990
'Dunbar Named Payload Commander for USML-1.' NASA Lyndon B. Johnson Space Center, Houston, Texas, 13 September 1990
'ASTRO flight grounded by new Shuttle leaks.' *Flight International*, 26 September 1990
'Budget cut threat to Soviet shuttle.' *Flight International*, 3 October 1990
'MBB carries out Eureca testing.' *Flight International*, 10 October 1990
'Soviet spacewalk to repair Kvant.' *Flight International*, 17 October 1990
'Discovery mission mooted for 1991.' *Flight International*, 24 October 1990
'The Hubble Space Telescope Optical Systems Failure Report.' NASA-Caltech Jet Propulsion Laboratory, Pasadena, California, November 1990
'Atlantis launch in November.' *Flight International*, 14 November 1990
'Cosmonauts fail to repair hatch.' *Flight International*, 14 November 1990
'Hydrogen dispersion systems improve Space Shuttle safety.' *Flight International*, 28 November 1990
'NASA Awards Commercial Middeck Augmentation Module Contract.' NASA Lyndon B. Johnson Space Center, Houston, Texas, 3 December 1990
'Shuttles to return to KSC runway.' *Flight International*, 5 December 1990
'Springer Retires from NASA, Marine Corps.' NASA Lyndon B. Johnson Space Center, Houston, Texas, 12 December 1990
'NASA Announces Crew Members for Future Shuttle Flights.' NASA Lyndon B. Johnson Space Center, Houston, Texas, 19 December 1990
'And the first prize isa journey into space.' *Flight International*, 2 January 1991
'NASA Awards Space Shuttle Orbiter Drag Chute Contract Mod.' NASA Lyndon B. Johnson Space Center, Houston, Texas, 4 January 1991
'Space Shuttle Orbiter Production Contract Modified.' NASA Lyndon B. Johnson Space Center, Houston, Texas, 9 January 1991
'Scientific value of Juno questioned.' *Flight International*, 23 January 1991
'August Shuttle team to include military reconnaissance expert.' *Flight International*, 30 January 1991
'NASA Awards Space Shuttle Orbiter 14-Inch Disconnect.' NASA Lyndon B. Johnson Space Center, Houston, Texas, 6 February 1991

'Discovery's Flight on STS-39 Delayed, Atlantis on STS-37 Next Up.' NASA Lyndon B. Johnson Space Center, Houston, Texas, 28 February 1991
'Cracks threaten Atlantis launch.' *Flight International*, 13 March 1991
'NASA Issues Modifications to Shuttle Manifest.' NASA Headquarters, Washington, DC, 21 March 1991
'New sleep regime for night workers.' *Flight International*, 3 April 1991
'Equipment Updates Enhance Endeavour, Orbiter Fleet.' NASA Lyndon B. Johnson Space Center, Houston, Texas, 5 April 1991
'Astronaut Gardner Named Commandant USAF Test Pilot School.' NASA Lyndon B. Johnson Space Center, Houston, Texas, 10 April 1991
'STS-39/Discovery Launch Date Set.' NASA Lyndon B. Johnson Space Center, Houston, Texas, 15 April 1991
'Shuttle Crew Assignments Announced.' NASA Lyndon B. Johnson Space Center, Houston, Texas, 19 April 1991
'Mir near miss as Progress docks.' *Flight International*, 24 April 1991
'Astronaut Mary Cleave Joins Environmental Project at Goddard.' NASA Lyndon B. Johnson Space Center, Houston, Texas, 2 May 1991
'NASA Selects Payload Specialists for Spacelab Mission.' NASA Lyndon B. Johnson Space Center, Houston, Texas, 2 May 1991
'Juno: The Jolly Folly?' *Flight International*, 15 May 1991
'ASTRO Mission to Refly.' NASA Lyndon B. Johnson Space Center, Houston, Texas, 20 May 1991
'KSC runway doubts as Shuttle lands badly.' *Flight International*, 22 May 1991
'New Transport Vehicle Allows More Timely Orbiter Egress.' NASA Lyndon B. Johnson Space Center, Houston, Texas, 28 May 1991
'Astronaut Lounge to Leave NASA.' NASA Lyndon B. Johnson Space Center, Houston, Texas, 30 May 1991
'Soviet cosmonaut logs record 541 space days.' *Flight International*, 5 June 1991
'GRO Spacecraft Grabs First Target of Opportunity.' NASA Headquarters, Washington, DC, 14 June 1991
'Soviets plan major event.' *Flight International*, 26 June 1991
'Astronaut O'Connor to Leave NASA.' NASA Lyndon B. Johnson Space Center, Houston, Texas, 28 June 1991
'Astronaut Coats to Leave NASA.' NASA Lyndon B. Johnson Space Center, Houston, Texas, 3 July 1991
'Atlantis "should have tougher tyres fitted".' *Flight International*, 24 July 1991
'Astronaut Class of 1990 Eligible for Flight Assignments.' NASA Lyndon B. Johnson Space Center, Houston, Texas, 29 July 1991
'Budget constraint cuts Soviet Mir missions.' *Flight International*, 31 July 1991
'Early Hubble Service Visit Not Scheduled.' NASA Headquarters, Washington, DC, 31 July 1991
'US and USSR Expand Space Co-operation.' NASA Headquarters, Washington, DC, 31 July 1991
'Soviet/US missions planned.' *Flight International*, 14 August 1991
'Atlantis arrives home safely.' *Flight International*, 21 August 1991

'No plans for early Hubble revisit.' *Flight International*, 21 August 1991
'NASA Announces Crew Members for Future Shuttle Flights.' NASA Lyndon B. Johnson Space Center, Houston, Texas, 23 August 1991
'Sharman mission faced post-launch abortion.' *Flight International*, 28 August 1991
'NASA Appoints Spacelab Payload Specialist.' NASA Headquarters, Washington, DC, 10 September 1991
'NASA Renames Gamma Ray Observatory in Honor of Compton.' NASA Headquarters, Washington, DC, 23 September 1991
'Payload Specialists for Tethered Satellite Mission Named.' NASA Lyndon B. Johnson Space Center, Houston, Texas, 26 September 1991
'Shuttle Discovery avoids space debris.' *Flight International*, 9 October 1991
'Seddon Named Payload Commander for SLS-2.' NASA Lyndon B. Johnson Space Center, Houston, Texas, 23 October 1991
'Ball nets Hubble COSTAR contract.' *Flight International*, 30 October 1991
'Endeavour launch delayed.' *Flight International*, 30 October 1991
'Single Soviet centre sought.' *Flight International*, 13 November 1991
'Cosmonauts to fly Chernobyl mission.' *Flight International*, 27 November 1991
'NASA Shuttle cuts threaten flights.' *Flight International*, 27 November 1991
'US reconnaissance expert is launched.' *Flight International*, 4 December 1991
'Payload Crew Named for Spacelab Life Sciences-2 Mission.' NASA Lyndon B. Johnson Space Center, Houston, Texas, 6 December 1991
'Earth observation mission clouded by rising pollution.' *Flight International*, 11 December 1991
'STS-44 hit by inertial measurement failure.' *Flight International*, 11 December 1991
Powers, C. Blake, Shea, Charlotte, and McMahan, Tracy, 'The First Mission of the Tethered Satellite System.' NASA George C. Marshall Space Flight Center, Huntsville, Alabama, 1992
'DARA confident of project despite Soviet upheaval.' *Flight International*, 8 January 1992
'Launch of final Mir modules delayed.' *Flight International*, 15 January 1992
'CIS tightens space management.' *Flight International*, 29 January 1992
'First Alenia Spacehab module arrives.' *Flight International*, 29 January 1992
'Shuttle Columbia to Stop Overnight in Houston.' NASA Lyndon B. Johnson Space Center, Houston, Texas, 4 February 1992
'Crew Assignments Announced for Future Shuttle Missions.' NASA Lyndon B. Johnson Space Center, Houston, Texas, 21 February 1992
'Bioreactor Team Earns NASA Inventor of the Year Honors.' NASA Lyndon B. Johnson Space Center, Houston, Texas, 6 March 1992
'Space Shuttle Crew Assignments Announced.' NASA Lyndon B. Johnson Space Center, Houston, Texas, 16 March 1992
'Russian Molniya launch lifts CIS spirits.' *Flight International*, 18 March 1992
'First Spacehab module is revealed.' *Flight International*, 15 April 1992
'Free-flying Spacelab is go.' *Flight International*, 8 May 1992
'Mir-core replacement delayed until 1997.' *Flight International*, 20 May 9192
'ESA eyes flights to CIS' Mir.' *Flight International*, 27 May 1992

'Spacewalking practice needed.' *Flight International*, 27 May 1992
'Astronaut Buchli to Retire and Leave NASA.' NASA Lyndon B. Johnson Space Center, Houston, Texas, 29 May 1992
'Astronaut Creighton to Retire and Leave NASA.' Lyndon B. Johnson Space Center, Houston, Texas, 25 June 1992
'Astronaut Hilmers Leaves NASA to Study Medicine.' NASA Lyndon B. Johnson Space Center, Houston, Texas, 21 July 1992
'Tethered Satellite Investigation Underway.' NASA Lyndon B. Johnson Space Center, Houston, Texas, 11 August 1992
'Astronaut Sullivan to Become Chief Scientist at NOAA.' NASA Lyndon B. Johnson Space Center, Houston, Texas, 14 August 1992
'Crew Assignments Announced for STS-58 and STS-61.' NASA Lyndon B. Johnson Space Center, Houston, Texas, 27 August 1992
'Payload Commander Named for IML-2 Mission.' NASA Lyndon B. Johnson Space Center, Houston, Texas, 2 September 1992
'STS-46 Space Shuttle Mission Report.' NASA Lyndon B. Johnson Space Center, Houston, Texas, October 1992
'STS-47 Space Shuttle Mission Report.' NASA Lyndon B. Johnson Space Center, Houston, Texas, October 1992
'Atlantis to Make East Texas Stops *En-Route* to California.' NASA Lyndon B. Johnson Space Center, Houston, Texas, 14 October 1992
'Mission Specialists Named for IML-2 Mission.' NASA Lyndon B. Johnson Space Center, Houston, Texas, 27 October 1992
'NASA Names Crew for STS-60 Mission with Cosmonaut.' NASA Lyndon B. Johnson Space Center, Houston, Texas, 28 October 1992
'Payload Specialist Selected for Second Life Sciences Mission.' NASA Lyndon B. Johnson Space Center, Houston, Texas, 29 October 1992
'Photo Opportunity With Cosmonauts.' NASA Lyndon B. Johnson Space Center, Houston, Texas, 5 November 1992
'Spacewalk Added to Shuttle Flight, More Expected.' NASA Lyndon B. Johnson Space Center, Houston, Texas, 25 November 1992
'Soyuz as Space Station Emergency Vehicle is Focus of Meeting.' NASA Lyndon B. Johnson Space Center, Houston, Texas, 30 November 1992
'Pre-Launch Mission Operation Report: Orbiting Retrievable Far and Extreme Ultraviolet Spectrometer-Shuttle Pallet Satellite (ORFEUS-SPAS).' Office of Life and Microgravity Sciences and Applications, Flight Systems Division, NASA Headquarters, Washington, DC, 1993
'Shuttle Crew to Teach Physics of Toys During Mission.' NASA Lyndon B. Johnson Space Center, Houston, Texas, 6 January 1993
'Endeavour mission ends successfully.' *Flight International*, 27 January 1993
'Mir cosmonauts switch duties.' *Flight International*, 3 February 1993
'Spacewalk Added to July Space Shuttle Flight.' NASA Lyndon B. Johnson Space Center, Houston, Texas, 3 February 1993
'Spacewalk Added to April Space Shuttle Flight.' NASA Lyndon B. Johnson Space Center, Houston, Texas, 17 February 1993

'Engine revisions set back Space Shuttle launches.' *Flight International*, 24 February 1993
'Mir spacewalk plans are revealed.' *Flight International*, 3 March 1993
'Backup Crew Member for STS-61, HST Maintenance Mission.' NASA Lyndon B. Johnson Space Center, Houston, Texas, 9 March 1993
'Astronauts toil for Hubble effort.' *Flight International*, 17 March 1993
'Astronaut named as Hubble back-up.' *Flight International*, 24 March 1993
'Valve failure holds Spacelab back.' *Flight International*, 31 March 1993
'NASA plans Atlas launch for April.' *Flight International*, 7 April 1993
'Discovery may fly-by the Mir 1.' *Flight International*, 14 April 1993
'Poliakov aims for new flight record.' *Flight International*, 28 April 1993
'Astronaut Seddon Injured During Training.' NASA Lyndon B. Johnson Space Center, Houston, Texas, 4 May 1993
'ESA proceeds with Mir missions.' *Flight International*, 19 May 1993
'Astronaut Story Musgrave Injured During Training.' NASA Lyndon B. Johnson Space Center, Houston, Texas, 1 June 1993
'Endeavour Space Shuttle grounded.' *Flight International*, 9 June 1993
'Mission Impossible?' *Flight International*, 9 June 1993
'Frenchman goes to Mir.' *Flight International*, 14 July 1993
'Payload Commanders Named for Future Shuttle Missions.' NASA Lyndon B. Johnson Space Center, Houston, Texas, 3 August 1993
'Patient doctor.' *Flight International*, 11 August 1993
'Shuttle shelters from meteor shower.' *Flight International*, 11 August 1993
'Russians at sea over Atlantis.' *Flight International*, 18 August 1993
'Crew Members Selected for STS-65.' NASA Lyndon B. Johnson Space Center, Houston, Texas, 23 August 1993
'Spacesuit fault halts TM-17 outing.' *Flight International*, 13 October 1993
'Discovery poised for fourth launch attempt.' *Flight International*, 1 September 1993
'Explosive deployment.' *Flight International*, 20 October 1993
'Cosmonauts forced to remain in space.' *Flight International*, 27 October 1993
'Flying saucer launch planned from Shuttle.' *Flight International*, 22 December 1993
'USA/Russia agree space deal.' *Flight International*, 5 January 1994
'Crew Selected for STS-67 Astronomy Mission.' NASA Lyndon B. Johnson Space Center, Houston, Texas, 10 January 1994
'Russia schedules ten Mir launches.' *Flight International*, 12 January 1994
'Soyuz has close encounter with Mir station.' *Flight International*, 26 January 1994
'Astronaut Linenger Joins STS-64 Crew.' NASA Lyndon B. Johnson Space Center, Houston, Texas, 28 February 1994
'Space lessons.' *Flight International*, 16 March 1994
'Improved Shuttle Tile to Fly on STS-59.' NASA Headquarters, Washington, DC, 30 March 1994
'US defence secretary visits Baikonur.' *Flight International*, 30 March 1994
'Russia agrees Baikonur lease terms.' *Flight International*, 20 April 1994
'SRL success prompts calls for new craft.' *Flight International*, 27 April 1994
'Cosmonauts tackle Shuttle training.' *Flight International*, 1 June 1994

'Crew Named for First Space Shuttle, Mir Docking Mission.' NASA Lyndon B. Johnson Space Center, Houston, Texas, 3 June 1994
'Astronaut chief to lead first Mir flight.' *Flight International*, 15 June 1994
'NASA claims Mir-1 input will cut costs.' *Flight International*, 6 July 1994
'Budget crisis causes Mir-1 postponement.' *Flight International*, 10 August 1994
'Progress docks with Mir at last.' *Flight International*, 14 September 1994
'Shuttle engine cleared.' *Flight International*, 21 September 1994
'Spektr module delayed.' *Flight International*, 5 October 1994
'Docking scare alarms Russia's Mir crew again.' *Flight International*, 19 October 1994
'Russia's Mir crew closes in on new space-endurance records.' *Flight International*, 2 November 1994
'Russian space programme funding reaches crisis point.' *Flight International*, 8 March 1995
'NASA considers retrieving UARS.' *Flight International*, 22 May 2001
'Last gasp for atmosphere satellite shut down on cost.' *Flight International*, 4 September 2001
Broad, William J., 'Shuttle Soars into Orbit to Test Device for Space Rescues.' *New York Times*, 10 September 1994
Burrough, Bryan (1998), *Dragonfly: NASA and the Crisis Aboard Mir*. London: Fourth Estate
Cassutt, Michael, 'Max Q Live: In space no one can hear you sing.' *Air & Space Magazine*, 1 March 2009
Cassutt, Michael, 'Secret Space Shuttles.' *Air & Space Magazine*, 1 August 2009
Clark, Phillip (1988), *The Soviet Manned Space Programme*. London: Salamander
Cosmo, M.L. and Lorenzini, E.C., *Tethers in Space Handbook* (Third Edition). Smithsonian Astrophysical Observatory, prepared for NASA George C. Marshall Space Flight Center, Huntsville, Alabama, December 1997
Date, Shirish, 'The Malarkey Milkshake: It's Not a Joke Any More.' *The Orlando Sentinel*, 2 May 1991
Dawson, Virginia P. and Bowles, Mark D. (2004), *Taming Liquid Hydrogen: The Centaur Upper Stage Rocket, 1958-2002*. Office of External Relations, NASA Headquarters, Washington, DC
Dunar, Andrew J. and Waring, Stephen P. (1999), *Power to Explore: The History of Marshall Space Flight Center, 1960-1990*. NASA George C. Marshall Space Flight Center, Huntsville, Alabama
Evans, Ben (2005), *Space Shuttle Columbia: Her Missions and Crews*. Chichester: Praxis
Evans, Ben (2009), *Escaping the Bonds of Earth*. Chichester: Praxis
Evans, Ben (2010), *Foothold in the Heavens*. Chichester: Praxis
Evans, Ben (2011), *At Home in Space*. Chichester: Praxis
Evans, Ben (2012), *Tragedy and Triumph in Orbit*. Chichester: Praxis
Froelich, Walter (1984), *Spacelab: An International Short-Stay Orbiting Laboratory*. NASA Headquarters, Washington, DC
Furniss, Tim, 'Shuttle design change to increase mission duration.' *Flight International*, 4 November 1989

Furniss, Tim, 'Infernal device.' *Flight International*, 26 September 1990
Furniss, Tim, 'Star Wars Shuttle.' *Flight International*, 20 February 1991
Furniss, Tim, 'Endeavour the latest to develop cracks.' *Flight International*, 27 March 1991
Furniss, Tim, 'Bent pipes in space.' *Flight International*, 24 July 1991
Furniss, Tim, 'Weightless exploration.' *Flight International*, 15 January 1992
Furniss, Tim, 'NASA changes main Endeavour engines.' *Flight International*, 15 April 1992
Furniss, Tim, 'NASA head alters Shuttle date.' *Flight International*, 6 May 1992
Furniss, Tim, 'Stretching the Shuttle.' *Flight International*, 27 May 1992
Furniss, Tim, 'Discovery is ready to deploy two satellites.' *Flight International*, 14 July 1993
Hall, Rex D. and Shayler, David J. (2003), *Soyuz: A Universal Spacecraft*. Chichester: Praxis
Heppenheimer, T.A. (1999), *The Space Shuttle Decision*. NASA Office of Policy and Plans, NASA Headquarters, Washington, DC
Jemison, Mae C. and Olsen, Patricia R., 'Executive Life: The Boss – What Was Space Like?' *New York Times*, 2 February 2003
Jenkins, Dennis R. (2001), *Space Shuttle: The History of the National Space Transportation System – The First 100 Missions*. Hinckley: Midland Publishing
Jones, Tom (2006), *Sky Walking: An Astronaut's Memoir*. New York: HarperCollins
Leary, Warren E., 'Woman in the News: A Determined Breaker of Boundaries – Mae Carol Jemison.' *New York Times*, 13 September 1992
Linenger, Jerry M. (2000), *Off the Planet: Surviving Five Perilous Months Aboard the Space Station Mir*. New York: McGraw-Hill
Lord, Douglas R. (1987), *Spacelab: An International Success Story*. Scientific and Technical Information Division, NASA Headquarters, Washington, DC
Morgan, Clay (2001), *Shuttle-Mir: The United States and Russia Share History's Highest Stage*. NASA History Series, Lyndon B. Johnson Space Center, Houston, Texas
Portree, David S.F. (1995), *Mir Hardware Heritage*. NASA Information Services Division, NASA Lyndon B. Johnson Space Center, Houston, Texas
Portree, David S.F. and Treviño, Robert C. (1997), *Walking to Olympus: An EVA Chronology*. NASA History Office, NASA Headquarters, Washington, DC
Reichhardt, Tony (ed.), *Space Shuttle: The First 20 Years*. London: DK Publishing, Inc. 2002
Sharman, Helen, with Christopher Priest (1993), *Seize the Moment*. London: Victor Gollancz
Shayler, David J. and Burgess, Colin (2007), *NASA's Scientist-Astronauts*. Chichester: Praxis
'Soviet Space Programs: 1981-87: Piloted Space Activities, Launch Vehicles, Launch Sites and Tracking Support'. Prepared at the Request of Hon. Ernest F. Hollings, Chairman, Committee on Commerce, Science and Transportation, United States Senate, February 1988
Tomalin, Terry, 'Flying in space to fly fishing.' *St. Petersburg Times*, 6 July 2007

Index

Adamson, Jim, 126–127, 145
Advanced Communications Technology Satellite, 336–347
Afanasyev, Viktor, 178–182, 188–190, 318, 375, 466, 471–477
Akers, Tom, 23, 25–29, 232, 239, 240–243, 245, 359, 363, 367–370, 372
Akiyama, Toyohiro, 179–181, 186
Allen, Andy, 248, 263, 269, 271, 273, 291, 395–396, 398–400
Apt, Jerome 'Jay', 74–84, 210, 277, 282, 284–285, 405, 407, 409
ASTRO-1, 17, 19, 23, 43–67, 112, 266, 305, 444–445, 448–451
Atlantis, 30–43, 70–85, 120–127, 135–149, 216–230, 261–276, 437–443, 475, 481
Artsebarski, Anatoli, 178, 183, 186, 188–190, 192–194, 200
Aubakirov, Toktar, 192, 194–196, 200, 467, 490
Avdeyev, Sergei, 458–460, 463

Bagian, Jim, 100–101, 105–106, 110–119, 348
Baker, Ellen, 248–249, 260
Baker, Mike, 125, 127, 291–297, 412, 423–430
Balandin, Alexander, 65, 172–176
Blaha, John, 103, 105–106, 120–128, 347, 350–356
Bluford, Guy, 86–99, 281, 297–302
Bolden, Charlie, 216–217, 221–224, 230, 282, 375–376, 383, 389–394
Bondar, Roberta, 203, 207–208, 211, 213, 296
Bowersox, Ken, 247–260, 363, 367
Brand, Vance, 20, 50, 52, 54–66, 72, 100
Brandenstein, Dan, 85, 136, 173, 211, 225, 231–245, 247, 323
Brown, Curt, 277, 281–286, 437, 440
Brown, Mark, 130, 132–135
Buchli, Jim, 130–135
Bursch, Dan, 338, 344–345, 412, 422, 426–427

Cabana, Bob, 23–28, 119, 297–303, 413–421
Cameron, Ken, 74–83, 316, 320–321, 411
Carter, Manley 'Sonny', 203–204, 209
Casper, John, 248–249, 304–309, 395–401
Challenger, 4–5, 15–17, 20
Chang-Díaz, Franklin, 73, 264–274, 319, 383–384, 388–394
Chiao, Leroy, 357, 413–419
Chilton, Kevin, 240–241, 305, 406–411
Cleave, Mary, 72, 203–204
Clervoy, Jean-François, 439–442, 457
Clifford, Michael 'Rich', 241, 297–302, 312, 405–410, 423
Coats, Mike, 41–42, 87, 92–99, 248, 291
Cockrell, Ken, 211, 316, 319–322
Columbia, 43–66, 100–119, 245–260, 286–296, 309–325, 347–355, 395–400, 412–421
Compton Gamma Ray Observatory, 67–70, 307, 445
Covey, Dick, 30–43, 363–373
Creighton, John 'J.O.', 19, 129–135
Culbertson, Frank, 32–43, 127, 336–345

Davis, Jan, 277–281, 283–284, 383–394
Defense Support Program, 135, 143, 146–149
DeLucas, Larry, 249, 253–260
Discovery, 1–29, 86–99, 128–134, 201–215, 297–302, 316–321, 336–346, 375–394, 430–435, 468, 470–471, 481

Duffy, Brian, 221–231, 327–336
Dunbar, Bonnie, 249–260, 387, 411, 474–475
Durrance, Sam, 44, 50–66, 445–448

Endeavour, 231–244, 277–285, 303–308, 326–335, 356–374, 401–411, 422–429, 443–452
European Retrievable Carrier, 269–276, 289, 320–321, 327, 332–333, 356, 470

Fettman, Marty, 349–354
Foale, Mike, 217–229, 316–320
Frimout, Dirk, 217, 221–231

Gaffney, Drew, 100–110, 352
Gardner, Guy, 52–66, 468
Gemar, Charles 'Sam', 32–34, 130–135, 241, 396–400
Gibson, Robert 'Hoot', 101, 211, 266, 277, 282–285, 291, 412, 481
Godwin, Linda, 74–84, 403–409
Grabe, Ron, 204, 210–211, 326–335
Gregory, Bill, 444, 446–451
Gregory, Fred, 136–149, 231, 281
Grunsfeld, John, 444–451
Gutierrez, Sid, 105–110, 112, 118–119, 407–411

Halsell, Jim, 413–421,
Hammond, Blaine, 87, 92–99, 431
Harbaugh, Greg, 87, 92–99, 304–309, 366–367
Harris, Bernard, 310–312, 322–325
Helms, Susan, 304–309, 312, 431–436
Hennen, Tom, 144–148
Henricks, Terence 'Tom', 137–149, 322–325
Hilmers, Dave, 4, 19, 209–211, 291
Hoffman, Jeff, 18–20, 44–67, 261–277, 356–374, 445
Hieb, Rick, 87, 89–99, 231–243, 413–421
Hubble Space Telescope, 5, 17, 30, 68, 75, 83, 245, 278, 308, 332–333, 345, 356–374, 393, 407, 412, 433, 445
Hughes-Fulford, Millie, 100–120

Inertial Upper Stage, 2–4, 26–29, 38–40, 124–125, 143, 146–147, 304, 307–308
Intelsat-603, 231–245, 307, 332
Ivins, Marsha, 118, 143, 248, 261–269, 291, 396, 399,

Jemison, Mae, 277, 280–286
Jernigan, Tammy, 102–103, 110, 118–119, 291–296, 328, 444–451
Jones, Tom, 24, 74, 102, 337, 403–412, 422–430

Kaleri, Alexander, 187, 192, 196, 199–200, 454, 457–459,
Kondakova, Yelena, 473, 478–481
Krikalev, Sergei, 155–159, 183, 186–200, 202, 217, 375–394, 453–454, 461

Laser Geodynamics Satellite, 287–295, 327
Lawrence, Wendy, 446–447, 451
Lee, Mark, 277–284, 430–435
Leestma, Dave, 44, 52, 221–229, 357
Lichtenberg, Byron, 217–230
Linenger, Jerry, 425, 431
Lounge, Mike, 4, 51–53, 59–67
Low, David, 126–127, 308, 327, 332–335
Lucid, Shannon, 120–124, 348, 352–353

MacLean, Steve, 291–296
McArthur, Bill, 350–353, 422
McMonagle, Don, 87, 92–96, 305–309, 437, 440, 443
Malenchenko, Yuri, 466, 473, 475–480
Malerba, Franco, 268–269
Manakov, Gennadi, 177–181, 460–466
Manarov, Musa, 169, 180–189, 192, 471, 481
Meade, Carl, 32–34, 210, 249, 256, 260, 430–434
Melnick, Bruce, 15, 23, 25, 29, 236, 240–242
Merbold, Ulf, 203, 207–208, 211, 263, 469–481
Mir space station,
  Base block, 154, 156, 181–182, 188–189, 194
  Buran and, 153, 172–174, 177–178, 186, 454–456, 461–462, 466
  Kvant-1 module, 154–156, 168, 170, 173, 175, 177–182, 189, 192–194, 460, 462–466, 476
  Kvant-2 module, 151, 155, 166, 168–169, 171–172, 174–179, 181, 188, 198–199, 457, 462–464, 467
  Kristall module, 151, 155, 169, 171–174, 178, 181, 189, 454, 460–466, 472, 474, 478, 480

Mir space station, *cont.*
 Spektr module, 456, 461–462, 466, 468, 478–479, 481
 Priroda module, 456, 461, 468
Mohri, Mamoru, 277–286, 414,
Mukai, Chiaki, 280, 413–419
Musabayev, Talgat, 192, 196, 473, 475–480
Musgrave, Story, 14, 137–148, 178, 241, 363, 366–372

Nagel, Steve, 70–85, 309–315, 321–326, 422
Newman, Jim, 308, 337–346
Nicollier, Claude, 208, 261–274, 363–368

O'Connor, Bryan, 103–119, 248
Ochoa, Ellen, 316–320, 437, 440
Oswald, Steve, 204, 210–213, 316, 319–321, 446–448, 451, 453

Parazynski, Scott, 439–440, 445
Parise, Ron, 44, 50–66, 445–448
Parker, Bob, 43–44, 47, 49–62
Payload Assist Module, 26–28, 39
Poleshchuk, Alexander, 460, 462–466
Precourt, Charlie, 322–324

Readdy, Bill, 204, 209–215, 236, 336–339, 342–344
Reightler, Ken, 130–135, 375, 383–384, 392
Richards, Dick, 3, 22–30, 33, 44, 52, 247–260, 431, 436
Ross, Jerry, 61, 68–85, 173, 266, 310–312, 322–326, 339, 453
Runco, Mario, 138, 143, 147–148, 304, 306–309, 412–413

Schlegel, Hans, 310–312, 322–324
Searfoss, Rick, 350–356
Seddon, Rhea, 100–119, 123, 209, 277, 284, 347–356
Sega, Ron, 383–392
Serebrov, Alexander, 151, 155, 159–160, 166, 168–173, 464, 466–467, 472, 478,
Sharman, Helen, 183–190
Shepherd, Bill, 15, 23–25, 291, 296
Sherlock, Nancy, 327–335
Shriver, Loren, 19, 267–276, 291, 347
Shuttle Pallet Satellite, 91–98, 343–345, 438, 440–443
 CRISTA-SPAS, 438, 440–443
 IBSS-SPAS, 91–98
 ORFEUS-SPAS, 343–345
Smith, Steve, 412, 424–427, 429
Solovyov, Anatoli, 65, 172–176, 453, 458–463
Spacehab, 326–335, 352, 383, 388–394, 471
Spacelab,
 ASTRO-1, 43–66
 ASTRO-2, 444–452
 ATLAS-1, 216–230,
 ATLAS-2, 316–321
 ATLAS-3, 437–443
 International Microgravity Laboratory-1, 201–215
 International Microgravity Laboratory-2, 412–421
 Lidar In-Space Technology Experiment, 430–435
 Spacelab-D2, 309–325
 Spacelab-J, 277–285
 Spacelab Life Sciences, 99–119, 347–355
 Space Radar Laboratory-1, 401–411
 Space Radar Laboratory-2, 422–429
 United States Microgravity Laboratory-1, 245–260
Spartan, 235, 289, 316–318, 432–437
Springer, Bob, 32–36, 42
Strekalov, Gennadi, 177–178, 180–181, 466, 475, 481
STS-35, 43–66
STS-37, 70–85
STS-38, 30–43
STS-39, 86–99,
STS-40, 100–119
STS-41, 1–29
STS-42, 201–215
STS-43, 120–127
STS-44, 135–149
STS-45, 216–230
STS-46, 261–276
STS-47, 277–285
STS-48, 128–134
STS-49, 231–244
STS-50, 245–260
STS-51, 336–346
STS-52, 286–296
STS-53, 297–302
STS-54, 303–308
STS-55, 309–325

STS-56, 316–321
STS-57, 326–335
STS-58, 347–355
STS-59, 401–411
STS-60, 375–394
STS-61, 356–374
STS-62, 395–400
STS-63, 383, 468, 470–471, 481
STS-64, 430–435
STS-65, 412–421
STS-66, 437–443
STS-67, 443–452
STS-68, 422–429
STS-71, 443, 475, 481,
Sullivan, Kathy, 88, 118, 145, 149, 204, 217–231, 362

Tanner, Joe, 437–438, 440, 443
Tethered Satellite System, 261–276
Thagard, Norm, 160, 203–211, 411, 453, 474–475, 478, 481
Thomas, Don, 413–421
Thornton, Kathy, 231–244, 356–374,
Thuot, Pierre 'Pepe', 231–244, 395–400
Titov, Vladimir, 160, 169, 178, 180, 217, 376, 383–384, 453, 461, 471, 481
Tognini, Michel, 439, 456, 458–459, 469
Tracking and Data Relay Satellite, 120–127, 303–308
Transfer Orbit Stage, 336–346

Trinh, Gene, 245–260,
Tsibliyev, Vasili, 464–467, 472, 474, 478

Ulysses, 1–29
United States Microgravity Payload, 286–296, 395–400
Upper Atmosphere Research Satellite, 128–134
Usachev, Yuri, 192, 196, 375, 466, 471–473, 476–477

Veach, Lacy, 86–99, 286–296
Viehböck, Franz, 192, 195–196, 469
Viktorenko, Alexander, 151, 155, 159–160, 166, 168–173, 192, 196, 199–200, 457–459, 473, 478–481
Volkov, Alexander, 155, 159, 187, 192, 195–202, 376, 472
Voss, Janice, 326–335
Voss, Jim, 135–149, 297–302

Wake Shield Facility, 375–394
Walker, Dave, 297–302
Walter, Ulrich, 309–325
Walz, Carl, 337–346, 412–421,
Wetherbee, Jim, 286–296
Wilcutt, Terry, 422–429
Wisoff, Peter 'Jeff', 326–335, 422–429
Wolf, Dave, 347–355

About the Author

Ben Evans is a space enthusiast and writer, based in Warwickshire, UK. He has written extensively for the magazines Spaceflight, Countdown and Astronomy Now and is presently the senior writer for AmericaSpace.com. He has written seven previous books for Springer-Praxis, including the first four volumes of this History series, one of which was a finalist for the 2010 Eugene M. Emme Award for Astronautical Literature.

GPSR Compliance

The European Union's (EU) General Product Safety Regulation (GPSR) is a set of rules that requires consumer products to be safe and our obligations to ensure this.

If you have any concerns about our products, you can contact us on

ProductSafety@springernature.com

In case Publisher is established outside the EU, the EU authorized representative is:

Springer Nature Customer Service Center GmbH
Europaplatz 3
69115 Heidelberg, Germany